ESTUARINE SHORES

ESTUARINE SHORES

Evolution, Environments and Human Alterations

Edited by

Karl F. Nordstrom
Institute of Marine and Coastal Sciences,
Rutgers University, USA

and

Charles T. Roman
National Biological Service,
University of Rhode Island, USA

JOHN WILEY & SONS

Chichester · New York · Brisbane · Toronto · Singapore

National 01243 779777
International (+44) 1243 779777
e-mail (for orders and customer service enquiries): cs-books@wiley.co.uk
Visit our Home Page on http://www.wiley.co.uk
or http:///www.wiley.com

Other Wiley Editorial Offices

John Wiley & Sons, Inc., 605 Third Avenue,
New York, NY 10158-0012, USA

Jacaranda Wiley Ltd, 33 Park Road, Milton,
Queensland 4064, Australia

John Wiley & Sons (Canada) Ltd, 22 Worcester Road,
Rexdale, Ontario M9W 1L1, Canada

John Wiley & Sons (Asia) Pte Ltd, 2 Clementi Loop #02-01,
Jin Xing Distripark, Singapore 129809

Library of Congress Cataloging-in-Publication Data

Estuarine shores : evolution, environments, and human alterations /
edited by Karl F. Nordstrom and Charles T. Roman.
p. cm.
Includes bibliographical references and index.
ISBN 0-471-96596-0
1. Estuaries. 2. Coast changes. I. Nordstrom, Karl F.
II. Roman, Charles T. (Charles True)
GC97.E79 1996
551.46'09 – dc20 96-12984
CIP

British Library Cataloguing in Publication Data

A catalogue record for this book is available from the British Library

ISBN 0-471-96596-0

Typeset in 10/12pt Times from the editor's disks by MHL Typesetting Ltd, Coventry
Printed and bound in Great Britain by Biddles Ltd, Guildford and King's Lynn
This book is printed on acid-free paper responsibly manufactured from sustainable forestation, for which at least two trees are planted for each one used for paper production.

To the memory of
Bill Carter

Contents

List of Contributors

Kenneth W. Able, *Marine Field Station, Institute of Marine and Coastal Sciences, Rutgers University, 800 Great Bay Boulevard, Tuckerton, NJ 08087, USA*

James R. Allen, *National Biological Service, 15 State Street, Boston, MA 02109, USA*

Edward J. Anthony, *Université du Littoral, Département de Géographie, URA 1688 CNRS, 2 Chaussée, 59140 Dunkerque, France*

Richard H. Burroughs, *Department of Marine Affairs, University of Rhode Island, Kingston, RI 02881, USA*

Russell Cole, *Marine Geosciences Group, University of Waikato, Hamilton, New Zealand*

Willem de Lange, *Marine Geosciences Group, University of Waikato, Hamilton, New Zealand*

J. P. Doody, *Joint Nature Conservation Committee, Monkstone House, City Road, Peterborough PE1 1JY, UK*

Mark S. Fonseca, *National Marine Fisheries Service, National Oceanic and Atmospheric Administration, Pivers Island, Beaufort, NC 28516-9722, USA*

Annick Garniel, *Kieler Institut für Landschaftsökologie, Bergstrasse 5, 24103 Kiel, Germany*

Dale D. Goehringer, *Biology Department, Woods Hole Oceanographic Institution, Woods Hole, MA 02543, USA*

Terry R. Healy, *Marine Geosciences Group, University of Waikato, Hamilton, New Zealand*

Brian L. Howes, *Biology Department, Woods Hole Oceanographic Institution, Woods Hole, MA 02543, USA*

Nancy L. Jackson, *Center for Policy Studies, New Jersey Institute of Technology, Newark, NJ 07102, USA*

Claire J.L. Jeuken, *Institute of Marine and Atmospheric Research, University of Utrecht, PO Box 80.115, 3508 TC Utrecht, The Netherlands*

Michael S. Kearney, *Department of Geography, University of Maryland, College Park, MD 20742, USA*

Reinhard Lampe, *Geographisches Institut, Ernst-Moritz-Arndt-Universität, Ludwig-Jahnstrasse 16, 0-2200 Greifswald, Germany*

Richard S. McBride, *Florida Marine Research Institute, 100 Eighth Avenue SE, St Petersburg, FL 33701, USA*

Ulrich Mierwald, *Kieler Institut für Landschaftsökologie, Bergstrasse 5, 24103 Kiel, Germany*

Carole C. McIvor, *National Biological Service, Arizona Cooperative Fish and Wildlife Research Unit, University of Arizona, Tucson, AZ 85721, USA*

Karl F. Nordstrom, *Institute of Marine and Coastal Sciences, Rutgers University, New Brunswick, NJ 08903, USA*

John S. Pethick, *Coastal Research Unit, Department of Geography, University of Cambridge, 62 Sidney Street, Cambridge CB2 3JW, UK*

Jonathan D. Phillips, *Department of Geography and Planning, East Carolina University, Greenville, NC 27858, USA*

Kenneth Pye, *Postgraduate Research Institute for Sedimentology, University of Reading, PO Box 227, Whiteknights, Reading RG6 2AB, UK*

Charles T. Roman, *National Biological Service, Cooperative Park Studies Unit, Graduate School of Oceanography, University of Rhode Island, Narragansett, RI 02882, USA*

Rodney A. Rountree, *National Marine Fisheries Service, National Oceanic and Atmospheric Administration, 166 Water Street, Woods Hole, MA 02543, USA*

Lawrence P. Rozas, *National Marine Fisheries Service, National Oceanic and Atmospheric Administration, 4700 Avenue U, Galveston, TX 77551, USA*

Douglas J. Sherman, *Department of Geography, University of Southern California, Los Angeles, CA 90089-0663, USA*

Kelly J. Smith, *Marine Field Station, Institute of Marine and Coastal Sciences, Rutgers University, 800 Great Bay Boulevard, Tuckerton, NJ 08087, USA*

J. Court Stevenson, *Horn Point Environmental Laboratories, Center for Environmental and Estuarine Studies, University of Maryland, Cambridge, MD 21613, USA*

John M. Teal, *Biology Department, Woods Hole Oceanographic Institution, Woods Hole, MA 02543, USA*

Jan H. van den Berg, *Institute of Marine and Atmospheric Research, University of Utrecht, PO Box 80.115, 3508 TC Utrecht, The Netherlands*

Ad J.F. van der Spek, *Geological Survey of The Netherlands, University of Utrecht, PO Box 80.115, 3508 TC Utrecht, The Netherlands*

José A. Vargas, *Centro de Investigación en Ciencias del Mar y Limnología, Universidad de Costa Rica, 2060 San Pedro, Costa Rica*

Peter K. Weiskel, *Water Resources Division, US Geological Survey, Marlborough, MA 01752, USA*

David A. Witting, *Marine Field Station, Institute of Marine and Coastal Sciences, Rutgers University, 800 Great Bay Boulevard, Tuckerton, NJ 08087, USA*

Colin D. Woodroffe, *Department of Geography, University of Wollongong, New South Wales 2500, Australia*

Preface

Estuarine Shores examines the spatial and temporal aspects of processes, sediments, landforms and biota and their interactions across the shallow subtidal, intertidal and frequently flooded supratidal portion of the estuary and nearby upland. The rationale for the book is to address the perceived lack of attention to interdisciplinary studies of linkages between and among the environments that define the shore. These environments include seagrass beds, mudflats, beaches, saltmarshes, mangroves, eroding uplands, stream banks and a wide variety of human-altered landforms and structures. The environments are becoming more fragmented through time as a result of human development, and the nature of the interaction between them is becoming more complex. *Estuarine Shores* addresses environmental interactions by highlighting the functional ecology of habitats, the interrelationships between habitats and the alteration of physical environments and habitats due to shoreline change and shoreline development.

The book includes 21 chapters (including case studies and reviews) by specialists in estuarine research from eight countries. It is organized into three parts – shoreline evolution, environments and processes, and human alterations and management – plus introductory and concluding chapters. The contributions, solicited specifically for this book, represent an attempt to achieve broad coverage across disciplines rather than treat a single discipline in detail. The conception of environmental linkages across time and space differs between chapters because of differences in the backgrounds and types of research conducted by the contributors. Although many of the authors found it difficult to define explicitly the nature of the linkages, particularly outside their disciplinary backgrounds, they underscored the value of adopting this perspective in future research programs. Thus, this book should be viewed more as the beginning of a comprehensive research program than the definitive results of this kind of program. As a result, we end the book with a series of questions that are designed to stimulate further research.

Karl F. Nordstrom and Charles T. Roman

Acknowledgements

We thank the authors for their rapid response to our requests for manuscripts and their patience during revisions and while waiting for the last few manuscripts to be completed. We are especially grateful to the authors for their willingness to address the specific themes that we wished to underscore in the book. We thought it best to concentrate on thematic issues and relax the requirement to adopt a common style throughout the book. As a result, English and American spellings and style have been used according to the preference of each author.

The chapters in this book have benefited from critical peer reviews, for which we are grateful to the following colleagues:

Gail M. Ashley
Rutgers University

J. Paul Barusseau
University of Perpignan

Bob Creese
University of Auckland

Henry Bokuniewicz
State University of New York at Stony Brook

Suzanne B. Bricker
National Oceanic and Atmospheric Administration

Richard H. Burroughs
University of Rhode Island

Virginia Carter
US Geological Survey

Harlan K. Dean
Harvard University and Johnson and Wales University

R.C. Dodd
Research Triangle Institute

Nancy L. Jackson
New Jersey Institute of Technology

Diane P. Horn
University of London

Kenneth L. Heck, Jr.
Marine Environmental Sciences Consortium, University of South Alabama

Michael J. Kennish
Rutgers University

George D. Klein
New Jersey Marine Sciences Consortium

Ronald T. Kneib
University of Georgia

Ashish J. Mehta
University of Florida

Richard A. Orson
Orson Environmental Consulting

Denise J. Reed
Louisiana Universities Marine Consortium

Peter S. Rosen
Northeastern University

Craig D. Taylor
Woods Hole Oceanographic Institution

Thomas A. Terich
Western Washington University

Chapter One

Environments, processes and interactions of estuarine shores

CHARLES T. ROMAN
National Biological Service

and

KARL F. NORDSTROM
Rutgers University

1.1 INTRODUCTION

Estuarine shores, as considered in this book, extend from nearshore subtidal areas, through the intertidal zone and landward to the limit of inundation by storms. Environments may include beaches, dunes, eroding uplands, stream channels, rocky platforms, saltmarshes, mangroves, seagrass meadows, algal beds, tidal flats, and a wide variety of human-altered landforms. Estuarine shoreline environments provide spawning and nursery habitat for marine fishes (McHugh 1980), and an essential habitat for migratory birds (Myers 1983) and can serve as filters to maintain water quality (Schubel and Kennedy 1984). These values, coupled with intense human development that is fragmenting estuarine shores, provide an impetus to focus on shore environments, their evolution, the abiotic and biotic processes shaping them and linkages among them. Other estuarine-related books and monographs provide insight to specific estuarine environments, such as saltmarshes (Chapman 1960; Ranwell 1972; Daiber 1982, 1986; Allen and Pye 1992), seagrass beds (McRoy and Helfferich 1977; Larkum et al. 1989), beaches (Nordstrom 1992), mangroves (Lugo and Snedaker 1974), tidal flats (Reise 1985) and shallow lagoons (Barnes 1980; McComb 1995) or relate to specific areas of interest, including physical processes (Dyer 1973; Officer 1976), geomorphology

Estuarine Shores: Evolution, Environments and Human Alterations. Edited by K. F. Nordstrom and C. T. Roman.

(Perillo 1995), chemistry (Burton and Liss 1976; Neilson and Cronin 1981) and ecology (Day et al. 1989). Other books provide comprehensive coverage of specific estuarine types (e.g. fjords, Syvitski et al. 1987) or diverse topics (Lauff 1967; Cronin 1975; Wiley 1976, 1978; Kennedy 1980, 1982, 1984; Wolfe 1986).

This book is different from the others because it concentrates exclusively on estuarine shores in terms of linkages (the way environments are connected through exchanges of sediments, nutrients and biota). In addition, physical and ecological processes that define the structure and function of these environments are analyzed spatially and temporally from an interdisciplinary perspective. This book is organized into three parts that investigate shoreline evolution, environments and processes, and human alterations and management. Contributions in Part One are case studies of the geomorphology and evolution of estuarine shores of different parts of the world. Part Two includes contributions on mudflats, estuarine beaches, salt marshes and seagrass beds, with a focus on the physical, chemical and biological processes that affect change and linkages among adjacent environments. The final section of the book (Part Three) is an explicit treatment of human impacts on estuarine shorelines with presentation of management and policy alternatives. The purpose of this introductory chapter is to provide an overview of estuarine shoreline environments, their evolutionary trends, and the processes that affect their dynamics and linkages.

1.2 ENVIRONMENTS, EVOLUTION AND PROCESSES

Physical processes are the primary determinants of the type, configuration and distribution of environments or morphologic units that will occur on the estuarine shore. These processes may include streamflow influencing sediment type and sediment transport patterns in the riverine portion of an estuary, and tidal flow that may be more important toward the estuary mouth. Shoreline exposure to waves is also paramount in determining the type of environment that will be established and maintained in a given location. In addition to these macro-scale processes (e.g. tides or surge levels, currents, salinity, wave energy), estuarine elevation gradients (e.g. bathymetry and adjacent upland topography), landscape features (e.g. headlands, coves, human structures) and biotic processes (e.g., primary production, bioturbation) greatly influence the variability of environments encountered along the shores of a typical estuary. The chapters in this book demonstrate the importance of these physical features and processes in defining the spatial distribution of estuarine environments, controlling temporal changes and linking the transport of sediments, nutrients and biotic assemblages among environments.

Figure 1.1 presents an idealized estuary with a variety of shoreline environments, along with key variables affecting the distribution and structure of the environments. The characteristics of estuarine shorelines differ greatly depending

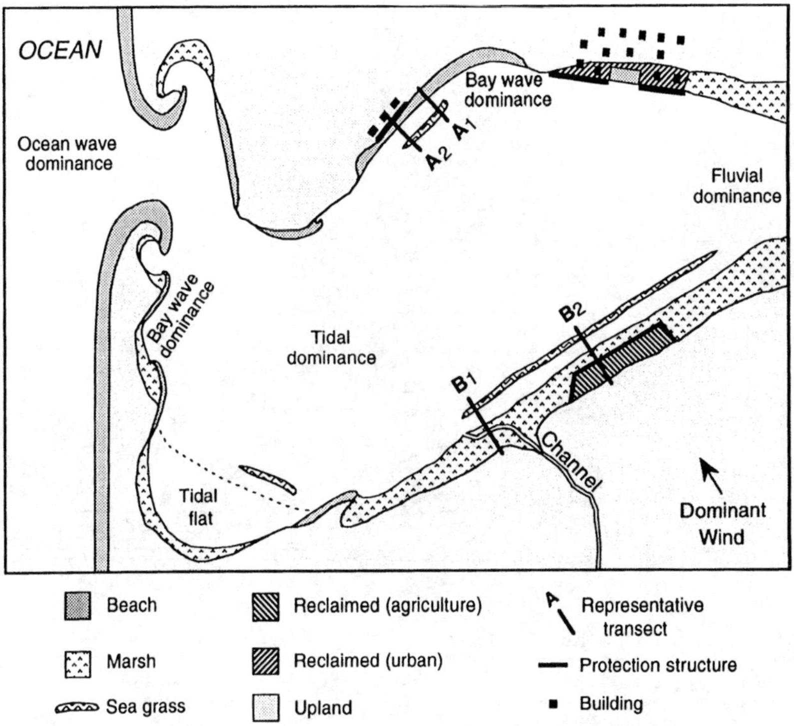

Figure 1.1 An idealized temperate estuary showing dominant shoreline environments and major physical forces. Representative cross-shore transects are portrayed in Figure 1.2

on geologic structure, climate type and the size and stage of evolution of the basins (see Chapters 2 to 6). We have presented an illustration of only one kind of estuary at one point in time. The figure depicts components of a drowned river valley, with significant riverine input and a barrier spit fronting the embayment. Shoreline environments of the idealized estuary are described below and portrayed in Figure 1.2 as linked systems.

Estuarine beaches, defined here as deposits of unconsolidated sediment (sand, gravel, or shell) freely reworked by waves and wave-induced currents, are often found along the shoreline where wave energies are adequate to erode unconsolidated materials from the coastal formations (Figures 1.1 and 1.2A(i)). Beaches also dominate the estuarine shoreline near the ocean inlet where sediments are delivered into the estuary by waves and flood-tide currents and at other locations where sand is plentiful on the bay side of the barrier, particularly associated with overwash platforms or former oceanside dunes (Nordstrom 1992). Small marsh-barrier beaches can form in less energetic areas as a result of erosion of marsh substrate and overwash across the marsh surface (Rosen 1980). Dunes may comprise a subenvironment of estuarine beaches that are sufficiently wide to

2A(i) Beach

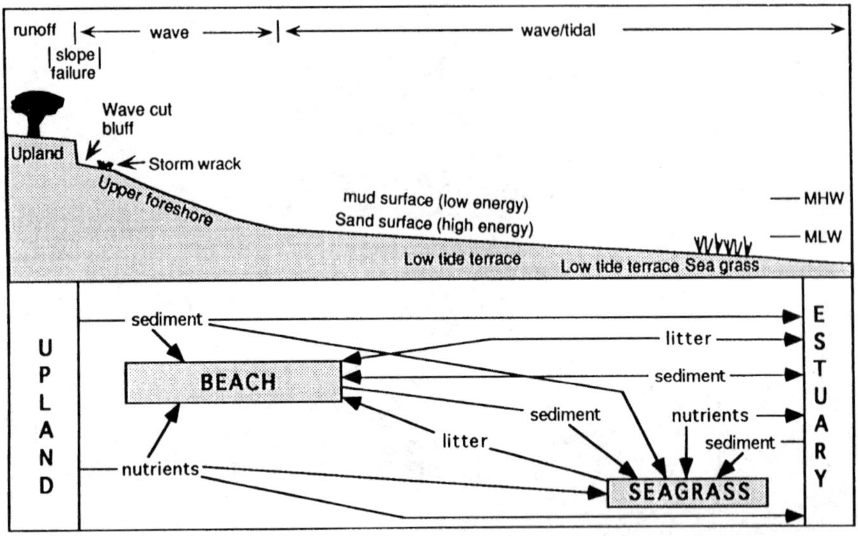

2A(ii) Bulkheaded beach

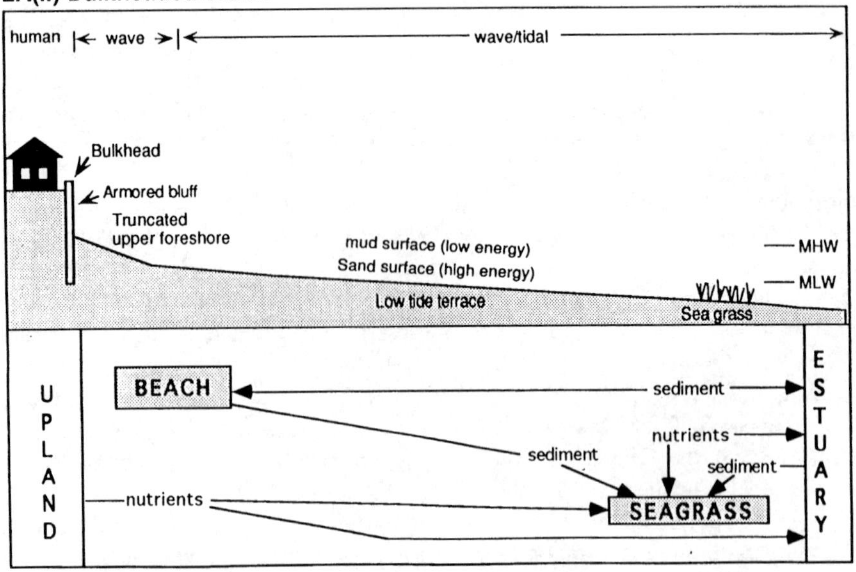

Figure 1.2 Profile view of representative transects depicted in Figure 1.1, showing dominant processes and flow diagram of linkages between estuarine shore environments, the upland, and deeper portions of the estuary. Sediment, nutrient and biotic (fishes and crustaceans) interactions are depicted

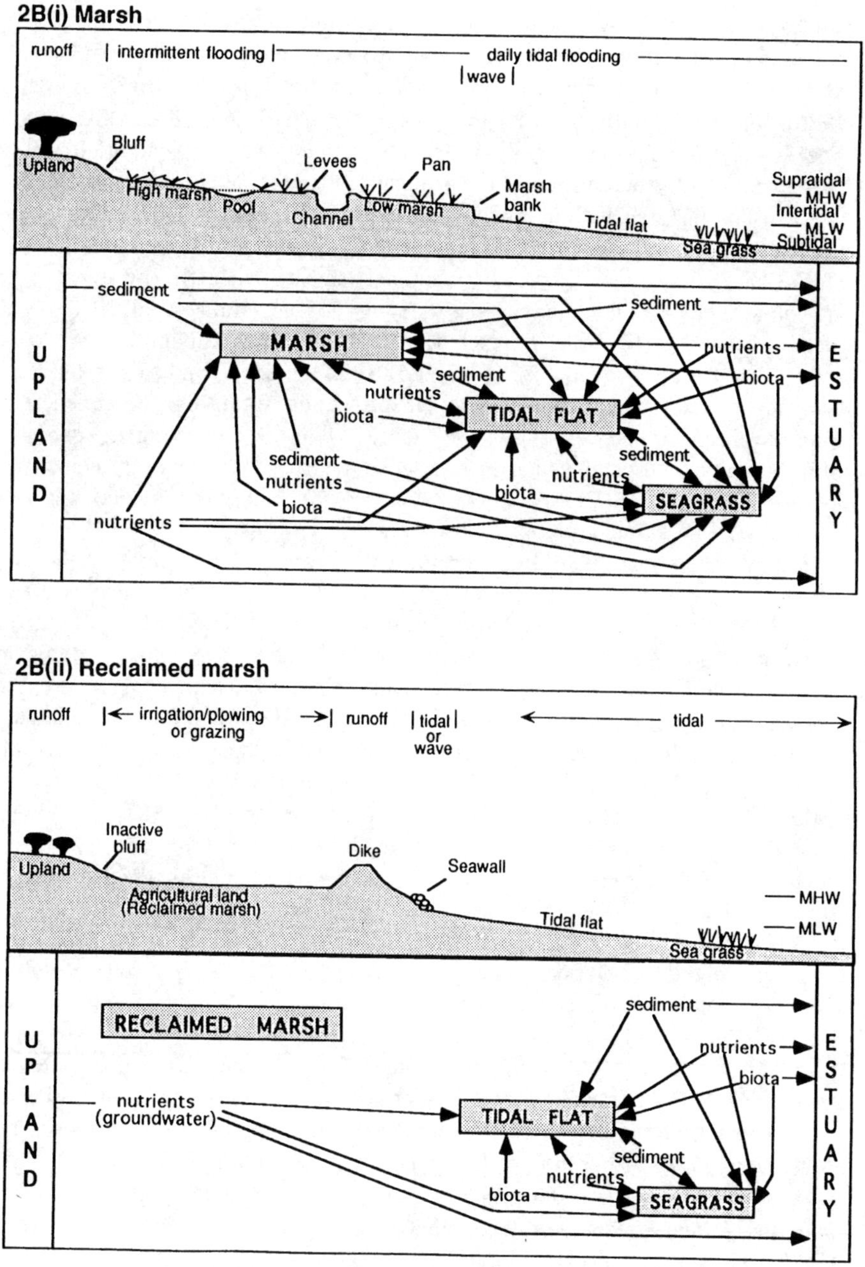

Figure 1.2 (*continued*)

provide a sediment source or where the shoreline is stable enough to allow ample time for slow accretion and prevent wave erosion (Nordstrom and Jackson 1994), although estuarine dunes are usually small and ephemeral (Nordstrom 1992).

Saltmarshes (Figures 1.1 and 1.2B(i)) are accretionary environments accumulating sediment and organic matter in response to sea level rise (Stevenson et al. 1986). Marshes initially form on low-energy shorelines and are characterized by fine sediments and low elevation gradients (Chapman 1960; Redfield 1972). Examples of distinctive morphological marsh units determined by physical gradients or features within estuaries include fluvial marshes, occurring on the upper estuarine margins of rivers; bluff-toe marshes that form at the base of coastal bluffs; back-barrier marshes; and transitional marshes where freshwater marshes are colonized by saltmarsh in response to rising sea level (Kelley et al. 1988). In our idealized estuary (Figure 1.1), marshes are formed on alluvium in the upper reaches, on the upwind side of the estuary and on the downwind side in the lee of bayside spits or headlands that provide protection from high wave energies. Mangrove ecosystems, not shown on Figure 1.1, are found along the shores of tropical and subtropical estuaries, while saltmarshes are found principally in temperate regions (Day et al. 1989). Like marshes, mangrove systems generally accumulate fine-grained, organic substrates.

Seagrass beds and tidal flats are located seaward of beaches, marshes and mangrove systems. Seagrass beds are found along estuarine shorelines worldwide (den Hartog 1970) and are most often associated with sheltered areas of relatively stable substrates. Their role in modifying currents, attenuating wave energies and trapping sediments is well documented (Ginsberg and Lowenstam 1958; Fonseca and Cahalan 1992; Fonseca, this volume Chapter 11). Intertidal flats are ubiquitous in estuaries although their dimensions and locations may differ dramatically. Particular attention has been given to the hydrodynamics of tidal flats and influences on cycles of erosion and deposition (Anderson et al. 1981; Pethick, this volume Chapter 8). Tidal flats are generally considered as barren habitats, although benthic microalgae (Reise 1985), macroalgae (e.g. Gordon et al. 1985; Vadas and Beal 1987) and a rich benthic fauna (Reise 1985; Vargas, this volume Chapter 15) are often conspicuous.

Human-altered shorelines are an integral, and often major, component of estuaries. Bulkheads and dikes are depicted along the shores of our idealized estuary (Figures 1.1 and 1.2A(i)) because of their extensive use. Other shore-parallel structures could include seawalls and revetments. Jetties, often in association with stabilization of inlets or protection of harbors, and groins are other types of ''hard'' shoreline structures. Artificially nourished beaches and constructed dunes represent ''soft'' human-altered shoreline forms. The literature on developed coasts is considerable because of the extent of shoreline alteration and the often dramatic effects that such alterations have on shoreline configuration, land surface elevation, bathymetry, drainage patterns, hydraulic regimes, sediment budgets, and ecological processes (Walker 1988; Nordstrom

1992, 1994; Garniel and Mierwald, this volume Chapter 16; Jackson, this volume Chapter 17).

1.3 ESTUARINE SHORE LINKAGES

Estuarine shoreline environments are not discrete landscape units but are closely linked by hydrological, geomorphological and ecological processes. Linkages among estuarine environments can be expressed as the transport of materials (e.g. shells from tidal flat to beach or sediment from eroding bluff to beach), cyclic coupling (e.g. marsh-tidal flat nutrient cycling), and utilization of multiple habitats by biota (e.g. movement of fishes between marsh surface, tidal channels and seagrass beds). Examples of linkages of sediment, nutrients and biota among estuarine shoreline environments are provided below and depicted in Figure 1.2. Despite a simplistic presentation, this figure demonstrates the complexity of estuarine shoreline linkages and also illustrates that with human development of the shoreline habitat linkages can be greatly simplified and the dimensions truncated.

The input of sediment to estuarine shores is from many sources, including erosion of upland bluffs, upland runoff, aeolian transport from ocean dunes, and transport from ocean inlets, estuarine mouths and deeper portions of estuaries. The interdependence of saltmarshes and mudflats through the cyclic exchange of sediments is one type of linkage between environments (Pethick 1992; Pethick, this volume Chapter 8). Channels serve as conduits for the exchange of sediments between marshes, mudflats and adjacent habitats (French and Stoddart 1992). The role of seagrass beds in determining the character of the shore landward of them is well documented (McRoy and Helfferich 1977). For instance, die-back of seagrass communities can have dramatic effects on nearshore sedimentation processes, especially on the sediment structure and form of estuarine beaches (Rasmussen 1973; Christiansen et al. 1981). The role of seagrass and marsh-derived litter in enhancing aeolian deposition and forming dunes (Hemminga and Nieuwenhuize 1990; Lampe, this volume Chapter 4) and in the accumulation of sediment on estuarine beaches (Nordstrom and Jackson 1992) is demonstrated. Despite their low frequencies, storms may account for much of the sediment exchange within and among estuarine shoreline habitats, including saltmarshes (Stumpf 1983; Reed 1989; Roman and Daiber 1989) and beaches (Nordstrom 1992).

Linkages among adjacent shoreline environments by estuarine-dependent fauna are evident on tidal, diurnal and seasonal time-scales, and can be a function of life history (Kneib and Wagner 1994; McIvor and Rozas, this volume Chapter 13; Able et al., this volume Chapter 14). For example, the saltmarsh mummichog (*Fundulus heteroclitus*) feeds on the marsh surface during high water and returns to creek habitats as the marsh drains (Weisburg et al. 1981). Because of this feeding strategy, it is suggested that mummichogs are an important energy link between the marsh and subtidal food webs (Valiela et al. 1977). The tidal to seasonal migrations of fishes and crustaceans among estuarine habitats and use of multiple habitats (Heck et al. 1989; Sogard and Able 1991; Deegan 1993; Kneib and Wagner 1994; McIvor and Rozas, this volume Chapter 13; Able et al., this vulume Chapter 14; Vargas, this volume Chapter 15) represent significant trophic linkages and transfers of energy and nutrients.

Nutrient linkages among estuarine shore environments are especially complex (Nixon 1980; Neilson and Cronin 1981) due to recycling, transformations, remineralization and fluxes. This book focuses on the role of saltmarshes in intercepting groundwater nutrients prior to transport to estuarine waters, thus providing a critical upland–marsh–estuary link (Howes et al., this volume Chapter 12). Groundwater can be a significant source of terrestrially derived nutrients to estuarine shorelines (Valiela et al. 1990). Regarding seagrass–marsh linkages, it has been argued that decomposing *Zostera* (eelgrass) wrack, when deposited on adjacent saltmarshes, is a process that retains nutrients within the seagrass bed–marsh complex, rather than losing them to other systems (Gallagher et al. 1984). Similarly, Welsh (1980) found a strong coupling of saltmarsh and adjacent mudflat. The mudflat–macroalgal environment tends to trap nutrients (particularly phosphate and ammonia) that are exported from marshes, thus conserving and recycling nutrients within the nearshore zone.

1.4 CONSERVATION AND MANAGEMENT OF ESTUARINE SHORES

Estuarine shores and watersheds are locations of major industrial ports, population centers, recreational sites and agricultural lands. Given these uses and associated impacts, the effective management and protection of estuarine shores become complex endeavors. Several chapters in this book identify impacts on estuarine shores and propose alternative solutions for the protection of natural shoreline functions and the restoration of degraded environments.

Nordstrom (1992) has reviewed the effects of groins, bulkheads, jetties and other structures on the geomorphology of estuarine shorelines, particularly beaches and dunes. Other authors focus on the restriction of tidal flow to saltmarshes caused by roads, bridges and causeways and evaluate the associated ecological effects (Roman et al. 1984) or address the widespread practice of diking saltmarshes for mosquito control, waterfowl habitat enhancement, agricultural reclamation and flood control (Daiber 1986). The impacts of human activities on shoreline geomorphology, hydrology and ecology can be dramatic and efforts are required to minimize impacts of future shoreline developments and restore altered shoreline environments where feasible.

Estuarine shorelines are dynamic features, and coastal managers should incorporate the concept of change into management and policy actions (Davidson et al. 1991). To protect natural elements of estuarine beaches, marshes, mangroves, tidal flats and other shore environments, development must be designed to maintain sediment budgets, minimize alteration of energy fields (e.g. waves, currents), and preserve the opportunity for the landward, seaward or longshore migration of environments. The policy of managed retreat for the United Kingdom (Doody, this volume Chapter 18) adopts this premise of change, whereby saltmarshes that had been diked and reclaimed for agricultural purposes are opened to tidal flow, effectively widening the estuarine shoreline, re-establishing the natural ability to respond to sea level rise, buffer storm wave energies, and restore ecological functions. The loss of estuarine environments and linkages implies that estuarine shoreline management programs should incorporate a policy of habitat restoration as a valuable management tool for preserving habitat diversity in the coastal zone (Zedler 1988; Thayer 1992; Roman et al. 1995).

Eutrophication of shoreline environments from development patterns is becoming increasingly evident in estuaries (Nixon et al. 1986; Valiela et al. 1992). In response to nutrient enrichment, studies have documented increased macroalgal growth in shallow lagoons and often a change of seagrass to macroalgae-dominated communities (Harlin 1995). These structural changes can have profound effects on the functional ecology, geomorphology and nutrient dynamics of estuarine shoreline environments. For decades, regulatory approaches were adopted and refined to control point-source inputs of nutrients to estuaries. More recently, regulatory efforts have been made to control the more diffuse non-point-source inputs from agricultural, urban and rural runoff. Adoption of best management practices, such as buffer strips (Phillips, this volume Chapter 20), technological advances in waste treatment (Burroughs, this volume Chapter 19), and responsible zoning requirements are feasible approaches for alleviating eutrophication of estuarine shoreline environments.

1.5 PROSPECT

This volume is dedicated to understanding the processes and linkages of environments fringing estuarine shorelines. Our purpose in organizing this book was not to provide comprehensive coverage of estuarine shoreline environments but to present examples that highlight the geomorphological, hydrological and ecological processes that define the structure and function of key estuarine habitats and demonstrate their interrelationships.

Contributions to this book focus on the evolution and geomorphology of estuarine shoreline environments (Chapters 2–6), on specific shoreline types and associated processes (Chapters 7–12), on shoreline habitat ecology (Chapters 13–15), and on policy and management implications (Chapters 16–20). We hope this volume will encourage scientists and coastal resource managers to view estuarine shorelines from an interdisciplinary perspective and examine the exchanges of sediment, nutrients, biota and energy that are fundamental to understanding and managing estuarine shores. We include a summary of research needs (Chapter 21), derived largely from concepts presented in the individual chapters, to assist in establishing a research agenda related to linkages among estuarine shore habitats.

REFERENCES

Allen, J.R.L. and K. Pye (eds.). 1992. *Saltmarshes: Morphodynamics, Conservation and Engineering Significance.* Cambridge University Press, Cambridge.

Anderson, F.E., L. Black, L.E. Watling, W. Mook and L.M. Mayer. 1981. A temporal and spatial study of mudflat erosion and deposition. *Journal of Sedimentary Petrology* 51: 729–736.

Barnes, R.S.K. 1980. *Coastal Lagoons: The Natural History of a Neglected Habitat.* Cambridge Univ. Press, Cambridge.

Burton, J.D. and P.S. Liss (eds.). 1976. *Estuarine Chemistry.* Academic Press, London.

Chapman, V.J. 1960. *Salt Marshes and Salt Deserts of the World.* Interscience Publishers, Inc., New York.

Christiansen, C., H. Christoffersen, J. Dalsgaard and P. Nornberg. 1981. Coastal and near-shore changes correlated with die-back in eel-grass (*Zostera marina*, L.). *Sedimentary Geology* 28: 163–173.

Cronin, L.E. (ed.). 1975. *Estuarine Research, Vol. I and II.* Academic Press, New York.

Daiber, F.C. 1982. *Animals of the Tidal Marsh.* Van Nostrand Reinhold Co., New York.

Daiber, F.C. 1986. *Conservation of Tidal Marshes.* Van Nostrand Reinhold, New York.

Davidson, N.C., D. d'A. Laffoley, J.P. Doody, L.S. Way, J. Gordon, R. Key, M.W. Pienkowski, R. Mitchell and K. Duff. 1991. *Nature Conservation and Estuaries in Great Britain.* Nature Conservancy Council, Peterborough, UK.

Day, J.W., Jr., C.A.S. Hall, W.M. Kemp and A. Yanez-Arancibia. 1989. *Estuarine Ecology.* John Wiley and Sons, New York.

Deegan, L.A. 1993. Nutrient and energy transport between estuaries and coastal marine ecosystems by fish migration. *Canadian Journal of Fisheries and Aquatic Sciences* 50: 74–79.

den Hartog, C. 1970. *The Seagrasses of the World.* North-Holland Publishers, Amsterdam.

Dyer, K.R. 1973. *Estuaries: A Physical Introduction.* John Wiley and Sons, London.

Fonseca, M.S. and J.A. Cahalan. 1992. A preliminary evaluation of wave attenuation by four species of seagrass. *Estuarine, Coastal and Shelf Science* 35: 565–576.

French, J.R. and D.R. Stoddart. 1992. Hydrodynamics of salt marsh creek systems: implications for marsh morphological development and material exchange. *Earth Surface Processes and Landforms* 17: 235–252.

Gallagher, J.L., H.V. Kibby and K.W. Shirvin. 1984. Detritus processing and mineral cycling of seagrass (*Zostera*) litter in an Oregon salt marsh. *Aquatic Botany* 20: 97–108.

Ginsberg, R.N. and H.A. Lowenstam. 1958. The influence of marine bottom communities on the deposition environment of sediments. *Journal of Geology* 66: 310–318.

Gordon, D.M., C. van den Hoek and A.J. McComb. 1985. An aegagropiloid form of the green alga *Cladophora montagneana* Kutz. (Chlorophyta, Cladophorales) from southwestern Australia. *Botanica Marina* 27: 57–65.

Harlin, M.M. 1995. Changes in major plant groups following nutrient enrichment, p. 173–187. *In* A.J. McComb (ed.), *Eutrophic Shallow Estuaries and Lagoons*, CRC Press, Boca Raton, Florida.

Heck, K.L., Jr., K.W. Able, C.T. Roman and M.P. Fahay. 1989. Fishes and decapod crustaceans of Cape Cod eelgrass meadows: species composition, seasonal abundance patterns and comparison with unvegetated substrates. *Estuaries* 12: 59–65.

Hemminga, M.A. and J. Nieuwenhuize. 1990. Seagrass wrack-induced dune formation on a tropical coast (Banc d'Arguin, Mauritania). *Estuarine, Coastal and Shelf Science* 31: 499–502.

Kelley, J.T., D.F. Belknap, G.L. Jacobson Jr. and H.A. Jacobson. 1988. The morphology and origin of salt marshes along the glaciated coastline of Maine, U.S.A. *Journal of Coastal Research* 4: 649–665

Kennedy, V.S. (ed.). 1980. *Estuarine Perspectives.* Academic Press, New York.

Kennedy, V.S. (ed.). 1982. *Estuarine Comparisons.* Academic Press, New York.

Kennedy, V.S. (ed.). 1984. *The Estuary as a Filter.* Academic Press, Orlando, Florida.

Kneib, R.T. and S.L. Wagner. 1994. Nekton use of vegetated marsh habitats at different stages of tidal inundation. *Marine Ecology Progress Series* 106: 227–238.

Larkum, A.W.D., A.J. McComb and S.A. Shepherd (eds.). 1989. *Biology of Seagrasses.* Elsevier Science Publishers, London.

Lauff, G.H. (ed.). 1967. *Estuaries.* Publ. No. 83, American Association for the Advancement of Science, Washington, DC.

Lugo, A. and S. Snedaker. 1974. The ecology of mangroves. *Annual Review of Ecology and Systematics* 5: 39–64.

McComb, A.J. (ed.). 1995. *Eutrophic Shallow Estuaries and Lagoons.* CRC Press, Boca Raton, Florida.

McHugh, J.L. 1980. Coastal fisheries, p. 323–346. *In* R.T. Lackey and L.A. Nielsen (eds.), *Fisheries Management.* Blackwell Scientific Publishers, Oxford.

McRoy, C.P. and C. Helfferich (eds.). 1977. *Seagrass Ecosystems.* Marcel Dekker, New York.

Myers, J.P. 1983. Conservation of migrating shorebirds: staging areas, geographic bottlenecks and regional movements. *American Birds* 37: 23–25.

Neilson, B.J. and L.E. Cronin (eds.). 1981. *Estuaries and Nutrients.* The Humana Press, New Jersey.

Nixon, S.W. 1980. Between coastal marshes and coastal waters – a review of twenty years of speculation and research on the roles of salt marshes in estuarine productivity and chemistry, p. 437–527. *In* P. Hamilton and K. MacDonald (eds.), *Estuarine and Wetland Processes.* Plenum Press, New York.

Nixon, S.W., C.A. Oviatt, J. Frithsen and B. Sullivan. 1986. Nutrients and the productivity of estuarine and coastal marine ecosystems. *Journal of the Limnological Society of South Africa* 12: 43–71.

Nordstrom, K.F. 1992. *Estuarine Beaches.* Elsevier Science Publishers, London.

Nordstrom, K.F. 1994. Beaches and dunes of human-altered coasts. *Progress in Physical Geography* 18: 497–516.

Nordstrom, K.F. and N.L. Jackson. 1992. Effect of source width and tidal elevation on aeolian transport on an estuarine beach. *Sedimentology* 39: 769–778.

Nordstrom, K.F. and N. L. Jackson. 1994. Aeolian processes and dune fields in estuaries. *Physical Geography* 15: 358–371.

Officer, C.B. 1976. *Physical Oceanography of Estuaries (and Associated Coastal Waters).* John Wiley and Sons, New York.

Perillo, G.M.E. (ed.). 1995. *Geomorphology and Sedimentology of Estuaries.* Elsevier Science Publishers, Amsterdam.

Pethick, J.S. 1992. Saltmarsh geomorphology, p. 41–62. In J.R.L. Allen and K. Pye (eds.), *Saltmarshes: Morphodynamics, Conservation and Engineering Significance.* Cambridge University Press, Cambridge.

Ranwell, D.S. 1972. *Ecology of Salt Marshes and Sand Dunes.* Chapman and Hall, London.

Rasmussen, E. 1973. Systematics and ecology of the Isef Fjord marine fauna (Denmark) with a survey of the eelgrass (*Zostera*) vegetation and its communities. *Ophelia* 11: 1–507.

Redfield, A.C. 1972. Development of a New England salt marsh. *Ecological Monographs* 42: 201–237.
Reed, D.J. 1989. Patterns of sediment deposition in subsiding coastal salt marshes, Terrebonne Bay, Louisiana: the role of winter storms. *Estuaries* 12: 222–227.
Reise, K. 1985. *Tidal Flat Ecology.* Springer-Verlag, Berlin.
Roman, C.T., W.A. Niering and R.S. Warren. 1984. Salt marsh vegetation change in response to tidal restriction. *Environmental Management* 8: 141–150.
Roman, C.T. and F.C. Daiber. 1989. Organic carbon flux through a Delaware Bay salt marsh: tidal exchange, particle size distribution, and storms. *Marine Ecology Progress Series* 54: 149–156.
Roman, C.T., R.A. Garvine and J.W. Portnoy. 1995. Hydrologic modeling as a predictive basis for ecological restoration of salt marshes. *Environmental Management* 19: 559–566.
Rosen, P.S. 1980. Erosion susceptibility of the Virginia Chesapeake Bay shoreline. *Marine Geology* 34: 45–59.
Schubel, J.R. and V.S. Kennedy. 1984. The estuary as a filter: an introduction, p. 1–11. *In* V.S. Kennedy (ed.), *The Estuary as a Filter.* Academic Press, Orlando, Florida.
Sogard, S.M. and K.W. Able. 1991. A comparison of eelgrass, sea lettuce macroalgae, and marsh creeks as habitats for epibenthic fishes and decapods. *Estuarine, Coastal and Shelf Science* 33: 501–519.
Stevenson, J.C., L.G. Ward and M.S. Kearney. 1986. Vertical accretion in marshes with varying rates of sea level rise, p. 241–259. *In* D.A. Wolfe (ed.), *Estuarine Variability.* Academic Press, Orlando, Florida.
Stumpf, R. P. 1983. The process of sedimentation on the surface of a salt marsh. *Estuarine, Coastal and Shelf Science* 17: 495–508.
Syvitski, J.P.M., D.C. Burrell and J.M. Skei (eds.). 1987. *Fjords: Processes and Products.* Springer-Verlag, New York.
Thayer, G.W. (ed.). 1992. *Restoring the Nation's Marine Environment.* A Maryland Sea Grant Book, College Park, Maryland.
Vadas, R.L. and B. Beal. 1987. Green algal ropes: a novel estuarine phenomenon in the Gulf of Maine. *Estuaries* 10: 171–176.
Valiela, I, J. Costa, K. Foreman, J. Teal, B. Howes and D. Aubrey. 1990. Groundwater-borne inputs from watersheds to coastal waters. *Biogeochemistry* 10: 177–198.
Valiela, I., K. Foreman, M. LaMontagne, D. Hersh, J. Costa, P. Peckol, B. DeMeo-Anderson, C. D'Avanzo, M. Babione, C. Sham, J. Brawley and K. Lajtha. 1992. Couplings of watersheds and coastal waters: sources and consequences of nutrient enrichment in Waquoit Bay, Massachusetts. *Estuaries* 15: 443–457.
Valiela, I., J.E. Wright, J.M. Teal and S.B. Bolkmann. 1977. Growth, production and energy transformations in the salt marsh killifish, *Fundulus heteroclitus. Marine Biology* 40: 135–144.
Walker, H.J. (ed.). 1988. *Artificial Structures and Shorelines.* Kluwer Academic, Dordrecht.
Weisberg, S.B., R. Whalen and V.A. Lotrich. 1981. Tidal and diurnal influence on food consumption of a salt marsh killifish *Fundulus heteroclitus. Marine Biology* 61: 243–246.
Welsh, B.L. 1980. Comparative nutrient dynamics of a marsh–mudflat ecosystem. *Estuarine and Coastal Marine Science* 10: 143–164.
Wiley, M. (ed.). 1976. *Estuarine Processes, Vol. I and II.* Academic Press, New York.
Wiley, M.L. (ed.). 1978. *Estuarine Interactions.* Academic Press, New York.
Wolfe, D.A. (ed.). 1986. *Estuarine Variability.* Academic Press, Orlando, Florida.
Zedler, J.B. 1988. Restoring diversity in salt marshes: Can we do it?, p. 317–325. *In* E.O. Wilson (ed.), *Biodiversity.* National Academy Press, Washington, DC.

Part 1

EVOLUTION OF ESTUARINE SHORES

The discussion in the preceding introductory chapter, and the depiction of estuarine environments in Figures 1.1 and 1.2 refer to the complexity of linkages between shoreline environments at near steady-state conditions. Estuarine shores are much more complex and dynamic than can be depicted in a brief overview, due to different geologic and latitudinal constraints and stages in their evolution. The chapters in this section provide a view of the variability in location, size, orientation and juxtaposition of environments at large spatial (basin or multiple basins) and temporal (centuries to millennia) scales. The chapters indicate that processes operate at different intensities and at different location sin the estuary through time, resulting in a complex mix of active and inactive forms. Thus, proper interpretation of these features requires a knowledge of past processes. Some of the contributions later in the book are also of basin scale, but they have been placed in subsequent sections because of their focus on a specific aspect of the basin (e.g. tidal channels in Chapter 7 and human alterations in Chapter 6).

The chapters in Part 1 include case studies of estuarine systems in five countries from different climate types in the northern and southern hemispheres. We have structured the chapters roughly according to the spatial scale of examination, beginning with the study of one estuary through time and ending with investigation of many estuaries of different origins and stages of evolution. Pye (Chapter 2) demonstrates how shoreline environments and their linkages vary in both cross-shore and longshore directions over a period of several centuries in an estuary in Great Britain. Anthony (Chapter 3) presents an examination of estuaries in Africa and extends the spatial scale of analysis to two different kinds of estuarine systems in a tropical zone. Lampe (Chapter 4) and Woodroffe (Chapter 5) document the many changes that have occurred in their estuarine systems during the Holocene, when differences in sea level caused dramatic changes in coastal sedimentation and landforms, and they identify the significance of geomorphic inheritance. Healy and co-authors (Chapter 6) introduce greater degrees of complexity by describing New Zealand estuaries of diverse origins, structural controls and processes.

Chapter Two

Evolution of the shoreline of the Dee estuary, United Kingdom

KENNETH PYE
Reading University

2.1 INTRODUCTION

Estuaries have long been the subject of intensive scientific interest in view of their importance to a wide range of human activities including navigation, port development and tidal power generation. Numerous studies of estuarine hydraulic behaviour, sediment transport regime and fluxes of nutrients and pollutant substances have been undertaken, but only relatively recently has significant attention been paid to the dynamics of estuarine shorelines and long-term development of estuary morphology (e.g. Allen 1989; Nordstrom 1992; Allen and Pye 1992; Pye and French 1993). In part this growth of interest arises from increasing concern about the effects on estuaries of environmental changes, both natural (e.g. due to sea level rise) and human-induced (e.g. due to tidal barrage construction, dredging, embanking and managed retreat). However, before predictions about morphological responses to future changes can be made with any confidence, there is a need to achieve an adequate understanding of present-day estuarine processes and morphology, including the factors that have influenced past morphological changes in estuaries.

This paper reviews the nature and causes of shoreline changes that have taken place in the Dee estuary, located on the boundary between northwest England and North Wales, UK, during the past 250 years. The Dee has been selected for this purpose because it provides a clear example of the way in which cross-shore variations in morphology, sediment characteristics and habitat type vary spatially within an estuary in response to gradients in energy regime, and how changes in

Estuarine Shores: Evolution, Environments and Human Alterations. Edited by K. F. Nordstrom and C. T. Roman.

one part of an estuary may produce spatially and temporally variable knock-on effects which continue for several decades or even centuries.

2.2 ENVIRONMENTAL SETTING AND PROCESS REGIME OF THE DEE ESTUARY

The Dee estuary is located in the southeastern corner of Liverpool Bay, in the southeastern Irish Sea (Figures 2.1 and 2.2). It has a maximum axial length, orientated NW–SE, of about 30 km and a maximum width near the mouth of 8 km. Before extensive reclamation and channel modification in the inner estuary, the estuary was almost rectangular in form, tapering slightly upstream towards the city of Chester, which was an important port in Roman times (AD 43–410). Some reclamation may have taken place as early as the Roman period, but much of the inner estuary between Chester and Connah's Quay (Figure 2.3) was reclaimed from the mid-18th century onwards. Extensive active saltmarshes which post-date most of the reclamations occupy much of the mid estuary, particularly on the northern (Wirral) shore, giving the unvegetated part of the estuary a funnel shape in plan (Figure 2.2).

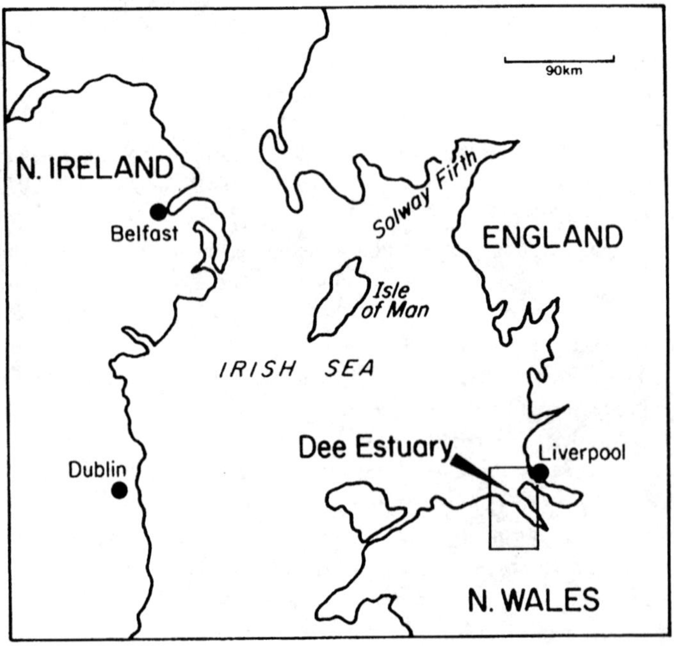

Figure 2.1 Location of the Dee estuary.

Figure 2.2 Satellite image of the Dee estuary and Wirral Peninsula, taken in 1985, showing the large area of active saltmarshes in the southeastern part of the estuary

The estuary occupies a fault trough which is underlain by Carboniferous sandstones, coals and shales and is flanked to the east by Triassic sandstones and conglomerates. The western flank comprises Carboniferous sandstones, coals, limestones and Palaeozoic mudrocks (Wedd et al. 1924). The rock floor beneath the estuary lies at an average of 50 m below Ordnance Datum Newlyn (ODN), with depressions reaching −80 m ODN, and in most places is covered by several tens of metres of glacial till, outwash deposits and Holocene alluvium. Glacial till and outwash also form a discontinuous drift cover on the adjoining areas. Gresswell (1964) suggested that the trough occupied by the Dee estuary represents a glacially over-deepened iceway formed in the late Pleistocene by ice advancing from the northwest, but erosion by sub-glacial or pro-glacial meltwater may also have played a part in its formation (Howell 1973). The lower Dee floodplain consists largely of sediments which were deposited in one or more pro-glacial lakes during the late Pleistocene, overlain by a varying thickness of Holocene alluvium (Wills 1912; Lambert 1988).

The River Dee and its main tributaries rise in the mountains of North Wales, the combined catchment area to Chester weir being 1816 km^2 (Lambert 1988). The river is regulated by a series of reservoirs and the total freshwater discharge (average 36.8 $m^3\ s^{-1}$; Turner et al. 1994) represents only about 14% of the combined estuarine discharge (Parr 1988). Chester weir, originally built in Norman times (11–12th century AD), creates a backwater effect for the upstream section of

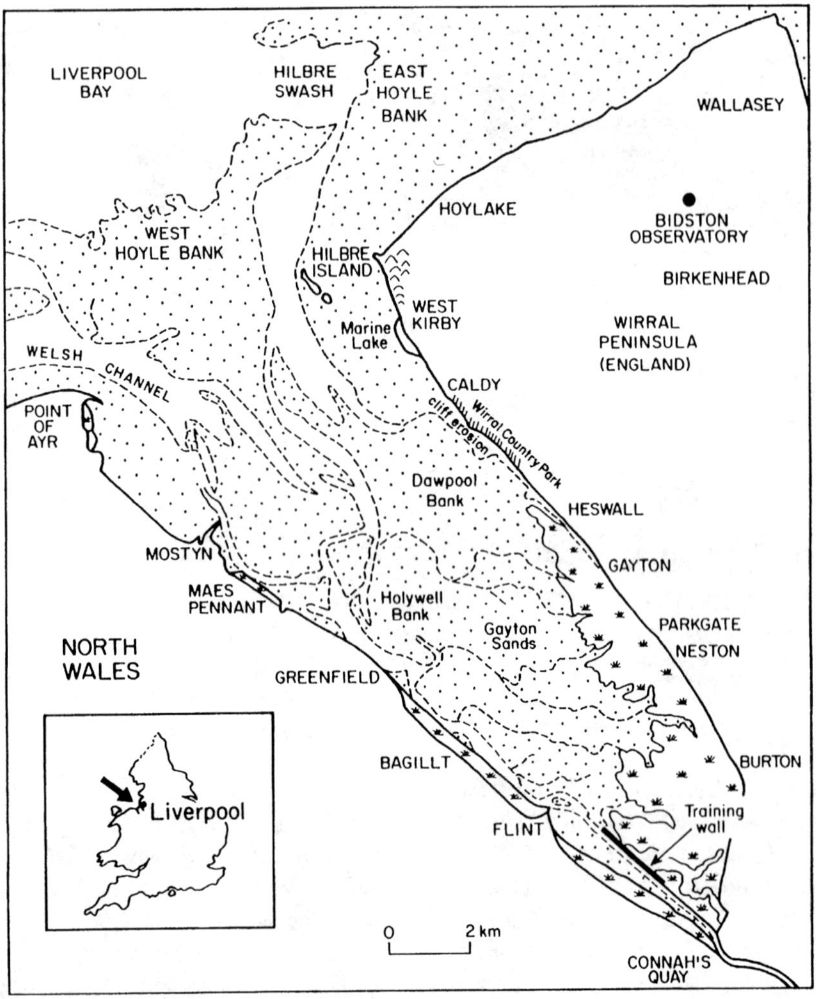

Figure 2.3 Major intertidal morphological features of the Dee estuary, based on Admiralty Chart 1978 (Admiralty 1991) and Ordnance Survey maps

the river and provides the normal limit for saline water influence. However, extreme high tides may exceed the height of the weir and can cause temporary reversals of flow up to 20 km upstream from Chester (Lambert 1988).

The sides of the modern estuary are almost straight and, particularly on the Wirral side, are marked by a low cliff or break of slope which is cut mainly into glacial till. Raised sand and gravel beach deposits occur at c. 4.0 m ODN in several places along the Welsh shore and near Burton on the Wirral shore, suggesting the possibility of a slightly higher energy regime (and possibly sea level) at some time earlier in the Holocene.

Table 2.1 Summary of tidal levels at stations in the Dee estuary. Heights are given in metres ODN. MHWS, mean high water spring; MHWN, mean high water neap; MLWN, mean low water neap; MLWS, mean low water spring; MSTR, mean spring tidal range; MNTR, mean neap tidal range — no data

Station	MHWS	MHWN	MLWN	MLWS	MSTR	MNTR
Hilbre Island	4.07	2.27	−1.83	−3.63	7.70	4.10
Mostyn Quay	4.00	2.20	—	—	—	—
Connah's Quay	3.95	2.25	—	—	—	—
Chester	4.60	2.60	—	—	—	—

(Data taken from Admiralty 1991;1993)

The estuary is macrotidal, mean spring and mean neap tidal ranges at Hilbre Island being 7.7 m and 4.1 m respectively (Table 2.1). Near the head of the estuary, at Connah's Quay, the mean spring tidal range is approximately 3.4 m and the mean neap tidal range is approximately 1.7 m. The tides are semi-diurnal. At the entrance to the estuary the flood tide is of shorter duration than the ebb (Figure 2.4), and average flood tidal current velocities are higher than the average ebb velocities. At Hilbre Island, an average spring flood tide floods for 6 h and 35 min and ebbs for 6 h 50 min. The tidal wave is progressively modified as it moves up the estuary, the slightly funnel-shaped morphology reinforcing the tendency for flood dominance. At Connah's Quay an average spring tide floods for only 2.5 h and ebbs for 9 h. On neap tides there are no tidal effects in the canalized channel above Connah's Quay, but during high spring tides a bore up to 1 m high may run up the channel as far as Chester weir (Lambert 1988). The flood-dominated velocity asymmetry, combined with the processes of "settling lag" and "scour lag" (Postma 1967), favour net accumulation of sediment within the estuary.

The highest predicted tides, exceeding 4.5 m ODN, occur around the time of the spring and autumn equinoxes (February–March and September–October). During the winter months storm surges can raise predicted high water levels by up to 2 m, although surges of >1.5 m are exceptional (Graaf 1978). In recent decades, major storm surges occurred in 1976, 1977, 1983 and 1990.

At low water of spring tides, extensive sand flats and banks are exposed within the estuary and in the neighbouring part of Liverpool Bay (Figure 2.3). The total intertidal area is approximately 13 000 ha, of which saltmarsh comprises more than 2100 ha (Davidson et al. 1991). These areas provide imporant feeding grounds for migrant birds (Prater 1981), and consequently the estuary has been designated as a site of major conservation importance. A management strategy for the estuary is currently in preparation (Jemmett 1994; 1995a,b).

The banks and channels in the outer estuary are continually changing, but at present there are two main deep water channels, the Welsh Channel and the Hilbre Swash Channel, separated by the West Hoyle Bank (Figure 2.3) which is important

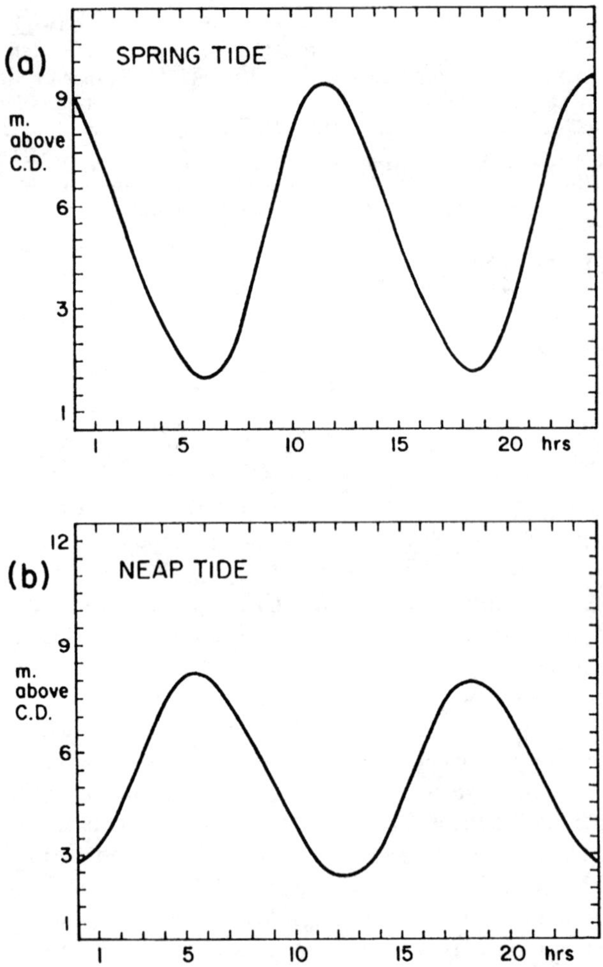

Figure 2.4 Typical tidal curves for (a) a spring tide and (b) a neap tide at Hilbre Island (after NERC 1992)

in restricting wave ingress into the estuary. Liverpool Bay and the Dee estuary are relatively sheltered from the prevailing southwesterly winds, and the dominant waves approach the estuary from the northwest and west (Figure 2.5). The northeastern shore between West Kirby and Heswall has the greatest exposure to wave action, although the northwestern shore between Point of Ayr and Mostyn periodically experiences moderate to high waves from the north-northwest. Wave data recorded at the Mersey Bar light vessel in the late 1950s indicated a mean annual wave height of <1 m, but deepwater waves with heights greater than 5 m can occur in winter (Draper and Blakey 1969). Nearshore wave recordings over a

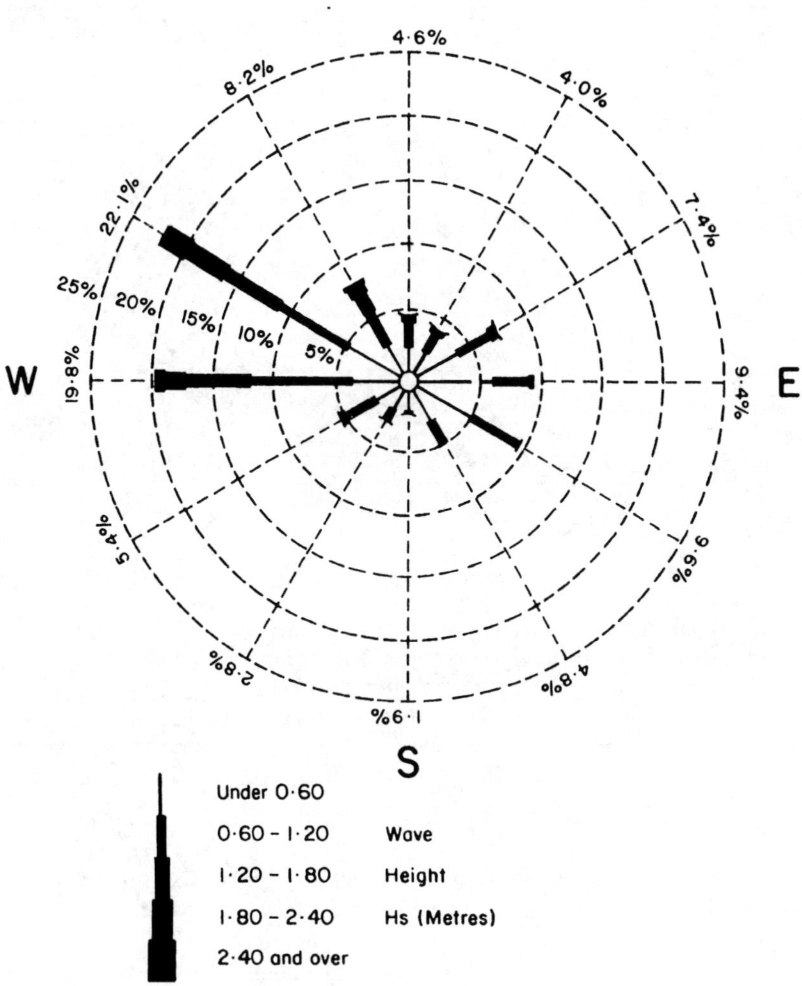

Figure 2.5 Wave rose for the North Wales coast, hindcast using HR Wallingford's HINDWAVE model and wind data for Squiresgate Airport, Blackpool, 1970–1990 (after Jelliman et al., 1991)

one-year period off Hilbre Island indicate a wave height of 1 m exceeded only 3–4% of the time, corresponding to a wave period of 4.0–4.5 s (EAU, 1987). Model studies suggest that wave direction within the Dee estuary is rapidly modified by the banks and channels at the estuary mouth, and that there is a significant refraction-induced reduction in height as waves penetrate the estuary (BMT 1988; Shoreline Management Partnership 1993).

The results of sea bed drifter studies and analyses of sand wave orientations on the sea floor suggest a strong net easterly to southeasterly movement of sediment

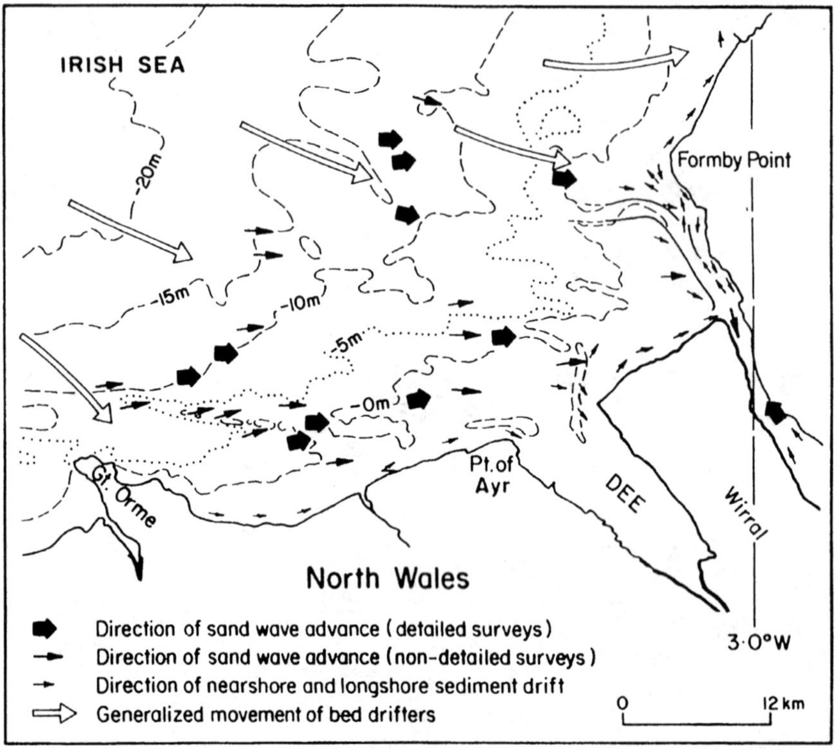

Figure 2.6 Directions of net sediment transport in the southeastern Irish Sea, based on evidence compiled by Sly (1966) and Hydraulics Research (1990)

along the North Wales coast towards Liverpool Bay and the Dee estuary (Sly 1966; Welsby and Motyka 1989; Hydraulics Research 1990; Figure 2.6).

Compared with the neighbouring Mersey estuary, which has been more intensively studied, relatively little is known about patterns of water mixing, residual current circulation and sediment transport within the Dee estuary. Preliminary field and laboratory studies, carried out by Hydraulics Research (1970) in connection with a proposed water storage scheme, suggested a pattern of net seaward surface movements and net landward movements near the bed, but the pattern is complicated by the banks and channels. A sampling exercise undertaken in December 1991 indicated that suspended particulate matter concentrations are low in comparison with other UK macrotidal estuaries such as the Mersey and the Severn (Turner et al. 1994), and there is no marked turbidity maximum. In this study, a dominance of sandy suspended sediment was observed throughout the estuary. Based on a consideration of geochemical evidence, Turner and colleagues estimated that <0.5 % of the sediment in the Dee estuary is derived from fluvial sources, the remainder originating from marine sources outside the estuary.

2.3 NATURE OF THE ESTUARINE SHORELINE

At the northwest corner of the Dee estuary lies the Point of Ayr, a complex of recurved shingle spits and cheniers, partly capped by dunes, which have formed in response to west to east littoral drift of sediment along the North Wales coast (Welsby and Motyka 1989). At the present time the dune frontage near the Point of Ayr lighthouse is eroding, and sand is being transported into the estuary by littoral drift. On the eastern side of the Point of Ayr, an area of sand ridges and saltmarsh suffered significant erosion during the first half of this century (Flint County Council 1958).

Between the Point of Ayr and Mostyn Quay the Welsh shore is backed by a railway embankment, largely reinforced by masonry walls, which encloses a strip of reclaimed saltmarsh. Up the estuary from Mostyn Quay, which is the only harbour in the estuary now capable of handling large sea-going ships, the shoreline is fronted by a mixture of made ground, composed largely of industrial waste and locally protected by concrete blocks, stone rip-rap and masonry walling, with residual areas of fringing saltmarsh (Figure 2.7). The saltmarsh edge shows widespread erosion and attempts have been made to provide protection using gabions, armourstone or by dumped building rubble (Figure 2.8). At Bagillt a licensed landfill site extends onto the saltmarsh and is surrounded by low embankments constructed of colliery waste. Elsewhere the active marshes are backed by grassed embankments, locally reinforced by concrete revetments, and areas of industrial waste material.

Along the northernmost part of the Wirral Peninsula shore, between Hilbre Point and West Kirby, unprotected dunes are fronted by a wide intertidal sand flat which forms the western end of East Hoyle Bank (Figure 2.3). Immediately to the north of West Kirby the dune frontage adjoining the Royal Liverpool Golf Links is protected by rock armour. A variety of types of seawall, revetments and rock armouring extend 2.8 km southwards along the frontage between the Marine Lake and Caldy Golf Club. The beach along this section of shore is predominantly sandy, wide and low angle (typically 2–3 deg), but from Caldy to Heswall the character of the intertidal zone is much more variable, consisting of a wave-cut platform eroded in glacial till which is partially overlain by a relatively steep (4–5 deg) upper beach consisting of coarse sands, mixed pebble–sand deposits and cobble lag deposits (Figure 2.9). The lower beach typically has a lower angle (2–3 deg) and is composed of finer sand with fewer pebbles. The beach narrows and steepens towards the south owing to the existence of a tidal channel which comes closer to the shore.

The shore between West Kirby and Heswall is backed by soft cliffs composed largely of glacial till, pro-glacial meltwater deposits and a thin capping of windblown sand and silt, with a few localised outcrops of Triassic sandstones and conglomerates (Slater 1929; Pitts 1983). The cliff section which is backed by the Wirral Country Park, together with a small area owned by the National Trust, is largely unprotected and is currently eroding, with numerous landslips evident following the winter high tides of 1993–94. The cliffs at the northern end of the

Figure 2.7 Unprotected eroding marsh near Bagillt on the Welsh shore of the estuary (August 1995)

Figure 2.8 Building rubble forming ad hoc protection at the edge of an eroding marsh near Bagillt on the Welsh shore of the estuary (August 1995)

Figure 2.9 Eroding cliffs of glacial till, fronted by an erosional wave cut platform and cobble lag deposit, on the Wirral Country Park frontage (view looking southeast, April 1994)

Wirral Country Park have a maximum height of about 18 m, but they become lower southwards towards Heswall and northwards towards Caldy. Erosion of the cliffs has occurred for at least 80 years through a combination of sliding, slumping, freshwater overflows at the cliff top and wave-induced toe erosion. Erosion during this period has been exacerbated by a landward movement of the tidal channel which runs parallel with the cliffline. Attempts to slow the rate of cliff retreat have been made in a number of places and have involved provision of cliff drainage systems, cliff re-profiling and toe protection (e.g. Pitts 1983). For example, works were carried out along the frontage of the Caldy Golf Club in 1956, 1983 and 1991, on the latter two occasions following severe storm surges in January 1983 and February 1990 (Fahy et al. 1993). The cliff section between the northern end of Caldy Golf Club and West Kirby, which is backed by residential property, is now entirely defended by rock armour, gabions, short sections of concrete revetment, and sandstone block walling (Figure 2.10).

Coast protection works, consisting of steel sheeting, concrete walls and armourstone revetments, have also been undertaken since 1930 between the Heswall sewage works and Gayton. The northern saltmarsh limit now lies just north of the sewage works (Figure 2.11), and to the south of this point the marsh provides a degree of shelter for the shoreline. The main belt of marsh is separated from the shore by a 10–20 m wide tidal channel (Figure 2.11), to landward of which is a 20–50 m wide belt of *Puccinellia–Aster–Limonium*-dominated fringing marsh (Figure

Figure 2.10 Soft cliffs composed of late Pleistocene glacial deposits now largely protected by sandstone walls, gabions and armourstone revetment (view from the northern end of Caldy Golf Club towards West Kirby, April 1994)

2.12). Landward of this fringing marsh is a high tide sandy beach, of width 10–15 m and composed of medium textured, moderately to well sorted sand (Figure 2.12). This is backed by a low wave-cut cliff which has been partly buried by 1–2 m of recent windblown sand. Progressive northward extension of the saltmarsh has led to gradual stabilization of the formerly eroding cliffline, followed by vertical accumulation of windblown sand to form low dunes.

The tidal channel which runs parallel to the shore extends as far as the northern end of Parkgate, but the high tide sand beach is present only to a point midway between Gayton and Parkgate. South of this location the saltmarsh, fringed by a zone of *Phragmites* reeds along the tidal limit, extends right up to a masonry seawall and promenade built in 1900 (Figure 2.13).

2.4 SHORELINE CHANGES IN THE HISTORICAL PERIOD

The Dee estuary has been silting up at least since the eleventh century. Around the second century AD, Chester was established as a Roman garrison and port at the head of a deep, sheltered estuary. In the twelfth century, problems with siltation were already being attributed to reduced scour following the construction of a weir to drive watermills at Chester. By 1450, an outport had been established at

Figure 2.11 Vertical air photograph taken in November 1986 showing the northern limit of accreting saltmarsh near Heswall sewage works (the sewage outfall pipe is labelled SO). The zone of fringing marsh and high tide sand beach are labelled fm and sb respectively. Reproduced by permission of Cambridge University Collection of Air Photographs; copyright reserved. Scale 1:5000

Shotwick, 8 km further downstream. Continued problems with silting led to construction of a new port (New Haven) at Neston in the mid-sixteenth century, but it too was eventually abandoned, becoming known as the Old Quay, while a New Quay was established 0.8 km further downstream near Parkgate. Between 1730 and 1830 Parkgate was of great commercial importance, providing the main ferry embarkation point for Dublin and Bagillt (Ashton 1920; Pearson 1985).

The earliest available charts, dating from the seventeenth and early eighteenth centuries, show the main low water channel of the Dee on the Cheshire side of the estuary and a wide belt of fringing saltmarsh along the Welsh shore (Figure 2.14A).

Figure 2.12 The high tide sand beach and fringing saltmarsh, north of Gayton on the Cheshire shore. The beach is backed by a low cliff cut in glacial till which is partly buried by windblown sand (April 1994)

The channel was evidently navigable as far as Parkgate at low water. The channel lay in approximately the same position in 1732 (Figure 2.14B), when Nathaniel Kinderley obtained Parliamentary authority to construct a straight channel, known as the New Cut, between Chester and Wepre Brook near Connah's Quay with the objective of providing improved navigation access to Chester. Construction of the New Cut embankments between 1732 and 1737 also allowed the marsh and intertidal flats between Chester and Shotwick to be reclaimed (Figure 2.9). The effect of these works was to shift the main channel of the river to the Welsh side and dramatically to reduce the tidal scour on the Cheshire shore between Burton Point and Parkgate. Boydell's map of 1771 (Figure 2.14B) shows that the main channel turned sharply to the right at the end of the New Cut towards Parkgate, from where it continued along the Cheshire shore to beyond Heswall, before swinging northwestwards again.

The embankment and training walls of the New Cut were extended several times during the nineteenth century, progressively forcing the main Dee channel away from Parkgate and further down the estuary (Figure 2.14C). This further enhanced sedimentation along the Cheshire shore between Burton Point and Heswall and allowed further reclamation of the saltmarsh between Shotwick and Burton Point between 1800 and 1870 (Figure 2.15). A bank was constructed from Burton Point towards the end of the New Cut in 1875. Although this bank was

Figure 2.13 Vertical air photograph taken in November 1986 showing the mature saltmarsh developed in front of the promenade at Parkgate. This marsh and its associated dendritic drainage system developed on a planar sand flat after 1947. Reproduced by permission of Cambridge University Collection of Air Photographs; copyright reserved. Scale 1:5000

breached by storms in 1877 it continued to provide shelter and to enhance mud accretion on the developing marsh behind it. In 1898 another embankment was built across the mid part of this marsh to convey the Neston–Hawarden railway, and the area on its landward side was then reclaimed. In 1916 the breaches in the 1875 wall were sealed, and the area behind reclaimed and subsequently developed as the site for a steelworks, paper mill and rifle range. Further reclamation at the seaward edge of the area took place in the mid-1980's when some $1 \times 10^6\,m^3$ of sandy sediment was pumped from the White Sands Gutter to raise the land level.

On the Welsh side of the estuary an area of approximately 1850 ha was

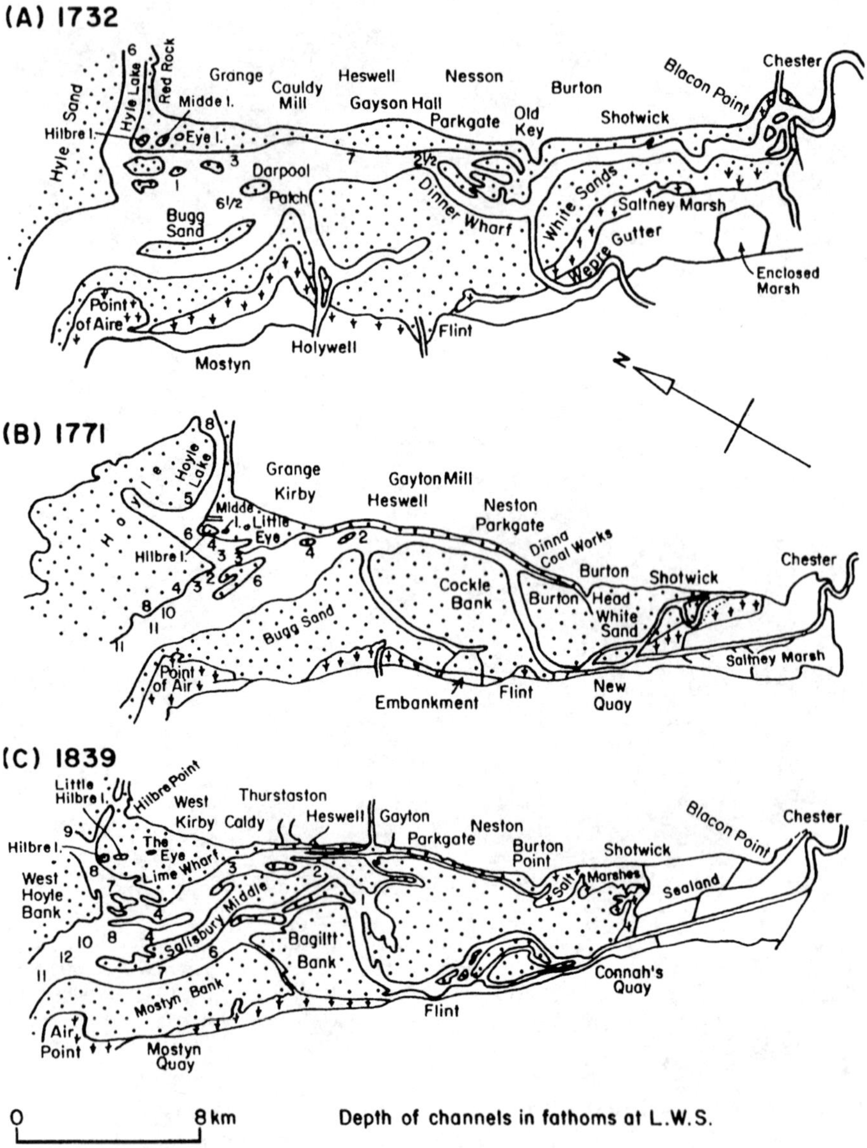

Figure 2.14 Intertidal morphology of the Dee estuary as indicated by charts published in (A) 1732, (B) 1771 and (C) 1839. Modified after Stopford (1951).

reclaimed before 1870, with further extensive reclamations around Connah's Quay, Flint and Mostyn after 1916. The total area of the estuary reclaimed for agricultural and industrial purposes since 1732 was estimated by Fahy and co-authors (1993) to be at least 4950 ha, or approximately 24% of the original area of the estuary.

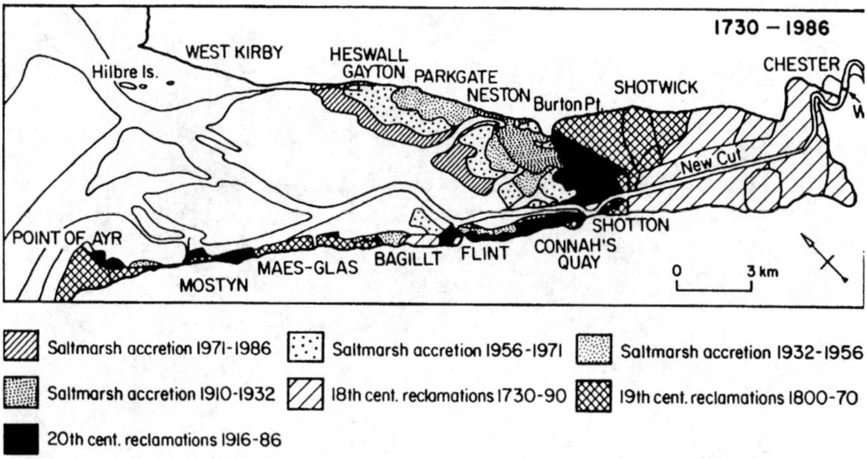

Figure 2.15 Reclamations in the Dee estuary 1730–1986 (modified after Davidson et al. 1991)

Movement of the main Dee channel towards the Welsh shore led to steepening of the foreshore and rapid saltmarsh erosion between Mostyn and Holywell. In 1834 the embankment along this section of coast was fronted by a belt of marsh more than 400 m wide and the main channel lay some 2 km offshore; by 1898 the channel had been forced to within 300 m of the embankment and only 25 m of saltmarsh remained. In 1904 the embankment at Holywell was breached and required repair by repeated slag tipping (Marker 1967).

The training of the upper Dee channel probably also caused changes in the banks and channels at the entrance to the estuary. As early as 1771 the momentum of the flow down the main low water channel had partially cut a new channel through the middle of the Hoyle Bank (Figure 2.14C), and by 1813 this channel completely separated the West Hoyle Bank from the East Hoyle Bank (Hydraulics Research Wallingford 1990; Shoreline Management Partnership 1990).

Despite the shift of the main low water channel, the shoreline north of Burton Point remained sandy throughout the late eighteenth and nineteenth centuries. Parkgate became a fashionable bathing resort after 1780 and a stone promenade was built during the period 1810–14. At this time there was still a relatively deep low water channel, and even as late as the First World War fishing boats were moored opposite the Parkgate promenande. However, progressive shallowing of the channel contributed to the discontinuation of the Irish ferry service in 1830, and the popularity of Parkgate as a resort diminshed, although it continued to be used as a bathing beach by the local population until the Second World War (Figure 2.16).

By 1938 saltmarsh was established at Moorside, 0.8 km south of Parkgate, but at Parkgate itself there was still a sandy beach covered by every high tide. The first *Salicornia* seedlings were recorded on Parkgate beach in 1939 and by 1947

Figure 2.16 Children paddling on Parkgate beach in the 1920s

saltmarsh had colonized the entire beach as far as the slipway at the northern end of the frontage (Marker 1967). In 1947 the width of the marsh was 450 m, increasing to 1200 m by 1967. Immediately adjacent to the seawall, a thickness of 523 mm of mud accumulated on top of the beach between 1938 and 1947 (58 mm a^{-1}); a further 238 mm accreted between 1947 and 1963 (average rate 15 mm a^{-1}) (Marker 1967). At the time of writing the marsh in front of the promenade is more than 1.2 km wide and is sufficiently high to allow growth of *Phragmites* reeds along the drift line (Figure 2.17).

The main native pioneer marsh species in the Dee are *Salicornia* spp, and *Puccinellia maritima*, but since the Second World War marsh growth has been heavily influenced by the spread of *Spartina* spp, which was first introduced to a site near Connah's Quay in 1928–29 (Webster 1930). The initial plantings were unsuccessful (Massey 1937), and it was not until after 1947 that extensive areas of *Spartina* marsh developed between Parkgate and Heswall (Marker 1967; Taylor and Burrows 1975). During the mid-1950s *Spartina* invaded the open coast recreational beaches at Hoylake on the north coast of the Wirral Peninsula, requiring investigation of its control by spraying (Taylor and Burrows 1968, 1975). The reasons for this rapid expansion are not entirely clear, but may be related to a decrease in mean wind speed and wave activity within the estuary which favoured greater stability of the intertidal sediments (Pye unpublished data). *Spartina* in the Dee is found between approximately Mean High Water Neap (MHWN) tide level (2.2–2.3 m OD) and Mean High Water Spring (MHWS) tide level (3.95–4.05 m OD). It appears to have a competetive advantage over other species at lower elevations and is able to spread rapidly to fill an essentially unoccupied niche when growth conditions are favourable (Gray 1992). Since the 1960s *Spartina* has been

Figure 2.17 The promenade and fronting marsh at Parkgate in February 1994

largely replaced by other species (including *Halimione, Aster, Limonium, Puccinellia, Festuca* and *Agrostis*) at Parkgate and Heswall as the marshes have grown higher in the tidal frame and *Spartina* has been out-competed. Today it occurs mainly as a fringe along the accreting northwestern margin of the saltmarsh zone (Figure 2.11; White 1982; Hill 1986, 1987; Gall 1988). The rate of northerly advance of the saltmarsh front has slowed in recent years, partly reflecting higher wave energy towards the mouth of the estuary, and partly a regional increase in mean wind speed (and, by inference, wave height; Pye unpublished data).

Studies of ^{137}Cs concentrations in the marsh sediments along the Wirral shore (Hutchinson and Prandle 1994; Hutchinson 1994) have provided additional information about spatial variations in vertical marsh accretion rates as a function of marsh age. The main source of ^{137}Cs in the Irish Sea is the Sellafield nuclear reprocessing plant, located 130 km to the north of the Dee estuary on the coast of Cumbria. Peak discharges of ^{137}Cs from Sellafield occurred between 1974 and 1978, with virtually no discharges before 1955 and after 1986. Analysis of sediment cores taken from the inner parts of Burton Marsh showed peak ^{137}Cs concentrations at a depth of <100 mm, indicating average accretion rates of <5 mm a^{-1} in the last 20 years. Cores from the upper and middle marshes at Parkgate showed peak concentrations at a depth of 180–200 mm, implying average sedimentation rates of approximately 10 mm a^{-1} over the same period. The youngest marshes near Heswall showed peak ^{137}Cs concentrations at 400–500 mm depth indicating vertical accretion rates of >20 mm a^{-1}. Considering the magnitude of both vertical and lateral marsh accretion, Hutchinson and Prandle (1994) estimated that net mud accretion on the Dee saltmarshes in the past 20 years amounts to approximately $6\times106\,m^3$. Since only a small fraction of this amount could have been supplied by recorded rates of lateral cliff recession between Caldy and Heswall (average about $0.3\,m\,a^{-1}$), or from fluvial sources, these authors concluded that the bulk of the material must have been transported into the estuary by tidal exchange processes.

On the basis of a comparison between the Admiralty survey of 1888 and an aerial survey of the estuary in 1965, Hydraulics Research (1970) estimated total net sediment accumulation in the estuary of $173\times106\,m^3$, at an average annual accumulation rate of about $2\times10^6\,m^3a^{-1}$. Analysis of volumetric changes in the banks of the outer estuary, downstream of Mostyn, based on sequential chart surveys, suggested average annual accretion during the last 130 years of about $0.5\times106\,m^3\,a^{-1}$ (BMT Ceemaid 1990). Of the total net accretion during the 130-year period, more than 85% occurred between 1929 and 1987. The reasons for the apparently enhanced accretion rate in the estuary during the last 50 years are not fully understood. Canalization, embanking and reclamation in the inner estuary may have had a significant effect by reducing the tidal prism and the average velocity of ebb tidal currents, while the rapid expansion of saltmarsh following the establishment of *Spartina* may have had a similar effect. A further factor may be the ending of channel migration in the inner and middle estuary. Before the position of the deep water channel in the mid and inner estuary was fixed, migration of the channels caused episodes of saltmarsh and intertidal flat erosion, providing a natural control on the saltmarsh area within the estuary. Following training of the channel, this natural sediment flushing mechanism became less effective, allowing progressive storage of sediments in the intertidal zone.

2.5 CONCLUSIONS

Shoreline changes in the Dee estuary during the past 200 years have, in large part, been determined by human activities in the inner estuary, particularly canalization, stabilization of the low water channel and land reclamation. The most important effects have been the initiation of saltmarsh and intertidal flat erosion along the Welsh shore, and the initiation of rapid intertidal flat and saltmarsh accretion on the Wirral shore. These contrasting erosion/accretion regimes are reflected by differing cross-shore patterns of the sediments and landforms on the two sides of the estuary. Important longshore variations are also evident and reflect the transition from a low energy environment at the estuary head to a relatively higher energy environment near the estuary mouth. In general, the Welsh shore is characterized by a relatively steep intertidal gradient, erosional intertidal flats and receding saltmarsh cliffs which have been modified to varying degrees by ad hoc coast protection works. The southern part of the Wirral shore is characterized by a wide intertidal zone of low gradient, slowly accreting intertidal flats and relatively stable, vertically accreting saltmarshes. The landward margin of the saltmarsh zone is experiencing a transition to brackish conditions, with local development of *Phragmites* and *Scirpus* reed communities. The saltmarshes are still actively spreading towards the northwest, but at a much reduced rate compared with the 1950s to 1970s. Longshore variations in the width and height of the saltmarsh, combined with increasing water depth and wave exposure towards the estuary mouth, produce a spatial sequence of shoreline characteristics which is also changing slowly with time. Beyond the northern limit of the saltmarsh, sand and mixed sand–pebble beaches are backed by unprotected soft cliffs which still experience periodic erosion, especially during storm surges. To the south of the northern saltmarsh limit, wave activity at the shoreline has been reduced sufficiently to bring cliff erosion to an end, and windblown sand derived from a high tide beach has accumulated as low dunes. Still further south, wave activity has been reduced by the greater height and width of the saltmarsh zone, allowing reed marsh vegetation to invade the former high tide beach. A similar spatial and temporal sequence of shoreline changes could be expected in other estuaries in which there has been rapid lateral and vertical growth of saltmarsh.

REFERENCES

Admiralty. 1991. *Great Ormes Head to Liverpool. Admiralty Chart 1978.* Admiralty, London.

Admiralty. 1993. *Admiralty Tide Tables* 1994. Hydrographer of the Navy, Taunton.

Allen, J.R.L. 1989. Evolution of saltmarsh cliffs in muddy and sandy systems: a qualitative comparison of British west-coast estuaries. *Earth Surface Processes and Landforms* 14: 85–92.

Allen, J.R.L. and K. Pye (eds.). 1992. *Saltmarshes – Morphodynamics, Conservation and Engineering Significance.* Cambridge University Press, Cambridge.

Ashton, W. 1920. *The Evolution of a Coastline. Barrow to Aberystwyth and The Isle of Man.* Edward Stanford, London, and W. Ashton, Southport.

BMT Ceemaid Ltd. 1988. *Flint By-Pass Wave Study.* Contract Report No. 4635500, prepared for Flint County Council.

BMT Ceemaid Ltd. 1990. *Wirral Dee Study.* Contract Report No. 46287, prepared for Wirral Metropolitan Borough Council.

British Geological Survey. 1984. *Liverpool Bay. Sheet 53N 04W. Sea Bed Sediment and Quaternary.* Ordnance Survey, Southampton.

Davidson, N.C., D. d'A. Laffoley, J.P. Doody, L.S. Way, J. Gordon, R. Key, C.M. Drake, M.W. Pienkowski, R. Mitchell and K.L. Duff. 1991. *Nature Conservation and Estuaries in Great Britain.* Nature Conservancy Council, Peterborough.

Draper, L. and A. Blakey. 1969. *Waves at the Mersey Bar Light Vessel.* National Institute of Oceanography Report A37. National Institute of Oceanography, Wormley.

EAU. 1987. Environmental Report 4A/87. Dee and Mersey Estuaries Environmental Background. Unpublished report prepared for Shell UK Exploration and Production, Environmental Advisory Unit, University of Liverpool.

Fahy, F.M., J.D. Hansom and D.P.N. Comber. 1993. Estuaries Management Plans. Coastal Processes and Conservation. Dee Estuary. Unpublished report prepared for English

Nature, Peterborough. Coastal Research Group, Department of Geography and Topographic Science, Glasgow University.

Flint County Council. 1958. *Flintshire County Development Plan.* Flint County Council.

Gall, A. 1988. Dee Estuary *Spartina* Monitoring 1979–1986. Unpublished report prepared for the Nature Conservancy Council.

Gilham, R.M. 1978. An ecological investigation of the intertidal benthic invertebrates of the Dee estuary. PhD dissertation. University of Liverpool.

Graaf, J. 1978. Abnormal sea levels in the northwest of England. *Dock and Harbour Authority* 58: 366–371.

Gray, A.J. 1992. Saltmarsh plant ecology: zonation and succession revisited, p 63–79. In J.R.L. Allen and K. Pye (eds.), *Saltmarshes – Morphodynamics, Conservation and Engineering Significance.* Cambridge University Press, Cambridge.

Gresswell, R.K. 1964. The origin of the Mersey and Dee estuaries. *Geological Journal* 4: 77–85.

Hill, M.I. 1986. Population studies of *Spartina anglica* C.E. Hubbard in the Dee Estuary. Ph.D. dissertation. University of Liverpool.

Hill, M. 1987. *Vegetation Change in Communities of* Spartina anglica *C.E. Hubbard in Salt Marshes of North-west England.* Chief Scientist Directorate Contract Surveys No 9, Nature Conservancy Council, Peterborough.

Howell, F.T. 1973. The sub-surface drift of the Mersey and Weaver catchments and adjacent areas. *Geological Journal* 8: 285–296.

Hutchinson, S.M. 1994. Distribution of ^{137}Cs in saltmarsh sediments in the Dee estuary, NW England. *Marine Pollution Bulletin* 28: 262–265.

Hutchinson, S.M. and D. Prandle. 1994. Siltation in the saltmarsh of the Dee estuary derived from ^{137}Cs analysis of shallow cores. 1993. *Estuarine Coastal and Shelf Science* 38: 471–478.

Hydraulics Research Ltd. 1970. *Dee estuary Scheme. General Description of Hydraulics Research Studies, With Summary of Results and Conclusions.* Report EX 527, Hydraulics Research Ltd. Wallingford.

Hydraulics Research Ltd. 1990. *Marine Sand Extraction in Liverpool Bay, a Study of the Effects of Dredging.* Report EX 2081, Hydraulics Research Ltd., Wallingford.

Jelliman, C., Hawkes, P. and A.H. Brampton. 1991. *Wave Climate Change and Its Impact on UK Coastal Management.* Report SR260, Hydraulics Research, Wallingford.

Jemmett, A.W.L. 1994. The Dee Estuary Review Executive Summary. Department of Leisure Services and Tourism. Metropolitan Borough of Wirral, Birkenhead, unpublished report.

Jemmett, A.W.L. 1995a. The Dee Estuary Strategy Consultation Draft, February 1995. Unpublished report.

Jemmett, A.W.L. 1995b. Integrated coastal zone management: lessons from the Dee estuary, U.K. p. 447–450. *In* M.G. Healy and J.P. Doody (eds.), *Directions in European Coastal Management.* Samara Publishing Ltd, Cardigan.

Lambert, A. 1988. Regulation of the River Dee. *Regulated Rivers: Research and Management* 2: 293–308.

Marker, M.E. 1967. The Dee estuary: its progresive silting and saltmarsh development. *Transactions of the Institute of British Geographers* 41: 65–71.

Massey, J.D. 1937. *Spartina townsendii* H.& J. Groves in the Liverpool District. *Northwest Naturalist* 12: 308–310.

NERC 1992. *United Kingdom Digital Marine Atlas.* Version 2.0. Proudman Oceanographic Laboratory, Bidston.

Nordstrom, K.F. 1992. *Estuarine Beaches.* Elsevier Science Publishers, London.

Parr, S.J. 1988. The Dee Estuary Site of Special Scientific Interest. Unpublished report

prepared for the Nature Conservancy Council.
Pearson, J. 1985. *Neston and Parkgate.* Countyvise Ltd, Birkenhead.
Pitts, J. 1983. Geomorphological observations as aids to the design of coast protection works on a part of the Dee estuary. *Quarterly Journal of Engineering Geology* 16: 291–300.
Postma, H. 1967. Sediment transport and sedimentation in the estuarine environment, p. 158–179. *In* G.H. Lauff (ed.), *Estuaries.* American Association for the Advancement of Science Publication No. 83, Washington DC.
Prater, A.J. 1981. *Estuary Birds of Britain and Ireland.* T. & A.D. Poyser, Calton.
Pye, K. and P.W. French. 1993. *Erosion and Accretion Processes on British Saltmarshes. Volumes I–V.* Cambridge Environmental Research Consultants Ltd, Cambridge.
Shoreline Management Partnership. 1990. River Dee Study. Unpublished report prepared for Dee Tidal Users Group.
Shoreline Management Partnership. 1993. Dee Tidal Users Group Annual Monitoring Report 1993. Unpublished report prepared for Dee Tidal Users Group.
Slater, G. 1929. The Dawpool section of the Dee Estuary, Cheshire. *Proceedings of the Liverpool Geological Society* 15, 134–143.
Sly, P.G. 1966. Marine geological studies in the Eastern Irish Sea and adjacent estuaries, with special reference to sedimentation in Liverpool Bay and the River Mersey. Ph.D. dissertation. University of Liverpool.
Stopford, S. 1951. An ecological survey of the Cheshire foreshore of the Dee estuary. *Journal of Animal Ecology* 20: 103–122.
Taylor, M.C. and E.M. Burrows. 1968. Chemical control of fertile *Spartina townsendii* (s.l.) on the Cheshire shore of the Dee estuary. 1. Field trials on the *Spartina* sward. *Weed Research* 8: 170–184.
Taylor, M. and E.M. Burrows. 1975. Studies on the biology of *Spartina* in the Dee estuary, Cheshire. *Journal of Ecology* 56: 795–809.
Turner, A., G.E. Millward and A.O. Tyler. 1994. The distribution and chemical composition of particles in a macrotidal estuary. *Estuarine Coastal and Shelf Science* 38: 1–17.
Webster, F. 1930. The River Dee reclamations and the effect upon navigation. *Transactions of the Liverpool Engineering Society* 51: 63–92.
Wedd, C.B., H.H. Thomas, W.C. Simmons and T.C. Cantrill. 1924. *The Geology of the Country Around Flint, Hawarden and Caergwrle.* Memoirs of the Geological Survey of England and Wales, Explanation of Sheet 108. HMSO, London.
Welsby, J. and J.M. Motyka. 1989. *A Macro Review of the Coastline of England and Wales. Volume 8. The North West. The Great Orme to the Solway Firth.* Report SR 207, Hydraulics Research Ltd, Wallingford.
White, D.A. 1982. *Dee Estuary Vegetation Monitoring 1971–1979.* Wales Field Survey Unit Report. Nature Conservancy Council, North Wales Region.
Wills, L.J. 1912. Late glacial and post-glacial changes in the lower Dee Valley. *Quarterly Journal of the Geological Society of London* 68, 180–198.

Chapter Three

Evolution of estuarine shoreline systems in Sierra Leone

EDWARD J. ANTHONY
Université du Littoral

3.1 INTRODUCTION

The coast of Sierra Leone, in West Africa (Figure 3.1), shows sharp longshore contrasts in process domains and patterns of Holocene coastal and shoreface sedimentation. Estuarine deposits, which cover an area of 2347 km^2 (38% of the area of Holocene coastal sediments in Sierra Leone) are particularly well developed between Sherbro Bay and the Scarcies estuary (Figure 3.1). These estuaries are a heritage of a dense drainage network in a humid tropical setting, and have been variably infilled since the eustatic highstand was achieved on this coast at around 5500 years BP. Estuaries are practically absent south of Sherbro Bay where extensive sand barrier systems have diverted river mouths into a common, now largely infilled, fluvio-lagoonal system (Figure 3.1).

In order to illustrate spatial gradients and linkages in estuarine shoreline evolution on this tropical coast, two examples, Sherbro Bay and the Scarcies estuary (Figure 3.1), representing two types of settings and degrees of estuarine infill, have been singled out for this study. This chapter is a review of results distilled from several reports on estuarine and coastal systems in Sierra Leone undertaken since 1980. It addresses two related aspects: (1) estuarine development patterns, involving progradation of tidal flats and their shorelines and infill of estuarine channels by sand derived from local, marine and upland sources, and (2) the identifiable hydrological, geomorphic and ecological gradients and linkages that characterize estuarine plains over space and time and that determine, in part, rates and patterns of estuarine shoreline progradation. The chapter also briefly

Estuarine Shores: Evolution, Environments and Human Alterations. Edited by K. F. Nordstrom and C. T. Roman.

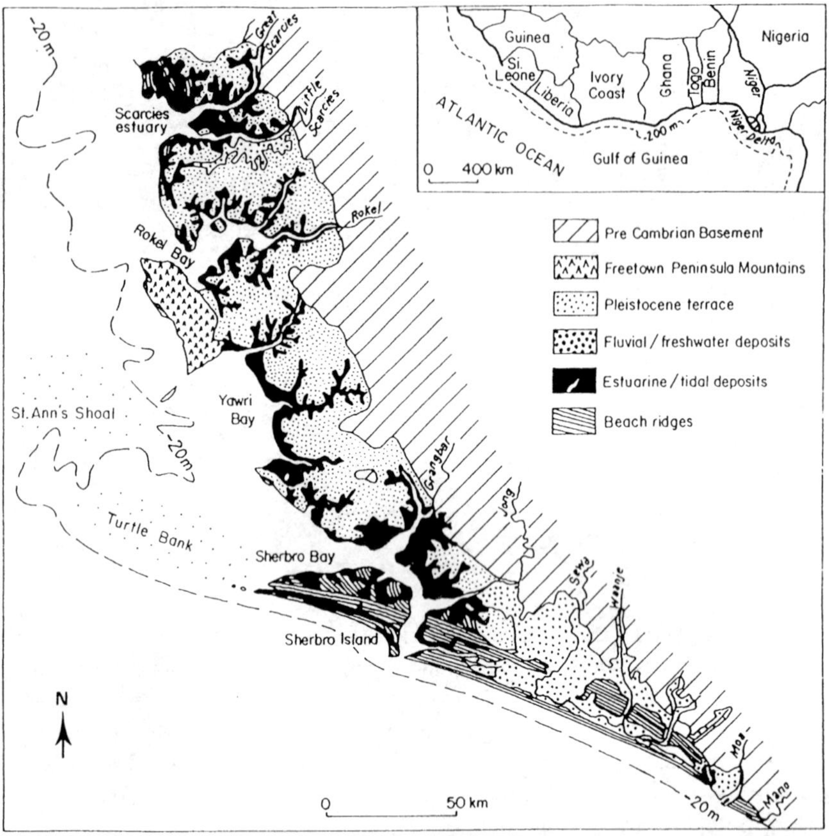

Figure 3.1 The coast of Sierra Leone, showing the well developed estuarine systems between Sherbro Bay and the Scarcies river mouths

speculates on the possible response of estuarine plains and shorelines to future sea-level rise. Estuarine plains in West Africa and their mangrove vegetation have long been shown to be highly productive ecosystems (e.g. Marius 1985). Vast areas of tidal flats liable to be affected by future sea-level rise have been used for decades as tidal ricelands while mature mangrove forests constitute a reservoir for fishing activities.

3.2 GENERAL SETTING OF ESTUARINE SYSTEMS

The coast of Sierra Leone is characterized by a seasonal tropical climate. Annual rainfall, virtually totally concentrated between May and November, varies from 4000 mm to 3000 mm, decreasing from south to north. Mean monthly temperatures range from a minimum of 24°C in August to a maximum of 28°C in March. The seasonal rainfall regime is paralleled by river flow regime, characterized by heavy discharge and sediment supply to the coast in the wet-season months.

The estuarine systems on this humid tropical coast occur in two types of settings: sheltered bays and open river mouths. The sheltered Rokel and Sherbro Bays are composite estuarine systems comprising several, more or less interlinked estuary mouths. Open-coast river-mouth estuaries debouch onto low-energy coastal embayments and may also be interlinked by more or less extensive, adjacent estuarine shores. Such muddy coasts show little spatial contiguity (Figure 3.1) as a result of marked variations in incident wave energy dissipation resulting from a very irregular inner shelf bathymetry (Anthony 1991a and b).

The degree of estuarine infill has depended essentially on inherited estuarine valley morphology and size, and on sediment supply from both the upland and the sea. While some rivers such as the Great and Little Scarcies in northern Sierra Leone and the Moa in southern Sierra Leone (Figure 3.1) have rapidly filled their mouths to evolve into deltaic estuaries or deltas that inject sediment onto the open coast, others are still deep and actively trap incoming sediment, including sands from the nearby shelf. These include Rokel Bay, a structurally controlled drowned river valley, and Sherbro Bay, formed in the Holocene behind beach-ridge barriers.

3.3 ESTUARINE MORPHOLOGY AND SEDIMENTARY ENVIRONMENTS

3.3.1 Sherbro Bay

The Sherbro estuarine complex, which covers an area of around 450 km^2, is fed by several rivers whose total catchment area is only 24 500 km^2, of which 65% is drained by the Sewa River (Figure 3.1). This estuarine system occupies the transitional coast between the tidally influenced, low-energy, muddy Yawri Bay and the major wave-formed beach-ridge barrier complex of southern Sierra Leone (Figure 3.1). Much of Sherbro Bay is protected from swell wave influences by these northward-prograded spits and beach-ridge barriers, which progressively closed off a series of embayments in southern Sierra Leone, over a distance of 200 km (Anthony 1983, 1991). Further protection from open swell is offered by the Turtle Bank, an extensive sand shoal platform at the main bay mouth.

Sherbro Bay is a good example of a largely unfilled estuarine system. The main bay channel is an unfilled disequlibrium embayment, in the sense of Swift (1976). Sediment inputs from terrestrial and marine sources have not been sufficient to fill the space created behind the Holocene beach-ridge barrier system. Adjacent to the main bay channel, Late Pleistocene sheetwash deposits, which outcrop inland as a terrace (Figure 3.1), form a shallow (6–10 m deep) substrate on which have aggraded Holocene muds. These have been derived from the upland and eroded by tidal channels from older muds underlying parts of the beach-ridge plain. These estuarine aggradational muds have been enriched by in situ organic inputs from mangroves.

The muds have formed a composite or coalesced estuarine plain that interlinks the river mouths debouching into the bay. On Sherbro Island, tidal sandflats corresponding to infilled lagoonal arms or formed from shoreface shoal sand incorporated within the prograded beach-ridge barriers (Anthony 1991b), are also a part of the Sherbro bay estuarine system. The estuarine plain abruptly merges inland into freshwater fluvial floodplains, except in parts of the northern flank of the bay where they merge into ill-drained pockets of freshwater swamps.

The estuarine plain is dissected by a dense network of tidal channels whose role in sediment transport processes is discussed below. The plain is fringed by low-gradient muddy intertidal and subtidal slopes alternating with 2–10 m wide, 2–100 km long estuarine beaches of well to very well sorted fine to medium quartz sand. The low-energy sandy estuarine beaches, sometimes colonized by mangroves, are also fronted by wide, low-gradient muddy intertidal to subtidal slopes. These beaches commonly occur along the bayward flanks of the beach ridges that bound the bay. Along much of the northern flank of the bay, the mudslopes are protected by the Turtle Bank and St Ann's Shoal, an inner shelf sand shoal complex (Figure 3.1). These two contiguous shoals control wave refraction patterns and dissipate much of the incident wave energy on this coast, accounting for extensive low-energy open-coast mangrove shorelines in Yawri Bay (Figure 3.1).

Within the main Sherbro Bay channel, sediment sources, channel hydrology and depositional processes have led to marked contrasts in patterns of infill. The bay channel exhibits two clear-cut sectors of advanced sandy infill, separated by relatively deep channel segments (Figure 3.2). The bay-mouth shoal massif, the Turtle Bank, is distinct from the inner bars and islets that have formed at the mouths of the various rivers. This bayward migrating sand massif is derived from the St Ann's Shoal (Anthony 1983; 1990; Anthony and Marius 1985), and is cut by a complex system of ebb and flood channels. It is similar in its formation and sediment dynamics to other bay-mouth shoals such as those of Chesapeake Bay (Colman et al. 1988).

The Turtle Bank area is dominated by clear salt water and bayward directed residual currents (Anthony 1990) and contrasts sharply with parts of the Sherbro Island flank and the northern, continental bay flank, which show, superficially, less saline but much more turbid water and seaward-directed residual flows (Figure 3.2). The concentration of clear flood-dominated water over the Turtle Bank area may reflect a topographically induced asymmetry in salt-wedge intrusion resulting from a marked difference in depth between the shallow northern continental flank that drains several rivers, and the hitherto deep southern flank, adjacent to the Sherbro Island beach-ridge barrier. The latter flank has been infilled in part by the marine-derived Turtle Bank sand body. This asymmetry effect may be strengthened by the Coriolis force across the wide bay mouth in spite of the low latitude. The Turtle Bank cannot be considered as a simple flood tide delta, as it shows well

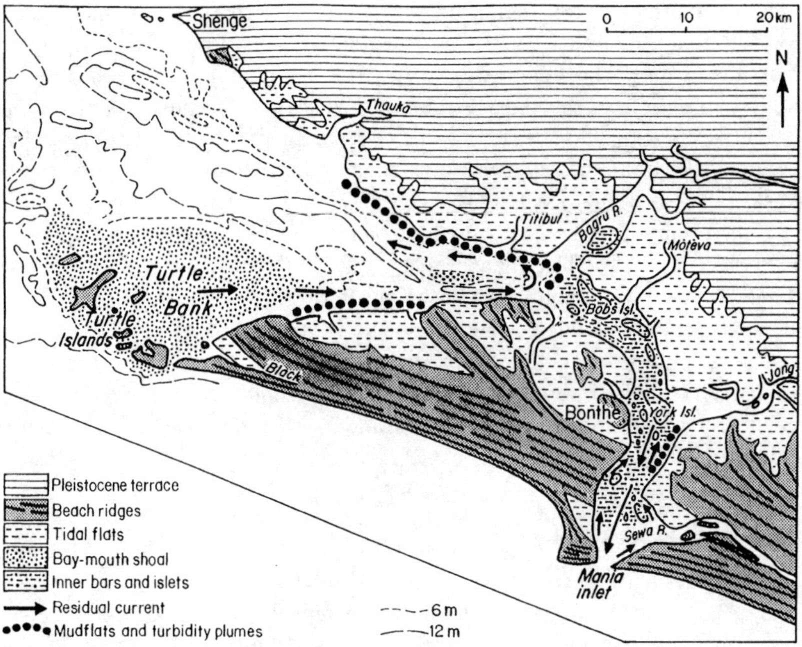

Figure 3.2 Morphology, residual current patterns, turbidity plumes and sedimentation patterns in the Sherbro Bay channel

developed ebb flows. A similar reasoning has been applied to the Chesapeake bay-mouth shoals by Colman and colleagues (1988). On the Turtle Bank, these flows are generated by ebb drainage of the Black River lagoon (Figure 3.2) and of the adjacent tidal flats on Sherbro Island. The Turtle Bank also comprises a number of islets representing swash platforms that have been built up into supratidal deposits by waves and mild aeolian activity.

The inner bars and islets (Figure 3.2) correspond to "central basin" infill deposits (in the sense of Dalrymple et al. 1992) formed at the mouths of the rivers, where friction-dominated flows occur in association with sudden deceleration as the river jets debouch into the bay. This central sector of the bay is narrow. The spring tidal range also progressively increases towards this sector in response to this narrowing, from around 2 m at the main bay entrance to 2.6 m at York Island (Sierra Leone Tide Tables) before diminishing rapidly up the mouths of the inflowing rivers. The bars, formed from sand eroded by the rivers from the adjacent beach ridges or from the Pleistocene terrace upland, considerably dissipate tidal energy, leading to a well mixed salinity and turbidity regime. Changes in bar morphology and location hinder navigation. Aggraded bars in this sector serve as a substrate for mud accumulation and sometimes show rapid mangrove colonization by *Avicennia africana*.

The circulation patterns south of York Island are extremely complex (Africa Pilot 1967) and involve mutually evasive tidal streams (Figure 3.2), separated by debris-lined fronts. This complexity arises from retardation of the ebb flow by the vast system of adjacent mangrove-colonized tidal flats, the presence of the Mania Inlet (Figure 3.2), whose flood jet is deflected to the flanks of the bay by the outgoing water from this bay, part of which ebbs through this inlet, and by the Sewa River discharge, which accounts for over 50% of the river discharge into Sherbro Bay. The Mania Inlet is an outgrowth of patterns of beach-ridge development and has persisted as a stable channel between Sherbro Island and the mainland throughout much of the formation of the beach-ridge barrier (Anthony 1991b). The outflowing jet through this channel is especially strong in the wet season when ebb flow is strengthened by the Sewa discharge. This strong ebb jet has had a fundamental influence on the geomorphic development of Sherbro Foreland, influencing wave refraction and the sediment drift cell patterns that control it (Anthony 1991b).

3.3.2 The Scarcies estuary

The Scarcies estuary occupies the mouths of the Great and Little Scarcies Rivers (Figure 3.3) which, together, drain a catchment of 25 500 km^2. It shows depositional environments typical of a largely infilled mixed fluvial/tide-dominated estuary and occupies a broad, mesotidal, low-energy embayment. The estuarine fill is up to 80 m thick in the river thalwegs (Japan International Cooperation Agency 1983), but is less thick over the palaeo-interfluves which consist of Pleistocene deposits that largely outcrop as a coastal terrace. The Scarcies estuary is the southern-most component of a dense open-coast estuarine system that stretches for over 1000 km from northern Sierra Leone to Guinea Bissau (Figure 3.4). The development and evolutionary pattern of the Scarcies estuarine shores are fairly representative of much of this tide-dominated coast of West Africa (Anthony 1993).

This coast is fronted by a relatively shallow and rather irregular nearshore zone, linked to a 150–200 km wide, low-gradient shelf. The shelf surface exhibits remnant barrier spits, barrier island–lagoon complexes, sea cliffs, river channels and major, drowned fine-grained deltas north of the Sierra Leone/Guinea border (McMaster et al. 1970). These preserved forms reflect a rapid rate of shoreface translation during the last transgression (Anthony 1989a; 1990). The inner shelf and nearshore zone considerably dissipate the Atlantic swell while favouring tidal range amplification. These hydrodynamic conditions have favoured the development of open, strongly tide-influenced river mouths and extensive tidal channel networks, while precluding the formation of beach-ridge barriers and lagoons which could trap fluvial sediments (Figure 3.4). The open nature of the estuaries has allowed for direct supply of mud from the upland to the coast while the prevailing low wave energy conditions have favoured muddy progradation. Many of these estuarine

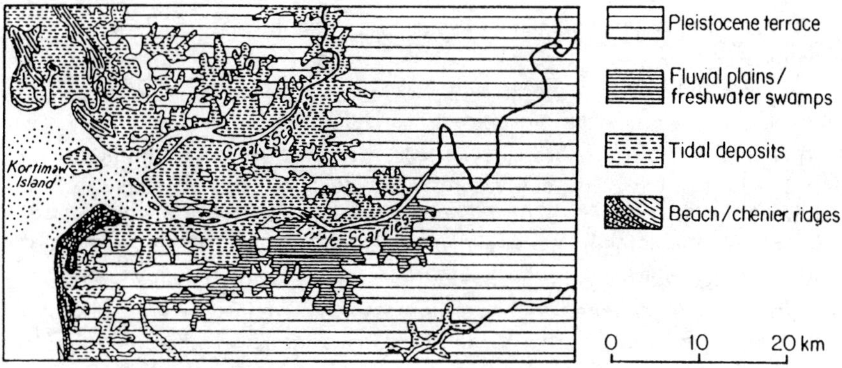

Figure 3.3 The composite Scarcies estuary

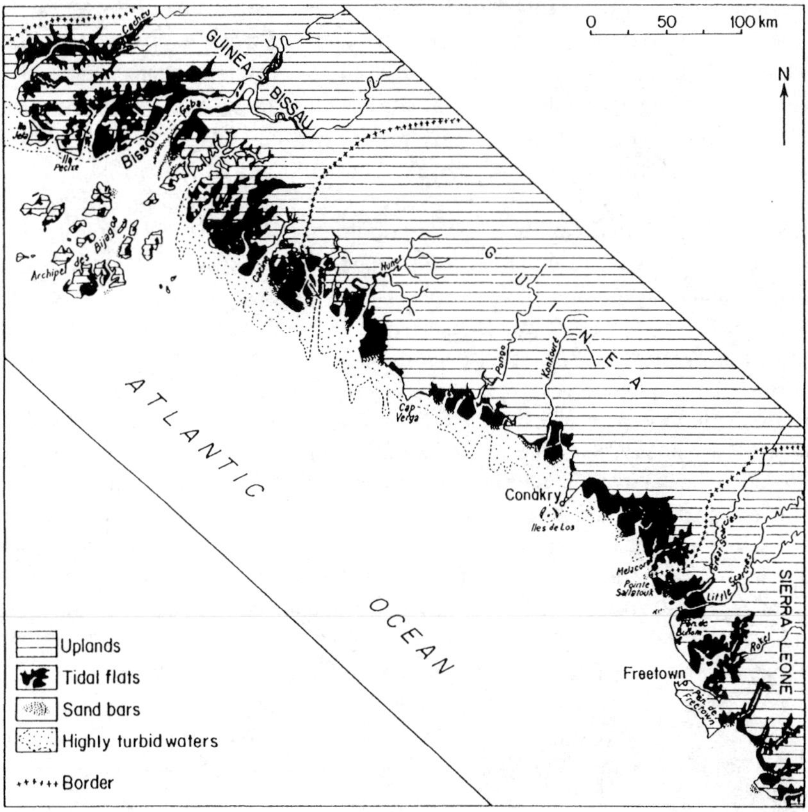

Figure 3.4 The open-coast estuarine systems of northern Sierra Leone, Guinea and Guinea-Bissau. Mud supplied by the numerous coastal rivers and by marine sources is distributed alongshore and especially transported southwards by large-scale coastal circulations

systems directly pass into upland terrace deposits, and fluvial floodplains are poorly developed.

Estuarine and fluvial plains in the Scarcies embayment extend along both the valley axes and the adjacent open-coast shoreline to the north (Figure 3.3), where progradation of up to 10 km has occurred. The mangrove cover on the estuarine plain has been extensively removed for rice cultivation. Bands of cheniers occur within this muddy plain north of the estuary (Figure 3.3). The vegetated tidal flats forming the estuarine plain are bounded seaward by gentle bare slopes (1:1000) of mud intermixed with very fine sand. These muddy foreshores are widely exposed at low tide and do not show any change in cross-shore gradient or grain-size except locally where the upper portion of the intertidal zone is occupied by 100–500 m wide low-energy beaches of fine to medium quartz sand sometimes intermixed with mud. This longshore alternation between fairly short (< 10 km long) segments of sand beaches occupying the higher part of the intertidal zone and much longer stretches of monotonous muddy shoreline is a characteristic feature of much of this part of the West African coast. Cheniers are locally formed presently (Anthony 1989b), and bound part of the shoreline north of the Scarcies estuary (Figure 3.3).

The composite Scarcies estuary-mouth, which closely approximates two adjacent trumpet-shaped meso-tidal estuaries (Figure 3.5), comprises a vast shallow sand platform cut by tidal channels. Little is known of the hydrology of this estuary and of the current circulation patterns. Parts of this platform, formed from sand reworked by tidal currents from poorly consolidated nearshore deposits (Anthony 1990), are representative of the upper flow regime flats (Figure 3.5) of Dalrymple and colleagues (1992). The extensive bar system on this platform results in well mixed dry season flows characterized by turbulence (Japan International Cooperation Agency 1983). The bars dissipate both tidal and wave energy, favouring sedimentation. Convergence between fluvial and tidal flow at this estuary-mouth zone also leads to significant fine-grained sedimentation over the shallower bars, many of which are colonized by mangroves. The two main river channels do not show the intermediate meandering reach described from tide-dominated estuaries (Dalrymple et al. 1992), probably because of the importance of river discharge, the relatively moderate tidal range and considerable estuary-mouth dissipation of tidal energy. The tidal range, which attains a mean spring value of 2.6 m at the mouths of the estuary, is rapidly attenuated upstream. In the Little Scarcies (a misnomer, as this river has a bigger catchment and a more infilled mouth than the Great Scarcies), the maximum limit of tidal intrusion upstream during the dry season occurs well beyond the estuarine plain and corresponds to the junction between the coastal plain and the metamorphic basement. In this river, the only one of the two with a freshwater floodplain (Figure 3.3), tidal waters upstream are confined to the main channel by river-built levees which isolate freshwater floodbasins from saline water. In the wet season, the significant increase in river discharge results in significant tidal flushing and in temporary stratified flow. The tip of the salt wedge associated with this high river discharge regime intrudes no

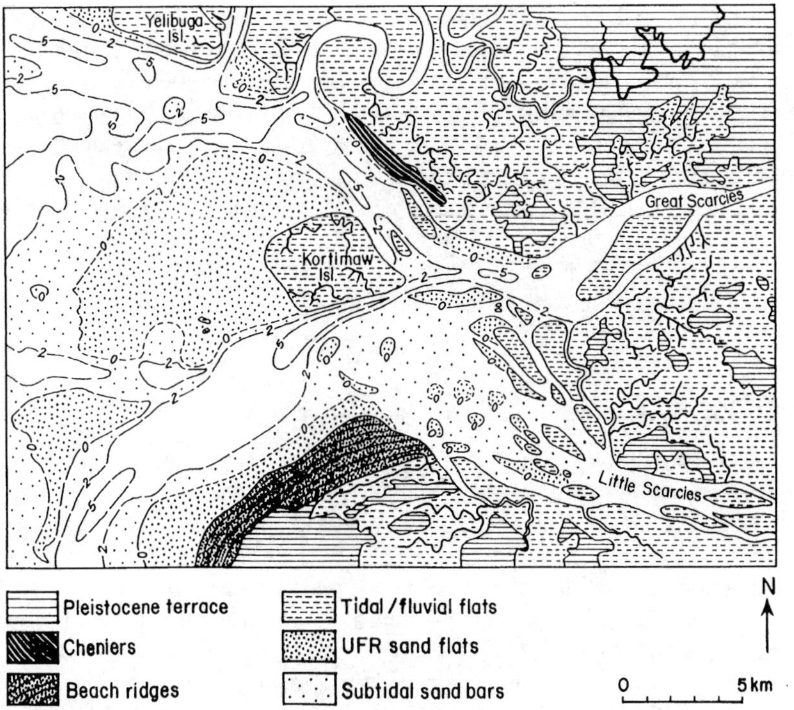

Figure 3.5 Morphology and bar deposits at the friction-dominated mouth of the Scarcies rivers. The prograded alluvial plain at the mouth of these rivers consists of vegetated tidal/fluvial flats that are subject to tidal flooding in the dry season. Tidal flushing and river dominance occur during the wet season. These flats are fringed by beaches of fine quartz sand that merge with upper flow regime sand flats and subtidal bars within the common mouth of the two rivers. These beaches give way within the main estuarine and secondary tidal channels and along much of the open coast north of the estuary mouth to low-gradient mud slopes

more more than around 5 km upstream (Japan International Cooperation Agency 1983).

3.4 ESTUARINE SHORELINE PROGRADATION: THE ROLE OF LONG-TERM HYDROLOGICAL, GEOMORPHIC AND ECOLOGICAL INTERACTIONS

Estuarine shoreline evolution depends not only on sediment supply from the upland and from marine or longshore sources but also on patterns and rates of sediment redistribution within and across the vegetated tidal flats forming the estuarine plains and within the main estuarine channels. The processes of sediment redistribution that may lead to tidal flat accretion and/or shoreline progradation involve long-term interaction and mutual adjustements between tidal flat ecology, hydrology, geomorphology and sedimentation. The accretion of vegetated tidal flats close to the high spring-tide level, for instance, appears to be a mechanism that results in net export of upland-derived mud to the lower intertidal and subtidal zones via dense tidal drainage networks. As accretion slows down, the upland-derived mud becomes increasingly redistributed towards the intertidal slopes fringing these flats, leading to their accretion and progradation. The gradients and linkages involved in these long-term processes, and their hydrological, geomorphological and ecological implications, are discussed in the following subsections.

3.4.1 Tidal flat accretion and progradation of estuarine shoreline systems

Estuarine shoreline progradation has depended on two modes of sediment supply and redistribution: (1) cross-shore supply of terrigenous muds towards the lower intertidal foreshore and subtidal zones by tidal channels, and (2) lateral dispersion of these muds by the local tidal and non-tidal circulation systems followed by sedimentation. In the case of interlinked open-coast estuarine systems, progradation may be significantly enhanced by longshore redistribution of terrigenous and marine muds transported by large-scale coastal currents and deposited on estuarine foreshores and their adjacent shallow subtidal zones.

3.4.1.1 Tidal channel hydrology and suspended sediment transport

The tidal export of terrigenous or locally derived muds towards the foreshore is conditioned, as stated above, by long-term substrate–hydrology relationships across tidal flats. Discussion of the processes that lead to long-term tidal flat accretion and development is beyond the scope of this chapter. Long-term rates of accretion on the flats are not available for the study area. Initial rapid sedimentation during and immediately after the Holocene eustatic transgression, probably aided by flood tidal lag in the manner described by Postma (1967), should be expected to diminish over time, as depths of flooding and duration of flooding decrease with accretion. Concomitantly, the tidal channel network on these accreting deposits should develop from a rudimentary system of rectilinear drains into the dense network of meandering creeks (Figure 3.6) that transport suspended sediment from tidal flats to

---

Figure 3.6 (*opposite*) The dense channel networks that serve as conduits for the ebb export of mud onto prograding foreshores and subtidal slopes in the Scarcies (top) and Sherbro estuarine systems. Rectilinear channels such as those of the Yelibuya Island mud bar are developing features while the meandering channels represent mature, near-equilibrium systems. Circled numbers refer to some of the transects, from the tidal channels to the tidal flats, whose maximum instantaneous velocities at mid-ebb tide are shown in Figure 3.7

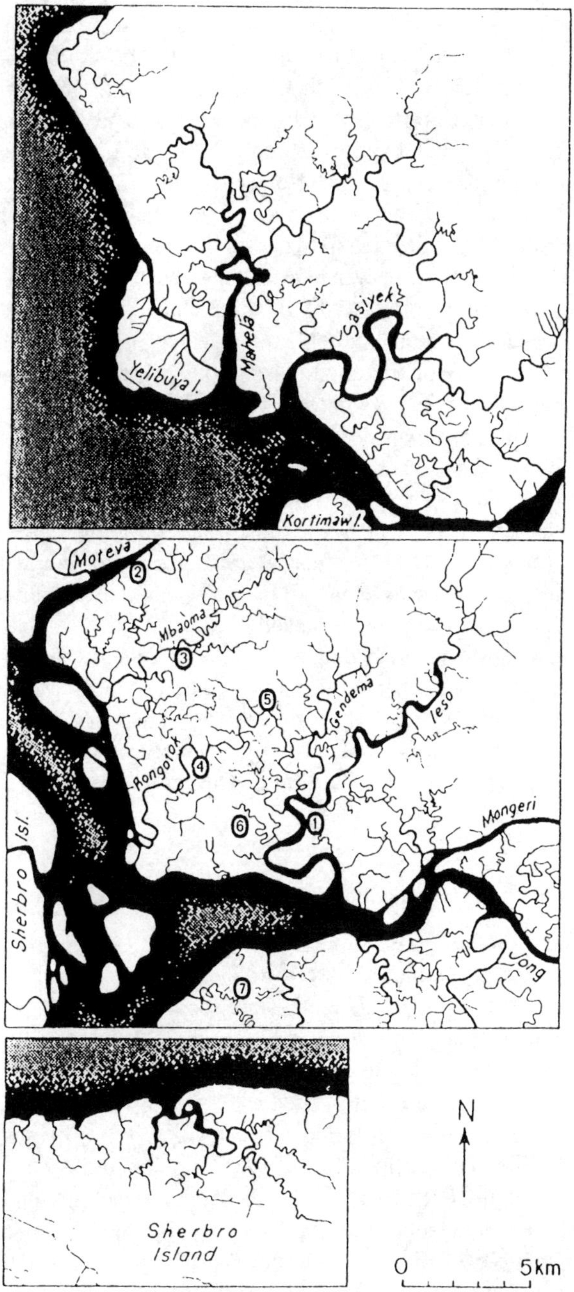
Yelibuya I.
Mahela
Sasiyek
Kortimaw I.
Moteva
Mbaoma
Gendema
Ieso
Rongotok
Mongeri
Jong
Sherbro Isl.
1
2
3
4
5
6
7
Sherbro Island
N
0
5km

the foreshore. This role of tidal channels in suspended sediment transport to the foreshore has been reported in a number of studies from mangrove swamps in Australia (e.g. Wolanski et al. 1980; Boto and Bunt 1981). The ebb flows in the tidal channel networks are commonly stronger than the flood flows, as a result of lesser frictional dissipation within the channels during falling tides and the overall downstream slope effect. Ebb flows in the main channels are also strengthened by river discharge from the upland and by rainfall discharge from the flats in the wet season. The duration of these ebb flows also exceeds that of the flood flows because of the retardation effect across the tidal flats. A marked gradient in current strength between vegetated flats and the channels has been identified by Wolanski and co-authors (1980) in their study. This gradient has also been observed in several short (20 m) transects from the tidal flats to the flanking channels (Figure 3.7) in Sherbro Bay (Anthony 1984). Maximum mid-ebb tide surface flows across the vegetated flats were weak, as a result of frictional dissipation and retardation over the rough substrate of mangrove roots, pneumatophores and leaf litter, but showed sharp velocity increases in the channels (Figure 3.7). On rare occasions, however, mangrove litter on the tidal flats resulted in local dam effects that led to maximum velocities of up to $0.3\,m\,s^{-1}$.

The main tidal channels, which evacuate the ebb flow, may exhibit high turbidity values (of up to $30\,g\,l^{-1}$). Much of this suspended sediment is supplied from upland, while the rest may come from the erosion of organic-rich concave meander banks lining the channels, especially where the local sediment supply is augmented by sand-sized bedload material that accumulates as sand bars in the channel thalwegs, leading to channel instability. The channels also export organic debris from the tidal flats. Particularly high turbidity values in the wet season (up to $40\,g\,l^{-1}$) in rivers draining the mainland north of Sherbro Bay may be due to the possible erosion of rutile and bauxite mining tailings in the upper catchments of these small rivers.

3.4.1.2 Longshore suspended sediment transport and sedimentation

Outflowing water from the vast complex of tidal flats in both Sherbro Bay and the Scarcies estuary forms permanent well defined turbidity plumes. Along the northern flank of Sherbro Bay, these plumes are separated from clear marine water by sharp debris-lined fronts. This structure suggests that the water and suspended matter flushed from the tidal flats are essentially trapped close to the shore by southwesterly winds. The turbidity plumes formed in Sherbro Bay stretch alongshore northwestwards towards the adjacent Yawri Bay in response to the residual, seaward-directed streams in the former bay (Figure 3.2). The long-term changes involved in tidal flat substrate–hydrology interactions and the ensuing estuarine shoreline progradation are schematically presented in Figure 3.8, which shows a representative profile across estuarine plains in Sierra Leone with a temporal sequence lasting several hundreds to a few thousand years.

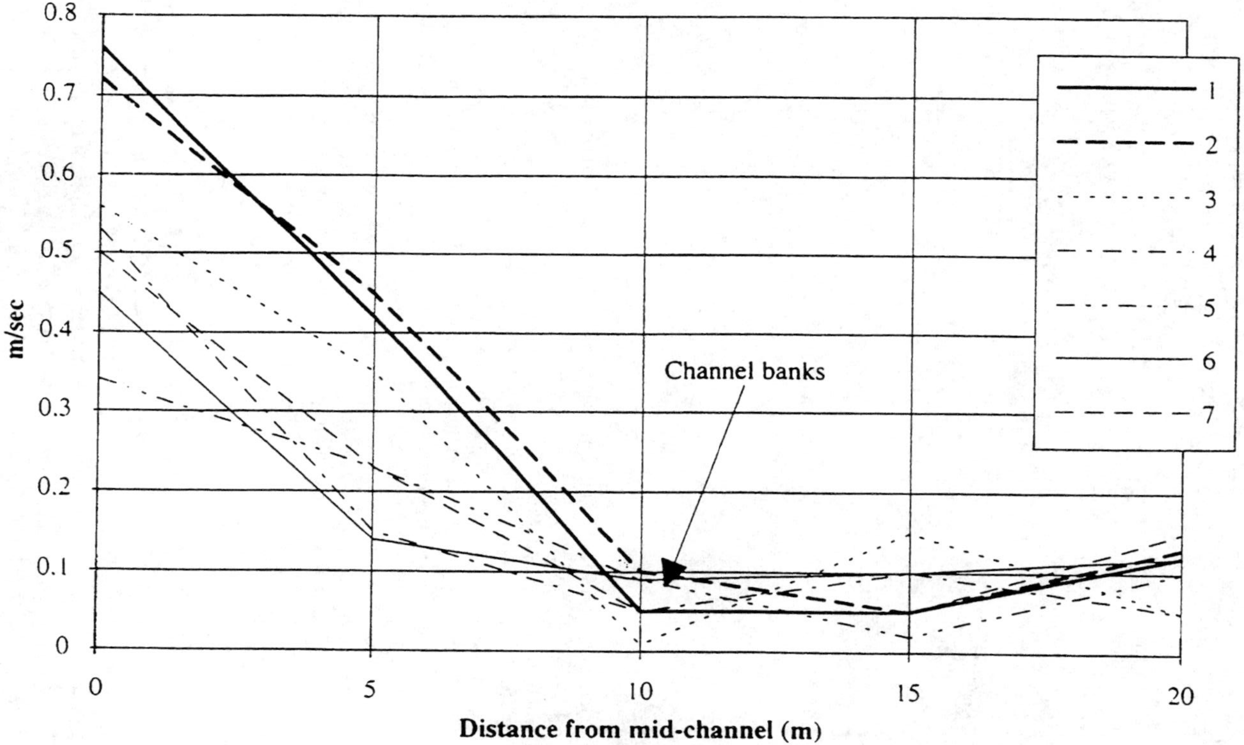

Figure 3.7 Maximum instantaneous current velocities at mid-ebb tide (generally three hours after the start of the ebb tide) for transects from the bounding channels to the mangrove-colonized flats. Note the sharp drop in velocities over the tidal flats. Velocities tend to be higher in the bigger tidal channels

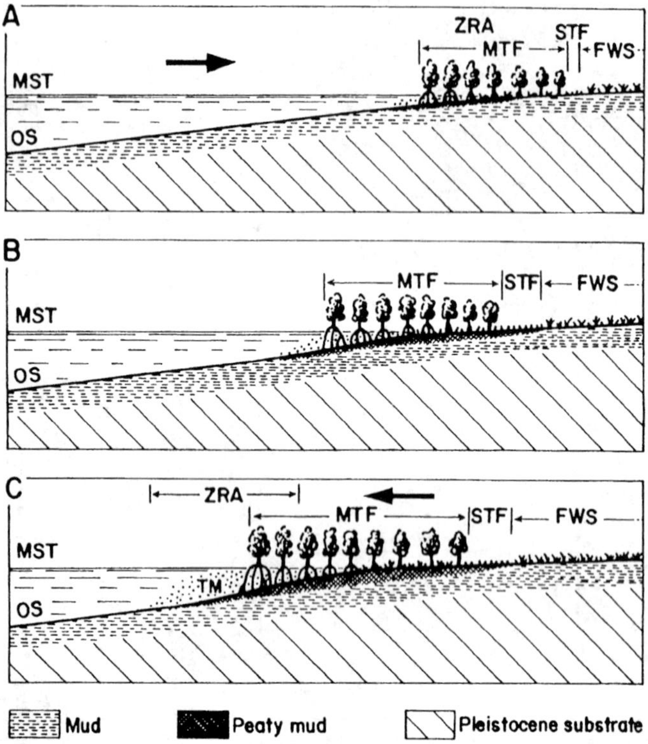

Figure 3.8 Schematic profile of long-term changes (order of hundreds to thousands of years) in tidal hydrology, mud dynamics, accretion and progradation across tidal flats of estuarine systems in Sierra Leone. MST = Mean high spring tide level; OS = Original surface; TM= Turbidity maximum; ZRA= Zone of rapid accretion; MTF = Mangrove-colonized tidal flats; STF = Supratidal flats (salt-tolerant grasses flooded only occasionally by spring tides; may not be present behind the mangrove flats); FWS = Freshwater swamps. Arrows refer to dominant mud transport directions by tides. Initial accretion may be favoured by flood-tidal asymmetry. As substrate elevation increases, increasingly ebb-dominated channelized flows lead to net export of terrigenous mud onto the tidal flat foreshore and subtidal zones, leading to a seaward shift of the rapid accretion zone. The exported mud forms turbidity plumes that are stretched alongshore to feed the accretion of sectors adjacent to the channel mouths, including the inner subtidal zone fronting low-energy sand beaches. Profile gradient is of the order of 1:1000

The turbidity plumes in Sherbro Bay also feed the depressions between beach ridges on Sherbro Island. These depressions have acted over time as particularly efficient mud sinks (Figure 3.9). Net accumulation in these sinks has probably been enhanced by reduced seaward flushing of suspended sediment during the wet season, compared with tidal flats on the continent affected by river flushing. In the long term, the turbidity plumes also nourish the aggradation of subtidal slopes

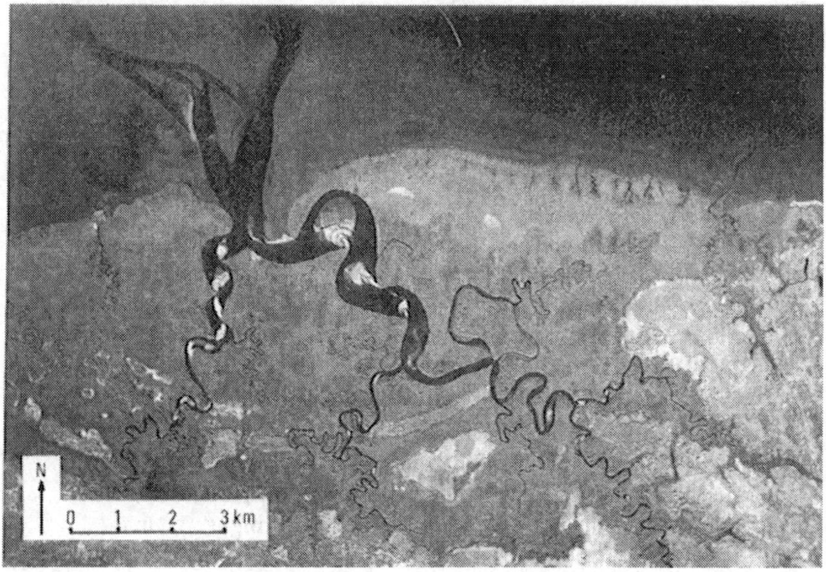

Figure 3.9 Aerial view of mangrove-colonized tidal flats and their extensive intertidal foreshore and subtidal mud slopes (north shore of Sherbro Island) in Sherbro Bay. These sediments have prograded bayward from a broad depression between beach ridges sets. Parts of a beach-ridge set show up on the right

adjacent to the intertidal shoreline (Figure 3.9), including areas where low-energy estuarine sand beaches constitute the shoreline. Within this bay, the longshore distribution of turbidity plumes, wave winnowing processes and the occurrence of beach ridge deposits combine to give a variable Sherbro Bay shoreline (Figure 3.2).

In addition to the feeding of longshore turbidity plumes by the cross-shore mode of sediment supply (Figure 3.8) described above, progradation of the Scarcies estuarine shoreline has also been favoured by large-scale longshore mud supply. The sediment discharge system that characterizes the West African coast north of the Scarcies estuary shows high concentrations of suspended material in the lower river mouths in the wet season (Diop 1990). These feed turbidity plumes that stretch for several hundreds of kilometres along the coastlines of northern Sierra Leone and Guinea. The estuarine systems are thus interlinked by a well developed open-coast, fine-grained sediment dispersal system similar to that described by Wright (1989) for other tropical estuarine and deltaic shorelines. Field observations, confirmed by aerial photographs, show that some of the suspended sediment load that enters the Scarcies estuarine system is derived from alongshore. Muds in suspension off this estuary are trapped in a 10–30 km wide nearshore zone (up to depths of 20 m) by moderate southwesterly breezes and coastal currents. This feature is more pronounced off the Guinea and Guinea-Bissau coasts (Rüe 1989;

Diop 1990; Bertrand 1993). The turbidity plumes are stretched southward from the Guinean estuary mouths by coastal currents driven by northerly winds in the winter months and by the north-south Guinea Current. In addition to terrigenous inputs, a significant proportion of these suspended sediment loads is composed of marine mud. An estimate of this source is given by the 10–30% of smectites comprising the clay fraction in mud from the Scarcies estuary (Anthony 1990). Smectites in this part of West Africa are considered as a marine-derived mineral. Their percentage in estuarine sediments increases northwards towards the Senegal and Cape Verde source areas, where they occur in equal amounts with kaolinite in Holocene and modern estuarine sediments (Marius and Lucas 1982).

Fine-grained sediments off the Scarcies estuary accumulate up to 30 km offshore (Longhurst 1959), contributing to wave energy dissipation while constituting a shallow nearshore foundation for progressive shoreline progradation. Closer inshore, the laterally transported muds feed the low-gradient muddy intertidal slopes. The low-energy estuary mouths and shoreline embayments, especially along the highly indented macrotidal Guinea-Bissau coast, are particularly efficient mud sinks that trap both the upland-derived and laterally drifting muds. In these settings, the muds form mudbanks that are akin, although on a somewhat smaller scale, to the fluid muds described from other high-sediment discharge tropical areas such as the Guyanas (e.g. Froidefond et al. 1988; Augustinus et al. 1989). The muddy aggrading slopes that bound these open-coast estuarine systems are cut by developing rectilinear tidal drains. Examples of such channels occur on Yeibuya Island (Figure 3.6), a vast mud bar that is being progressively colonized by mangroves. Similar rectilinear channels occur as abundant tidal runoff drains on rapidly accumulating mud bars elsewhere, including in Guinea, and in French Guyana. These drains contrast with the meandering channels (Figure 3.6) incised into mature, mangrove-colonized organic-rich substrates.

The importance of mud slopes on the open-coast shorelines of estuarine deposits north of the Scarcies reflects not only the significant suspended sediment loads that enter this coastal system but also the considerable dissipation of waves as they propagate across the shallow low-gradient nearshore zone and encounter the mud banks. In this muddy setting, cheniers have been formed shoreward of bathymetric windows between stable mudbanks where efficient wave energy transmission has favoured the sand-sorting processes associated with the construction of these ridges. Over the past three decades, more active chenier formation, including along the shoreline north of the Scarcies estuary, has been observed from comparisons of aerial photographs (Rüe 1989; Anthony 1990; Bertrand 1993). This seems to be due mainly to lesser protection by mud banks as fine-grained sediment supply to this coast has waned in response to greater regional climatic aridity and considerable reduction in river discharge. Waning river discharge also implies reduced seaward flushing of mud trapped in estuaries, and on the Guinea coast, Rüe (1989) has even described estuaryward migration of shore-attached mud banks, leading to coastal retreat and erosion during the wet-season months when wave energy increases.

3.4.2 Tidal flat accretion and ecological change in estuarine systems in Sierra Leone

The processes of estuarine shoreline development described above also affect tidal flat ecology in ways that are not clear. The accretion of tidal flats up to the limits of high spring-tide flooding leads to the establishment of a delicate substrate–hydrology–ecology equilibrium that may be perturbed by changes in several parameters such as mean sea level, tide range, river discharge and subsidence. Perturbation of this delicate equilibrium is illustrated by the ill-drained pockets of freshwater swamps, not linked to seaward prograding river floodplain deposits. These freshwater swamps are common to both the Scarcies and Sherbro estuarine systems. They are generally colonized by grasslands of *Eleocharis dulcis*, sometimes mixed with *Eichornia natans* and *Nymphoides indica* and overlie organic-rich acid muds dominated by *Rhizophora* peats (Jordan 1964; Dijkerman 1969). The muds show evidence of oxidation, thus suggesting phases of drying out that resulted in marked soil geochemical changes and the formation of *sulfaquents* i.e. acid-sulphate soils (Anthony and Marius 1985) that are harmful to mangroves. The swamps occupy depressions that are flooded by high river discharge in the wet season. They are now protected from tidal influences both by tidal wave attenuation across the estuarine plains and by levees marginal to the river channels. The mangrove swamps they overlie are evocative of the Big Swamp phase described from tropical estuaries in Australia (Woodroffe et al. 1989), and attributed by these authors to river-induced aggradation beyond the upper limits of tidal flooding. The buried mangrove swamps of Sierra Leone are most probably the product of similar fluvial sedimentation, but other possible factors such as long-term changes in tide range, or in seasonal, wind-driven sea-level fluctuations, must not be overlooked.

While the relationship between tidal flooding, fluvial sedimentation, soil oxidation and mangrove elimination to the benefit of supratidal and freshwater plants appears certain, the successional changes, within the intertidal zone, induced by accretion and progradation are not clear. Mangrove zonation and succession patterns in estuaries in West Africa have been proposed by Jordan (1963), who recognized *Rhizophora racemosa* as a dominant pioneering species, followed by *Avicennia africana.* Clear zonation and succession patterns in estuaries in Sierra Leone are not readily identifiable, except in two morpho-sedimentary environments: young developing bars within the main estuarine channels and some of the bigger tidal channels, on which the pioneer species is commonly *Avicennia*, and tidal channel banks, where an outer fringe of *R. racemosa* commonly occurs, followed by an inner mixed zone within the tidal flats. The zonation pattern in the latter morpho-sedimentary environment suggests that as the channel meander belts migrate, the two aforementioned habitat belts migrate in consequence (Figure 3.10). As the external belt of pure stands of high *R. racemosa* shifts along the meander point bars attached to the convex channel banks, it is succeeded by mixed stands comprising shrubby specimens of *R. racemosa, R.*

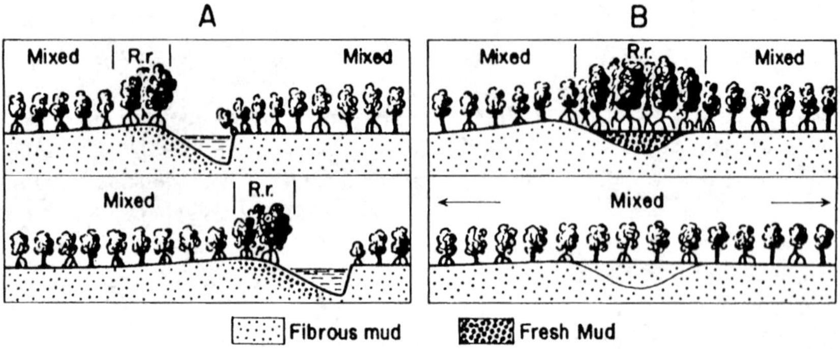

Figure 3.10 Schematic long-term shifting of mangrove belts in response to the progressive migration of tidal channel meanders across the tidal flats (A), and of their silting up (B). R. r. = *Rhizophora racemosa*; Mixed = *Rhizophora racemosa* + *Rhizophora mangle* + *Avicennia africana.* The temporal sequence is of the order of several hundred years

mangle, A. africana and, occasionally, *Laguncularia racemosa. R. racemosa* probably maintains itself in the outer belt because of its territorial advantage, easier implantation of its long propagules over soft muddy substrates, better drainage conditions, and shading out of its competitors. As the *Rhizophora* belt shifts, the old generations of this species are deprived of fresh sediment supplies, become decadent and are replaced by, or form mixed stands with the other species.

In estuarine systems such as those described in this chapter, where sediment supply has not been massive enough to induce rapid changes in tidal flat and tidal channel shape and morphology (see Section 3.5 below), these changes probably involve long time spans (order of several hundred years) that probably exceed the lifespan of a generation of mangroves. Mangroves are generally long-living organisms but the ages of stands or individual trees are difficult or impossible to determine (Woodroffe 1990). Once tall mature *Rhizophora* stands disappear under the ecological changes described above, the renewal of this species has been shown to be rather poor on organic-rich substrates (Rosevear 1947). The geochemical characteristics, notably salinity, measured by anion and cation content, and chemical composition, of muddy, mangrove-colonized substrates in Sherbro Bay show extremely variable values (Anthony and Marius 1985), and meaningful correlations cannot be established with mangrove species distribution. Habitat characteristics and the attendant species distributions are in reality conditioned by a multitude of auto-correlated factors that include dispersal and thriving mechanisms proper to each species, as well as substrate morphology, stability and grain-size characteristics, substrate and soil water geochemistry, water depth, salinity and circulation patterns, all of which commonly exhibit more or less marked gradients across estuarine plains. While the mechanisms probably involved in generating zonation have been thoroughly reviewed by Snedaker (1982), very little work has been done on the relationships between mangrove successions and substrate geochemistry, a measure of whose complexity in the tropical estuarine settings of West Africa has been given by Marius (1985).

3.5 THE PRESENT GEOMORPHIC STATUS OF ESTUARINE SYSTEMS IN SIERRA LEONE AND THEIR RESPONSE TO FUTURE SEA-LEVEL RISE: A PRELIMINARY ASSESSMENT

The estuarine plains of West Africa are of immense economic interest, and vast areas of their tidal flats are extensively used as ricelands and for salt extraction while their mangroves constitute valuable fishing grounds and provide wood for both fuel and fish traps in some areas. The economic value of these estuaries is likely to gain impetus in the future. In this regard, identifying the present geomorphic status of these plains, and monitoring their possible future evolution, are important. Aerial photographs of both the Scarcies and Sherbro estuarine systems show very little visible geomorphic or ecological change in the past 30 years, apart from moderate changes in shoreline position along the Scarcies estuary and removal of mangroves for rice cultivation in this estuary (Anthony 1990). The little geomorphic change together with the well developed mangrove stands characterizing Sherbro Bay starkly contrast with some of the marked physical and bio-geochemical changes that have been recorded in estuarine systems further north along the West African coast (e.g. Marius 1985; Barusseau et al. 1985; Diop 1990; Bertrand 1993). These changes are basically due to increased aridity, the effects of which have not been directly detected in estuarine systems in Sierra Leone (Anthony 1990). In the more extreme cases in Senegal, the drought that has affected the northern parts of West Africa has led to large-scale elimination of mangrove swamps (Marius 1985), while drastically modifying the estuarine tidal and sediment dynamics (e.g. Barusseau et al. 1985). The general impression from the well developed mangrove flats in Sierra Leone and their stable tidal meanders is that these systems are in a state of equilibrium or near-equilibrium. Incoming mud brought by rivers is evacuated to the foreshore areas, unlike in deltaic plains where marked short-term changes in sediment supply, involving switching and silting up of channels, lead to rapidly changing environmental gradients (e.g. Allen 1965; Thom 1967).

Sea-level rise should be envisaged as the major future environmental constraint on estuarine systems and mangroves in West Africa (Verstraete 1989; Anthony 1990). It is not clear how a mangrove coastline will respond to sea-level change. Such a response will depend on shoreline topography, rate of sea-level rise and sediment supply (Woodroffe 1990; Ellison and Stoddart 1991). The relationship between the rate of sea-level rise and the rate of sediment supply is manifestly a critical element in the maintenance of mangroves. There are no sea-level or reliable long-term sedimentation data (based on dated peat) available from Sierra Leone. The analysis of tidal gauge data from two sites in West Africa (Takoradi, Ghana, and Santa Cruz, Tenerife) by Verstraete (1989) suggests respective annual sea-level rises of 3.4 mm (±0.6 mm) and 2.6 mm (±0.5 mm). Rates of substrate change (accretion or erosion), averaged over a period of 6 to 9 years using thin stainless steel rods implanted on both mangrove-colonized and bare fringing mudslopes, range from minimal values of –0.5–1 mm a^{-1} on the former, to 0–4 mm a^{-1} on the latter (Anthony 1984).

Both the rates of sea-level rise quoted above and rates of accretion recorded on tidal flats must be viewed with caution. The former have been determined from stations several hundreds of kilometres away that may not be tectonically stable, while the latter represent partial, short-term records of accretion that are sometimes difficult to obtain correctly in the presence of surficial accumulations of leaf litter over the tidal flats (Anthony 1984). The relationship between mangrove swamps and sea level may be estimated using the rate of autochthonous versus allochthonous sediment inputs suggested by Ellison and Stoddart (1991). Although present accretion rates may be lower than rates of sea-level rise, this might simply reflect equilibrium between tidal flooding and sediment accretion, with excess mud from upland being evacuated to the foreshores. Both the high turbidity levels observed in suspended sediment inputs to the foreshores in the wet season and the relatively low organic matter content of mangrove soils suggest dominantly allochthonous sediments. The low organic matter content is indexed by the loss on ignition values ($m = 18\%$) of over 75 sediment samples from mangrove-colonized tidal flats in Sierra Leone. Thus mangroves may not have to be in the fragile situation of having to provide autochthonous, organic sediments, to maintain themselves as sea level rises.

As far as the evolution of the estuarine shoreline systems themselves is concerned, sea-level rise may lead to a drop in the transport of inorganic mud to the foreshores fringing the tidal flats, favouring accretion of the latter. Sea-level rise should lead to an increase in the frequency and duration of tidal flat flooding. In terms of tidal processes, this may reduce the ebb-tidal effect of suspended sediment transport to the foreshores, with ebb currents diminishing in strength. Pronounced sea-level rise may in fact enhance the flood-tidal lag effect of Postma (1967) mentioned in Section 3.4.1.1, thus favouring widespread sedimentation over the tidal flats. North of the Scarcies estuary, sea-level rise may lead to more active chenier formation, as nearshore wave energy levels increase (Anthony 1990). This increase may result in estuarine shoreline erosion in places while more common cheniers may provide backbarrier areas of muddy aggradation.

3.6 CONCLUSIONS

The coast of Sierra Leone shows extensively developed estuarine systems that reflect a dense drainage network related to a humid tropical setting. Sheltered bays have provided a setting for the development of composite estuaries, while low wave energy and mesotidal conditions have favoured the development of more or less interlinked open-coast estuarine systems. The evolutionary patterns of two systems, those of Sherbro Bay and of the Scarcies estuary, have been briefly highlighted. The Scarcies is a now largely infilled estuarine system exhibiting attributes of a mixed, tide and river-dominated estuary, while Sherbro Bay is a still largely unfilled estuarine complex. Cross-shore and alongshore gradients and linkages are expressed over both estuarine systems and are marked by long-term hydrological, geomorphic and ecological change. Accretion of inner estuarine tidal flats, for instance, is shown to lead to net mud export by strengthened ebb currents towards the estuarine fore-shores where turbidity plumes are maintained, feeding the aggradation and progradation of the estuarine shorelines. While such cross-shore mud supply has been important in both estuarine systems, longshore mud supply by large-scale coastal currents has played a significant role in the progradation of open-coast estuarine shorelines, such as those of the Scarcies, as of the vast estuarine systems of the Guinea coasts. Progradation has, in turn, had repercussions on long-term tidal channel dynamics and on the ecology of mangrove swamps. Changes in estuarine ecology, for instance, imply complex bio-physical and bio-geochemical trans-formations that need to be further explored on the basis of a more multi-disciplinary approach involving botanists, geomorphologists, hydrologists, geochemists and soil scientists. The chapter has briefly speculated on the possible response of these low-lying, economically valuable, estuarine shorelines to future sea-level rise.

The themes discussed in the chapter call for further research on the relationship between processes within estuarine plains and channels and shoreline evolution in both profile and plan view. The most important line of further research appears to be that of the hydrological processes involved in sediment redistribution patterns between estuarine plains and their shorelines and the changes these processes undergo as estuarine plains accrete and as the main estuarine channels become infilled, or conversely, as sea level rises. Of particular interest is the progressive switch from flood-tidal asymmetry to ebb-tidal asymmetry across tidal flats in response to rapid accretion, and vice versa, in response to sea-level rise. Another important research need, particularly geared to the extensively interlinked muddy open-coast estuarine and deltaic systems of West Africa and of other humid tropical coasts, concerns the dynamics of longshore fine-grained sediment dispersal systems. These dispersal systems are important to the progradation of future wetlands that are valuable to the economies of several tropical countries. Both of these research needs require the acquisition of detailed field data, while the study of the dynamics of longshore sediment dispersal systems would require the extensive use of satellite-acquired remote sensing data.

REFERENCES

Africa Pilot. 1967. Volume 1. Hydrographer of the Navy, London.
Allen, J.R.L. 1965. Coastal geomorphology of Eastern Nigeria: beach ridge barrier islands

and vegetated tidal flats. *Geologie en Mijnbouw* 44: 1–21.
Anthony, E.J. 1983. Holocene geomorphic evolution of the coast of southern Sierra Leone. *Bulletin de l'Association Sénégalaise pour l'Etude du Quaternaire Africain (ASEQUA)*. 68/69: 35–45.
Anthony, E.J. 1984. Les unités géomorphologiques et pédologiques de la côte nord de la Sierra Leone: Etude des rapports entre la géomorphologie et la pédologie de la mangrove. Centre d'Etudes de Géographie Tropicale, Bordeaux.
Anthony, E.J. 1989a. Translation du littoral ouest-africain de 18000 ans B.P à l'actuel: l'exemple de la Sierra Leone. *CNRS Bulletin* 36: 255–258.
Anthony, E.J. 1989b. Chenier plain development in northern Sierra Leone, West Africa. *Marine Geology* 90: 297–309.
Anthony, E.J. 1990. *Environnement, géomorphologie et dynamique sédimentaire des côtes alluviales de la Sierra Leone, Afrique de l'Ouest.* Revue d'Analyse Spatiale Quantitative et Appliquée, No. 27–28, Presses CRDP, Nice.
Anthony, E.J. 1991a. Coastal progradation in response to variations in sediment supply, wave energy and tidal range: examples from Sierra Leone, West Africa. *Géodynamique* 6: 57–70.
Anthony, E.J. 1991b. Beach-ridge plain development: Sherbro Island, Sierra Leone. *Zeitschrift für Geomorphologie* N.F. Supplement Band 81: 85–98.
Anthony, E.J. 1993. Holocene coastal and shoreface sedimentation in West Africa. *In*: Quaternary Coastal Evolution: Models, Processes and Local to Global Factors. PIGC Projet 274 Oostduinkerke. *Extended Abstract Volume* 3–6.
Anthony, E.J. and C. Marius. 1985. Géomorphologie, sédiments et sols de la Baie de Sherbro (Sierra Leone méridionale). *Cahiers ORSTOM – Série Pédologie* 21: 97–108.
Augustinus, P.G.E.F., L. Hazelhoff and A. Kroon. 1989. The chenier coast of Suriname: modern and geological development. *Marine Geology* 90: 269–281.
Barusseau, J.P., E.S. Diop and J.L. Saos. 1985. Evidence of dynamics reversal in tropical estuaries: geomorphological and sedimentological consequences (Salum and Casamance Rivers, Senegal). *Sedimentology* 32: 543–552.
Bertrand, F. 1993. Contribution à l'Étude de l'environnement et de la dynamique des mangroves de Guinée. Données de terrain et apport de la Télédétection. Collection Etudes et Thèses, ORSTOM, Paris.
Boto, K.G. and J.S. Bunt. 1981. Tidal export of particulate matter from a northern Australian mangrove system. *Estuarine Coastal and Shelf Science* 13: 247–255.
Colman, S.M., C.R. Berquist Jr. and C.H. Hobbs III. 1988. Structure, age and origin of the bay-mouth shoal deposits, Chesapeake Bay, Virginia. *Marine Geology* 83: 95–113.
Dalrymple, R.W., B.A. Zaitlin and R. Boyd. 1992. Estuarine facies models: conceptual basis and stratigraphic implications. *Journal of Sedimentary Petrology* 62: 1130–1146.
Dijkerman, J.C. 1969. Soil resources of Sierra Leone, West Africa. *African Soils* 14: 185–206.
Diop, E.S. 1990. La Côte Ouest-Africaine – du Saloum (Sénégal) à la Mellacorée (République de Guinée). Collection Etudes et Théses, ORSTOM, Paris.
Ellison, J.C. and D.R. Stoddart. 1991. Mangrove ecosystem collapse during predicted sea-level rise: Holocene analogues and implications. *Journal of Coastal Research* 7: 151–165.
Froidefond, J.M., M. Pujos and X. André. 1988. Migration of mud banks and changing coastline in French Guiana. *Marine Geology* 84: 19–30.
Japan International Cooperation Agency. 1983. The feasibility study of the Rhombe Swamp. Agricultural Development Project. Ministry of Agriculture and Natural Resources, Freetown, Sierra Leone.
Jordan, H.D. 1963. The vegetation of mangrove swamps in West Africa. *Agronomie Tropicale* 18: 796–797.

Jordan, H.D. 1964. The relation of vegetation and soil to development of mangrove swamps for rice growing in Sierra Leone. *Journal of Applied Ecology* 1: 209–212.

Longhurst, A.R. 1959. Benthos densities off tropical West Africa. *Journal du Conseil International pour l'Exploitation de la Mer* 25: 21–28.

Marius, C. 1985. Mangroves du Sénégal et de la Gambie: Ecologie – Pédologie – Géochimie – Mise en valeur et d'aménagement. Collection Etudes et Théses, ORSTOM, Paris.

Marius, C. and J. Lucas. 1982. Evolution géochimique et exemple dîaménagement des mangroves au Sénégal (Casamance). *Oceanologica Acta* 11: 151–159.

McMaster, R.L., T.P. Lachance and A. Ashraf. 1970. Continental shelf geomorphic features off Portuguese Guinea, Guinea and Sierra Leone. *Marine Geology* 9: 203–213.

Postma, H. 1967. Sediment transport and sedimentation in the estuarine environment. pp. 158–179. *In* G.H. Lauff (ed.), *Estuaries.* American Association for the Advancement of Science, Washington, DC.

Rosevear, D.R. 1947. Mangrove swamps. *Farm and Forest* 8: 23–30.

Rüe, O. 1989. Mobilité des forêts de mangrove sur les côtes de Guinée. *CNRS Bulletin* 36: 205–207.

Sierra Leone Tide Tables. Government Printing Department, Freetown, Sierra Leone.

Snedaker, S.C. 1982. Mangrove species zonation: why? p. 111–125. *In* D.N. Sen and K.S. Rajpurohit (eds.), *Tasks for Vegetation Science*, Junk Publishers, The Hague.

Swift, D.J.P. 1976. Coastal sedimentation, p. 225–310. *In* D.J. Stanley and D.J.P. Swift (eds.), *Marine Sediment Transport and Environmental Management.* Wiley, New York.

Thom, B.G. 1967. Mangrove ecology and deltaic geomorphology, Tabasco, Mexico. *Journal of Ecology* 55: 301–343.

Verstraete, J.M. 1989. Le niveau de la mer le long des côtes de l'Afrique de l'Ouest. *UNESCO Comaraf Bulletin* 4: 5–43.

Wolanski, E., M. Jones and J.S. Bunt. 1980. Hydrodynamics of a tidal creek–mangrove swamp system. *Australian Journal of Marine Freshwater Research* 31: 431–450.

Woodroffe, C.D. 1990. The impact of sea-level rise on mangrove shorelines. *Progress in Physical Geography* 14: 483–520.

Woodroffe, C.D., J. Chappell, B.G. Thom and E. Wallensky. 1989. Depositional model of a macrotidal estuary and floodplain, South Alligator River, Northern Australia. *Sedimentology* 36: 737–756.

Wright, L.D. 1989. Dispersal and deposition of river sediments in coastal seas: models from Asia and the Tropics. *Netherlands Journal of Sea Research* 23: 493–500.

Chapter Four

Shoreline changes in the bodden coast of northeastern Germany

REINHARD LAMPE
Ernst-Moritz-Arndt-University of Greifswald

4.1 INTRODUCTION

This chapter is designed to identify the significance of temporal and spatial scales in shaping an estuarine shoreline by incorporating historical and modern dynamic processes into an evolutionary description of an estuary that has undergone considerable change in water level and wave energy. The study area is the bodden coast of eastern Germany. Glacial activity during the Pleistocene and subsequent variations in water level, wave energy and coastal erosion are examined through time. The spatial significance of active processes and geomorphic inheritance is examined in terms of position in the estuary. Examples are provided of coastal change associated with both natural and anthropogenic causes to reveal the complex interaction between hydraulic, geomorphic and biotic variables in determining the character of the shoreline.

The post-Pleistocene morphogenesis of the Baltic coastline is the result of the coincidence of four modes of activity:

1. Glacial and fluvioglacial erosion and deposition during the Pleistocene that determined the type, stratification conditions and morphology of the rocks forming the coast.
2. Isostatic rebound in the formerly glaciated areas, starting after the ice age and producing high elevations in the north and lower elevations in the south, resulting in regressing coasts (coasts of emergence) in Scandinavia and the

Estuarine Shores: Evolution, Environments and Human Alterations. Edited by K. F. Nordstrom and C. T. Roman.

Baltic countries and transgressing coast (coast of submergence) on the southern shore of the Baltic (Kolp 1991).
3. Post-glacial re-shaping of the coastline during stages of development of the Baltic Sea, especially during the Litorina transgression, resulting in a rapid flooding of older coastlines, particularly in the southern Baltic (Kliewe and Janke 1982).
4. Coastal formation processes following the Litorina transgression under the influence of contemporary hydrographical and meteorological conditions (Lampe 1987).

The interplay between these complexes along the German Baltic coast influenced the development of different types of coast, including: (1) the firth coast in eastern Schleswig-Holstein; (2) the large-bay coast between the islands of Fehmarn and Poel; (3) the smoothed coast between Poel and the Fischland; and (4) the smoothed bodden coast from the Fischland to the Polish border (Figure 4.1).

4.2 THE GENESIS OF THE SMOOTHED BODDEN COAST

The smoothed bodden coast is a rare and perhaps unique type of coast. It is found in the area beyond the so-called ice divide of Jasmund (NE Rügen). The term ''bodden'' is the local name for a shallow lagoon behind the exposed Baltic coast. The term ''haff'' is used where there is a strong river water input.

Tectonic uplift in this coastal region caused upper cretaceous chalk deposits to rise above the Pleistocene base of the surroundings to a height of up to 200 m. This high area divided two glacier streams from one another in the late Pleistocene. The glacial streams reached beyond the Arkona basin into the region of the lower Oder valley and the area of today's Darss/Zingst bodden chain. The Baltic and Oder bay glaciers, which oscillated strongly as they withdrew, left behind a relatively diversified relief. This relief is assumed to follow a morphology present prior to the Weichsel period because of the weak shaping force of late Weichsel inland ice. To the rear of these ice-rim areas, exaration processes caused the development of several rows of flat glacier-tongue basins arranged along the margin, which probably remained sealed with detached ice for a long period towards the end of the late ice age (Kliewe 1987). A number of fairly large ice-barrier lakes formed between and above the detached ice; these may have been related to the first stage in the development of the post-ice age Baltic Sea, the Baltic Ice Lake.

As the inland ice receded farther and the water level of the Baltic Ice Lake fell as a result, the area of the present coast dropped to a dry level of about 40 m below sea level. The loss of the link between the lake and the ocean through the central Swedish depression (Degerfors), as a consequence of the isostatic elevation of Scandinavia, led to the rapid water level rise of the so-called Ancylus Lake from about 9300 BP onwards. This occurred as a result of the fresh water inflow from the

Baltic tributaries and the final melting of the remaining ice caps of Fennoscandia. By about 8900 BP the Ancylus Lake had reached its highest level of about 8 m below sea level but was able to penetrate only into the deepest parts of the glacier tongue basins beyond the Jasmund ice shed (North Rügen Bodden and Greifswald Bodden). The subsequent rapid retreat of this freshwater lake, as a result of its breaking through the morain ridge of the Darss sill (its natural dam), constituted the beginning of another dry-land period in the present-day coastal region. Not until the Litorina transgression, beginning around 7900 BP (when the Danish Baltic region flooded due to rising sea level in the ocean) did this phase come to an end. Initially, the rise in the water level attained a speed of up to 2.5 $m\,c^{-1}$ and it then slowed down to 0.3 $m\,c^{-1}$ by about 5700 BP (Kliewe and Janke 1982; Figure 4.2).

The rapid rise in sea level caused: (1) a marked erosion of the coasts; (2) sedimentation of some of the eroded material in the embayments of the present day boddens; and (3) formation of an archipelago with many fossil cliffs that are still visible today. Only when the transgression began to progress more slowly did the longshore drifting and shoreline smoothing become more pronounced. Between 6000 and 3500 BP a large number of spits and bars were formed. These features attached themselves to the areas of former glacial deposition and cut off the flat basins behind them from the open sea. It is assumed that the build-up of the spits and bars was completed by about 3500 BP, although isolated inlets (Seegats) still existed up to the 16th century and some even until the 19th century. Most of the inlets were closed artificially. As glacially shaped embayments flooded by the sea and later largely separated from it, the boddens have a complex genesis. They are recent formations during the Holocene.

The more energetic outer (Baltic Sea) coast developed into a smooth coast (in plan view) but with an alternation between steep (Pleistocene till and sands) and flat shores composed of Holocene sands. The inland (bodden) coast stagnated at an early stage of coastal smoothing and remained irregular in plan view because of a marked decrease in fetch distance and the consequent weak wave energy for moving sediment. Only 340 km of the 1470 km long coast of Mecklenburg/West Pomerania are exposed to the Baltic Sea; the other 1130 km are along the boddens and haffs. There is a close link between land and sea in the bodden coastal area, with a great diversity of landscape elements and a rich endowment of valuable ecological habitats.

The mouths of the rivers retreated towards the land as the Litorina transgression continued and the erosion base rose. At the beginning of the Ancylus period, the Oder river still ended in the Arkona basin 44 m below sea level. The eastern arm of the Dzivna River ended in the southern part of the Bornholm basin at almost the same depth (Kolp 1983). Today their mouths lie 50–100 km farther inland. The retreat of the river mouths during the transgression and the isolation of the embayments of the boddens into which they then emptied caused changes in the hydrography of the boddens so that they became large mixed basins between sea and river water, forming shallow, polymictic, tideless estuaries.

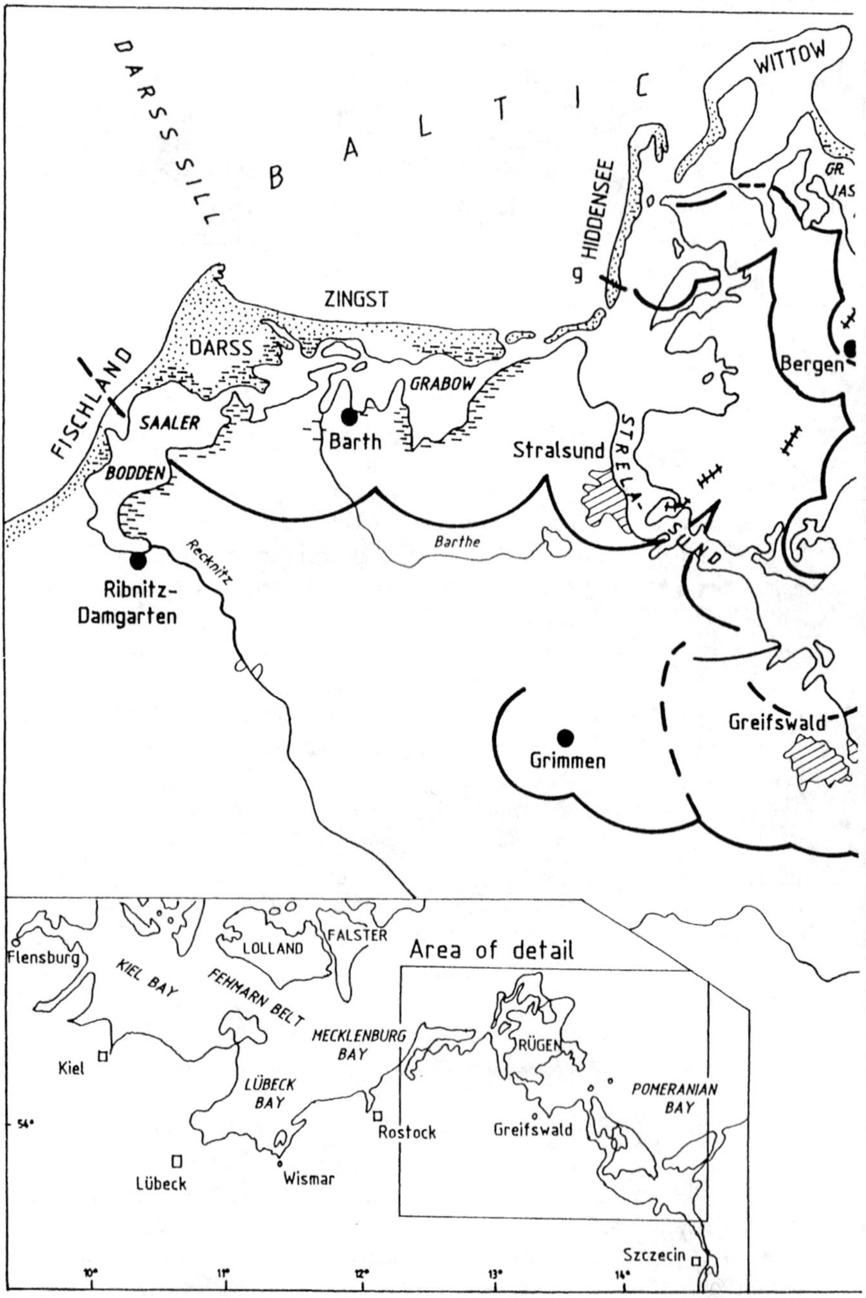

Figure 4.1 (*and opposite*) Position of the West Pomeranian bodden landscape with late Pleistocene and Holocene morphogenetic entities related to its formation (moraine ridges, spits, bars, marine sand plains and peat lowlands)

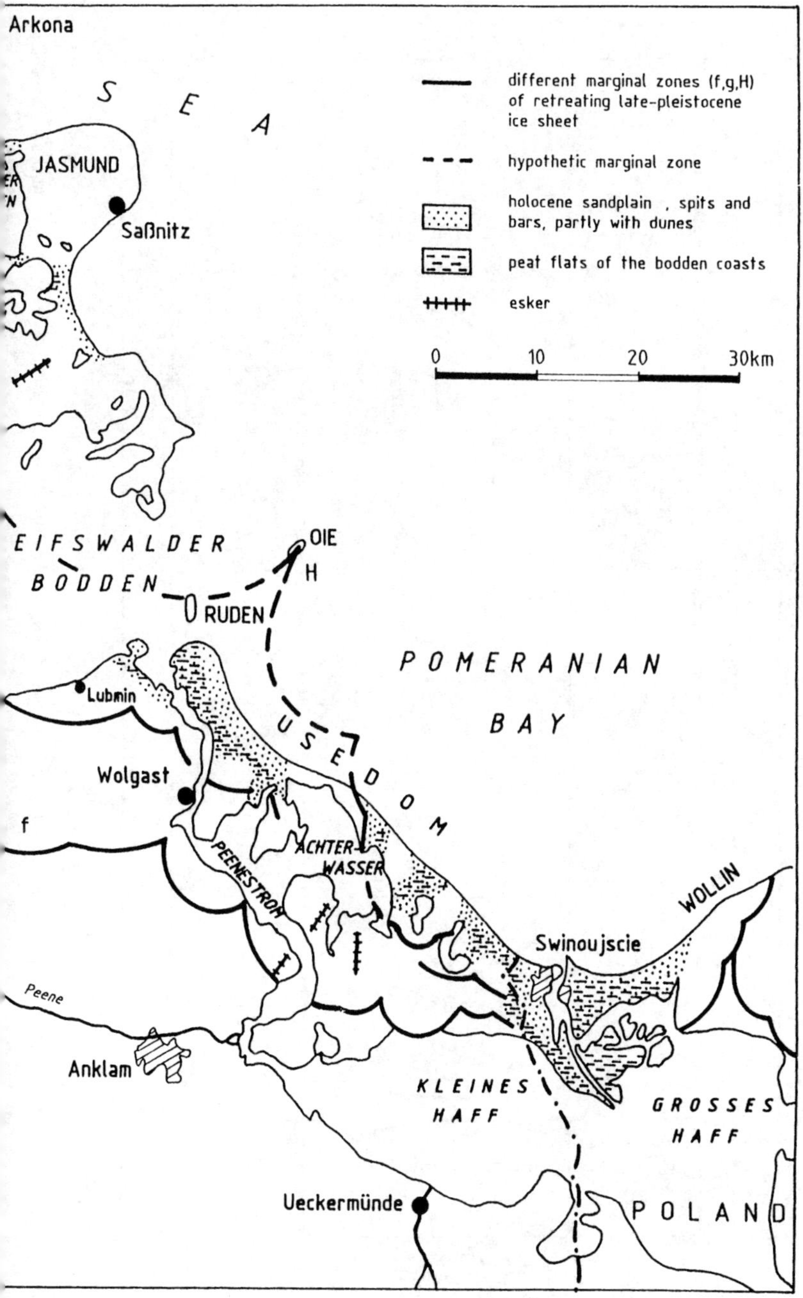
Arkona
S E A
JASMUND
Saßnitz
different marginal zones (f,g,H) of retreating late-pleistocene ice sheet
hypothetic marginal zone
holocene sandplain , spits and bars, partly with dunes
peat flats of the bodden coasts
esker
0
10
20
30km
EIFSWALDER
BODDEN
OIE
H
RUDEN
POMERANIAN
BAY
Lubmin
USEDOM
Wolgast
f
ACHTER-
WASSER
PEENESTROM
WOLLIN
Swinoujscie
Peene
Anklam
KLEINES
HAFF
GROSSES
HAFF
Ueckermünde
POLAND

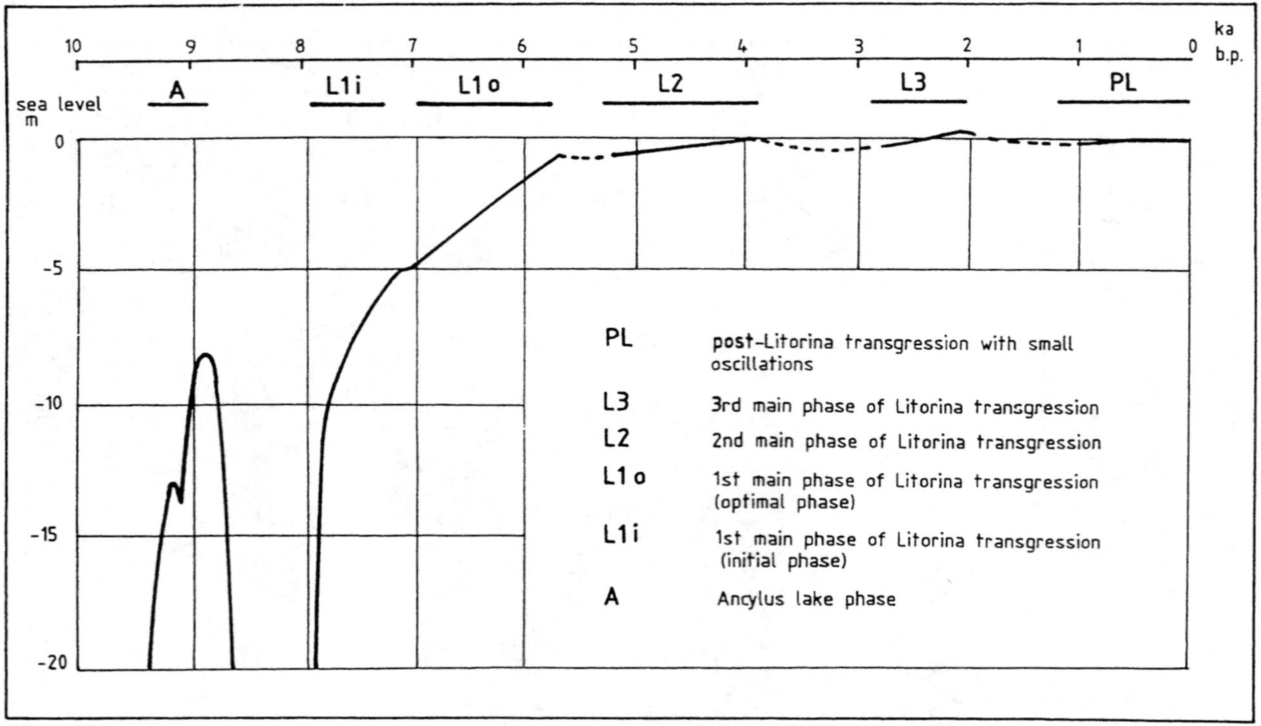

Figure 4.2 Graph of the Holocene rise in sea level along the coast of West Pomerania (simplified from Kliewe and Janke 1982)

4.3 CHANGES IN HYDROGRAPHY FOLLOWING ISOLATION OF THE BODDENS FROM THE BALTIC

The cutting off of the boddens from the Baltic as a result of the evolution of spits and sand bars caused profound changes in the hydrographic conditions leading, in turn, to the development of morphodynamics and an ecology peculiar to these areas. The alteration of the hydrography resulted mostly from the narrowing of the sea area (causing reduction in wave energy) and the limited exchange of water with the Baltic (causing desalination, eutrophication and attenuation of fluctuations in water level).

4.3.1 Wave influence on the bodden coasts

The limited sea area and shallow water of the boddens are the reasons why the coastal forming processes (including cliff erosion, sand transport parallel and perpendicular to the coast, spit formation) are less intense than along other coasts composed of unconsolidated or poorly consolidated sediments. The wave movement of the bodden waters is more like that of large shallow lakes than of other waters on the edge of the Baltic Sea (e.g. firths).

Perspective on wave energy may be gained using Krylov's procedure (Weiss et al. 1978) that expresses wave energy as deep water wave energy per metre of coast per year. The maximum along the outer coast is 13 MWh m^{-1} a^{-1} with an average of 6–8 MWh m^{-1} a^{-1} at Cap Arkona (Wiemer and Gurwell 1991); along the bodden coast, the maximum is 1 MWh m^{-1} a^{-1} (at peninsula Gross Zicker, Greifswald Bodden), with an average of 550 kWh m^{-1} a^{-1} in the Greifswald Bodden, c. 150 kWh m^{-1} a^{-1} in the Achterwasser, and less than 100 kWh m^{-1} a^{-1} in the Darss/Zingst bodden waters (Lampe 1987).

The vector of wave propagation (which determines the size and direction of wave and current processes), can be divided into the shore-normal and the shore-parallel components to estimate the intensity and direction of sediment transport along the coast. For the bodden coast, the angle between the approach direction of the wave resultant and the shore-normal is usually small, i.e. the component parallel to the coast is usually also small. Longshore sediment transport is therefore frequently of less importance than on the outer coast, and shore-normal energy becomes relatively more important. One possible cause of this phenomenon is the rounded form of many bodden waters which prevents a greater concentration of energy delivered in one direction.

4.3.2 Water exchange between the boddens and the Baltic Sea

The smoothing of the outer coast made the links between the Baltic Sea and the boddens increasingly long and narrow, although complete closure did not occur due to compensatory currents. Variations in the water level caused by the wind in the

Table 4.1 Morphological parameters of the bodden waters (selected from Correns 1979).

Region	Area (km²)	Volume ($10^6 m^3$)	Average depth (m)	Maximum depth (m)
Bodden south of Darss and Zingst	196.8	397.1	2.0	12.0
Bodden between Hiddensee and Rügen	171.3	300.5	1.8	7.6
Rügen inland bodden	58.6	553.5	3.5	10.3
Strelasund	64.4	252.5	3.9	16.0
Greifswald Bodden	510.2	2960.0	5.8	13.5
Peenestrom	163.9	429.3	2.6	16.0
Kleines Haff	277.2	1025.6	3.7	7.8
Grosses Haff	409.7	1556.9	3.8	

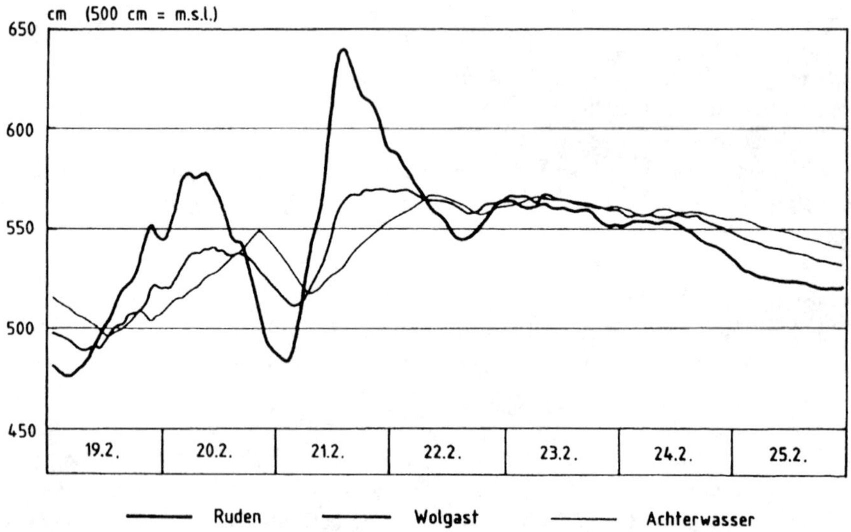

Figure 4.3 Variations in water level at various gauge stations on outer and inner (bodden) coast as example of the weakening and phase shift in water level changes (measurement period 19 February to 25 February 1993; sites of the stations in Figure 4.1)

tideless Baltic could only be transferred to the inner coastal waters in a weakened form, with decreasing force and a greater time lag, due to the greater distance from the opening link (Correns 1979, Figure 4.3). As a result, the boddens farthest inland have the lowest salinity and the least variation in water level. Since the degree of variation in the water level has an effect on wave energy and coastal erosion, this results in a marked decline or complete elimination of coastal shaping through waves and currents far from the inlets. Processes of phytogenic sedimentation and an increasingly limnetic form of the shore increase as the water becomes less salty. Parallel to the decreasing salt content towards the mouth of the rivers in the boddens there is also a drastic increase in the amount of eutrophication. Because of its stimulating effect on the growth of phytoplankton, eutrophication is revealed in reduction in the visible depth, inhibition of the growth of macrophytobenthos and the sedimentation of mud that is rich in nutrients (Lampe 1993).

4.3.3 Coast formation by erosion, accretion and phytogenic sedimentation

The primary cause of the shifting of shore sediment is the rapidly increasing orbital movement of the water particles in the waves as the depth decreases. As a result of wave asymmetry (short steep crest, long flat trough), the coarser content of the sediment begins to shift landwards. The marked dependency of this process on the depth is critical for grading and sorting of heterogeneous sediment. Where the sediment is relatively coarse grained, a water depth of 3–4 m means that only fine sand is transported; at a depth of 0.5 m the gravel may also be transported. As a result of the constantly changing wave conditions, the grain fractions shift to their equilibrium locations on the shore. In the process, they form three zones of sediment accumulation (the beach, the sand riff area and the mud area farther from the shore) which are separated by the abrasion or lag sediment zones. This sedimentation sequence was first described and demonstrated by Pratje (1939), and has been confirmed by other scholars working in the outer coasts of the western and central Baltic (Seibold et al. 1961; Flemming and Wefer 1973) and in the West Pomeranian bodden waters by Gusen (1978) and Lampe (1987) (Table 4.2). Along the bodden coasts, there is hardly any tendency for this zonation to be altered by wave movement because of the limited wave energy. This is expressed in the characteristic biotopes that develop in the individual shore-perpendicular zones (Figure 4.4).

Table 4.2 Characteristics of sediment zones

Zones of marine sediments	Nearshore lag sediment zone	Nearshore sand zone	Farshore lag sediment zone	Farshore sand zone	Mud zone
Kinds of sediment	Coarse sands, gravels, blocks, partly sand	Fine and medium sands	Coarse sand, gravel, partly fine sand	Fine and medium sands	Silt, mud, sandy mud
Sorting	Very poor	Well	Medium to very poor	Very well	Poor
Depths (m)					
Western Baltic	0–6	6–10	10–15	15–17	>17
Central Baltic	0–10	10–20	20–40	40–50	>50
Greifswalder B.	0–0.5	0–2.5	2–4	4–7	>7

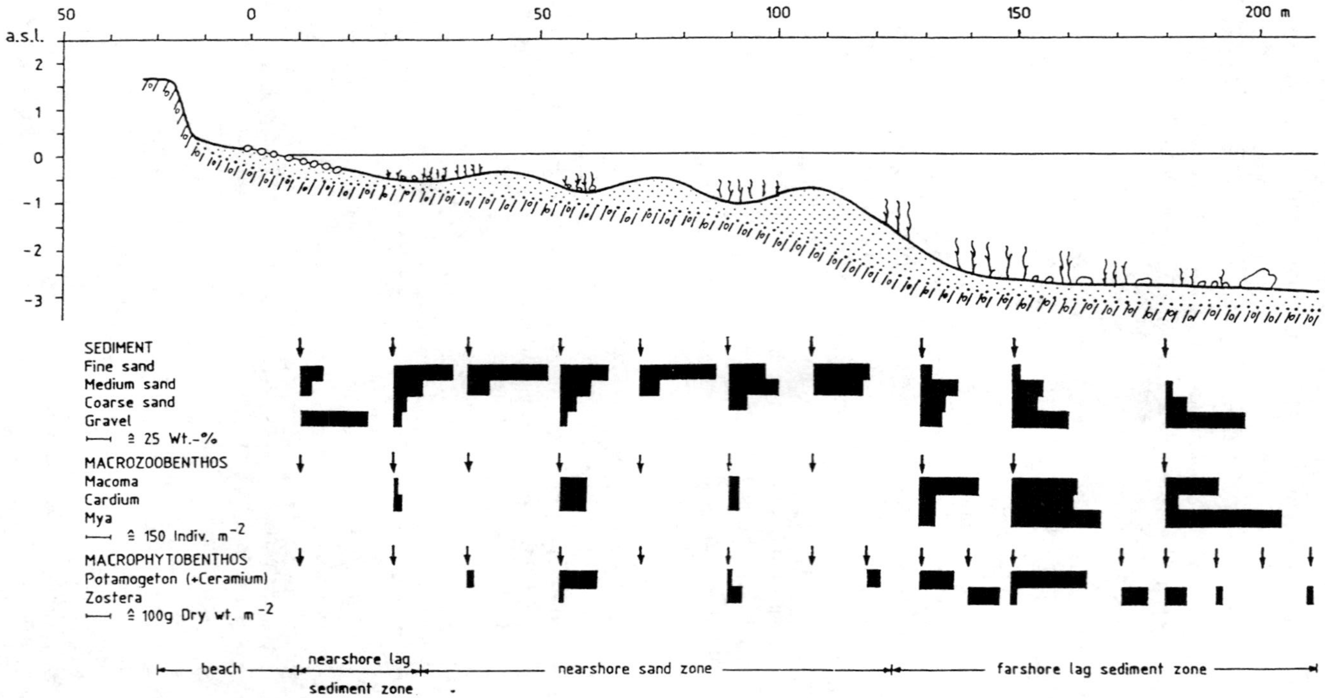

Figure 4.4 Close-up photograph of the sediment, macrozoobenthos and macrophytobenthos zones of a coastal profile in the western Greifswald Bodden (after Lampe 1987)

Other landscape elements are used to divide the coast in the longshore direction. Coastal segments characterized by abrasion, transport or accumulation are grouped together as relatively closed but balanced systems or coastal cells (Clayton 1980; Gurwell et al. 1982). It is difficult to determine the hydrodynamics along the coast directly in most cases, so the identification of coastal character must be based on morphological and lithological features. Restriction of the analysis to an investigation of morphological forms can provide misleading results. It may be that quasi-fossil forms will be preserved because structures that were created at higher energy levels may only be reshaped slowly by later processes occurring at a lower energy level. Examples of quasi-fossil forms include the abrasion terraces found at various points where they could not form under recent wave conditions and were formed when the boddens were still bays of the Baltic. The correlation between morphology, lithology and hydrodynamics is more or less inapplicable at these points, so that morphological and lithological studies cannot be conducted without knowing both the present and past processes.

As a result of the limited effect of wave movement and the higher surge level, the cross-shore extent of the coastal changes in the boddens is generally far less than along the outer coast. The lack of a threat of major coastal changes is responsible for the small number of references in the literature; the sparse documentation concentrates mainly on ascertaining land losses resulting from exceptionally high storm floods. Thus an estimate of the coastal regression in the past is possible only for a few places, especially where there were major landslides, where farmsteads were endangered (Gomolka 1971) or at old measurement points. Gomolka (1987) and Janke and Lampe (1993) have described to what extent it is possible to reconstruct changes in the coastline with the help of historical maps. The average regression amounts to between 0.10 and $0.20\,\mathrm{m\,a^{-1}}$ (cf. example 2, below) for the active sections of the Greifswald Bodden. Along the coasts of the other boddens, the average regression is $<0.10\,\mathrm{m\,a^{-1}}$.

Averages of coastal erosion for long periods can be obtained with the help of fossil submarine cliffs. Cliffs came into existence at various stages of stagnation in the Litorina transgression; they were then drowned when the water level rose further. These cliffs can be identified today at various points in the Greifswald Bodden, the Achterwasser and in the area of the Peenestrom (Gusen 1978; Lampe 1987). For these places it can be concluded that the regression was between 75 and 300 m over the past 6800 years. When compared with contemporary measurements, these figures lead one to assume long periods of limited regression or stagnation.

Large amounts of deposition are not expected considering the small amount of sediment available from coastal erosion. The relatively great amount of cross-shore transport is also a hindrance to the development of accretional forms and the shortage of sand on the few beaches in existence. Recent formations like spits and bars are rare and of limited dimensions despite the irregular shoreline configuration. Fossil forms are much more frequent; these point to greater exposure to waves and currents before the closing of the sand barriers (Lampe 1987, 1992).

The development of the coasts is stagnant, at present, as a consequence of the limited wave energy and the low intensity of erosion and accumulation processes. Shaping by the flow of lithogenic material is replaced by the accumulation of mainly autochthonous phytogenic material; the shoreline is fringed by stands of reeds of varying breadths and with few species, of which reed grass (*Phragmites australis*) greatly predominates. Wave energy serves here as the limiting factor to further expansion in the seaward direction, whereas trophic properties and the salinity of the waters decisively affect the composition and the development of the vegetation within the reed. The use of such areas in the past and the present also has direct modifying effects.

4.4 EXAMPLES OF PROCESSES AND FORMS OF COASTAL EVOLUTION

Three examples will be given to illustrate the following: (1) the principles of coastal genesis; (2) changes in the sediment budget of a ''balance system'' with reference to technogenic and anthropogenic measures; and (3) sedimentation in the stagnant regions of the bodden coasts under the influence of varying degrees of agricultural exploitation and increase in nutrients.

4.4.1 The genesis of the bodden waters to the south of the Fischland–Darss–Zingst peninsula

This peninsula and the Darss/Zingst bodden chain form the westernmost element of the bodden smooothed coast comprising predominantly glacial and marine landscape units (Figure 4.5). The basic structure consists of dome-shaped moraine cores at different heights that stretch from Dierhagen moraine core via the upper Fischland, the Schifferberg and other, lower lying cores on the Zingst peninsula to the moraine core below the Grosser Werder. The almost flat plate of the Alt-Darss, at a level of up to 8 m above sea level and composed of sands of a late glacial melt water lake, also functioned as a core (Alt-Darss). During the Litorina transgression, with an initial inundation rate of $>2.5\ \text{mm a}^{-1}$, the former coastline was displaced so fast that the friable material was mainly redistributed under the sea, and no more than small, short-lived spits occasionally formed. The coastal landscape was drowned, and an archipelago developed. Both the Fischland core and the Alt-Darss must have been considerably more extensive at that time, although this cannot be reliably reconstructed. During the more recent Litorina period, the embayments of the present-day boddens were then flooded and thus became open bays of the Baltic Sea. Active cliffs came into existence along their mainland shores. From about 6000 BP onwards, the transgression process slowed down to an inundation rate of $<1.5\ \text{mm a}^{-1}$ and less material was brought in from the destruction of the island cores, creating conditions for a widespread coastal smoothing. Since about 4000

BP, the material eroded from the Fischland has been deposited as sand ridges in front of the inactive cliff of the Alt-Darss, forming the Neu-Darss (described in detail by Kolp (1982)). The whole 5.5 km long section of the Neu-Darss between Esper Ort and Darsser Ort has taken shape within the past 1200 years, indicating the rapid speed of its growth. The links between the other island cores also narrowed increasingly as spits were formed (Ribnitzer Stadtwiesen, Vor-Darss, Sundische Wiese and Bock) but without losing the links with the sea. The Permin, Loop, Hundsbeck, Hundetief and Straminke inlets were still open, in part, until the 15th century (the Prerowstrom was open until 1874). The inlets were navigable, in part, until they were closed naturally or artificially for protection against flooding. On their bodden side, some surge deltas (similar to flood tide deltas on tidal coasts) developed because inflowing water reached much higher flow velocities during the rapid rise in the water level of the Baltic during storm floods than in the subsequent outflow phase, when the water level fell more slowly. Surge deltas of this type can be found wherever a narrow link between the Baltic and a bodden is combined with a large supply of sediment from neighbouring erosion areas and include Schmidtb?lten in the Prerowstrom area, Kirr Island on the Straminke (Figure 4.5), Vierendehlgrund to the south of the Gellen/Hiddensee, Hahnentiefschaar/Hiddensee, Swina mouth/Usedom.

With decreased sediment input as a consequence of a more or less stagnant sea level, the smoothed coast is increasingly subject to erosion. The Fischland cliff has been cut back by 0.75 m a^{-1} over 100 years, corresponding to a movement of about 6 $m^3 m^{-1} a^{-1}$ of sediment. This amount is not sufficient to saturate the sand flow along the west coast of the Neu-Darss, so a retreat of the shoreline by about 0.65–0.95 m a^{-1} can be observed. Accretion is ascertainable only near Darsser Ort with an average of 9.4 m a^{-1}. This growth in a northerly direction leads, in turn, to a marked shortage of sediment, especially near Zingst on the north coast of the peninsula. Here the erosion amounts to about 0.60–0.70 m a^{-1}. However, it is difficult to obtain data that truly reflect the natural process because coastal protection measures have been in use for over 100 years. Initially, dunes were constructed in an attempt to stop erosion; later groynes were built for this purpose. Beach nourishment has been conducted in the most recent period to counteract the steepening of the shore. A total of $1296.2 \times 10^3 m^3$ of sand were deposited in the area of Zingst between 1968 and 1990 (Weiss 1991). If the rise in sea level were to speed up again as a result of climatic changes, without the introduction of massive protective measures, the bodden chain may open up again from the west. Compensatory flows between the boddens and the Baltic would be established mainly through the inlets; the strong currents in the present openings to the east between the Werder islands and between the islands of Bock and Hiddensee would decrease, and these openings would rapidly close because of the large supply of sediment.

The evolution of the bodden smoothed coast in the Fischland–Darss–Zingst region provides a classic example of cyclical development determined by progress or stagnation during a transgression and influenced by the transport of material

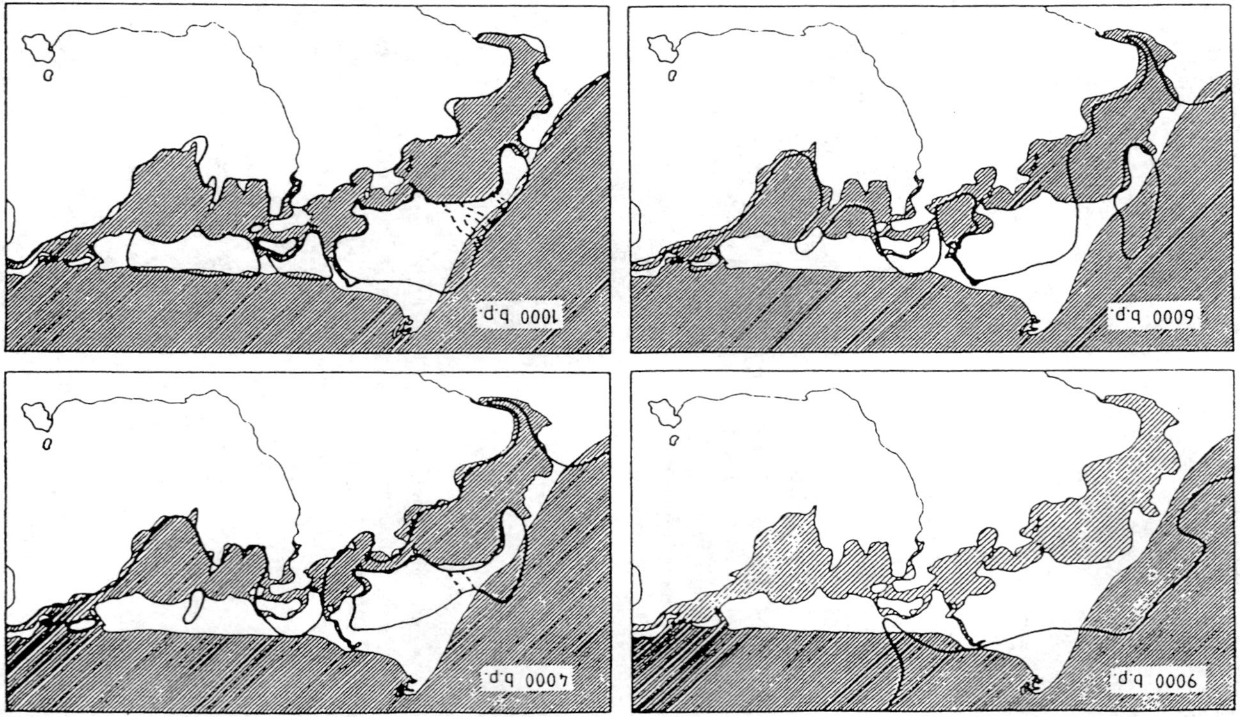

Figure 4.5 Shift in the coastline during the development of the Fischland–Drass–Zingst peninsula (upper four figures) and topography, geological construction units and recent rate of coastline shift calculated from analysis of historical maps (*next page*) (after Janke and Lampe 1993)

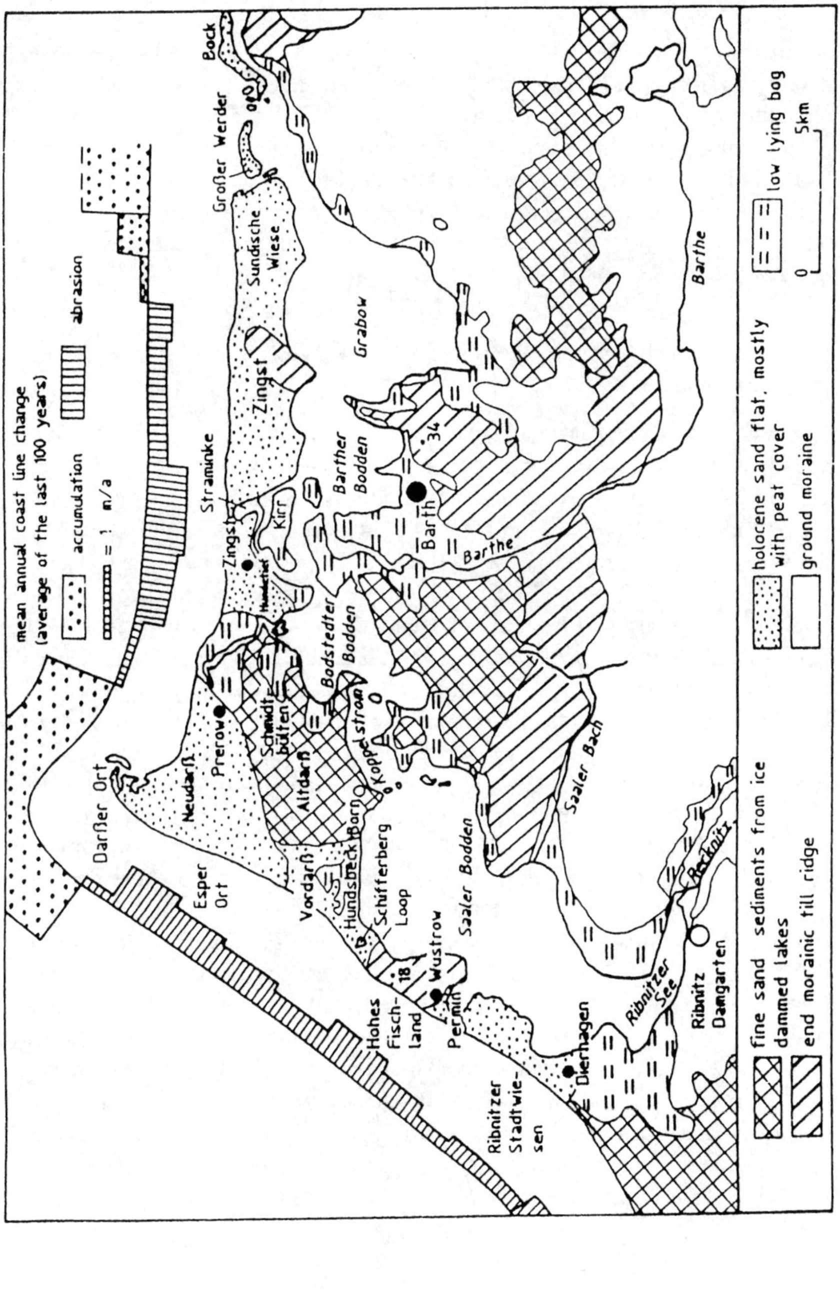

mean annual coast line change
(average of the last 100 years)
accumulation
abrasion
= 1 m/a
Darßer Ort
Esper Ort
Neudarß
Prerow
Zingst
Straminke
Kirr
Zingst
Sundische Wiese
Großer Werder
Bock
Grabow
Barther Bodden
Barth
34
Barthe
Bodstedter Bodden
Schmidt bülten
Altdarß
Koppelstrom
Vordarß
Hundsbeck
Barn
Schifferberg
Loop
Saaler Bodden
Saaler Bach
18
Wustrow
Hohes Fisch-land
Permin
Dierhagen
Ribnitzer See
Ribnitz Damgarten
Recknitz
Ribnitzer Stadtwie-sen
fine sand sediments from ice dammed lakes
end morainic till ridge
holocene sand flat, mostly with peat cover
ground moraine
low lying bog
0
5km

generally to the east. The different stages are: (1) flooding during rapid sea level rise, causing marked cutting back of the coasts and sedimentation of the eroded material in the basin and raising of the submarine areas of future spits; (2) flooding of the landward basins (the present day boddens) and formation of active cliffs along the internal shores; (3) reduction in the speed of the transgression, growth of spits and bars along the line of sediment flow (north or east) and maintenance of inlets by compensatory currents; (4) narrowing of fetch distances in the boddens with marked attenuation of the hydrodynamics and inactivation of the cliffs, starting with phytogenic sedimentation, artificial closing of some inlets and eastward shift of the opening to the Baltic; and (5) erosion of the sand bars that close off the boddens in the west due to reduced supply of sediment, resulting in reopening of inlets to the Baltic in the west and formation of a smoothed section in the east where links closed.

4.4.2 Natural and technologically induced changes on the south coast of the Greifswald Bodden

The Greifswald Bodden, with an area of 510 km^2, an average depth of 5.8 m, and a wide opening to the Pomeranian Bay, has the greatest wave and current energies. Maximum wave heights have been measured at 1.8 m with wind forces of 10 Bft. Coastal changes take place more quickly here than in other boddens and are relatively well documented on the more highly populated south coast.

The south coast of the Greifswald Bodden is characterized by considerable variety. In the eastern part from Struck Island to the nuclear power station Nord (no longer functioning) it is composed of salt grassland, part of which is protected by dikes and mainly belonging to the nature conservation area ''Peenemünder Haken, Struck and Ruden''. Under the peat, there is a vast amount of sand from a late glacial lake. This sand is also found in the shallow water area of the Freesendorf Shoal. From the outflow channel of the former power station (km 4, Figure 4.6) to the village of Lubmin, the ice lake sands rise gradually to 12 m above sea level (the so-called Lubmin heath) and form an active cliff that feeds an accumulation area to the west of the jetty protecting the outflow channel. The shoreline here has moved seaward considerably since the jetty was built at the end of the 1960s. Farther westward, glacial till emerges above the basin sands. From km 6.5 the till forms a stretch of cliff 2–6 m high; the cliff is active to varying degrees and extends to the area north of the village of Gahlkow. Up to the Gahlkow Shoal (km 12.5) the coast is flat with dunes (which can also be found in places in the inactive steep coastal areas on either side). From the Gahlkow Shoal to the beginning of the nature conservation area ''Lanken'' the coast consists of a holocene shore ridge with dunes in front of a marshy lowland. The diversity in coastal morphology is reflected in a corresponding diversity of independent biotopes, and the Freesendorf Marsh, the Grosses Holz near Loissin and the shore ridges of the ''Lanken'' have been designated nature preserves.

The natural accumulation areas in the east and west and erosion between Vierow and the Lubmin heath can still be identified, but engineering works have divided the coast into a series of smaller systems, generally increasing erosive tendencies. The contemporary geo-ecological inventory and dynamics of the coast can be seen in detail in Figure 4.6. It is possible to distinguish four separate sections with different coastal dynamics.

The first section includes the Freesendorf Shoal and the salt grassland of Struck Island and the Freesendorf Marsh. The wide swing of the 2 m isobath in the Freesendorf Shoal area does not indicate extensive accumulation under present conditions; the shallow water region is formed by the same late glacial basin sands as were deposited along the coast. The cover of recent sediment is no deeper than 0.5–1 m (Gusen 1988). The low sand cliff and the thin peat of the salt grassland are generally subject to erosion. Historical maps reveal an average retreat of the coast of about $0.7\,\mathrm{m\,a^{-1}}$ since 1694. After construction of the outflow channel, the erosive tendency in the western part of this section increased because the contribution of sands from the Lubmin heath was arrested.

The second section stretches from the outflow channel (Figure 4.6) to km 4.6 where an active cliff begins. The 400 m long jetty has trapped sand, resulting in a marked seaward swing to the shoreline from the west to the channel entrance; the dunes also formed largely after the jetty was built (Gomolka 1971).

The third section is characterized by extensive erosion and runs from km 4.6 to km 10.3. There are small areas of accumulation only on the two sides of the solid Seabridge; otherwise erosion predominates, especially to the east. With a wave action of approximately $720\,\mathrm{kWh\,m^{-1}\,a^{-1}}$ the coastal losses in the past were about $0.4\,\mathrm{m\,a^{-1}}$. Groynes were built to the east of the pier as early as the 1930s; a steel mantle was built along the cliff near the village; and in 1988 an artificially nourished beach of $80\,\mathrm{m^3\,m^{-1}}$ was placed between km 5.75 and km 7.75. These protection works had a generally positive effect, compensating for the marked lack of sediment and stopping coastal retreat. The neighbouring areas benefited from this additional fill material to some extent, especially between km 7.75 and km 8.75. The originally active cliff was rendered inactive and an artificial dune was established at the cliff foot to develop as a sand deposit. The boulder field that existed at Lubmin prior to the nourishment is still covered by fill material to a large extent (marked on Figure 4.6 by a dotted line) and is now only conspicuous from km 7.75 onwards. Part of this boulder field, the Teufelsstein (devil's stone), is a granite block measuring $32\,\mathrm{m^3}$. The stone is an excellent marker for measuring erosion, and it reveals a loss of $0.3–0.4\,\mathrm{m\,a^{-1}}$. The eastern part of this area now benefits from sediment provided by the beach nourishment in Lubmin which has at times hidden the dynamic character of the shoreline. At km 9.5 the trend of the coast is interrupted by the quay of Vierow Harbour. After the harbour was built, cliff erosion was so great on both sides that it was necessary to build two longitudinal stone structures. A change in the general longshore sediment transport can be observed in the surroundings of the harbour. The long-term average

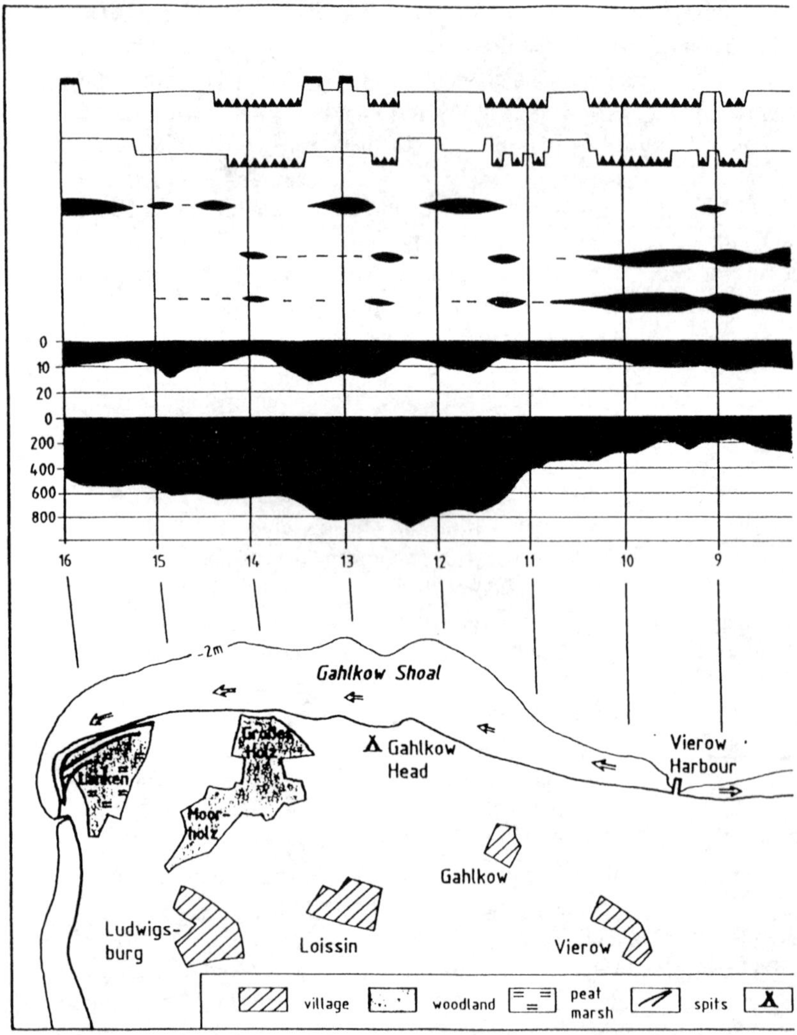

Figure 4.6 (*and opposite*) Coastal dynamics and geo-ecological inventory of the south coast of the Greifswald Bodden

indicates an easterly longshore transport east of the harbour and a westerly transport to the west. These trends can be explained as a consequence of changing fetch distances and the resulting change in longshore power of the waves. Also, the protective effect of the shallow Gahlkow Shoal against westerly winds must be taken into account. The net transport away from the harbour is accompanied by a marked sediment deficit.

In the fourth section, from km 10.75 to the western end of the segment, there is

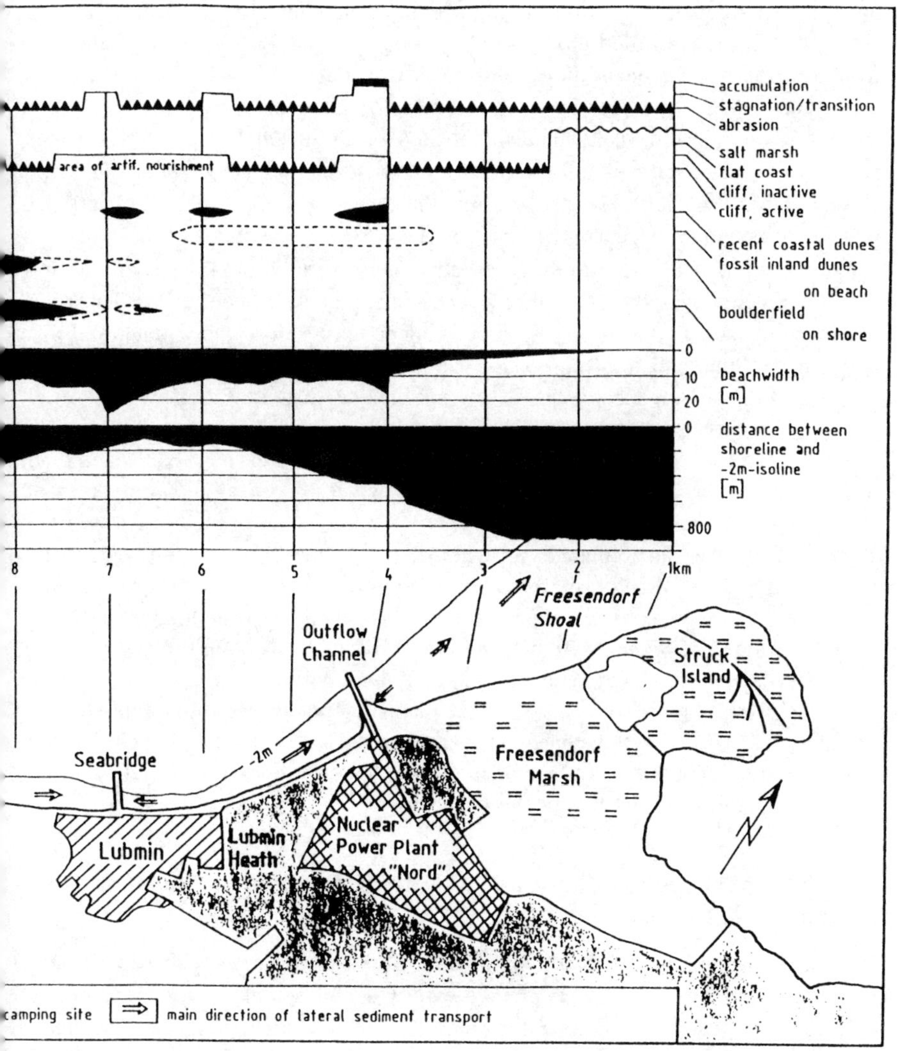

an increasing tendency for sediment to accumulate, creating wider beaches and dunes and resulting in the inactivation of the cliffs. Erosion now predominates only along the low cliff in front of the Grosses Holz. At the western end, this section terminates with the beach ridges and dunes of the Lanken.

With a few insignificant exceptions, the dunes along the south coast of the Greifswald Bodden are the only ones along the inner coastal waters. Dunes develop only where large amounts of sand are available for drifting and they are high enough that frequent high floods cannot destroy incipient forms. Normally these conditions are not fulfilled along the low bodden coasts that have little sand and tend towards phytogenic sedimentation. Only in the Greifswald Bodden are the waves and fluctuations in the water level strong enough to form beaches of sufficient width. Here, plant matter deposited on the upper beach by high water, serves as the starting point for dune formation. Eelgrass (*Zostera marina*) rots slowly and is especially effective in creating dunes by forming long barriers that trap sand that would otherwise be lost from the beach. The volume of these barriers sometimes amounts to several thousand cubic metres (Subklew 1987). The barriers become covered by a thin drift; after a short time, plants take root on this substrate, which is fertilized by the slowly disintegrating eelgrass below; these plants stabilize the primary dune and trap more sand. Dune formation on the south coast occurs because of the coincidence of the following: (1) a wide area of shallow water, rich in plants, that creates sufficient plant material; (2) exposure both to a wind from the southwest and storm floods from the northeast that makes it possible for plant material to be washed up frequently and extensively; (3) a large supply of sand from coastal erosion in neighbouring areas.

Dune formation also depends on whether beach vegetation is removed. In the past, the collection of eelgrass was often an important secondary source of income for fishermen, who sold the plant material as fertilizer, insulating material or mattress filling. Between 1880 and 1896, 5000 t were transported inland annually from Greifswald Harbour. The commercial exploitation only stopped in 1955 (Subklew 1987). Now, the decaying remnants of plants are often removed for aesthetic reasons to make the beach more attractive for tourists. No thought is given to the consideration that this prevents the formation of primary dunes and adversely affects the ecological balance of the beach.

4.4.3 The phytogenic sedimentation coast in the Grabow and the Barth Bodden

In contrast to the more energetic Greifswald Bodden, the largest part of the rest of the bodden coast (about 75%) is marked by phytogenic sedimentation, especially in the form of peat bogs. Seaward advance of reeds, predominantly *Phragmites australis* and a few other reed and marsh plants that are salt tolerant, results in deposits of peat and mud layers a few decimetres thick, covering the Pleistocene or holocene sediment. Marine sands and muds are frequently washed up, creating coastal-flood moors.

The actual form of the sedimentation fringe is dependent on several factors including: (1) the inclination of the shore, incoming wave energy and coastal exposure, which determine the likelihood of growth on the seaward side and the entry of marine clastic material; (2) the trophic state and salinity of the offshore bodden water by which salts and nutrients are transported into the shore region during regular flooding; (3) the supply of fresh water and material from the hinterland with nitrate from agricultural exploitation that contributes to the eutrophication of the reed area on the landward side; and (4) the type and intensity of use of this fringe (Voigtland 1983).

The supply of water and nutrients and different degrees of salinity result in a characteristic zoning parallel to the coastline that is revealed in different communities of plants in the bodden reed banks within the various zones *Phragmites* also develops different stalk types (shore, high, low and spring reeds) which differ significantly in length, stalk density, productivity and can be used to characterize the consumption of material specific to each zone (Figure 4.7).

In many places over the past centuries, the *Phragmites* reed belts have been transformed by extensive grazing into salt grassland. The grass peat formed by grazing is not much more than 800 years old. The peat accumulation in the present day sedimentation area started as the Litorina transgression came to an end and took the form of only partially disintegrated reed peat with a high silt and sand content (transgression peat). There was a transition to deposits of highly disintegrated slimy peat muds which points to a regression phase 2000 to 1200 BP (Figure 4.2). To some extent, a sediment gap must also be taken into account. With the start of the post-Litorina transgression, around 1000 BP, the development of the salt grassland began with the formation of salt meadow peat (Jeschke and Lange 1992). Grazing cattle played a decisive part in their genesis: the cropping of the biomass by the cattle was an obstacle to the dominance of the K-strategist, *Phragmites australis*. Simultaneously, trampling by animals worked some of the stalks into the sediment, and the greater compactness of the soil slowed down the decomposition of this organic matter. Thus, grazing resulted in a suppression of the potentially dominant species, in compression of the peat and, when associated with frequent flooding, in the change to a grassland that could tolerate salt (Jeschke 1987).

With intensification of agricultural production, especially between 1960 and 1989, many areas subject to flooding and sedimentation were either diked, drained and turned into grassland for intensive exploitation or completely abandoned. In both cases, this entailed profound changes in the salt grassland ecosystems. Intensively used areas were thenceforth characterized by sown grassland, subsidence and mulling phenomena and the disintegration of the network of flooding channels, while the abandoned surfaces were recolonized by *Astero-Phragmitetum* and the drier sites by *Agropyron repens* grass. At present, a progressive retreat of the reed banks analogous to that of the freshwater reeds can be observed as a result of the ongoing eutrophication (Lippert 1990).

On the basis of studies of vegetation ecology, Slobodda (1989) was able to identify a series of six types of phytogenic-sedimentation coast which correspond to a sequence from initial to strongly altered sedimentation coasts depending on different relationships with the hinterland. Within each series, the different degrees of anthropogenic modification determine the natural features of the area. He distinguishes between: (1) initial sedimentation coasts; (2) sedimentation coasts in

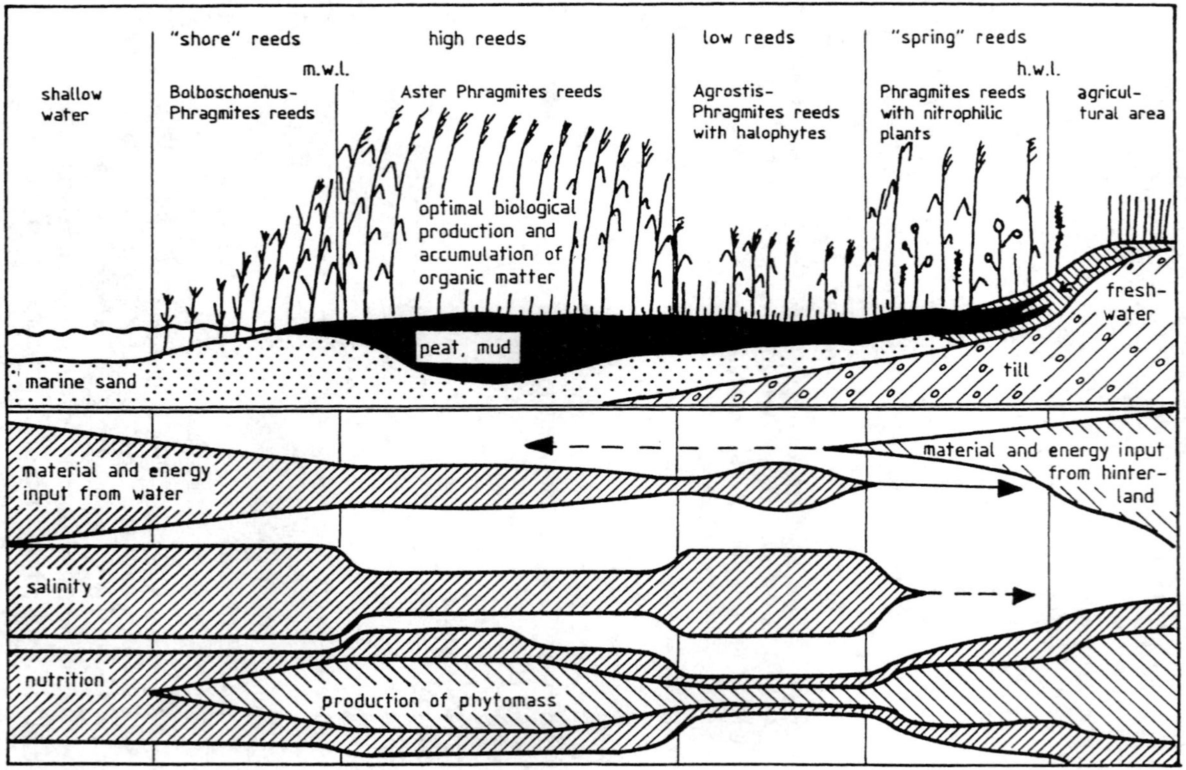

Figure 4.7 Outline sketch of the structure and zoning of sedimentation reed bank on a bodden coast including the intensity of material and energy contributions (after Slobodda 1989)

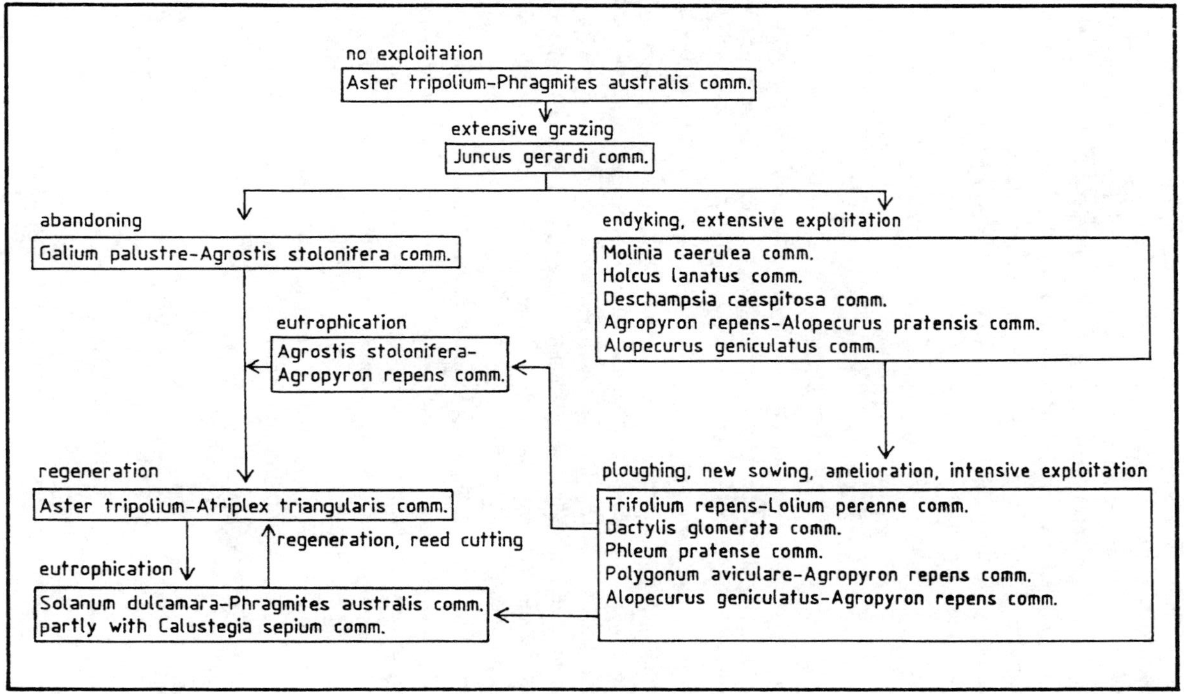

Figure 4.8 Basic characteristics of vegetation development on vegatation sites influenced by salt along the bodden phytogenic sedimentation coasts (after Lampe et al. 1987)

front of high moraine plates with fossil cliffs; (3) sedimentation coasts in front of low moraine plates and marine sand plains; (4) sedimentation coasts in front of marshy lowlands; (5) sedimentation coasts in front of marshy lowlands with very little exposure; and (6) salt grassland coasts as anthropogenic–zoogenic modification of sedimentation coasts.

Figure 4.8 provides an example of the basic characteristics of the change in vegetation within the bodden margins with the sequence of plant communities under the influence of various forms of exploitation to the regeneration of a potential natural reed vegetation (Lampe et al. 1989). The figure indicates both the variety of alterations to the vegetation and the complexity of interactions that occur as a result of human activities.

4.5 CONCLUSIONS

Reduction in wave energy and sedimentation associated with waves and wave-induced currents following closure and partial closure of the boddens has increased the likelihood of phytogenic sedimentation, while vegetation has become increasingly limnetic as the water has become less salty. There has been an increase in the amount of eutrophication, stimulating the growth of phytoplankton, inhibiting the growth of macrophytobenthos and increasing deposition of mud that is rich in nutrients. Quasi-fossil forms (abrasion terraces and inactive cliffs) are preserved because structures that were created at higher energy levels are reshaped slowly by subsequent processes that occur at lower energy levels. Coastal genesis is associated with changes in the sediment budget, given both technogenic and anthropogenic effects. The latter include increasing nutrients as a result of runoff from agricultural fields, grazing of marshes, diking and draining (or abandoning) land, nourishing beaches, cleaning beaches and implementing protection projects. These alterations cause changes in vegetation that lead, in turn, to changes in sedimentation and alterations in the character of the estuarine shoreline. The correlation between morphology, lithology and hydrodynamics has become less conspicuous with the reduction in energy levels, indicating that knowledge of both present and past processes is required adequately to reconstruct estuarine shoreline landscapes and interpret the most influential processes.

REFERENCES

Clayton, K.M. 1980. Beach sediment budgets and coastal modification. *Progress in Physical Geography* 4: 471–486.

Correns, M. 1979. Der Wasserhaushalt der Bodden- und Haffgewässer der DDR als Grundlage für die weitere Erforschung ihrer Nutzungsfähigkeit zu Trink- und Brauchwasserzwecken. Ph.D. dissertation. Humboldt University, Berlin.

Flemming, B. and G. Wefer. 1973. Tauchbeobachtungen an Wellenrippeln und Abrasionserscheinungen in der westlichen Ostsee südlich Bokniseck. *Meyniana* 23: 9–18.
Gomolka, A. 1971. Der Uferrückgang bei Freesendorf. *Wissenschaftliche Zeitschrift der Universität Greifswald, mathematisch-naturwissenschaftliche* Reihe 1/2: 19–25.
Gomolka, A. 1987. Untersuchungen über geomorphologische Veränderungen an Boddenküsten in den letzten drei Jahrhunderten unter besonderer Berücksichtigung des Greifswalder Boddens. Ph.D. dissertation. University of Greifswald.
Gurwell, B., D. Weiss and E. Zielisch. 1982. Beitrag zur Charakterisierung von physiographischen Einheiten und Bilanzsystemen. *Zeitschrift für geologische Wissenschaften* 10: 1347–1355.
Gusen, R. 1978. Geologisch-morphologische Untersuchungen auf der Schorre von Neu-Reddevitz (Greifswalder Bodden) – ein aktuogeologischer Beitrag zur Untersuchung des strukturellen und texturellen Baurezenter Flachwassersedimente. Thesis. University of Greifswald.
Gusen, R. 1988. Sedimentverteilung und geologischer Bau der Schorre vor Lubmin (Südküste des Greifswalder Boddens). *Zeitschrift für angewandte Geologie* 34: 86–90.
Janke, W. and R. Lampe. 1993. *Bericht zur holozänen Küstenentwicklung sowie zur subrezenten und rezenten Küstendynamik der Halbinsel Fischland-Darss-Zingst.* ILG Bureau of Coastal Studies, Greifswald.
Jeschke, L. 1987. Vegetationsdynamik des Salzgraslandes im Bereich der Ostseeküste der DDR unter dem Einfluss des Menschen. Hercynia N.F. 24: 321–328.
Jeschke, L. and E. Lange. 1992. Genese der Küstenüberflutungsmoore der vorpommerschen Boddenküste, p. 208–215. In K. Billwitz, K.-D. Jäger and W. Janke (eds.), *Jungquartäre Landschaftsräume.* Springer, Berlin.
Kliewe, H. 1987. Genetische und stratigraphische Merkmale von Küstenniederungen im Bereich der südbaltischen Boddenausgleichsküste. *Petermanns Geographische Mitteilungen* 131: 73–81.
Kliewe, H. and W. Janke. 1982. Der holozäne Wasserspiegelanstieg der Ostsee im nordöstlichen Küstengebiet der DDR. *Petermanns Geographische Mitteilungen* 126: 65–74.
Kolp, O. 1982. Entwicklung und Chronologie des Vor- und Neudarsses. *Petermanns Geographische Mitteilungen* 126: 85–94.
Kolp, O. 1983. Die schrittweise Verlagerung der Odermündung von der Bornholmmulde bis in die Oderbucht in folge holozäner wasserstandsanderunger in südlichen Ostseeraum. *Petermanns Geographische Mitteilungen* 127: 73–87.
Kolp, O. 1991. Die Bedeutung der isostatischen Kippbewegung für die Entwicklung der südlichen Ostseeküste. *Zeitschrift für geologische Wissenschaften* 9: 7–22.
Lampe, R. 1987. Morphodynamisch-lithologische Kennzeichnung und Kartierung ausgewählter Schorrebereiche der inneren Seegewässer der DDR. Ph.D. dissertation. University of Greifswald.
Lampe, R. 1992. Morphologie und Dynamik der Boddenküsten Vorpommerns. *Geographische Rundschau* 44: 632–638.
Lampe, R. 1993. Environmental state and material flux in western part of the Oder river estuary – results and consequences. *Petermanns Geographische Mitteilungen* 137: 275–282.
Lampe, R., A. Gomolka, Th. Leipe, S. Slobodda and R. Voigtland. 1987. Ergebnisse der Untersuchung von Struktur und Dynamik von Boddenküsten. *Wissenschaftliche Zeitschrift der Universität Greifswald, mathematisch-naturwissenschaftliche Reihe* 2/3: 99–111.
Lippert, K. 1990. Zur Nutzung der Naturressource Schilf an den Boddengewässern der DDR. *Naturschutzarbeit in Mecklenburg-Vorpommern* 33: 17–21.
Pratje, O. 1939. Die Sedimentation in der südlichen Ostsee. *Annalen der Hydrographie und maritimen Meteorologie* 67: 209–221.

Seibold, E., R.F. Dill and E. Walger. 1961. Tauchbeobachtungen und petrographische Untersuchungen zur Sedimentumlagerung in der Kieler Aussenförde. *Meyniana* 11: 82–96.

Slobodda, S. 1989. Landschaftsökologische Kennzeichnung und Typisierung von Bodden-Verlandungssaumufern an den inneren Seegewässern der DDR unter Einbezug vegetationsökologischer Untersuchungen. Ph.D. dissertation. University of Greifswald.

Subklew, H.-J. 1987. Das Angespül des Greifswalder Boddens. Acta hydrochimica et hydrobiologica 15: 65–78.

Voigtland, R. 1983. Wasserchemismus und Biomasseproduktion in Schilfröhrichten und deren Beeinflussung durch Grund- und Boddenwässer. *Wissenschaftliche Zeitschrift der Universität Greifswald, mathematisch-naturwissenschaftliche Reihe* 1/2: 113–119.

Weiss, D., B. Gurwell, K. Hackert et al. 1978. Beeinflussung der Seegangs- und Sedimentdynamik vor den Steilufern der DDR-Küste. Water Management Board ''Coast''. Warnemünde. (Unpublished report.)

Weiss, D. 1991. Küstenschutzbauwerke an der Ostseeküste von Mecklenburg-Vorpommern. *Wasser und Boden* 43: 17–26.

Wiemer, R. and B.R. Gurwell. 1991. Die Ostseeküste in Mecklenburg-Vorpommern. *Wasser und Boden* 43: 6–10.

Chapter Five

Late Quaternary infill of macrotidal estuaries in northern Australia

COLIN D. WOODROFFE
University of Wollongong

5.1 INTRODUCTION

Clastic coastal depositional environments can be conveniently classified in relation to the extent to which river, tide and wave processes dominate within them (Boyd et al. 1992). Such a classification has been widely applied to deltas (Galloway 1975; Wright 1985; Orton and Reading 1993) but may also be extended to describe estuaries. Dalrymple and colleagues (1992) have discriminated wave- and tide-dominated estuaries, and Cooper (1993) has demonstrated that there is a range of river-dominated estuaries (Figure 5.1).

Estuaries are a sink for sediment which may be derived either from riverine input, or through transport from seaward by tidal or wave processes (Guilcher 1967; Nichols and Biggs 1985; Dronkers 1986). In addition to the processes of sediment transport, estuaries may also differ as a result of other boundary conditions. In particular their morphology and the manner in which they infill, reflect geological inheritance, and relative sea-level change with respect to that prior topography (Roy et al. 1995; Chappell and Woodroffe 1995).

Modern estuaries have formed through marine inundation of previously subaerial valleys as a result of sea-level rise since the last glaciation. In the broadest sense all estuaries are rias, or drowned river valleys, typical of the submerging shorelines described by Johnson (1919). Eustatic sea-level changes, resulting in baselevel change, were an important component of the geographical

Estuarine Shores: Evolution, Environments and Human Alterations. Edited by K. F. Nordstrom and C. T. Roman.

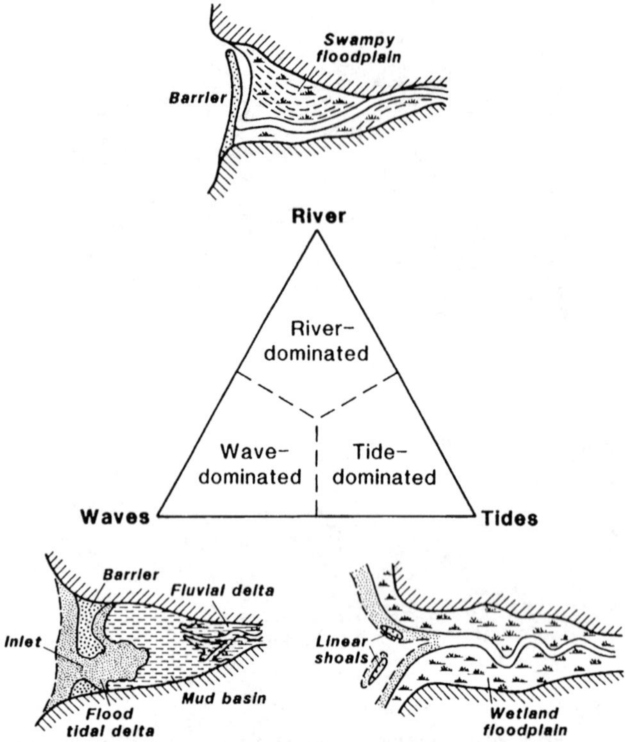

Figure 5.1 A ternary diagram illustrating the continuum of river-, wave- and tide-dominated estuary types. Morphological characteristics of wave- and tide-dominated estuaries after Dalrymple et al. (1992), and of river-dominated estuaries after Cooper (1993)

cycle proposed by W.M. Davis (1902). Although Davis' concepts of denudation chronology, comprising peneplanation and rejuvenation, are no longer widely accepted, sea-level variations are recognised as an important control on sediment accumulation over geological time. In recent decades there have been considerable advances in the correlation of global sea-level changes as a result of seismic stratigraphy, and specifically through the concepts of sequence stratigraphy (Vail et al. 1977; Posamentier and Vail 1988).

Sequence stratigraphy is the study of lithological relationships of genetically related strata in a chronostratigraphic framework, bounded by surfaces of erosion or non-deposition or their correlative conformities (Van Wagoner et al. 1990). This approach has only recently been applied to lithofacies relationships in macrotidal estuaries in three separate instances (Bay of Fundy: Dalrymple et al. 1992; Gironde estuary: Allen and Posamentier 1993; Daly River: Chappell 1993). Each of these interpretations (Figure 5.2), though differing in detail, recognises a basal lowstand systems tract, a transgressive systems tract and a highstand systems tract of

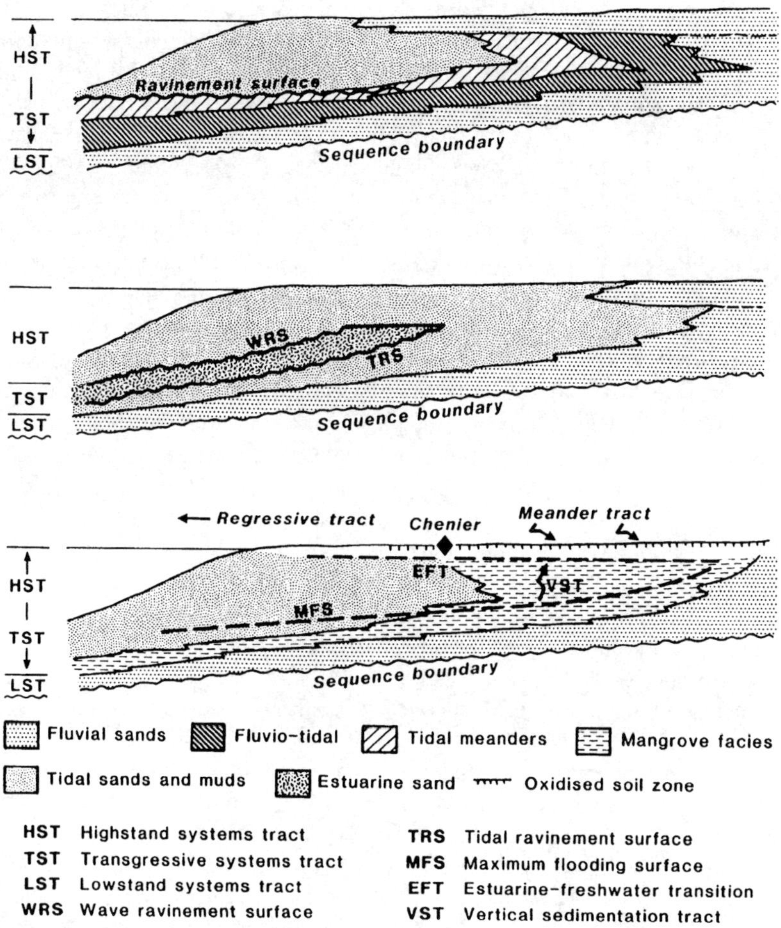

Figure 5.2 Sequence stratigraphic relationships for three macrotidal estuarine infills: (a) Bay of Fundy, North America, after Dalrymple et al. (1992); (b) Gironde estuary, France, after Allen and Postamentier (1993); and (c) Daly River, northern Australia, after Chappell (1993)

sediments, and is based upon the interpretation that sediments have infilled these estuaries with a landward onlapping sediment wedge during periods of sea-level rise, but that the surface details of estuarine infill have occurred during a subsequent period of sea-level stability. This view puts prime emphasis on sea level as the main control over sedimentation, and infers a Type 1 sequence in the terminology of Van Wagoner and co-authors (1990), whereby sea-level fall leads to rejuvenation of fluvial channel networks, which initially incise, but then aggrade during the lowstand in response to increased accommodation space. This interpretation may apply on steep, narrow coasts, which may also show Type 2 sequences where river mouths discharge over the continental shelf break at lower sea level, leading not to incision, but to the stacking of deltaic systems in a shelf margin systems tract (Suter et al. 1987). However, the cut and fill interpretation of incision and subsequent infill during sea-level cycles has recently been questioned, and the importance of other factors, such as tectonism, sediment supply and environmental change, has been recognised (Miall 1991; Schumm 1993; Wescott 1993).

Storage of sediment within floodplains is a complex response to a variety of factors. Floodplains are constructed both through lateral accretion as a result of channel migration and vertical, overbank, accretion (Nanson and Croke 1992), and more rapid storage of sediment in floodplains may be possible during periods of sea-level rise when there is greater accommodation space, than during stillstands (Wright and Marriott 1993). Whereas accommodation space is important in enabling sedimentation to occur, the significance of variations in sediment supply in relation to continental shelf sedimentation has been stressed by Swift and Thorne (1991) and Thorne and Swift (1991), and Schlager (1993) has emphasised that environmental change is a major control on sediment supply. Although incision of channels into the exposed continental shelf at lower sea level can provide a source of sediment, basins cannot fill in entirely through this self-cannibalisation (Galloway 1989). Galloway has favoured a tripartite continuum of sedimentation response to eustatic sea-level change, basin subsidence and sediment supply.

There appears to be good reason to question whether a Type 1 response will characterise the majority of rivers over the course of a eustatic sea-level cycle. The widest acceptance has come from studies in the Gulf of Mexico. Fisk (1944), for instance, assumed that the Mississippi had incised when sea level was lower. Although up to 40 m of incision during glacial periods has been identified for several rivers draining into the Gulf of Mexico (Thomas and Anderson 1992), incision does not appear to have extended beyond Baton Rouge (370 km upstream) up the Mississippi River (Autin et al. 1991), and many of the smaller rivers deposited Type 2 sequences, characterised by shelf margin deltas on the shelf edge (Suter and Berryhill 1985; Suter 1995).

Rivers can adjust to changes of sea level without undergoing incision, by adjusting their pattern, roughness or channel geometry (Wescott 1993). Schumm (1993) summarised experimental and theoretical studies which show that whether or not rivers incise appears to depend upon the rate and direction of sea-level change, geological controls (such as erodibility), and geomorphological controls such as the presence of levees which may constrain channels and channel pattern. Braided channels are likely to incise if sea level falls, whereas sinuous channels in uncohesive sediments are likely to adjust their channel pattern (Schumm 1993).

The timing and manner of estuary drowning can thus differ substantially, controlled by the pattern of past and ongoing relative sea-level change and local or regional tectonics. Relative sea-level change is an important boundary condition which differs so significantly between estuaries in the glacioisostatically adjusting high latitudes, particularly of the Northern Hemisphere, that it may often obscure the influence of other boundary conditions. Australia, on the other hand, represents a relatively stable platform (Aubrey and Emery 1986), with little active tectonism, and a subdued hydroisostatic adjustment to Quaternary ice and ocean volume changes (Nakada and Lambeck 1989). A comparatively simple sea-level history applies to adjacent estuaries which therefore differ less in this, than with respect to other boundary conditions. This situation favours the examination of other extrinsic and intrinsic factors which may result in differences between adjacent estuarine systems.

The northern coast of Australia is flanked by a broad continental shelf, the Sahul Shelf, which, like the Sunda Shelf to the north, was exposed when the sea was lower during the Quaternary glaciations. Tidal range is generally large along this north Australian coast (5–6 m), resulting in a sequence of macrotidal estuaries along the Top End of the Northern Territory and the north coast of Western Australia. Detailed stratigraphic investigation of several of these, together with some observations on the shelf, imply that the morphodynamics of adjacent estuaries, which will have experienced similar sea-level histories, can be different. Although sea level has been the prime control on the pattern of sedimentation, comparison of adjacent systems within northern Australia allows some insight into the role of other boundary conditions.

This chapter examines tidal and fluvial influences on sequences of channel relocation and infill that shape estuarine shoreline environments and influence ecological changes on the floodplains. The chapter outlines the Late Quaternary history of sedimentation in macrotidal estuaries in northern Australia, and compares this with stratigraphy that might be expected with a Type 1 sequence. It demonstrates that relative Holocene sea-level history, which has been broadly similar for estuaries around the Australian coast, has been a major control. Lowstand systems tracts, characterised by fluvial sediments, underlie extensive transgressive systems tracts characterised by rapid vertical sedimentation, terminating in a widespread swamp phase as sea level stabilised. However, the details of morphodynamic development of adjacent systems during the Holocene highstand differ, both regionally, indicating the importance of extrinsic controls such as climate and sediment supply, and also between adjacent systems, indicating the importance of intrinsic geomorphological controls.

5.2 REGIONAL SETTING

Seasonal rivers drain the north and northwest coasts of Australia. Those draining from the Proterozoic sandstone Arnhemland Plateau, into van Diemen Gulf, in the Top End of the Northern Territory are often bedrock-controlled in their upper reaches, but cross broad lowland plains with extensive tidal and freshwater wetlands. Littoral tidal wetlands, dominated by mangrove forests, are subject to inundation by salt water, but much of the area is presently dominated by seasonally inundated freshwater wetlands.

These wetlands are an important but fragile resource. They are productive, rich in wildlife, sustain a well established commercial fishing industry and rapidly growing recreational fisheries, support grazing by cattle and an increasingly managed buffalo population. They also feature prominently in an expanding tourist industry (Whitehead et al. 1990). On the other hand, they are threatened by feral animals (buffalo, pigs); by exotic weeds (*Mimosa pigra, Salvinia molesta*); and by saline intrusion, through extension of tidal creeks into the freshwater wetlands (Knighton et al. 1992).

The southern shore of van Diemen Gulf is an area of the wet–dry tropics. The region receives 1300–1400 mm of precipitation annually; rainfall is pronouncedly monsoonal with most falling in the wet season, December–March. The larger rivers (i.e. Mary, South Alligator, East Alligator) rise on the Arnhemland plateau, which is composed of sandstones of the Middle Proterozoic Kombolgie Formation. For much of their course the rivers flow over the lateritic lowlands associated with a late Tertiary erosional surface termed the Koolpinyah land surface, that has truncated Lower Proterozoic metasediments (schists, gneisses and carbonate rocks, intruded by granite and dolerite) which are widely covered with Cainozoic alluvium, and lateritic and ferricrete deposits. The river floodplains and coastal and estuarine plains are Holocene in age (Figure 5.3).

Tidal range in van Diemen Gulf is around 6 m at springs, reaching a maximum range of 7.8 m in Darwin Harbour. The low gradient of the plains and the large tidal range result in tidal influence penetrating more than 100 km up several of the river systems in the dry season, with river water close to sea water salinity at the tidal limit at the end of the dry season. During the wet season the saline water is flushed out and the estuaries are dominated by fresh but tidally reversing water, with a salt wedge at the river mouth (Woodroffe et al. 1986).

5.3 QUATERNARY SEA-LEVEL CHANGE IN THE AUSTRALIAN REGION

During the Quaternary, the volume of the world's oceans has fluctuated, with sea-level fall as water has been incorporated into polar and high-latitude ice sheets, and sea-level rise as these have subsequently melted. Reconstruction of sea-level changes over the most recent interglacial–glacial–postglacial cycle is shown in Figure 5.4a, based upon the deep-sea oxygen isotope record, calibrated against dated fossil reef shorelines on the rapidly uplifting coast of the Huon Peninsula, Papua New Guinea (Chappell and Shackleton 1986).

The shoreline over the broad Sahul and Sunda Shelves has undergone rapid lateral translation as well as vertical movements coincident with variations in ocean volume. Radiocarbon dates of 17 000–18 000 years BP have been determined on

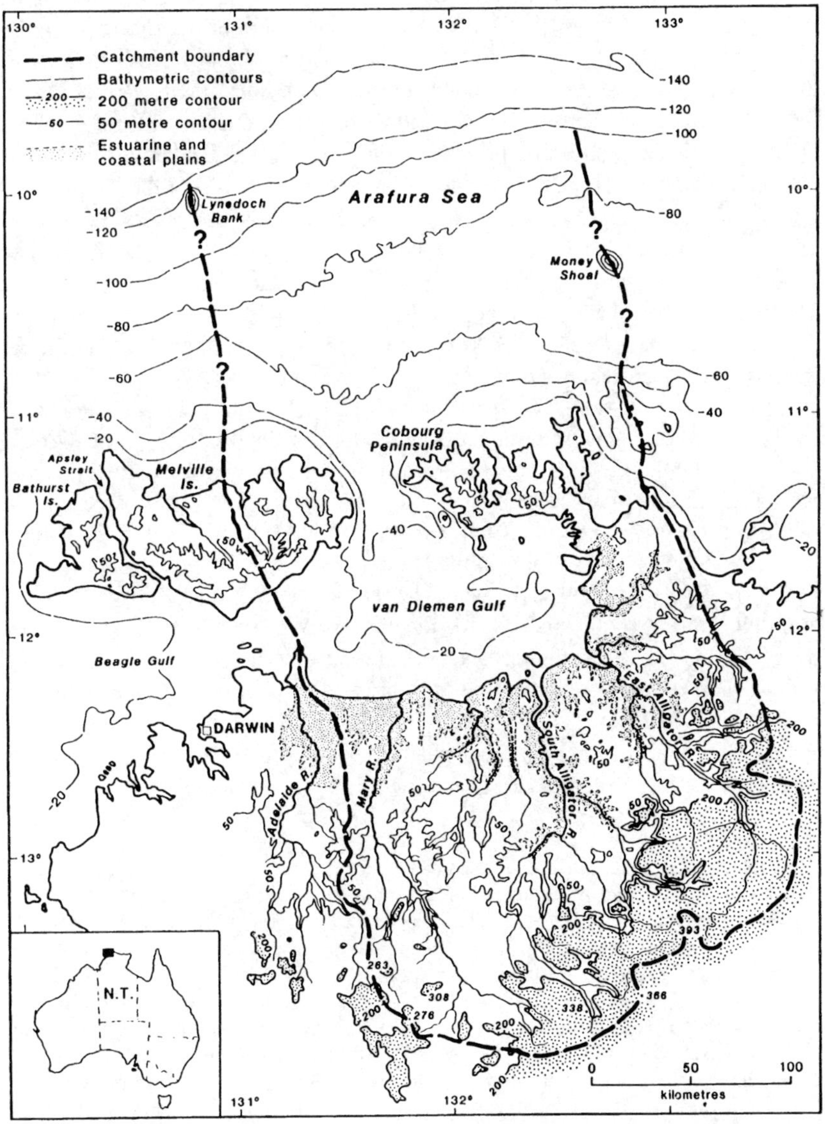

Figure 5.3 Location of the Adelaide, Mary, South Alligator and East Alligator Rivers, and the topography of their catchments, and the probable extent of the catchment of the river that drained the area when the sea was lower at the peak of the last glacial maximum. Note the relatively shallow gradient that the river would have had as it drained across the shelf

intertidal sediments, including beachrock, indicating that in northern Australia, the sea was 150–165 m below its present level at the peak of the glaciation (Jongsma 1970; Veeh and Veevers 1970). The probable exposure of the shelves is indicated in Figure 5.3. The shoreline must have migrated backwards and forwards across this shelf as the sea level fluctuated. Former shorelines, preserved as cliff lines, notches, terraces or scarps, have been identified from the floor of the Arafura Sea, at no fewer than eight different depths in the range 120–200 m (Jongsma 1974), and similar evidence of former shorelines has been described from the Timor Sea (van Andel and Veevers 1965, 1967; van Andel et al. 1967; Veevers 1969; Veevers 1971). Seismic traverses across the Arafura Sea indicate a series of sub-parallel reflectors, the uppermost of which (S1) is considered to mark the break between Pleistocene and Holocene sediments. Holocene sediments are inferred to have been deposited at $<0.1\ \mathrm{mm\,a^{-1}}$ 100 km north of the modern shoreline, and at around $2\ \mathrm{mm\,a^{-1}}$ 20 km offshore (Jongsma 1974).

The pattern of sea-level rise during the postglacial transgression has been reconstructed in detail from reefs off Barbados (Fairbanks 1989). In the Australian region, where there is not the same record of ongoing hydroisostatic adjustment to ice-melt, sea level can be shown to have reached a level close to present around 6000 years BP (Thom and Chappell 1975). The record of transgression from drilling the lowest reef terrace on the Huon Peninsula, corrected for uplift, is shown in Figure 5.4c (Chappell and Polach 1991), and agrees very well with the record from the stable Abrolhos Islands off the coast of Western Australia (Eisenhauser et al. 1993). There are similarities in the timing of sea-level change derived from radiocarbon dating of shells from stratigraphic studies on the New South Wales coast (Thom and Roy 1985) and from the record of transgression shown by dating mangrove remains in central Queensland (Grindrod and Rhodes 1984). Envelopes of curves for these records are also shown in Figure 5.4c, and are compared with

---

Figure 5.4 (*opposite*) A summary of Late Quaternary sea-level history in the Australian region. (a) The pattern of sea-level change over the last 140 000 years as reconstructed from oxygen-isotope analysis on deep-sea cores, and radiometric dating of fossil coral reefs on the Huon Peninsula (after Chappell and Shackleton 1986). (b) The maximum elevation at which the shoreline considered to mark the peak of the Last Interglacial is found at locations around Australia (data from Murray-Wallace and Belperio 1991): location of key Holocene sites: 1, Huon Peninsula; 2, Northern Territory sites; 3, Abrolhos Islands; 4, Central Queensland; and 5, Southeast Australia. (c) Summary of Holocene sea-level envelopes from around Australia. The sea-level envelope for Southeast Australia is based upon shell dates from New South Wales (after Thom and Roy 1985); that for the Northern Territory is based upon mangrove dates from van Diemen Gulf (after Woodroffe et al. 1987); results from Central Queensland are based upon mangrove dates from Hinchinbrook Island (after Grindrod and Rhodes 1984); Huon data represent dates on a coral core corrected for uplift (after Chappell and Polach 1991); and Abrolhos data come from cores on those islands (after Eisenhauser et al. 1993). The generalised curves for Karumba and the Great Barrier Reef (GBR) are based upon Chappell et al. (1982)

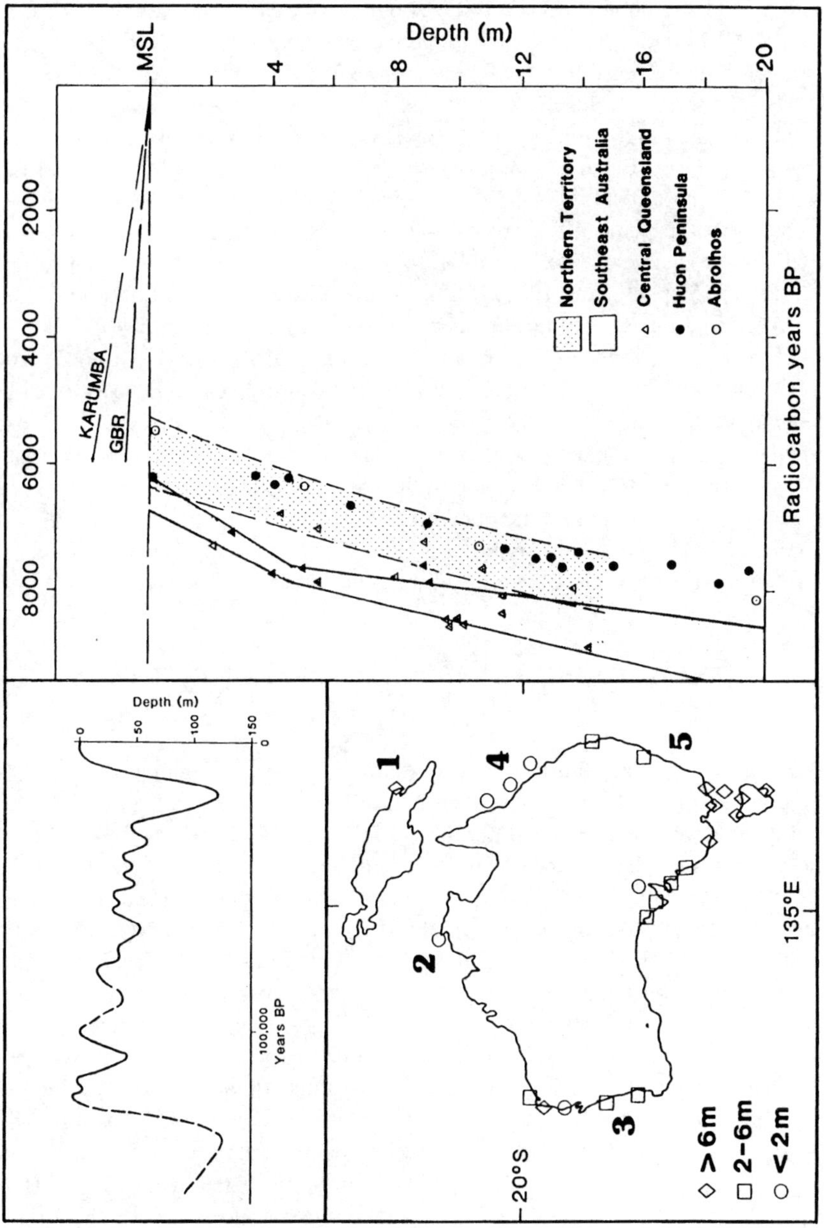

MSL
Depth (m)
4
8
12
16
20
2000
4000
6000
8000
KARUMBA
GBR
Northern Territory
Southeast Australia
Central Queensland
Huon Peninsula
Abrolhos
Radiocarbon years BP
Depth (m)
0
50
100
150
0
100,000
Years BP
1
2
3
4
5
135°E
20°S
> 6m
2–6m
< 2m

dates on mangrove remains from estuarine plains in the Northern Territory (Woodroffe et al. 1987), which are described in greater detail below.

Generally, there is little evidence for ongoing uplift or subsidence around the coast of Australia, although there is variation in the height at which the Last Interglacial shoreline is observed (See Figure 5.4b, data from Murray-Wallace and Belperio 1991). Sea level relative to the Australian coast appears to have reached a level close to present 6500–6000 radiocarbon years BP. For much of the Australian coastline there is evidence that the sea was higher than present in the mid-Holocene, although the height that was reached differs from place to place. From the central Great Barrier Reef, sequences of microatolls record the gradual fall of sea level from a level around 1 m above present 5500 years ago (Chappell 1982). Details of this fall differ around the Queensland coast, attributed to subtle hydroisostatic adjustments, with a fall from around 2 m above present, detectable at Karumba (Chappell et al. 1982; 1983). Geophysical modelling allows prediction of expected hydroisostatic adjustments and shows some agreement with observations of Holocene emergence (Nakada and Lambeck 1989). However, in the Northern Territory and Western Australia, the pattern of mid and late Holocene sea-level change remains inadequately defined. Thus in the van Diemen Gulf region, modelling predicts emergence (Nakada and Lambeck 1989), but observations (Woodroffe et al. 1987) do not support this. However, both the large tidal range and potential compaction of sediment complicate the reconstruction of sea level, which can be summarised as comprising a period of postglacial marine transgression up until around 6000 years BP, followed by a period of relative stability.

5.4 STRATIGRAPHY AND DEVELOPMENT OF ESTUARINE AND COASTAL PLAINS

The descriptions that follow are based upon detailed stratigraphy, pollen analysis and radiocarbon dating of more than 130 drill holes from the plains of the South Alligator River (Woodroffe et al. 1986; 1989), similar detail from around 60 drill holes from the Mary River (Woodroffe and Mulrennan 1993), and 20 drill holes from the Adelaide River (Woodroffe et al. 1993). Paleoecological reconstruction has been largely on the basis of pollen spectra, which allows discrimination of mangrove and freshwater wetland habitats (Grindrod 1988). Organic fragments from cores have been radiocarbon dated at the Australian National University Radiocarbon Dating Laboratory, and ages are reported as conventional radiocarbon dates, in radiocarbon years BP. Carbonate samples (principally shell) have not had an environmental correction applied as there is uncertainty about the value that is appropriate in these estuarine environments (Woodroffe et al. 1986). Thermoluminescence (TL) dates, where reported, are on quartz grains (coarse-grain technique) and have been calculated on the basis that acquired TL depends upon length of burial and environmental radiation dose rate, with a residual correction for incomplete bleaching determined from modern surface samples (see Nanson et al. 1993).

In each of these estuarine systems the present cross-shore distribution of environments consists of nearshore, subtidal shelly sand and mud, grading into the modern shoreline, which is often dominated by a fringe of mangroves 100–300 m wide and discontinuous chenier ridges (Figure 5.5). The mangrove forests comprise a seaward zone of *Avicennia* and *Sonneratia*, a zone of tall *Rhizophora* (15–20 m tall), a zone of *Ceriops* (5–10 m tall), and a landward zone of stunted *Avicennia*, backed by samphire and supratidal saline mudflat. The shoreline is backed by a prograded coastal plain, comprising sedge/grassland, with discontinuous shelly chenier ridges. Within the estuarine embayments, there is a vertically accreted estuarine plain that contains a freshwater clay floodplain, with a sedge/grassland vegetation (see Figure 5.5d).

The coastal plain has laterally prograded over the past 6000 years under the influence of shallow-marine and tidal sedimentation. The estuarine plains on the other hand have been dominated by both tidal and fluvial processes, and have accreted principally through vertical sedimentation (Figure 5.5d). The coastal and estuarine plains are differentiated by a discontinuous sandy ridge which marks a shoreline at around 6000 years BP (a TL date of 7500±800 was determined on sand 0.5 m beneath its surface on the eastern margin of the Mary River plains, and radiocarbon ages of 4950±80 and 4990±330 years BP have been determined on sediments beneath the ridge on the South Alligator and Adelaide plains respectively).

In the South Alligator River plains, drilling has shown that there is a pre-Holocene surface at a depth of 12–14 m, consisting of either oxidised sands and gravels, or a stiff mottled clay paleosol, which probably represents colluvial or alluvial deposits. These sediments correspond to the lowstand systems tract. Overlying this are bluish-gray, fine sandy or silty clays with organic fragments, often identifiable as mangrove wood. These sediments contain abundant mangrove pollen, and are referred to as mangrove facies. Mangrove facies has been demonstrated to underlie a large proportion of the estuarine plains (Figure 5.5a) and may be inferred to occur over much of the area not directly sampled (Woodroffe et al. 1985; 1993). In the case of the South Alligator plains, mangrove facies is not encountered within the meander tract (defined as that part of the plains across which the fluvio-tidal channel has meandered and marked at its outermost by the occurrence of infilled paleochannels). The meander tract comprises laminated sand and mud associated with lateral accretion point bars, and massive channel fill clay plugs within paleochannels.

More limited drilling within the Adelaide and Mary River plains also indicates widespread mangrove facies beneath the estuarine plains (Figures 5.5b and 5.5c). These mangrove sediments were deposited both during the early Holocene transgression, and the mid-Holocene stillstand. Initial inundation of the prior valley, floored by pre-Holocene sediments, occurred around 8000 years BP in the

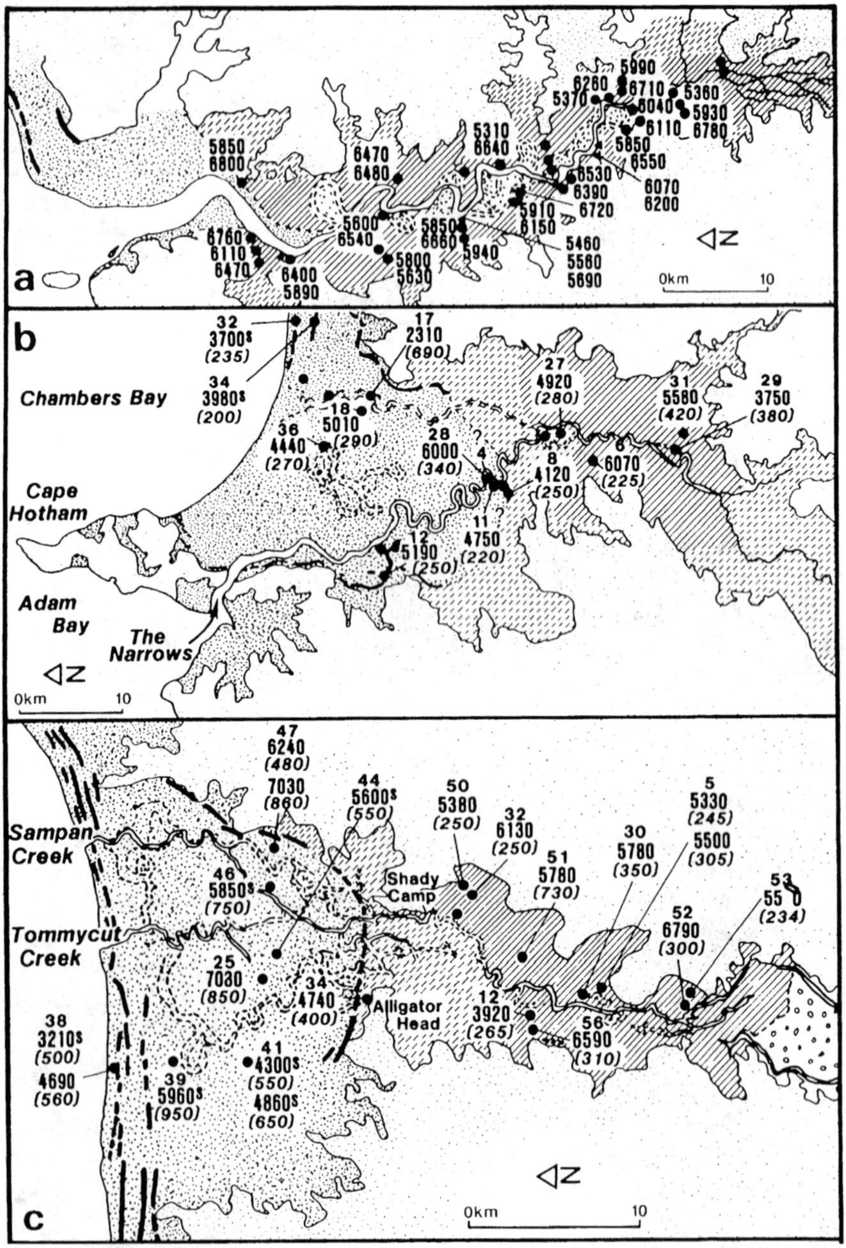

Figure 5.5 The extent of estuarine and coastal plains, and radiocarbon dates (years BP) on near-surface sediments (depths (cm) are given in brackets where samples are from deeper units), from rivers draining into van Diemen Gulf. (a) South Alligator River; (b) Adelaide River; (c) Mary River; (d) schematic cross-section of stratigraphy of plains of the Mary River (after Woodroffe et al. 1993)

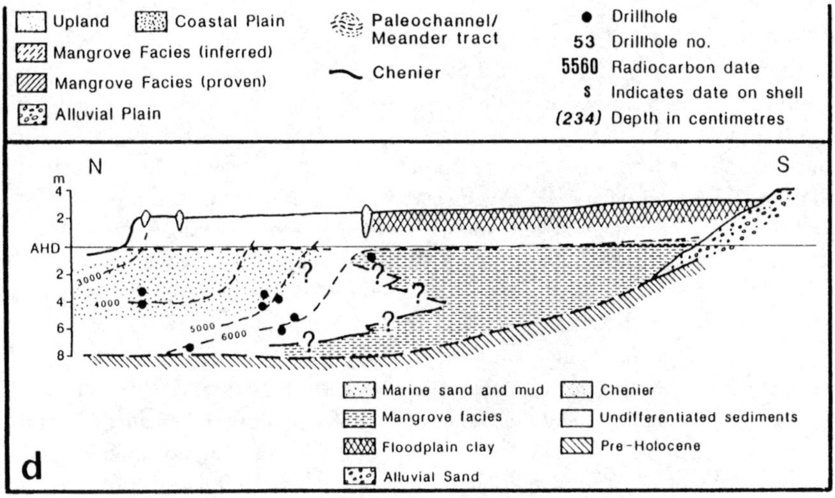

Figure 5.5 *(continued)*

South Alligator (which has a near-horizontal floor around 12 m below AHD – Australian Height Datum, which approximates Mean Sea Level). Initial inundation occurred only slightly later in the Adelaide and Mary River systems which have floors at around 8 m below AHD (see Woodroffe et al. 1993). Dates record the continued rapid rise of sea level until at least 6800 years BP when it was within 5 m of present level. Mangrove facies is particularly organically rich 2–5 m beneath the plains surface, and radiocarbon dates in the range 6800–6500 years BP are typical on this layer, interpreted as the development of widespread mangrove forests, termed the big swamp phase (Woodroffe et al. 1985), triggered by deceleration of sea-level rise.

During the trangressive phase, the cross-shore distribution of environments is likely to have been less complex than that which is seen at present, and which results from 6000 years of comparative sea-level stability. Basal transgressive mangrove facies indicate that as the sea was rapidly rising there was a littoral fringe of mangrove forest which retreated landward, and marine incursion into valleys occurred (Figure 5.5d).

5.5 CHANGING ENVIRONMENTS DURING THE HOLOCENE HIGHSTAND

The changing ecology of the mangrove forests of the South Alligator plains has been determined from pollen analysis of cores and pits throughout the estuary (Chappell and Grindrod 1985; Russell-Smith 1985; Woodroffe et al. 1985; Grindrod 1988). Continued sedimentation was accompanied by successional change from Rhizophoraceous forest to stands of *Avicennia*, subsequently replaced by non-halophytic vegetation of grasses and sedges (Poaceae and Cyperaceae). Continued vertical accretion of sediments during the big swamp phase, but without continued increase in accommodation space after sea level had stabilised, meant that sediment could only build up to the highest tidal flooding level. At this point much of the plains ceased to be inundated and most mangrove forests were eliminated by 4000 years BP. Freshwater clays accumulated thereafter.

The details of change appear remarkably consistent from cores throughout the estuarine plains of the South Alligator River, and are also reflected in the detailed palynological reconstructions from the Magela Plain, a part of the estuarine plains of the East Alligator River (Clark and Guppy 1988; Wasson 1992) and the Mary River plains (Woodroffe and Mulrennan 1993).

There have been three dominant sedimentation patterns since the big swamp phase: (1) coastal progradation resulting in extensive coastal plains; (2) vertical accretion of floodplains and the development of freshwater wetlands; and (3) channel migration and abandonment, resulting in lateral-accretion, estuarine channel deposits on some estuarine plains and paleochannels of different morphologies. The nature of each process, and its relative significance, have differed from one river system to another and are examined below.

5.5.1 Coastal plain progradation

The coastal plain has built out from the 6000-year shoreline. The rate of progradation, examined in detail by radiocarbon dating of shell fragments from beneath the coastal plains on the South Alligator and Mary plains (Woodroffe et al. 1986, 1989, 1993; Woodroffe and Mulrennan 1993), appears to have been greatest between 5000 and 3000 years BP. Since about 3000 years BP, sandy chenier ridges have formed, but little net progradation of the coastline appears to have occurred, except for some minor buildout near Point Stuart during the last few hundred years (Clarke et al. 1979; Lees 1987).

The major rivers, the South and East Alligator Rivers with catchments of $>10\,000\,km^2$, appear to have maintained a large estuarine funnel throughout progradation of the marginal coastal plains. Extensive estuarine paleochannel systems on the Adelaide and Mary River plains indicate that the location of these estuarine channels has switched. In the case of the Adelaide River an estuarine paleochannel formerly drained into Chambers Bay (Figure 5.5b), but is now largely infilled. The course of the Adelaide River has avulsed to occupy a highly meandering former fluvial course, through the bedrock-constrained Narrows, into Adam Bay. A radiocarbon date from wood within the channel fill of the paleochannel of 2310±80 years BP indicates that abandonment and infill of this former estuary is likely to have been within the most recent period of chenier formation. Plugging of the former estuary by coastal sedimentation may have resulted in estuarine channel avulsion (Woodroffe et al. 1993).

During the same period, the Mary River occupied consecutively two meandering estuarine paleochannels (Figure 5.5c). These filled with sediment, so that the fluvial discharge of the Mary River largely dissipated into extensive swamplands and evaporated with relatively minor discharge to the sea through lesser, avulsive distributaries which diverted the fluvial flow. The former estuarine system of the Mary appears to have infilled with sediment driven tidally upstream, after fluvial discharge was diverted (Woodroffe et al. 1993). Radiocarbon dates support the chronology of successive phases of channel activity and infilling. The western paleochannel appears contemporaneous with the more landward cheniers and was infilled by 3000 years BP, whereas the eastern paleochannel appears contemporaneous with the more seaward cheniers and to have infilled by 2000–1500 years BP (Woodroffe and Mulrennan 1993). Over the past 50 years, tidal creeks have again been cutting back from the coast and fluvial and tidal channels are once more connected (Knighton et al. 1992).

The pattern of coastal sedimentation has thus been broadly similar along southern van Diemen Gulf, but its impact on the extent of tidal systems has been considerably different. The largest rivers have maintained their courses to the sea; the smaller tidal systems to the east of the region, (i.e. the Wildman and West Alligator Rivers) also seem to have maintained channels to the sea. In contrast, sediment buildup appears to have diverted the course of the Adelaide River, which found an alternative, inherited route to the sea. Similar buildup impounded the Mary River, temporarily excluding tidal influence from the entire plain and permitting the extension of freshwater wetlands.

5.5.2 Floodplain accretion

The present surface of the estuarine plains has evolved since the disappearance of mangrove forests at the end of the big swamp phase. A similar big swamp phase can be identified from limited augering on the plains flanking the semi-arid Fitzroy and Ord Rivers (Jennings 1975; Thom et al. 1975). In the Top End, vertical accretion of clays has occurred beneath freshwater grass-sedgelands as a result of wet season floods. In the drier Western Australian environments, plains have not accreted to this extent and are dominated by saline mudflats (Wright et al. 1973; Woodroffe 1988).

The South Alligator plains are convex in form, declining in elevation away from a broad muddy "levée" flanking the river, and elevations close to the upland margin are often below the water levels reached by high tide in the tidal river. The Mary River plains are also low-lying, much being below the elevation reached by the highest tides at the coast, but inundation is prevented by broad levées (flanking paleochannels) and chenier ridges (Woodroffe and Mulrennan 1993). The elevation of the estuarine plain generally increases with distance from the coast. This effect is most pronounced along the South Alligator River, and least pronounced on the Adelaide and Mary Rivers.

5.5.3 Channel migration and abandonment

There are numerous paleochannels on the estuarine plains of Top End rivers, indicating former channel positions, but drilling has demonstrated that these have developed in different ways (Chappell and Woodroffe 1985, 1995). Four channel morphologies can be recognised within the tidal reach of the South Alligator: (1) a tapering estuarine funnel (0–30 km from the mouth); (2) a sinuous meandering segment (30–47 km); (3) a cuspate meandering segment within which meanders contain cusps or horns on the inner meander bends (47–73 km); and (4) an upstream segment in which the channel is largely straight but with occasional right-angle bends (73 km to tidal limit). Although much of the mid-plains is presently dominated by cuspate meanders, resembling the estuarine meanders described by Ahnert (1960), paleochannels on the adjacent plains are sinuous. Drilling within the meander tract has revealed laminated sand and mud, deposited in a channel margin environment, recording lateral migration of the channel. Figure 5.6 illustrates one example where there is a nested sequence of sinuous paleochannels, but the river is presently characterised by cuspate meanders. The probable evolution of this sequence is also shown in Figure 5.6, based upon several drill holes and radiocarbon dates (Woodroffe et al. 1986). This river appears to have been at its most sinuous 4000 years ago, and dating of former point bar sequences, as well as comparison of sequential aerial photographs, indicate a maximum rate of channel migration of the order of $1\ \mathrm{m\,a^{-1}}$.

By contrast, the Adelaide River, though presently dominated by a series of exquisitely sinuous meanders (Figure 5.7), shows few meander cutoffs on the estuarine plain. Drilling has not revealed laminated sand and mud channel-margin deposits, as on the South Alligator, presumably because there has not been a supply of sand, making discrimination of a meander tract much harder. Echo-sounding profiles of these meanders have indicated that they are incised into the basal pre-Holocene surface, and implies that these are inherited features which have been reoccupied by the modern river (Vertessy 1990). The former tidal channel of the Adelaide River, identifiable as a prominent anabranch on the plains, was dominated by broad estuarine meanders, of cuspate form. Its present meander form would thus appear to reflect reoccupation of a fluvial channel that incised into the basal sediments.

On the Mary River, there have been several episodes of channel abandonment and reoccupation on the former estuarine plain, but these are less marked than those preserved on the coastal plain and described above.

The extent to which a tidal channel migrates, the rate at which it migrates and the proportion of the floodplain it reworks are closely interrelated, and appear to depend upon the relative magnitude of tidal flows in the dry season and river-dominated floods in the wet season (Chappell and Woodroffe 1995). On the Daly River, with a much larger catchment, and an abundant sand supply, there are extensive scroll plains marking the former meander locations of what is evidently an actively migrating river channel. Rates of meander migration of up to 40 m a^{-1} have been inferred, and 15% of the floodplain has been reworked by the river (Chappell 1993). Paleochannels may be preferentially reoccupied, as in the case of scroll bars on the Daly or tidal creek extension in the Mary system. Channel abandonment following meander cutoff leads to ecological changes on the plains as the locus of tidal influence shifts to the new channel location with probable minor adjustments to tidal amplitude and equilibrium channel geometry.

5.6 SEA LEVEL AND OTHER BOUNDARY CONDITIONS

Sea-level change has evidently been a major control on the infill of the estuarine systems flanking van Diemen Gulf. The pre-Holocene embayment has been flooded as a result of the postglacial marine transgression and incursion. A major phase of estuarine development in northern Australia 8000–6000 years ago, coincided with delta initiation worldwide (Stanley and Warne 1994). The transgressive system tract of sediment, with a maximum flooding surface experienced early in the infill, has been dominated by vertical sedimentation tract (cf. Daly River, Chappell 1993). A similar stratigraphy is observed in semi-arid estuarine systems in Western Australia. However, in the semi-arid setting the plains contain extensive saline mudflats, whereas in the wet-dry tropical Top End setting, convex muddy floodplains have formed as a result of wet season inundation.

The extent to which this stratigraphy corresponds to the simple Type 1 sequence stratigraphic model of channel incision during phases of low sea level deserves further investigation in this low-gradient, broad-shelf setting. Figure 5.3 indicates topography along the South Alligator River, and its presumed extension across the shelf at times of low sea level, based upon shelf topography. In this setting the gradual gradient indicates rapid translation of the shoreline across the shelf as sea level has fluctuated. The river appears to have flowed out of the van Diemen Gulf depression (for which there are few detailed bathymetric data available) through Dundas Strait. However, instead of a longitudinal profile which steepens as the river drains across the shelf, the shallow gradient of the shelf implies that the river had a gradual gradient in its lower reaches during glacial times. Incision is less likely under these circumstances, and the river may have adjusted by alteration of channel pattern.

The age of the pre-Holocene sediments over which estuarine fill rests is central to an understanding of the lowstand systems tract. Generally these are sandy, often highly oxidised gravelly or muddy sands. Recent thermoluminescence dating of sands beneath the estuarine sediments at Kapalga yielded an age of 76.8 ± 10.1 ka (D. Price, unpublished data). Other dates from the sand sheets veneering the uplands indicate that these sediments have continued to accumulate through the Holocene.

Nanson and colleagues (1993) investigated the alluvial history of the Magela

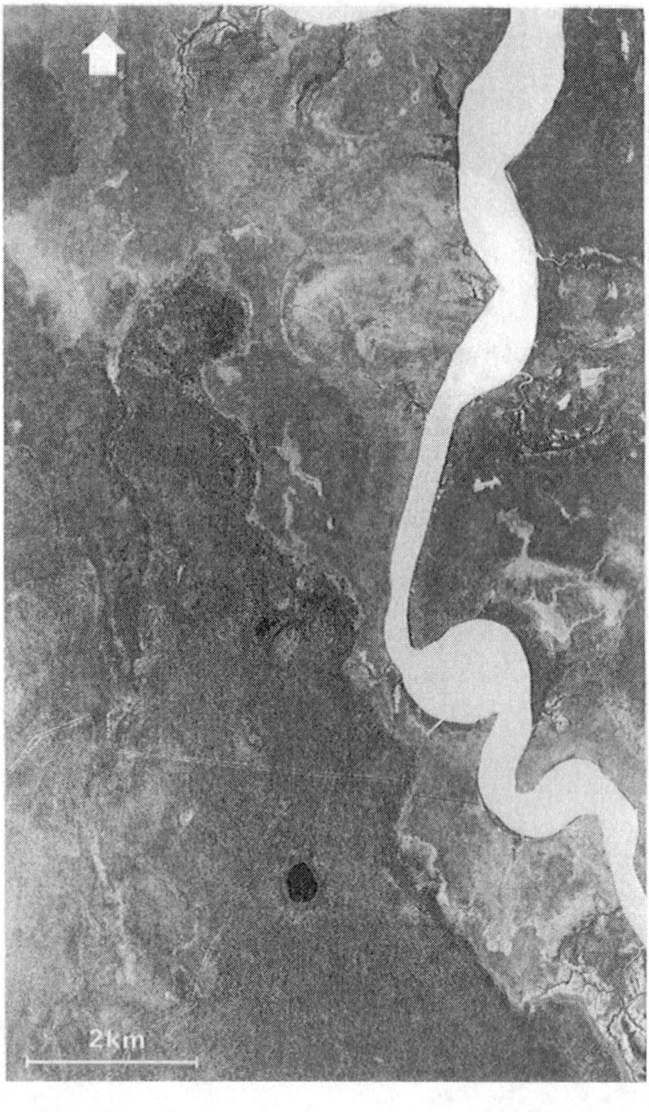

Figure 5.6 Vertical aerial photograph of the cuspate meanders that characterise the South Alligator River, and schematic block diagram interpretation of the Holocene evolution of the area. The aerial photograph (1987) shows a segment of the plains (reproduced by courtesy of the Department of Lands Planning and Environment, Northern Territory). The laminated channel sediments are considered to represent point bar deposits that have been deposited as the channel migrated. Two phases of active channel migration are indicated, resulting in the nested sequence of sinuous meanders. The area was dominated by mangroves 6800–6700

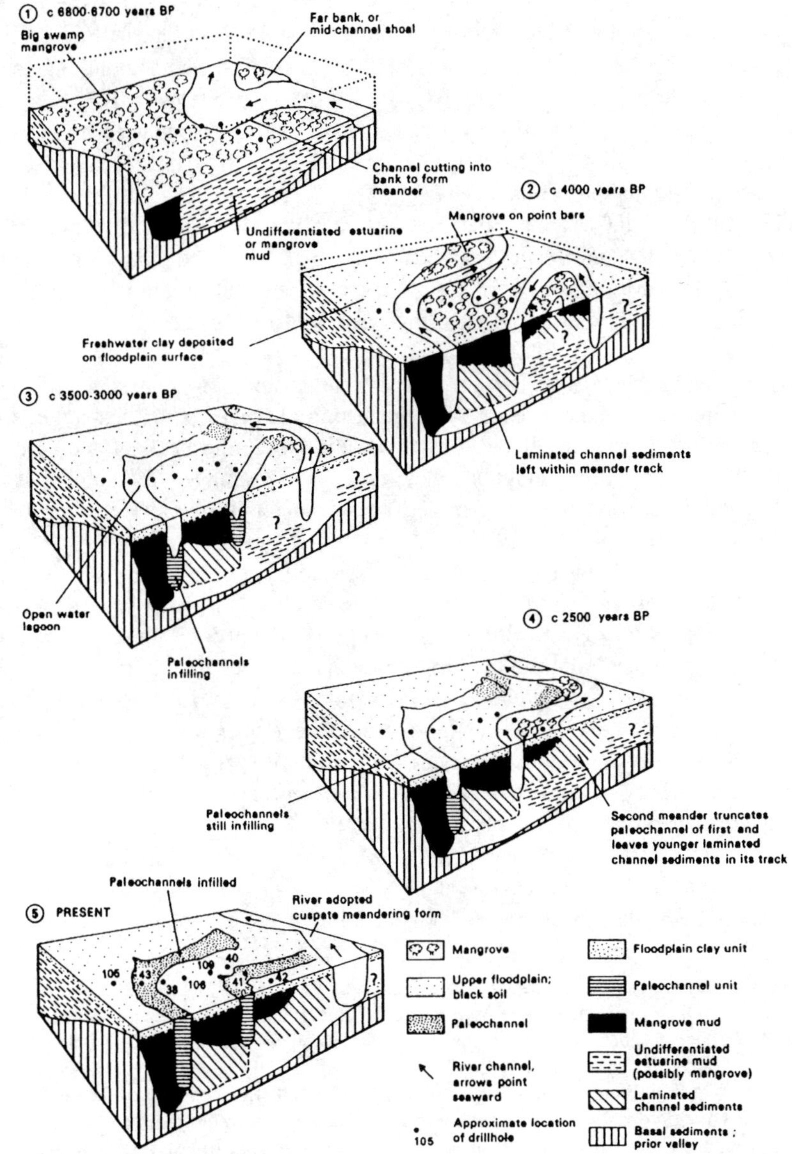

years BP as the sea level was still rising; by 4000 years BP a discrete channel had migrated to the position indicated by the westernmost meander. This river course was cut off, and the new channel migrated to occupy the inner meander location by around 2500 years BP, as indicated by radiocarbon dating of mangrove remains from the point bar of this former channel. The final stage of cutoff and the adoption of a cuspate meandering form appear to have occurred in the past 2500 years. Further details can be found in Woodroffe et al. (1986)

Creek system, a tributary of the East Alligator River, which in turn would have been a part of the proto-Alligator River when it drained across the shelf. Their TL dating indicates sedimentation back to > 300 000 years BP, with apparent phases of sedimentation 30 000–40 000 years BP and 65 000–80 000 years BP. Sand is presumed to have been derived from weathering of the sandstone escarpment and gulleying of sand aprons. They attribute incision to both sea level (and hence baselevel) control and to climate change. They record in detail the pattern of Holocene infill in the Magela sand bed system over the last 7500 years. This sequence records downstream progradation which is continuing with delta progradation into Mudginberri Billabong. Such a sequence implies downfilling of the valley with alluvium, not backfilling as might be expected as a result of sea level change (cf. Schumm 1993, Figure 12).

Thermoluminescence dates from the Kapalga and Magela sites suggest sedimentation, rather than incision, during the initial stages when sea level was lower. However, it remains unclear as to conditions at the peak of the glacial (18 000–20 000 years BP), though subdued terraces at Gulungul (flanking Magela Creek on the East Alligator River system) may span this period (Nanson et al. 1993).

The nature of the changes in climate which would have been experienced in this region over this period is far from clear. Greater aridity is to have been expected with the translation of the coastline over 200 km to the north because: (1) there is presently a pronounced rainfall gradient over such a distance inland from the coast; (2) there would have been greater continentality (Colhoun 1991; Nanson et al. 1992); and (3) there would have been fewer tropical cyclones which presently bring considerable rain to the region (Webster and Stretton 1978). Evidence has been presented for drier conditions than present during the peak of the glacial, from the northeastern Australian and Southeast Asian region (Verstappen 1975; Wyrwoll et al. 1986; Kershaw 1988; Torgersen et al. 1988). These conditions are likely to have led to a decrease in river discharge and sediment load, which in turn may have meant altered channel geometry. Channels have been encountered in seismic

---

Figure 5.7 (*opposite*) Vertical aerial photograph of the sinuous meandering segment of the Adelaide River, and schematic representation of the Holocene evolution of the plains. The aerial photograph (1987) shows a segment of the plains (reproduced by courtesy of the Department of Lands Planning and Environment, Northern Territory). The evolution of the plains has involved considerable progradation from the shoreline 6000 years ago, during the big swamp phase (after Woodroffe et al. 1993). For much of that time the river appears to have emptied into van Diemen Gulf via a palaeochannel on the eastern margin of the plains. A radiocarbon date of 2310 years BP (core 17, see Figure 5.5b) on infill within that palaeochannel indicates that sometime since 3000 years BP the river has adopted its present sinuous course to the sea. The incision of the present meanders into a Pleistocene surface, and the presence of terraces marking this surface in echo-sounded profiles across many of the meanders, has led to the suggestion that this meandering course has reoccupied an earlier fluvial channel pattern (see Vertessy (1990) for details)

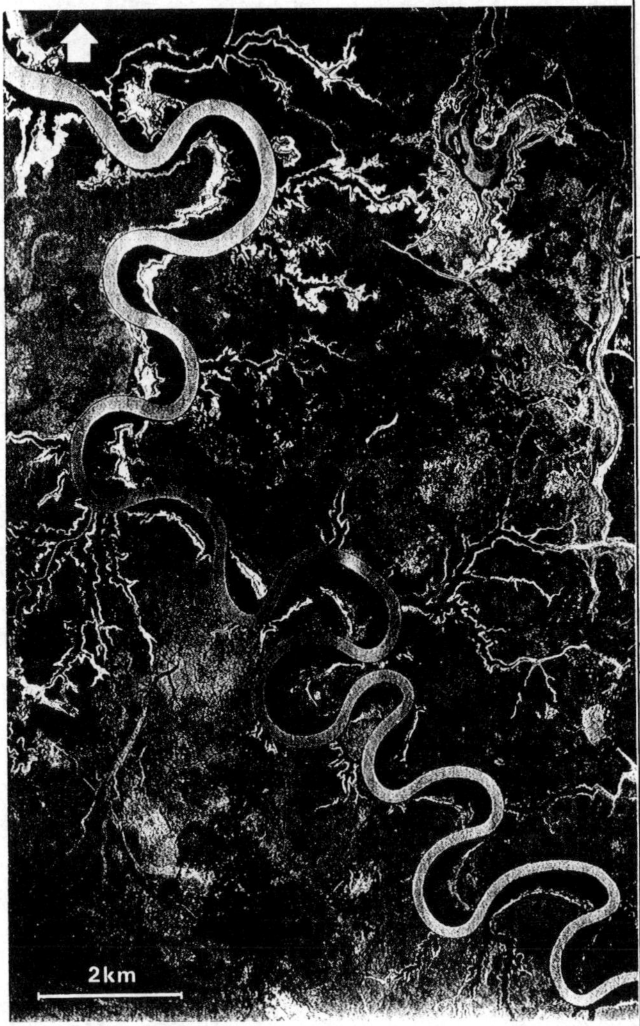
2km

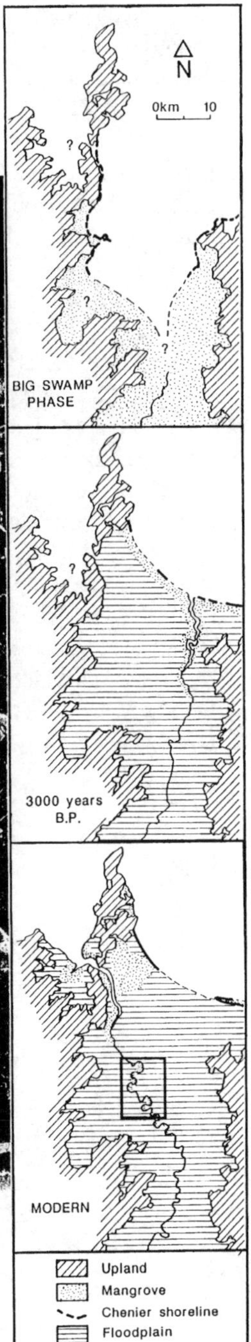
N
0km 10
BIG SWAMP PHASE
3000 years B.P.
MODERN
Upland
Mangrove
Chenier shoreline
Floodplain

traverses on the shelf, but there is little known about their form or pattern. If the sinuous form of the present Adelaide River has indeed reoccupied a lowstand channel and is inherited (Figure 5.7), then the lowstand channels may have been very sinuous.

The transgressive surface at the base of the transgressive systems tract (TST) is the clearest of bounding surfaces, and probably corresponds to the S1 reflector identified in seismic traverses on the shelf. The TST contains the thickest sediment column because of the continual availability of accommodation space as predicted by Wright and Marriott (1993).

The TST consists of mangrove facies deposited through vertical sedimentation and is widespread across estuarine systems in northwestern Australia. As pointed out by Chappell (1993) in the case of the Daly, this does not appear to have terminated in a maximum flooding surface (MFS), as is the case in the Gironde and Bay of Fundy systems. These systems, which experience higher wave energy show distinct ravinement surfaces. Instead, in the northern Australian estuaries the TST is bounded by an estuarine–freshwater transition (EFT), which is hard to distinguish as both estuarine and freshwater wetlands are characterised by mud, but can be discriminated using pollen analysis (see Figure 5.2). The highstand systems tract comprises the alluvial floodplain muds above the EFT, but also the meander tract and the regressive coastal plain sequence.

Sea level has been a major control on the evolution of north Australian estuaries. The extent to which rivers incised when the sea level was lower is still unclear; there may have been marked changes in channel pattern to compensate for altered lower valley gradients, and alluvial sediment aggradation may have been a response to altered climatic condtions. Each estuarine system underwent rapid sediment deposition during the transgressive phase, as accommodation space became available. The extent of channel reworking of the plains, and the thickness of the highstand systems tract, is not the same for all rivers, but appears to depend upon intrinsic factors such as the relative balance between fluvial input and tidal flows.

REFERENCES

Ahnert, F. 1960. Estuarine meanders in the Chesapeake Bay area. *Geographical Review* 50: 390–401.

Allen, G.P. and H.W. Posamentier. 1993. Sequence stratigraphy and facies models of an incised valley fill: the Gironde estuary, France. *Journal of Sedimentary Petrology* 63: 378–391.

Aubrey, D.G. and K.O. Emery 1986. Australia – an unstable platform for tide gauge measurements of changing sea levels. *Journal of Geology* 94: 699–712.

Autin, W. J., S. F. Burns, R. T. Miller, R. T. Saucie, and J. I. Snead. 1991. Quaternary geology of the lower Mississippi valley, p. 547–582. *In* R. B. Morrison (ed.), *Quaternary nonglacial geology: coterminous U.S.* (The Geology of America Volume k-2). Geological Society of America, Boulder, Colorado.

Boyd, R., R.W. Dalrymple and B.A. Zaitlin. 1992. Classification of clastic coastal depositional environments. *Sedimentary Geology* 80: 139–150.

Chappell, J.M.A. 1982. Evidence for smoothly falling sea levels relative to North Queensland, Australia, during the past 6000 years. *Nature* 302: 406–408

Chappell, J. 1993. Contrasting Holocene sedimentary geologies of lower Daly River, northern Australia, and lower Sepik-Ramu, Papua New Guinea. Sedimentary Geology 83: 339-358.

Chappell, J. and J. Grindrod. 1985. Pollen analysis: key to past mangrove communities and successional changes in North Australian coastal environments, p. 225–236. *In* K. Bardsley, J.D.S. Davie and C.D. Woodroffe (eds.), *Coasts and Tidal Wetlands of the Australian Monsoon Region.* North Australia Research Unit, ANU Press, Darwin.

Chappell, J. and H. Polach. 1991. Post glacial sea level rise from a coral record at Huon Peninsula, Papua New Guinea. *Nature* 349: 147–149.

Chappell, J.M.A. and N.J. Shackleton. 1986. Oxygen isotopes and sea level. *Nature* 324: 137–140.

Chappell, J. and C.D. Woodroffe. 1985. Morphodynamics of Northern Territory tidal rivers and floodplains, p. 85–96. *In* K. Bardsley, J.D.S. Davie and C.D. Woodroffe (eds.), *Coasts and Tidal Wetlands of the Australian Monsoon Region.* North Australia Research Unit, ANU Press, Darwin.

Chappell, J. and C.D. Woodroffe. 1995. Macrotidal estuaries, p. 187–218. *In* R.W.G. Carter, and C.D. Woodroffe (eds.), *Coastal Evolution.* Cambridge University Press, Cambridge.

Chappell, J.M.A., E.G. Rhodes, B.G. Thom and E. Wallensky. 1982. Hydro-isostasy and the sea level isobase of 5500 BP in North Queensland, Australia. *Marine Geology* 49: 81–90.

Chappell J.M.A., A. Chivas, E. Wallensky, H.A. Polach and P. Aharon. 1983. Holocene palaeo-environmental changes, central to north Great Barrier Reef inner zone. *BMR Journal of Australian Geology and Geophysics* 8: 223–235.

Clark, R.L. and J.C. Guppy. 1988. A transition from mangrove forest to freshwater wetland in the monsoon tropics of Australia. *Journal of Biogeography* 15: 665–684.

Clarke, M.F., R.J. Wasson and M.A.J. Williams. 1979. Point Stuart chenier and Holocene sea levels in Northern Australia. *Search* 10: 90–93.

Colhoun, E. A. 1991. Climate during the last glacial maximum in Australia and New Guinea; evidence inferred from biogeographical and geomorphological data. *Australia and New Zealand Geomorphology Group Special Publication*, 2.

Cooper, J.A.G. 1993. Sedimentation in a river dominated estuary. *Sedimentology* 40: 979–1017.

Dalrymple, R.W., B.A. Zaitlin and R. Boyd. 1992. Estuarine facies models: conceptual basis and stratigraphic implications. *Journal of Sedimentary Petrology* 62: 1030–1043.

Davis, W.M. 1902. Base-level, grade, and peneplain, p. 381–412. *In*: *Geographical Essays*, 18, Ginn, Boston.

Dronkers, J. 1986. Tide-induced residual transport of fine sediment, p. 228–244. *In* J. van de Kreeke (ed.), *Physics of Shallow Estuaries and Bays, Lecture Notes on Coastal and Estuarine Studies.* Springer, Berlin.

Eisenhauser, A., G.J. Wasserburg, J.H. Chen, G. Bonani, L.B. Collins, Z.R. Zhu, and K.H. Wyrwoll. 1993. Holocene sea-level determination relative to the Australian continent: U/Th (TIMS) and ^{14}C (AMS) dating of coral cores from the Abrolhos Islands. *Earth and Planetary Science Letters* 114: 529–547.

Fairbanks, R.G. 1989. A 17,000-year old glacio-eustatic sea level record: influence of glacial melting rates on the Younger Dryas event and deep-ocean circulation. *Nature* 342: 637–642.

Fisk, H. N. 1944. *Geological Investigation of the Alluvial Valley of the Lower Mississippi River.* Mississippi River Commission, Vocksburg, Missouri.

Galloway, W.E. 1975. Process framework for describing the morphologic and stratigraphic evolution of deltaic depositional systems, p. 87–98. *In* M.L. Broussard (ed.), *Deltas*. Houston Geological Society, Houston, Texas.

Galloway, W.E. 1989. Genetic stratigraphic sequences in basin analysis I: architecture and genesis of flooding-surface bounded depositional units. *American Association of Petroleum Geologists Bulletin* 73: 125–142.

Grindrod, J. 1988. The palynology of Holocene mangrove and saltmarsh sediments, particularly in northern Australia. *Review of Palaeobotany and Palynology* 55: 229–245.

Grindrod, J. and E.G. Rhodes. 1984. Holocene sea level history of a tropical estuary: Missionary Bay, North Queensland, p. 151–178. *In* B. G. Thom (ed.), *Coastal Geomorphology in Australia*. Academic Press, Sydney.

Guilcher, A. 1967. Origins of sediments in estuaries, p. 149–157. *In* G. H. Lauff (ed.), *Estuaries*. American Association for the Advancement of Science Publication, Washington.

Jennings, J.N. 1975. Desert dunes and estuarine fill in the Fitzroy estuary, north-western Australia. *Catena* 2, 215–262.

Johnson, D.W. 1919. *Shore Processes and Shoreline Development*. Prentice Hall, New York.

Jongsma, D. 1970. Eustatic sea level changes in the Arafura Sea. *Nature* 228: 150–151.

Jongsma, D. 1974. Marine Geology of the Arafura Sea. *BMR Bulletin* 157.

Kershaw, A.P. 1988. Australasia, p. 237–306. *In* B. Huntley and T. Webb (eds.), *Vegetation History*. Kluwer, Dordrecht.

Knighton, A.D., C.D. Woodroffe and K. Mills. 1992. The evolution of tidal creek networks, Mary River, Northern Australia. *Earth Surface Processes and Landforms* 17: 167–190.

Lees, B.G. 1987. Age structure of the Point Stuart chenier plain: a reassessment. *Search* 18: 257–259.

Miall, A.D. 1991. Stratigraphic sequences and their chronostratigraphic correlations. *Journal of Sedimentary Petrology* 61: 497–505.

Murray-Wallace, C.V. and A.P. Belperio. 1991. The Last Interglacial in Australia – a review. *Quaternary Science Reviews* 10: 441–461.

Nakada, M. and K. Lambeck. 1989. Late Pleistocene and Holocene sea-level change in the Australian region and mantle rheology. *Geophysical Journal* 96: 497–517.

Nanson, G. C. and J. C. Croke. 1992. A genetic classification of floodplains. *Geomorphology* 4: 459–486.

Nanson, G.C., D.M. Price and S.A. Short. 1992. Wetting and drying of Australia during the past 300 ka. *Geology* 20: 791–792.

Nanson, G.C., T.J. East and R.G. Roberts. 1993. Quaternary stratigraphy, geochronology and evolution of Magela Creek catchment in the monsoon tropics of northern Australia. *Sedimentary Geology* 83: 277–302.

Nichols, M.M. and R.B. Biggs. 1985. Estuaries, p. 77–173. *In* Davis, R. A. Jr (ed.), *Coastal Sedimentary Environments*. Springer, New York.

Orton, G.J., and H.G. Reading. 1993. Variability of deltaic processes in terms of sediment supply, with particular emphasis on grain size. *Sedimentology* 40: 475–512.

Posamentier, H. W. and P. R. Vail. 1988. Eustatic controls on clastic deposition II – sequence and systems tracts models, p. 109–124. *In* C.K. Wilgus, B.S. Hastings, C.G.S.C. Kendall, H.W. Posamentier, C.A. Ross, and J.C.Van Wagoner (eds.), *Sea-level Changes: An Integrated Approach*. Society of Economic Palaeontologists and Mineralogists Special Publication, Tulsa.

Roy, P.S., P. Cowell, M. Ferland and B.G. Thom. 1995. Wave-dominated coasts, p. 121–186. *In* R.W.G. Carter and C.D. Woodroffe (eds.), *Coastal Evolution*. Cambridge University Press, Cambridge.

Russell-Smith, J. 1985. A record of change: studies of Holocene vegetation history in the

South Alligator River region, Northern Territory. *Proceedings of the Ecological Society of Australia* 13: 191–202.

Schlager, W. 1993. Accommodation and supply – a dual control on stratigraphic sequences. *Sedimentary Geology* 86: 111–136.

Schumm, S.A. 1993. River response to baselevel change: implications for sequence stratigraphy. *Journal of Geology* 101: 279–294.

Stanley, D.J. and A.G. Warne. 1994. Worldwide initiation of Holocene marine deltas by deceleration of sea-level rise. *Science* 265: 228–231.

Suter, J. R. 1995. Deltaic coasts, p. 87–120 *In* R.W.G. Carter and C.D. Woodroffe (eds.), *Coastal Evolution*. Cambridge University Press, Cambridge.

Suter, J.R., and H.L. Berryhill. 1985. Late Quaternary shelf-margin deltas, northwest Gulf of Mexico. *American Association of Petroleum Geologists Bulletin* 69: 77–91.

Suter, J.R., H.L. Berryhill, and S. Penland. 1987. Late Quaternary sea-level fluctuations and depositional sequences, southwest Louisiana continental shelf, p. 199–219. *In* D. Nummedal, O.H. Pilkey, and J.D. Howard (eds.), *Sea-level Fluctuation and Coastal Evolution*. Society of Economic Palaeontologists and Mineralogists Special Publication, Tulsa.

Swift, D.J.P. and J.A. Thorne. 1991. Sedimentation on continental margins. I. A general model for shelf sedimentation, p. 3–31. *In* D.J.P. Swift, G.F. Oertel, R.W. Tillman and J.A. Thorne (eds.), *Shelf Sand and Sandstone Bodies: Geometry, Facies and Sequence Stratigraphy*. Blackwell Scientific Publications, Oxford.

Thom, B.G. and J. Chappell. 1975. Holocene sea levels relative to Australia. *Search* 6: 90–93.

Thom, B.G. and P.S. Roy. 1985. Relative sea levels and coastal sedimentation in southeast Australia in the Holocene. *Journal of Sedimentary Petrology* 55: 257–264.

Thom, B.G., L.D. Wright, and J.M. Coleman. 1975. Mangrove ecology and deltaic-estuarine geomorphology, Cambridge Gulf-Ord River, Western Australia. *Journal of Ecology* 63: 203–222.

Thomas, M.S. and J.B. Anderson. 1992. Eustatic controls on the facies architecture of the Trinity/Sabine incised valley system, Texas continental shelf. *Society of Economic Palaeontologists and Mineralogists Special Paper*, 52.

Thorne, J.A., and D.J.P. Swift. 1991. Sedimentation on continental margins. II. Application of the regime concept, p. 33–58. *In* D.J.P. Swift, G.F. Oertel, R.W. Tillman and J.A. Thorne (eds.), *Shelf Sand and Sandstone Bodies: Geometry, Facies and Sequence Stratigraphy*. Blackwell Scientific Publications, Oxford.

Torgersen, T., J. Luly, P. De Dekker, M.R. Jones, D.E. Searle, A.R. Chivas and W.J. Ullman. 1988. Late Quaternary environments of the Carpentaria Basin, Australia. *Palaeogeography Palaeoclimatolgy Palaeoecology* 67: 245–261.

Vail, P.R., R.M. Mitchum, R.G. Todd, J.M. Widmier, S. Thompson, J.B. Songree, J.N. Bubb and W.G. Hatelid. 1977. Seismic stratigraphy – applications to hydrocarbon exploration. *American Association of Petroleum Geologists Memoir* 26: 49–212.

van Andel, T.H. and J.J. Veevers. 1965. Sub-marine morphology of the Sahul Shelf, Northwestern Australia. *Geological Society of America Bulletin* 76: 695–700.

van Andel, T.H. and J.J. Veevers. 1967. Morphology and sediments of the Timor Sea. *BMR Bulletin* 83.

van Andel, T.H., G.R. Heath, T.C. Moore and G.F.R. McGeary. 1967. Late Quaternary history, climate and oceanography of the Timor Sea, Northwestern Australia. *American Journal of Science* 265: 737–758.

Van Wagoner, J.C., R.M. Mitchum, K.M. Campion and V.D. Rahmanian. 1990. *Siliclastic Sequence Stratigraphy in Well-logs, Cores, and Outcrops: concepts for high-resolution correlation of time and facies*. American Association of Petroleum Geologists, Tulsa.

Veeh, H.H. and J.J. Veevers 1970. Sea level at −175 m off the Great Barrier Reef 13,600 to 17,000 years ago. *Nature* 226: 536–537.

Veevers, J.J. 1969. Palaeogeography of the Timor Sea Region. *Palaeogeography Palaeoclimatology Palaeoecology* 6: 125–140.

Veevers, J.J. 1971. Shallow stratigraphy and' structure of the Australian continental margin beneath the Timor Sea. *Marine Geology* 11: 209–249.

Verstappen, H.T. 1975. On paleoclimates and landform development in Malaysia. *Modern Quaternary Research in Southeast Asia* 1: 3–35.

Vertessy, R. 1990. Morphodynamcs of macrotidal rivers in far northern Australia. Unpublished Ph.D. thesis, Australian National University.

Wasson, R.J. (ed). 1992. Modern sedimentation and late Quaternary evolution of the Magela Creek plain. *Research Report, Supervising Scientist for the Alligator Rivers Region* 6.

Webster, P.J. and N.A. Stretton. 1978. Late Quaternary ice age climates of tropical Austalasia: interpretations and reconstructions. *Quaternary Research* 10: 279–309.

Wescott, W.A. 1993. Geomorphic thresholds and complex response of fluvial systems – some implications for sequence stratigraphy. *American Association of Petroleum Geologists* Bulletin 77: 1208–1218.

Whitehead, P.J., B.A. Wilson, and D.M.J.S. Bowman. 1990. Conservation of coastal wetlands of the Northern Territory of Australia: The Mary River Floodplain. *Biological Conservation* 52: 85–111.

Woodroffe, C. D. 1988. Changing mangrove and wetland habitats over the last 8000 years, northern Australia and Southeast Asia, p. 1–33. *In* D. Wade-Marshall and P. Loveday (eds.), *Northern Australia: Progress and Prospects, Vol 2, Floodplains Research.* ANU Press, Darwin.

Woodroffe, C.D., B.G. Thom and J. Chappell. 1985. Development of widespread mangrove swamps in mid-Holocene times in northern Australia. *Nature* 317: 711–713.

Woodroffe, C.D., J.M.A. Chappell, B.G. Thom and E. Wallensky. 1986. *Geomorphological Dynamics and Evolution of the South Alligator Tidal River and Plains, Northern Territory.* Mangrove Monograph. No. 3, NARU, Darwin.

Woodroffe, C.D., J.M.A. Chappell, B.G. Thom and E. Wallensky. 1989. Depositional model of a macrotidal estuary and floodplain, South Alligator River, Northern Australia. *Sedimentology* 36: 737–756.

Woodroffe, C.D. and M.E. Mulrennan. 1993. *Geomorphology of the Lower Mary River Plains, Northern Territory.* North Australia Research Unit and Conservation Commission of the Northern Territory, Darwin.

Woodroffe, C.D., M.E. Mulrennan and J. Chappell. 1993. Estuarine infill and coastal progradation, southern van Diemen Gulf, northern Australia. *Sedimentary Geology* 83: 257–275.

Woodroffe, C.D., B.G. Thom, J.M.A. Chappell, E. Wallensky, J. Grindrod and J. Head. 1987. Relative sea level in the South Alligator River Region, North Australia, during the Holocene. *Search* 18: 198–200.

Wright, L.D. 1985. River deltas, p. 1–76. *In* R. A. Davis Jr (ed.), *Coastal Sedimentary Environments.* Springer, New York.

Wright, L.D., J.M. Coleman and B.G. Thom. 1973. Processes of channel development of a high-tide range environment: Cambridge Gulf-Ord River Delta, Western Australia. *Journal of Geology* 81: 15–41.

Wright, V.P. and S.B. Marriott. 1993. The sequence stratigraphy of fluvial depositional systems: the role of floodplain sediment storage. *Sedimentary Geology* 86: 203–210.

Wyrwoll, K.-H., N.L. McKenzie, B.J. Pederson and I.J. Tapley 1986. The Great Sandy Desert of northwestern Australia: the last 7000 years. *Search* 17: 208–211.

Chapter Six

Geomorphology and ecology of New Zealand shallow estuaries and shorelines

TERRY R. HEALY, RUSSELL COLE and WILLEM DE LANGE
University of Waikato

6.1 INTRODUCTION

The islands comprising New Zealand possess extensive indented coastlines exceeding 10 000 km in length. Taking a broad definition of estuaries to include both estuarine and lagoonal environments, as defined by Cameron and Pritchard (1963), McLay (1976) identified some 301 estuaries around the coastline of New Zealand and its outlying islands (Figure 6.1). They range in size from a few to over 15 000 ha, although most are less than 1700 ha.

The wide diversity of physical characteristics of the estuary types is a function of geological structure and lithology, climate and tidal range. Arising from its location on an active plate margin, the country contains a variety of lithologies, geological structures and topography (Figure 6.2), including axial ranges of low-grade metamorphosed greywacke, active neogene volcanism, soft cenozoic sediments, and tectonic uplifts in the North Island, to alpine glaciated ranges and outwash gravel plains in the South Island. The country is largely mountainous in the South Island, and mountainous-to-hilly in the North Island.

The elongated north–south orientation and varying topography transecting the zone of mid-latitude westerlies in an oceanic setting, result in contrasting climatic characteristics. The primary parameters that affect estuaries include the catchment rainfall, temperatures, winds (which impact on circulation and wave stirring), and evaporation. The mid-latitude westerlies bring frequent depressions which

Estuarine Shores: Evolution, Environments and Human Alterations. Edited by K. F. Nordstrom and C. T. Roman.

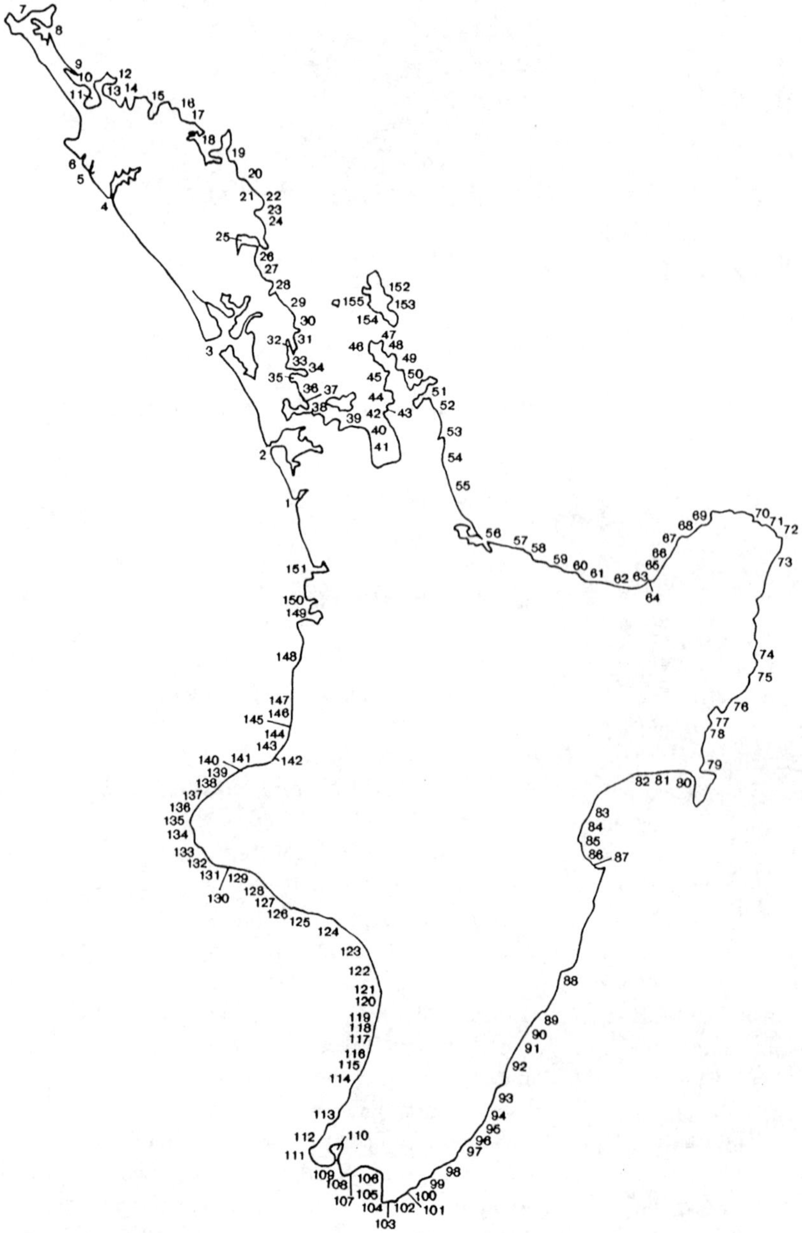

Figure 6.1 Locations of New Zealand estuaries (see McLay 1976 for names)

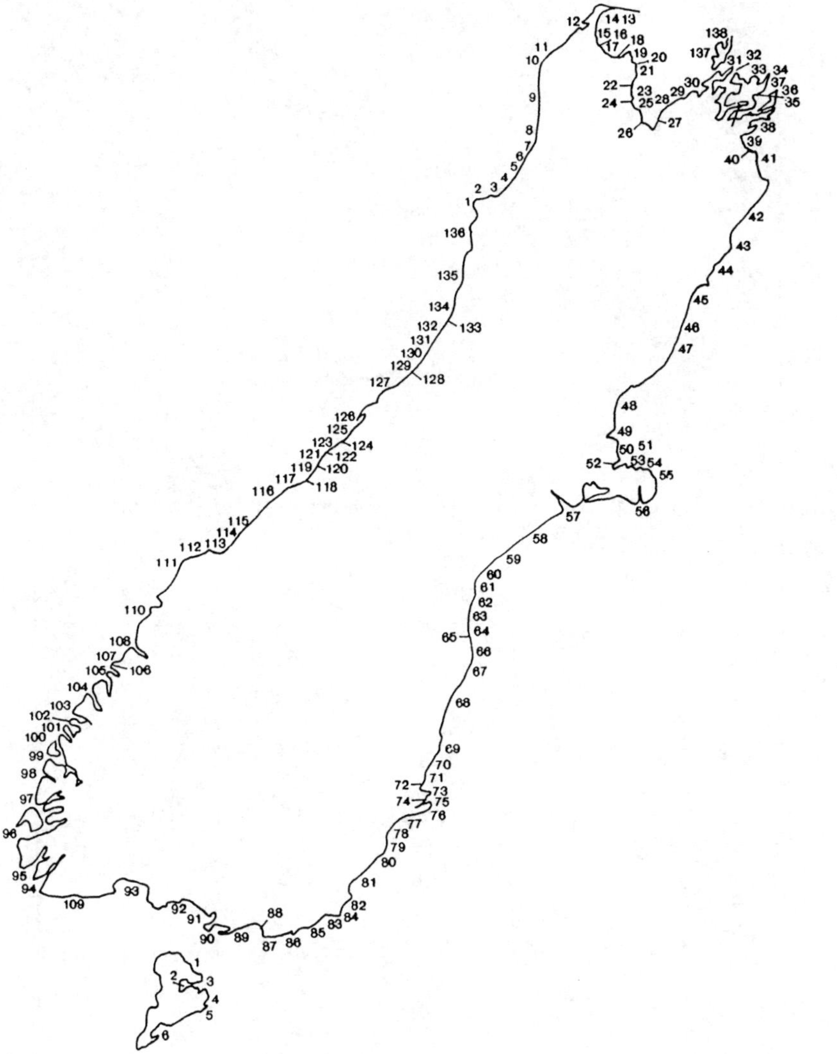

Figure 6.1 *(continued)*

occasionally interact to precipitate considerable orographically induced rainfall, so catchment flooding and erosion deliver considerable sediment to the estuaries and the littoral zone.

Tidal range is of fundamental importance to circulation in the estuaries. Typically the M2 tidal constituent is the most important influence (Heath 1977), with tides being semi-diurnal and typically ranging from about 2 m on the east coast to 3–4 m on the west coast (Figure 6.3).

From the interaction of the catchment structure, topography, geology, tectonic

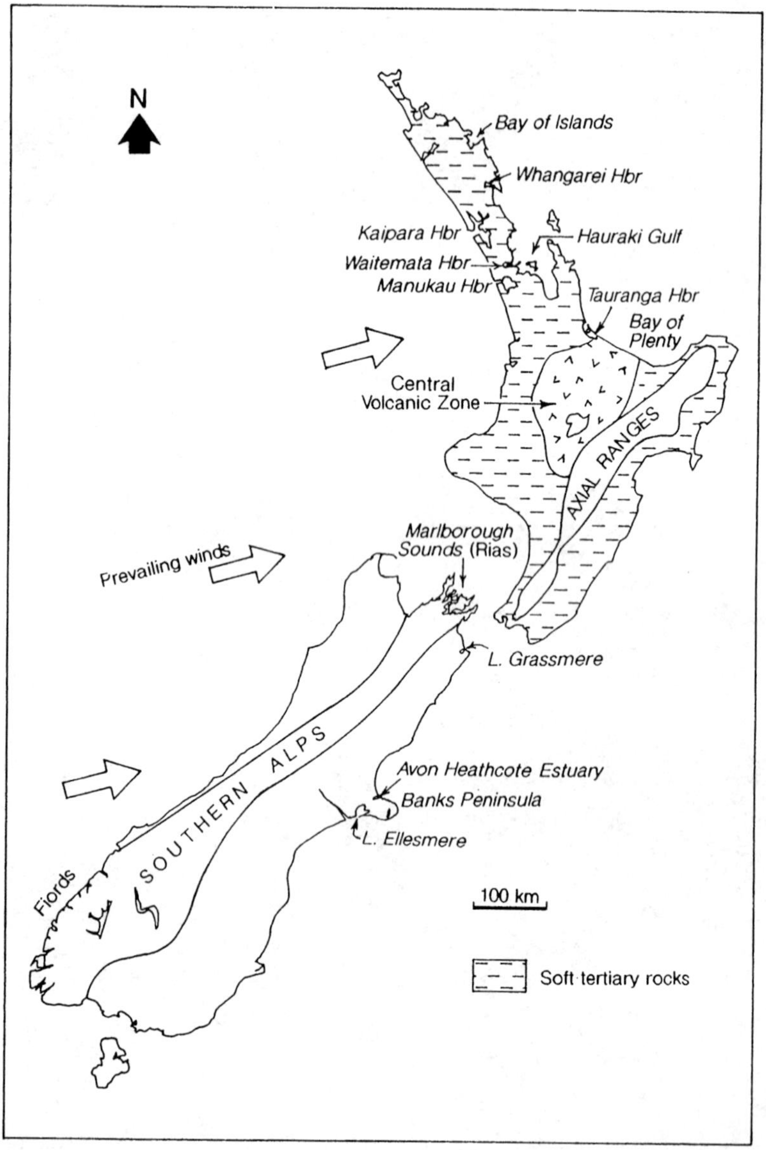

Figure 6.2 Lithology structure and climatic influences on estuarine type and evolution in New Zealand

history, hydrology and sedimentation, varying climatic parameters and tidal ranges, a wide range of coastal environments and estuarine types have evolved (Hume and Herdendorf 1988; Healy and Kirk 1992; Hume et al. 1992). Perhaps the most remarkable characteristic of New Zealand estuaries in comparison with their

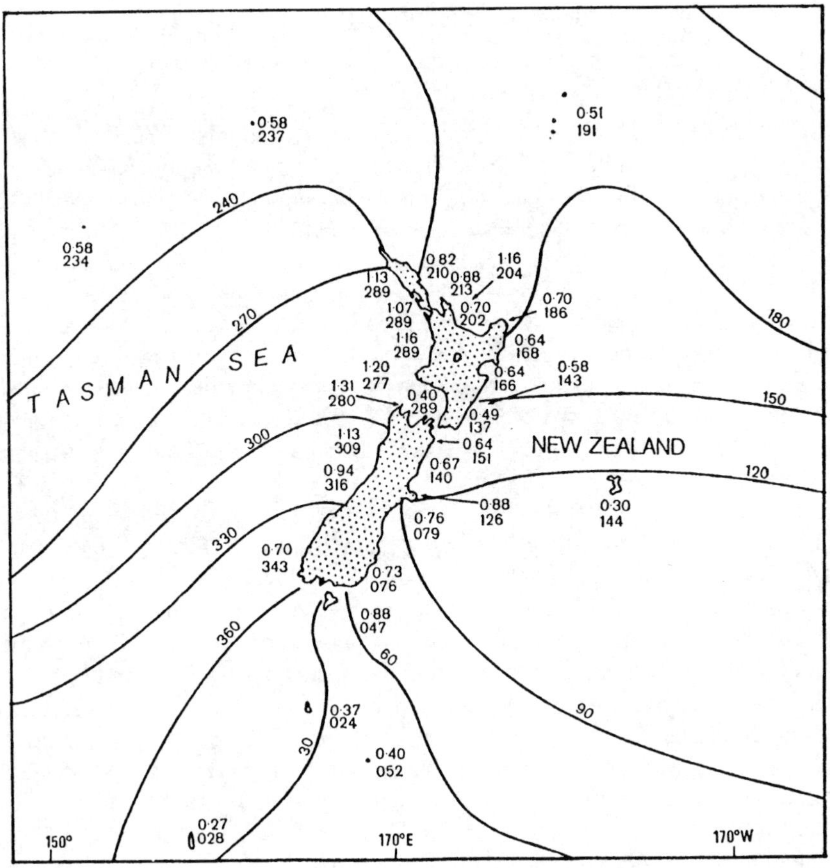

Figure 6.3 Observed M2 principal lunar and semi-diurnal tidal amplitude (m) and phase angle (°), adapted from Heath (1977)

counterparts on stable continental margins is the high rate of active infilling (Hume and Gibb 1987; Hume et al. 1989; Hume and Dahm 1991; Sheffield et al. 1995).

In this chapter we:

1. Identify the estuary types in this special geological context of an island environment on an active plate margin based on estuary morphology.
2. Present data on the hydrodynamics of shallow intertidal zones and beaches of estuarine lagoons.
3. Demonstrate the interactions and linkages between the biology and physical characteristics.
4. Present two cases of human interference with the natural estuarine regime that have impacted the sedimentation and ecology of the shallow sub- and intertidal zones and shorelines of an estuarine lagoon, namely in Tauranga Harbor and Maketu Estuary in the Bay of Plenty.

6.2 PREVIOUS ESTUARINE RESEARCH

6.2.1 Introduction

McLay (1976) summarized the types of investigations into New Zealand estuaries, and suggested a number of different classifications based on population densities, water quality, stratification and topography. He reported that most studies were concerned with either ecological or water quality surveys. A bibliography of scientific studies of New Zealand mainland estuaries and other sheltered waters lists 429 studies up to 1975, of which a substantial proportion are biological (Estcourt 1975). Hume and co-authors (1992) reviewed coastal oceanographic and sedimentological studies related predominantly to estuarine research; Putnam (1977) and Schiel (1991) reviewed marine biological and ecological studies in New Zealand; and Creese and Cole (1995) reviewed conservation issues relating to estuaries. Although these include non-estuarine studies, they offer a wider context for this review.

Further descriptions of the hydrodynamics, geomorphology and fauna and flora of estuaries may be found in university journals such as *Tane* (17 entries from estuarine environments) and *Mauri Ora* (three entries from estuarine environments). A 1990 Index to Articles in *Notornis* contained 152 North Island and 45 South Island entries listed under ''Estuaries and harbors''. There is also a vast unpublished literature of environmental impact audits, reports and statements, local body reports, and theses from universities; these will not be dealt with extensively. The most comprehensive study of biology in sheltered waters to date has been the National Institute for Water and Atmosphere (NIWA) studies of Manukau Harbor from 1989 to 1994 (Thrush and Roper 1988; Pridmore et al. 1990a,b, 1991; Roper et al. 1988, 1991; Thrush et al. 1989, 1991, 1992, 1994a, b; Cummings et al. 1993, 1995; Hewitt et al. 1993).

6.2.2 Hydrodynamics and geomorphology

Prior to the 1970s there were only a few studies of the hydrodynamics and geomorphology of New Zealand estuaries. Those early studies were primarily concerned with the developments and operations of port facilities (Hume et al. 1992) and were mainly undertaken for engineering purposes.

In the early 1970s increasing concern about the impacts of land use practices on the water quality, ecology and sedimentology of estuaries led to a series of environmental investigations, for example the Waitemata Harbor Study (Gregory and Thompson 1973). Most investigations tended to concentrate on either ecological or sedimentological aspects, and there was only limited consideration given to the interactions between ecology, sediments and physical processes.

Estuarine ecological studies include those on the Avon-Heathcote (summarized in Knox and Kilner 1973), which were mainly ecological community studies investigating the sediments in the course of wider investigations. Some similar studies were undertaken for some of the estuaries and harbors around the Hauraki Gulf in the North Island (Morton and Miller 1973; Grace 1970; Larcombe 1971).

A systematic investigation of the surficial sedimentology was made for numerous estuaries of the central North Island, for example, Sherwood (1973) for Raglan Harbor, Marks (1975) for Whangapoua Harbor, Davies-Colley (1976) for Tauranga Harbor, Richmond (1977) for Ohiwa Harbor, Millar (1980) for Whangarei Harbor, and White (1979) for Waikareo Estuary in Tauranga Harbor. All of these studies described the spatial characteristics of the surficial sedimentology, producing maps of textural parameters and composition, the latter usually based on the bulk XRD analysis, typically culminating in a standard facies analysis.

A pioneering study that integrated the sedimentology and the patterns of estuarine macro-benthic ecology was that by Murray (1978) for Maketu Estuary. Apart from the patterns of sediment distribution, Murray was also able to present the broad-scale distribution of the main benthic macrofauna relating to both the sediments, salinity structure and hydrodynamics (Figure 4). This work has subsequently been widely quoted as a baseline work against which to measure change when the Kaituna River is partially re-diverted back into the Maketu estuary.

The first comprehensive harbor study that involved intensive field observations was that for Whangarei Harbor, a large barrier spit-enclosed estuarine lagoon, carried out between 1979 and 1983. Investigations included both numerical and physical models, simultaneous 1-D and 2-D models and the world's first sediment transport model in a tidal inlet (Black 1983). For calibration and verification of the models, as well as general understanding of the inlet system, the extensive field studies included hydrographic surveys, underwater photography, side-scan sonar surveys, underwater video observations and sediment threshold experiments, tide gauges and continuous recording current meters as well as lagrangian drogue tracking, and wave refraction studies (Healy et al. 1987). From these extensive studies evolved some fundamental science on the importance of full tidal cycle velocity vector residuals as indicators of sandy sediment transport pathways and morphodynamics of a tidal inlet (Black 1983; Black and Healy 1982, 1986, 1988; Black et al. 1989). Although not specifically integrated in the study, the broad-scale macrobenthic ecology was reported from the underwater photography survey. Subsequent work on ecology was carried out by Dickie (1984a,b,c), May (1984), and Venus (1984).

A second major harbor study, the most detailed and extensive in a New Zealand estuary to date, was the Tauranga Harbor Study of 1983–85. Tauranga Harbor is also a large estuarine lagoon with a tidal inlet type entrance. Apart from the field studies mentioned above, detailed attempts at suspended sediment monitoring were also undertaken. A major first from this study, involving numerical hydrodynamic and sediment transport models, was the application of the sediment transport modeling at the design stage to determine the optimum length of bridge span relative to causeway length, minimizing scour and shallow estuary morphodynamic impact, for the then planned Tauranga Harbor bridge (Black 1984a,b).

Subsequently a number of hydrodynamic model studies have been undertaken for New Zealand port harbors, but not on the scale of the first two. They include Auckland and the Waitemata Harbor (Greig and Proctor 1988; Proctor and Greig 1989), Manukau, Otago, and Nelson, and relate to port developments and harbor management plans, while Wellington Harbor, a structural estuary, has been modeled for the effect of potential tsunami waves, in particular pertaining to the washover and flooding of the lower parts of the downtown city area. Most of these studies were undertaken with the focus on the physical issues, and they did not relate to the biology of the harbors and estuaries.

Since the 1980s, a related set of studies within, as well as outside, the estuaries has been modeling of effluent dispersion from outfalls for the purpose of environmental impact assessment (Wood et al. 1994). These studies typically have related the hydrodynamics to the microbiology and impacts on the macrobenthos (Roper et al. 1988, 1991; Pridmore et al. 1990a,b).

6.3 PHYSIOGRAPHY OF NEW ZEALAND ESTUARIES

Based on physiography, Healy and Kirk (1992) identified five major categories for New Zealand estuaries:

1. Drowned river valleys.
2. Barrier-enclosed estuaries and estuarine lagoons.
3. River-mouth estuaries.
4. Structurally and tectonically influenced estuaries.
5. Glacially excavated valleys or fjords.

6.3.1 Drowned river valleys

Topographically, New Zealand consists predominantly of hilly-to-mountainous terrain. Where the valleys intersect with the coastlines and deposition at the valley heads has been insufficient to infill the valleys to the stage of alluvial plains, the classical drowned river valley type estuary occurs. Because of the very active erosion and sedimentation processes characteristic of New Zealand, the classical drowned river valley estuaries tend to be relatively rare and large-scale features, but they do not exhibit large drainage catchments or large fresh water river input compared to tidal prism. Examples include the Waitemata Harbor, around which the city of Auckland has become established, Mahurangi Harbor and the Bay of

Figure 6.4 (*opposite*) Association between shallow sediments and macrobenthic communities for Maketu Estuary, after Murray (1978)

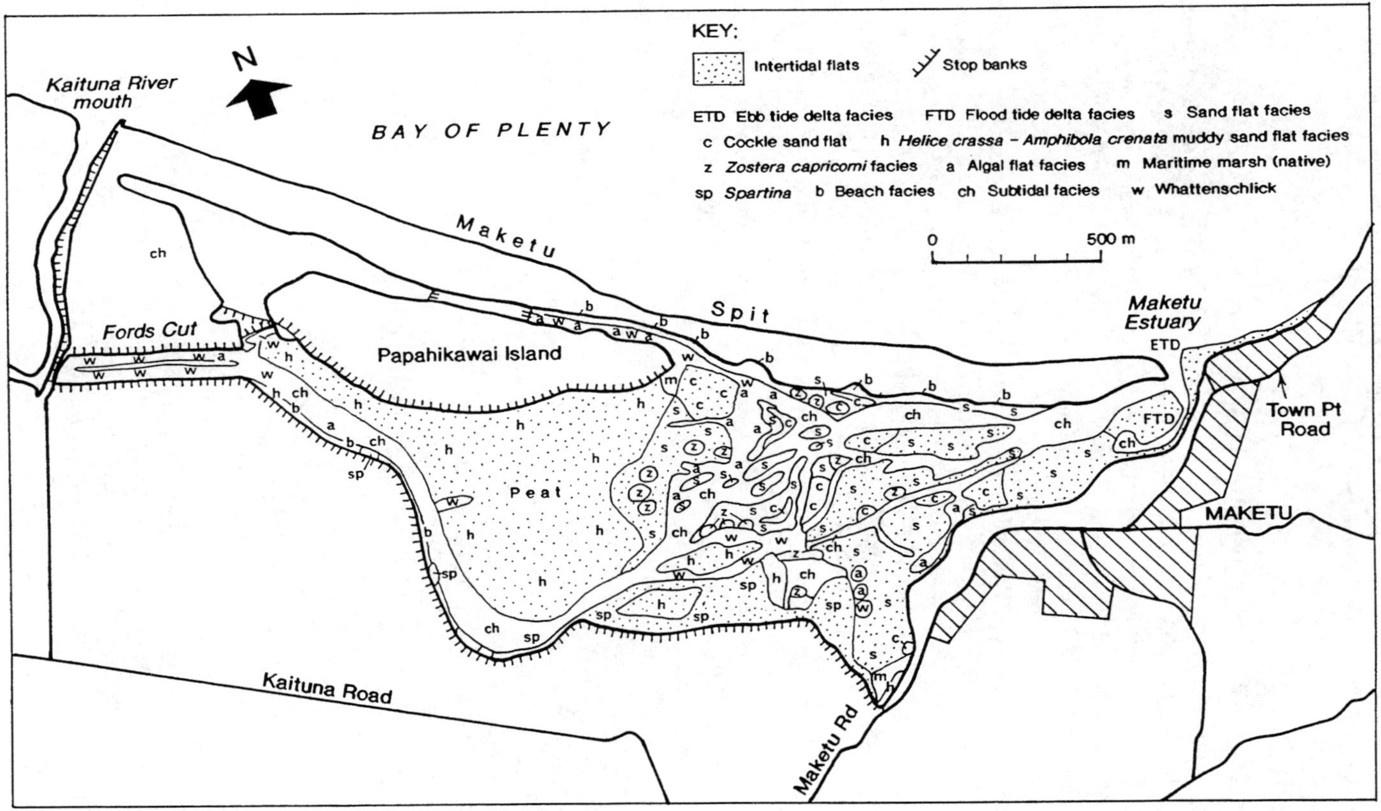

KEY:
Intertidal flats
Stop banks
ETD Ebb tide delta facies
FTD Flood tide delta facies
s Sand flat facies
c Cockle sand flat
h Helice crassa – Amphibola crenata muddy sand flat facies
z Zostera capricorni facies
a Algal flat facies
m Maritime marsh (native)
sp Spartina
b Beach facies
ch Subtidal facies
w Whattenschlick
0
500 m
N
Kaituna River mouth
BAY OF PLENTY
Maketu Spit
Fords Cut
Papahikawai Island
Peat
Maketu Estuary
Town Pt Road
MAKETU
Kaituna Road
Maketu Rd

Islands. Such drowned river valleys may contain smaller scale estuaries within their confines, such as drowned tributary valleys, barrier-enclosed estuarine lagoons, and river or stream mouth estuaries.

6.3.2 Barrier-enclosed estuaries and estuarine lagoons

Estuaries and lagoons enclosed by barrier systems, usually but not exclusively Holocene sandy spits, are termed ''barrier-enclosed''. Former writers (Johnson 1919; Cotton 1942; Pritchard 1967; Dyer 1973; Pethick 1984) rather loosely termed them ''bar built'' estuaries, but in most cases such terminology is genetically incorrect; the barrier systems did not originate from ''bars'' in the modern coastal process sense.

In the typical New Zealand setting, barrier-enclosed estuaries and estuarine lagoons comprise a Holocene sandy barrier spit across the mouth of a drowned river valley, enclosing a shallow lagoonal estuary linked via a tidal inlet system to the open ocean. The inlets are dynamic equilibrium systems, and susceptible to rapid shoreline change around the inlets. The lagoonal nature of the estuaries is evident from the extensive intertidal flats exposed at low tide within these estuaries, typically about 70% of the estuarine surface area (Davies-Colley and Healy 1978; Richmond et al. 1984; Burton and Healy 1985; Hume and Herdendorf 1990, 1992). For the case of Tauranga Harbor, a globally unique geomorphic estuarine lagoon system enclosed by two barrier tombolos and a barrier island (Figure 6.5), the stratigraphy within the estuarine lagoon is quite well known from the numerous cores drilled as part of the port development over the past decades (Davis and Healy 1992). Since the mid-1980s, Hume from the National Institute of Water and Atmosphere (NIWA) has led a program of investigation into the morphological and sedimentary characteristics of New Zealand barrier-enclosed estuaries, focusing in particular on the inlet characteristics and sedimentation processes (Hume and Herdendorf 1988, 1990, 1992).

New Zealand barrier-enclosed estuaries are not exclusively formed from Holocene sand spit deposits. Some large west coast North Island estuarine lagoons are enclosed by large dune systems of Pleistocene age, and in this sense are polycyclic (the Manukau and Kaipara harbors, for example). In the South Island particularly there are a number of gravel barrier systems, such as at Lake Ellesmere and Lake Grassmere (Kirk 1975; Pickrill 1977), while the barrier system at Nelson is composed of a 16 km long boulder bank spit.

Within large shallow and biogenically productive nearshore zones, as in the Firth of Thames at Miranda, or within estuarine lagoons such as Manukau and Kaipara harbors, the surfeit of dead bivalve (e.g. cockle *Austrovenus stutchburyi*, pipi *Paphies australis* and oval trough *Mactra ovata*) shells can be transported and concentrated by wave action into storm berm chenier ridges, and these ridges can build up to supratidal berms by storm surge, and under longshore drift wave action create small barrier systems enclosing an estuarine lagoon (Figure 6.6).

Figure 6.5 Tauranga Harbor, an example of a barrier-enclosed estuarine lagoon, the most common estuary type in northeast New Zealand. The barrier system here comprises two Holocene barrier tombolos and barrier island. The foreground right is part of a Pleistocene inner barrier

Figure 6.6 Tuapiro Spit is a recurved chenier ridge barrier spit, formed from molluscan bivalve shells of *Paphies australis, Austrovenus stutchburyi* and *Mactra ovata*, and located within the confines of northern Tauranga Harbour. Seaward is the flood tidal delta and Matakana barrier island. Extensive intertidal flat covered with *Zostera* and *Ulva* is evident

6.3.3 River mouth estuaries

These tend to be of small scale and are often found within the larger drowned river valleys. Locally they exhibit the classical funnel-shaped estuary form (Dyer 1973) and the intertidal and subtidal sediments are characterized by muddy deposition due to flocculation in the major mixing zone between the fresh and saline waters. In New Zealand, these estuaries have often become infilled as alluvial flats because of the rapidity of landscape erosion processes, facilitated by adjacent catchments containing soft erodible tertiary rocks or recent volcanic tephras producing high river sediment transport loads. Additionally, around the central North Island coasts, rapid tectonic uplift (Chappel 1975; Pillans 1986) has facilitated sedimentation infill of the river valleys. Frequently New Zealand river mouth estuaries may also be partially blocked by barrier spits, as at the entrance to the Whakatane (Figure 6.7), Kaituna, Mokau, Waipoua and Waimakariri river mouths.

6.3.4 Structurally and tectonically influenced estuaries

As is to be expected for coastal environments evolved on an active plate margin, some strong tectonic and structural influences may be reflected in the estuarine form. Examples in New Zealand include structural grabens, tectonically uplifted coastal river valleys, eroded calderas and rias.

Figure 6.7 Holocene sandy barrier spit enclosing the river mouth estuary of Whakatane River. Normal river flow is about 50 cumecs

6.3.4.1 Structural grabens

An example is the Firth of Thames (Naish 1990; Van Leeuwe 1991), a large linear horst–graben–horst feature which also fits the definition of a semi-enclosed sea (Healy and Harada 1991). This north–south aligned structural depression is bounded on east and west sides by fault escarpments and has considerable fluvial input from the larger inflowing rivers, resulting in extensive muddy deposition and suspended sediment loads with well developed lutoclines.

6.3.4.2 Tectonically uplifted coastal river valleys

An example is the Waitara Estuary in north Taranaki (Mathews 1982), part of the south central North Island that is undergoing rapid tectonic uplift. As a result, the estuarine development at the mouth of the Waitara River is inhibited, so that the estuary has remarkably low width:depth ratios, the estuary frequently being little wider than the normal river at its mouth, and high fresh:saline water ratios.

6.3.4.3 Eroded Calderas

A number of volcanic craters occur in the coastal zone, and where they have evolved as calderas, the sea has eroded into and flooded the collapsed crater to form an estuary. Examples are Lyttelton and Akaroa harbors on Banks Peninsula, while the Panmure Basin estuary is an eroded maar or tuff ring, which after erosion from the sea now forms an estuary. These estuaries have low fresh:saline water ratios.

6.3.4.4 Rias

Rias are tectonically down-sunken valleys. Classic examples of rias are the Marlborough Sounds in the northern South Island. These occur here on a regional basis in metasediment greywackes, and apart from the downsinking have also been affected by transverse faulting, so the rias tend to follow linear alignment. The rias are deep, narrow and elongated, with relatively small catchment areas and active sedimentation restricted to the uppermost headwater reaches. In these upper reaches typically there is some sedimentation, which is relatively minor compared to the area of the entire ria. Fresh:saline water ratios are low.

6.3.5 Glacially excavated fjords

Fjords occur in southwest New Zealand, carved into granodiorites of the oldest rocks in the country. These fjords have the general features outlined by Dyer (1973), while their circulation and suspended particulate matter have been discussed by Pickrill (1987). Typically there is some minor infill sedimentation at the headwaters of the fjord, while the sills at the entrance restrict the circulation. Because of their wild and scenic isolation and steep sides in hard paleozoic crystalline rocks, the shoreline dynamics change only over a long time-scale.

Although five broad morphological categories can be defined, many estuaries, except the fjords of south Westland and the rias of Marlborough, exhibit features of more than one category. For example, almost all estuaries in New Zealand show some sedimentation and morphological features related to drowning of pre-existing river valleys resulting from the Late Neogene post-glacial transgression. This is especially evident in the upper reaches of the estuaries where the river input of both fresh water and terrigenous sediments produces the typical upper estuarine deposition of fine muddy sediments.

In terms of the relative abundance and geographical distribution of the estuary types, the glacial fjords are restricted regionally to south Westland. Large-scale drowned river valley types are relatively rare, and occur at Auckland and in Northland. Structurally influenced types are also relatively rare, except that rias occur regionally as the Marlborough Sounds at the northern tip of the South Island. Otherwise barrier-enclosed types are the most common types, and river mouth types next in abundance.

6.4 ECOLOGY OF NEW ZEALAND ESTUARINE SHORES

The identities of organisms on estuarine shores around New Zealand have been gathered together most fully by Morton and Miller (1973). That study contained comprehensive descriptions of the fauna and flora of estuaries from Northland (Waiwera) to Otago, both from original observations and from reviews of other studies. More recent popular accounts have added to those descriptions (Cometti and Morton 1985; Stephenson 1986; Bradstock 1989; Walsby 1990), including long-term changes in the flora and fauna (Dromgoole and Foster 1983; Owen 1992). Descriptions of the estuarine shores of many areas are found in government reports and impact statements (Healy 1980; Burns et al. 1990; Davidson 1990; Davidson and Moffat 1991; Environment Bay of Plenty 1993). Species lists are given in those studies, but the different sampling times and techniques, and intensities of sampling in space and time do not permit a comprehensive analysis.

Schiel (1991) noted the under-representation of descriptive studies of soft shores in the *New Zealand Journal of Marine and Freshwater Research.* Most articles in the primary scientific literature identify habitats on the basis of tidal height and/or vegetation, and describe the organisms within them. Mangrove forests, salt marshes, and eelgrass beds have been described in many qualitative accounts, but there are few quantitative descriptions of habitat types.

6.4.1 Organisms of estuaries

It is difficult to study the flora and fauna of an area if taxa cannot be identified. The taxonomies of several conspicuous animal and plant groups remain to be analyzed in detail. This situation has been complicated by additions to the flora and fauna by a number of human activities (Luckens 1974; Dinamani 1971, 1981; Adams 1983; Dromgoole and Foster 1983; Hay and Luckens 1987; Partridge 1987; Creese and Cole, 1995). The fauna of the various habitats have been identified in some areas, quantified in fewer of those areas, and the dynamics of interactions assessed in hardly any areas (Table 6.1). The studies vary greatly in the taxonomic level to which organisms were identified, the nature of the sampling (qualitative or quantitative), and their spatial extent and duration. Most of the studies are descriptions of one estuary at one time, frequently limited to a particular taxonomic group, for example Webb's studies of fish in the Avon-Heathcote estuary (Webb 1972, 1973a,b,c,d). The most conspicuous permanent residents of many estuarine areas are bivalves and gastropods and those groups have been relatively well studied.

6.4.2 Spatial variability

There are many descriptions of biological gradients across intertidal shores, for example, McArdle and Blackwell (1989), as well as from marine to the limit of saline influence (e.g. Hayward and Hollis 1993). Partridge and Wilson (1987a,b; 1988a) conducted both physiological and transplant experiments to interpret the structure of salt marshes in Otago, concluding that competitive interactions were important in determining the lower distributional limits of the various plants.

Roper and colleagues (1992) suggested that post-metamorphic migration could be important in bivalve population dynamics in the Manukau Harbor, where the most comprehensive investigations have been done (Pridmore et al. 1990a,b, 1991; Thrush et al. 1991, 1992, 1994a,b; Cummings et al. 1995). Those studies examined spatial patterns of community composition, pollution gradients and pesticide impact, disturbances, and recolonization of sediments.

Experimental investigations of the use of seagrass beds by harpacticoid copepods have been conducted in Pauahatanui Inlet (Hicks 1984, 1986, 1988, 1989; Bell et al. 1988; Bell and Hicks 1991). Hicks (1984) found extremely high densities of a harpacticoid copepod in troughs of persistent sediment ripples, while observations by the authors (R. Cole, unpublished) found the distributions of molluscs and echinoderms to be influenced by sediment channels on Centre Bank, Tauranga Harbor.

The subtidal regions of New Zealand estuaries have been little described. The distributions of several intertidal bivalves have been shown to extend subtidally, for example, the estuarine clam, or New Zealand pipi, *Paphies australis* (Grace 1966 1970), the cockle *Austrovenus stutchburyi* (Larcombe 1973), and the wedge shell,

Table 6.1 Summary of the nature, extent, and duration of major biological studies conducted in New Zealand estuaries 1965–1994

Study (Chronological Order)	ID/Fauna	Quantitative	Experimental	Spatial extent (m^2) or site distance apart (km)	Temporal extent (months)	Gradient
Grace 1966	✓	✗	✗	~1	–	✓
Estcourt 1967	✓	✗	✗	~80	24	✗
Grace 1967	✓	✗	✗	~1	–	✓
McKnight 1968	✓	✗	✗	–	–	✗
McKoy 1970	✓	✓	✗	<3	–	✗
Webb 1972	✓	✓	✗	~4	12	✗
Webb 1973 a, b, c, d	✓	✓	✗	~4	12	✗
Fincham 1974	✓	✓		1 site	1 day, 1, 12	✗
Knight 1974	✓	✓	✗	7 km	24	✗
Luckens 1974	✓	✗	✓	2 sites, 100 km	2	✗
Grange 1977	✓	✓	✗	50 km	1	✗
Stanton et al. 1977	✓	✗	✗	3 km	3	✗
Grange 1979	✓	✓	✗	50 km	1	✗
Henriques 1980	✓	✓	✗	55 km	<36	✗
Gillespie and Mackenzie 1981	✓	✓	✗	~3 km	–	✗
Hubbard and Partridge 1981	✓	✗	✗	~14 km	276	✗
Juniper 1981	✗	✗	✗	–	2	✗
Rainer 1981	✓	✓	✗	19 km	18	✓
McClatchie et al. 1982	✓	✗	✓	4 m^2	1	✗
Woodroffe 1982	✗	✓	✗	1 site	12	✗
Lee and Partridge 1983	✓	✓	✗	30 km	108	✗
Read 1984	✓	✓	✗	~200 m	24	✓
Richmond et al. 1984	✓	✗	✗	–	–	✓
Hicks 1986	✓	✓	✓	~200 m	1 day, 6	✗
Roper 1986	✓	✓	✗	Single site	12	✗
Partridge 1987	✓	✗	✗	~1500 km	–	✗
Partridge and Wilson 1987a	✗	✗	✓	Lab	1–5	✗
Partridge and Wilson 1987b	✗	✗	✓	Lab	8	✗
Bell et al. 1988	✓	✓	✗	1000s of km	60	✗
Hicks 1988	✓	✓	✗	100s of m	6	✗
Partridge and Wilson 1988a	✓	✓	✗	200 km	–	✓
Partridge and Wilson 1988b	✓	✗	✓	25 km	7	✓
Roper et al. 1988	✓	✓	✗	~18 km	–	✓
Thrush 1988	✓	✓	✓	150 m^2	1	✗
Davidson 1989	✓	✓	✗	~500 m	9	✗
Dobbinson et al. 1989	✓	✓	✓	~1 km	9	✓
Hicks 1989	✓	✓	✓	~1 km	1 day, 1	✗
Thrush et al. 1989	✓	✓	✗	9000m^2	1	✗
Pridmore et al. 1990a	✓	✓	✗	~18 km	1 day	✓
Bell and Hicks 1991	✓	✓	✓	~1 km	3–5 days	✗
Pridmore et al. 1991	✓	✓	✓	300m^2	3	✗
Thrush et al. 1991	✓	✓	✓	800m^2	0.5	✗
Hicks 1992	✓	✓	✗	Single site	1 day	✗
Thrush et al. 1992	✓	✓	✓	36m^2	1	✗
Thrush et al. 1994a	✓	✓	✓	10 800m^2	6, 14	✗
Thrush et al. 1994b	✓	✓	✗	9000m^2	48*	✗

* At the time their MS was written – duration now approaching 7 years

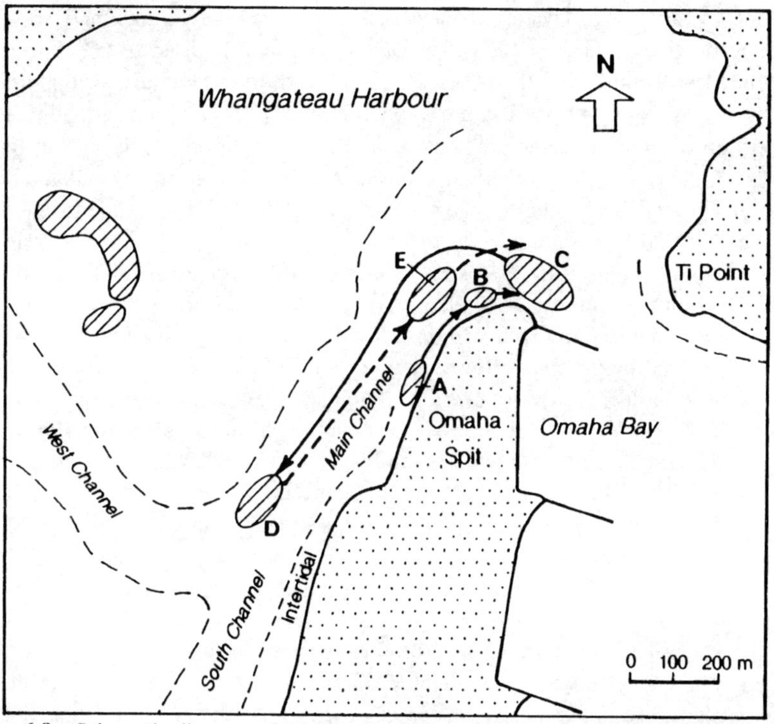

Figure 6.8 Schematic diagram of the hypothesized movement of pipi (*Paphies australis*) at the entrance to Whangateau Harbor, northeastern New Zealand (after Hooker 1995). A: The main recruitment site at mid-tide level; B: the area of high-density juvenile pipi at low tide mark; C: shallow subtidal area of high-density juvenile and adult pipi; D: subtidal area at the far extent of the pipi bed, also with a bimodal population size structure; E: exclusively adult pipi in deepest area of main channel. The subtidal bed is continuous from D to C; dark arrows indicate movements of pipi due to mucus drifting and the dotted arrows indicate movements by other mechanisms, such as passive movement with sediments, and attaching to debris. The dashed line indicates the edge of the channel, hatching indicates land

Macomona liliana (Grange 1979). *Paphies australis* (Grace 1970) is thought to drift between juvenile intertidal populations and dense subtidal populations (Figure 6.8: Hooker 1995). Hooker (1995) found that pipi survived when transplanted to non-estuarine sites.

Many studies have attempted to relate the fauna of areas to the sediment composition, but we could find no experimental investigations. Cassie and Michael (1968) analyzed the intertidal fauna and sediments from Karore Bank, Manukau Harbor, and identified a bivalve component of the fauna that could be associated with coarse sediments, and a crab- and polychaete-dominated assemblage associated with fine sediments. An early example of quantitative investigation of sediment–organism associations was by McKoy (1970), who analyzed sediments from four sites in Wellington Harbor, and suggested that filter feeders, deposit feeders and carnivores occurred most in sandy muds, silty muds and very sand muds respectively. Knight (1974) found two assemblages in Lyttelton Harbor, one associated with sand, and the other with mud; a third assemblage of opportunists occurred regardless of substratum type. Grange (1977) demonstrated that the proportion of deposit feeders increased, and the proportion of suspension feeders decreased, with decreasing grain size on intertidal shores in the Manukau Harbor. He noted that similar patterns had been found globally. However, Pridmore and colleagues (1990b) suggested that the relationships between sediment grain size and assemblage composition were not consistent in their studies, either in space or time. A subtidal survey was also undertaken by Grange (1979), and four assemblages identified: (1) a *Microcosmus/Notomithrax* assemblage, associated with coarse sediments in shallow water; (2) a *Halicarcinus/Bugula* assemblage, associated with deeper (>7 m) areas in channels, again with coarse sediments; (3) an *Amalda/Myadora* assemblage associated with shallow areas of medium to fine sand; and (4) a *Fellaster/Pagurus* assemblage associated with fine sand. Grange (1979) noted that Powell's (1937) study identified assemblages very similar to those in his study. In the same harbor Henriques (1980) suggested that mangroves and eelgrass increased the diversity (number of species) of benthic sediments, whereas *Gracilaria* had little effect on the fauna. In distinction, Rainer (1981) identified subtidal assemblages within Otago Harbor, and suggested that broken mollusc shell was important in determining the presence of macroalgae. He also related high species diversity with stable fine sand with shell, and lowest diversity with unstable fine sand and unconsolidated silts. It is clear that detailed experiments are needed to resolve whether sediments control the distribution of organisms in New Zealand estuaries.

McDowall (1976) noted that estuaries served as paths for migration for some marine and numerous freshwater species. Some planktonic organisms may remain in estuaries because of a combination of vertical migration and two-layered current flows (e.g. Fortier and Leggett 1983; Cummings et al. 1995). Roper (1986) suggested that the current patterns in the small Whangateau Estuary, north of Auckland, were such that larval vertical migration would not lead to organisms being retained within the estuary; hydrology might be more suitable for this to occur in larger estuaries. Ritchie (1976) argued that mangroves provided important nursery areas for other organisms. Morrison (1988) identified estuaries as recruitment sites for parore *Girella tricuspidata.*

6.4.3 Temporal variability

Most published studies of estuarine shoreline ecology to date have been of 2 years duration or less (Table 6.1). Many focus on "seasonal" variability, but few extend into multi-year comparisons. Longer term studies are necessary to identify seasonality, as distinct from simple temporal change.

Deely (1992) gives an account of major changes in the conspicuous flora such as eelgrass *Zostera* spp., and introduced cordgrasses *Spartina* spp. for the Avon-Heathcote estuary. Mason and Ritchie (1979) report a general decline in the amount of *Zostera* present in Whangarei Harbor, and observation suggests that it diminished greatly between 1966 and 1971. Knox (1983) – though mentioning *Zostera* beds in two of three categories of "vital areas" – does not describe its abundance in the Waitemata Harbor. Larcombe (1973) suggested that *Zostera* in the Waitemata had declined greatly due to disease. The area covered by *Zostera* in the Avon-Heathcote estuary declined between 1910 and the 1950s, but more recently increased again (Knox and Kilner 1973). The latter study also mentions increases in the green algae *Ulva lactuca* and *Enteromorpha ramulosa*, and documents the expansion of *Spartina townsendii* within the estuary over a ten-year period. Outbreaks of *Ulva lactuca* populations have also been recorded from Tauranga Harbor (Environment BoP 1993) and Ohiwa Harbor (Daniel 1984). Daniel (1984) and Crisp and co-authors (1990) summarize changes in the distribution of mangroves in several Bay of Plenty estuaries, noting that a small mangal vanished from Opotiki Harbor between 1945 and 1966, while in Ohiwa Harbor and the Waiaua Estuary there had been expansions of mangal since the 1940s. Mangal loss from Opotiki was suggested to be due to sedimentation.

The NIWA Manukau Harbor study is the only long-term investigation of estuarine organisms in New Zealand that the authors are aware of. It involves continued monitoring of six 9000 m^2 sites at 2-month intervals (Thrush et al. 1994b). The polychaete *Heteromastus filiformis* has undergone major population declines at one site in February and March, in three of four years of sampling. Recruitment of the bivalve *Macomona liliana* was found to be highly variable among years. Thrush and colleagues (1994a) used this database to simulate the effects of differing sampling intensities on the capacity to detect pattern; their conclusion was that low sample numbers could be highly misleading.

6.4.4 Hard shores

Morton and Miller (1973) present descriptions of the reef fauna in the Waitemata and Kaipara Harbors, emphasizing the importance of sediment and nutrient inputs from land. Batham (1956, 1965, 1969) and McKnight (1968) described rocky shore ecology in Fiordland and on Stewart Island. A series of studies has examined the black coral fauna of the reefs of Fiordland, dealing with taxonomy and growth (Grange et al. 1981; Grange 1985; Grange and Singleton 1988). Estuarine reefs may be sites of recruitment for fish of open coasts (Morrison 1988; Gillanders and Kingsford 1993). Estuarine aquaculture structures and wharf piles acting as artificial reefs, are colonized by a number of animals, including the 11-armed starfish *Coscinasterias calamaria*, which is a predator of aquaculture species (Poore 1968). Forrest (1991) suggested that oyster racks increased sedimentation locally. Reefs in Tauranga Harbor are usually occupied by many small triplefin and labrid fishes; groups of the pinnid bivalve *Atrina zelandica* may act as reefs there also. Sediments may cover and uncover reefs with changes in harbor morphology and this probably contributes to instability in their faunas.

6.4.5 Faunal interactions on estuarine shores

6.4.5.1 Herbivory

Estuarine soft shores in New Zealand have few macrophytes. *Zostera* is the only seagrass taxon in New Zealand. It evidently forms beds that provide a nursery for many organisms, and this phenomenon has been most intensively studied in Pauatahanul Inlet (Hicks 1984, 1986, 1988, 1989; Bell et al. 1988; Bell and Hicks 1991). Observations suggest that these beds may also affect water flow, and *Zostera* beds are often characterized by sediment mounds, presumably accumulated there due to changes in flow rate. A recent investigation of herbivores of *Ulva lactuca* (Gregor 1995) in Tauranga Harbor suggested that there were few herbivores that could reduce the volume of *Ulva* during the intermittent problem blooms. The major factors controlling the distribution of *Ulva* were thought to be nutrient supply, with winds, waves, and current patterns combining to transport it on to the beach. Dobbinson and colleagues (1989) found that cockles *Austrovenus stutchburyi* grew faster, were in better condition, and had reduced mortality low on the shore compared to high on the shore in Otago Harbor. The experiments necessary to discriminate between the potential explanations of (1) reduced food supply due to decreased submersion time, and (2) food depletion from organisms downshore have not been carried out.

6.4.5.2 Predation

The role of predation on the ecological communities in the shallow sub-, intertidal and beach zones of New Zealand estuaries has been best studied in Manukau Harbor. Thrush and co-workers (1991) found that feeding by eagle rays *Myliobatuis tenuicaudatus* turned over approximately 10 m^2 of sandflat per day, and had similar effects on assemblages dominated by polychaetes and shellfish. A subsequent study (Thrush et al. 1994b) suggested that bird predation had an important effect on the abundance of the wedge shell *Macomona liliana* after 6 months, and that elevated densities of *M. liliana* occurred in ray and bird exclusions after 14 months. Other abundant predators such as whelks, crabs and starfish have unknown effects. Dense populations of the starfish *Coscinasterias calamaria* occur in many harbors, such as Whangateau and Tauranga, frequently associated with beds of pipis, *Paphies australis* (Grace 1970; Cook 1989; R. Cole, unpublished data). The predator and prey together may act as biological reefs, to which small labrid and triplefin fishes again recruit. Few manipulative experiments have been done on the effects of predators in estuaries and to some extent this may be because the cages necessary to exclude grazers and predators are likely to produce artifacts of hydrodynamic boundary layer effects (Eckman 1979, 1987; Hulberg and Oliver 1980). The duration of impacts on the infauna likely reflects the local hydrological and sedimentary regimes. For example, Thrush and colleagues (1991) found that pits formed by ray feeding rapidly infilled again with sediment, while Thrush and co-workers (1992) found that adult infauna facilitated recolonization of defaunated sediments.

6.5 INTERRELATIONS OF HYDRODYNAMICS, SEDIMENTS AND ECOLOGY OF NEW ZEALAND SHALLOW SUB-TIDAL AND ESTUARINE SHORES

6.5.1 Estuarine beach form and sediments

A typical beach profile for the common type of sandy estuarine lagoon is given in Figure 6.9. The estuarine beaches typically slope in a ramp form at about 6° to intersect with a broad intertidal flat (Figure 6.10). The beach may be surmounted by low accumulations of sand, sometimes incipient foredunes. Intertidal flats extend out to a channel, and may frequently become colonized with *Zostera* beds. Sediments for the North Island estuarine lagoons are composed of a mixture of quartz, feldspar, titanomagnitite heavy mineral, and biogenic detrital shells, the latter normally comprising the gravel component. Substantial amounts of terrigenous gravels are rare in most northern sandy Holocene barrier-enclosed estuaries. For the central and northeastern North Island estuaries, pumice and volcanic glass are significant components of the sediments, deriving from the active acid volcanic hinterland.

Texturally the beach sediments are typically composed of moderately sorted medium-to-fine sands, but often with shelly gravel. On the intertidal flats the texture tends to finer sands, but with poorer sorting, reflecting a small percentage of silts, as well as significant biogenic components of shelly gravel.

Eroding estuarine beaches tend to possess a narrower beach face, with a much higher component of shelly gravel as a "lag" deposit, as well as surficial concentrations of titanomagnetite heavy mineral. The intertidal flat becomes very shell–gravel lagged, particularly from the predominant estuarine bivalves such as cockles *Austrovenus stutchburyi*, wedge shell *Macomona liliana* and pipi *Paphies australis*.

Observation shows that small-scale but significant beach and dune erosion is typical for sectors of most shallow estuarine lagoon shorelines (Figure 6.10), and eroding estuarine lagoon shorelines are known in Ohiwa, Tauranga, Tairua, Whangapoua, Whangateau, Mangawhai, Manukau and Kaipara harbors, to name but a few. The process is also recognized overseas (Jackson and Nordstrom 1992).

The primary mechanism responsible for the erosion process is the action of short "choppy waves" generated within the estuary by winds. In New Zealand, the

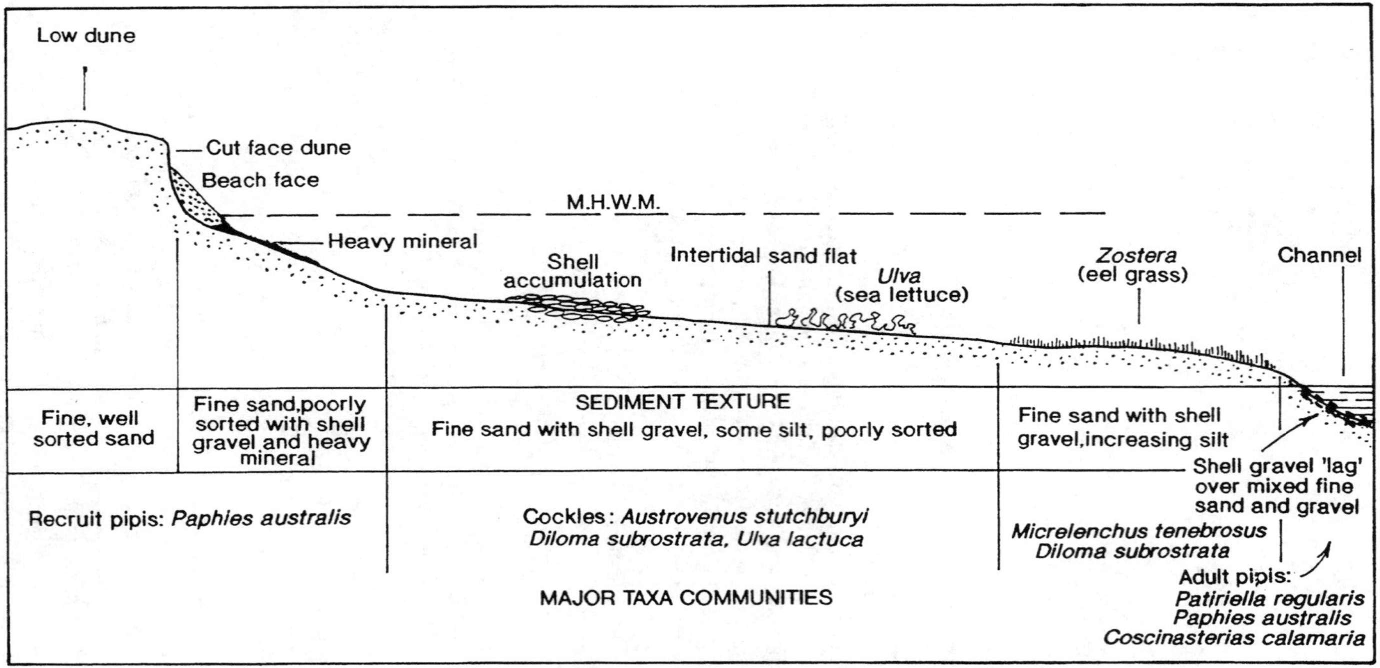

Figure 6.9 Typical beach profile, sediment textures and ecological associations for estuarine lagoons

Figure 6.10 Shoreline erosion on predominant wind-facing side (west) of northeast New Zealand estuarine lagoons. Note accumulations of heavy mineral in the upper beach, shell lag and *Zostera* beds further out on the intertidal flat

prevailing wind is from the west, and the west-facing shorelines are those that tend to be the most susceptible to erosion. The erosion of within-estuary beach and dune is facilitated at high tides, and by wind stress-induced surge occurring within the estuary. Typically for northeast New Zealand, the west-facing shoreline is the landward face of the barrier spit (Figure 6.10) and frequently such areas are under subdivision development. Rates of erosion may average $0.5\,\mathrm{m\,a^{-1}}$ from wave chop erosion. Examples of New Zealand eroding beaches in an estuarine lagoon were investigated by de Lange (1988), de Lange and Healy (1990b) and Healy (1994). For the latter case, from available survey data it was ascertained that the sediment budget of such beaches is about $0.4\,\mathrm{m^3\,m^{-1}}$ loss per year for a beach facing the predominant westerly winds with a fetch of 1–3 km across an estuarine lagoon (Figure 6.10).

6.5.2 Hydrodynamics

Relatively few studies exist on the hydrodynamics – currents and waves – impacting on estuarine shorelines. The only substantial New Zealand shallow subtidal and estuarine beach process hydrodynamic investigations are by Black (1983) and de Lange (1988). Black's work was carried out as part of the Whangarei Harbor study, while de Lange's (1988) study followed the intensive Tauranga Harbor

Study of 1983–84. This latter involved numerical sediment transport modeling which demonstrated that sediment transport in the estuarine lagoon occurred predominantly on the tidal deltas and in the deeper tidal channels. Tidal current measurements were also taken (Black 1984a; Healy 1985; Barnett 1985; de Lange and Healy 1990b), and at the several sites over the shallow and intertidal flats current speeds rarely exceed the threshold value for medium sand (0.3 m s-1). However, field observations suggest that an appreciable amount of sediment may also be transported over the intertidal flats where currents were too weak to entrain sediment. Accordingly an investigation was undertaken to define the wave climate applicable to shallow estuarine lagoons, and to ascertain whether wave-current action on the extensive intertidal flats would contribute significantly to the total sediment transport within the harbor. For this study a capacitance type wave probe and data logger were deployed. Distributions of the wave heights and periods are given in Figure 11a from de Lange and Healy (1990a). It was found that the longer period (>3 s) waves can be approximated by a Rayleigh distribution, but not the shorter period waves. The resulting average wave spectrum displays two main peaks (Figure 6.11b). The first, accounting for about 70% of the wave energy, represents external wave energy filtered through the harbor entrance, and about 10% of the spectral energy was attributed to short wind-generated waves of H = 0.15–0.3 m and T = 2–3 s. Generally such waves are capable of moving medium sand down to a depth of 1.2 m, which for a harbor such as Tauranga with a tidal range of ~2 m involves about 70% of the estuarine lagoon area (Figure 6.12).

6.5.3 Assemblages of estuarine lagoon beaches and intertidal flats

Here we focus on northern estuaries; the more southern ones may have similar bivalve faunas, but the vegetation is different (Deely 1992). As for the open coasts, the beach sector is relatively sparsely inhabited. The major benthic organisms may be recruit pipis *Paphies australis*, but crabs, gastropods and adult bivalves seldom occur there. Morton and Miller (1973) do not identify any isopods as being characteristic of the upper reaches of harbor shores, though they do occur there (R. Creese, Leigh Marine Laboratory, personal communication).

The intertidal flats carry a much richer flora and fauna: the flora may include macrophytes such as *Zostera* and *Ulva* (Figure 6.6), while the fauna comprises bivalves such as pipi *Paphies australis*, cockles *Austrovenus stutchburyi, Mactra ovata*, oysters *Crassostrea gigas* and *Tiostrea chilensis*, mussels *Perna canaliculus*, wedge shell *Macomona liliana, Nucula hartvigiana*, gastropods such as *Cominella* spp., *Maoricolpus, Micrelenchus* and *Amphibola*, crustaceans such as *Helice crassa, Pontophilus australis*, and squillids. Echinoderms such as sand dollars *Fellaster zelandica* and heart urchins *Echinocardium cordatum* may also occur toward the lower shore reaches. Feeding scars may mark visits by benthic-feeding fish such as *Pagrus auratus* and *Myliobatis tenuicaudatus* during high tide.

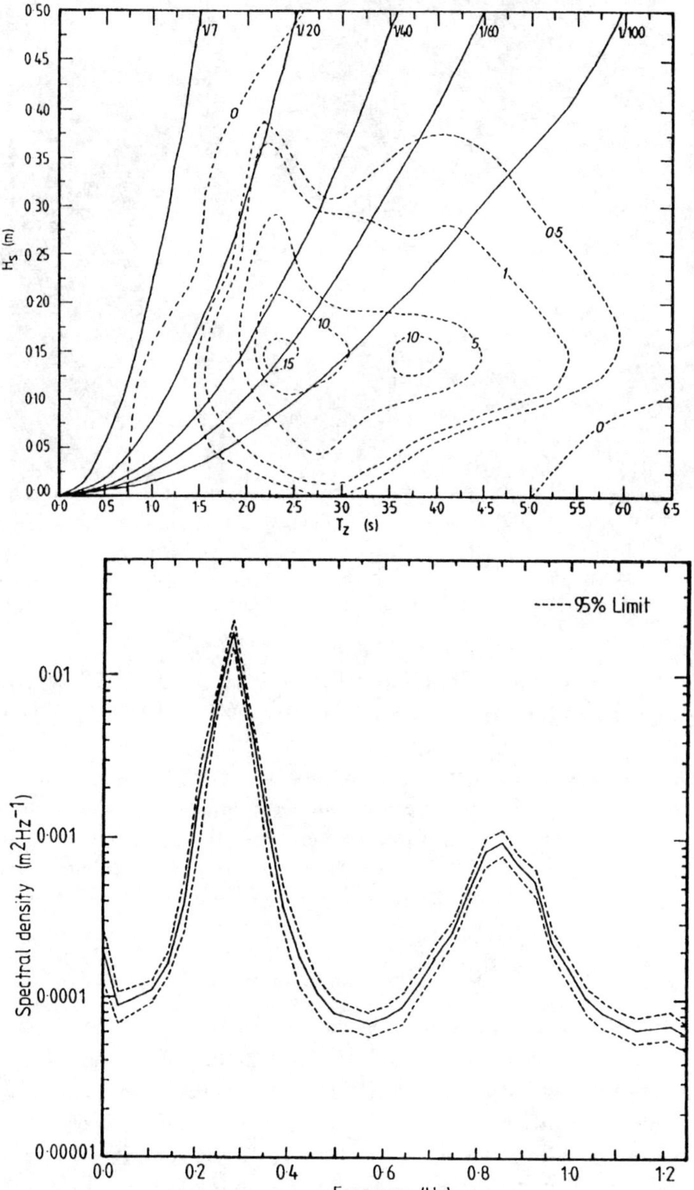

Figure 6.11(a) Distribution of significant wave heights (H_s) and periods (T_z) expressed as percentages of 35 000 waves measured by capacitance wave probe near the shoreline in 11 water depth, Tauranga Harbor. The dashed lines are percentage time contours, and the solid lines represent constant wave steepness derived from linear theory (after de Lange 1988). (b) Average wave spectrum for Pilot Bay, Tauranga Harbor, derived from 265 wave spectra (after de Lange 1988)

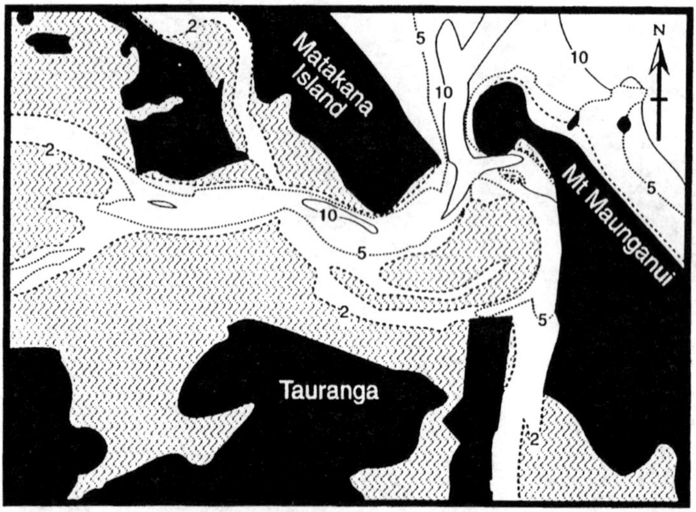

Figure 6.12 Southern compartment of Tauranga Harbor. The wavy area represents the shallow subtidal and intertidal flats and shoals influenced by wave stirring of the bottom sediments, based on the data presented in Figures 6.11a,b. The depth of wave stirring coincides neatly with the break in slope between the intertidal flats and the tidal channels

6.6 EXAMPLES OF HUMAN-INDUCED IMPACTS ON THE SHORELINE SEDIMENTATION AND ECOLOGY OF ESTUARINE LAGOONS

Examples of human impacts on estuarine shoreline abound. Two examples, which have been subjected to intensive study, are reported below. One relates to the changes consequent upon diversion of a significant river flow out of the estuarine lagoon, depriving it of much of its freshwater input; the other relates to maintenance of a beach downdrift from a port development.

6.6.1 Impact of Kaituna River diversion from Maketu Estuarine Lagoon

Prior to 1958 the Kaituna River flowed into the Maketu estuarine lagoon (Figure 6.13) enclosed by a long Holocene barrier spit comprising a single frontal dune. Owing to flooding problems in the catchment the engineering decision was made to divert the river out of the estuary. At the time of river diversion the estuary contained considerable areas of marshy wetland in a brackish environment, but after diversion, and with efforts to develop agriculture on the privately owned marshes, virtually all (95%) of the wetland disappeared as the estuary became predominantly saline. Major morphological changes occurred at the inlet entrance, as might be expected (Murray 1978; Burton 1987; KRTA 1986). The flood tidal delta enlarged massively for the size of the inlet, which has become unstable in location and since diversion has lost navigability for small craft except at high tide.

Figure 6.13 Diversion of the Kaituna River (foreground) out of the Maketu estuary. Since diversion the wetland area within the estuary declined drastically and the inlet entrance became unstable. There are now (1995) moves to restore the river partially to the estuary

Analysis of several tidal gaugings suggests that between 1972 and 1985 intertidal prism had been reduced as much as 34%. Over the past decades the reduction in tidal prism has evolved by gradual infilling of the estuarine lagoon by incursion of marine sands through the entrance and by episodic overtopping and breaching of the spit during storm events (Rutherford et al. 1989).

In recent years the issue of whether the river should be rediverted back into the estuary has been discussed, mainly on grounds of cultural preferences of the Maori people – considerations required under New Zealand's Resource Management Act (1991) – but also from the point of view of re-establishing the wetland. The difficulty with rediversion is that the river water is now of such poor quality in terms of BOD and coliform loads, that it would violate the existing higher water quality classification presently pertaining to the estuarine waters.

6.6.2 Maintenance of a beach downdrift from a port

Port operations and developments may affect adjacent beaches within the port or surrounding estuary. Examples of such impacts in New Zealand include Mission Bay beach at Auckland (Priestley and Consedine 1991), Pilot Bay beach at Tauranga (de Lange 1988; de Lange and Healy 1985; 1990b), Ngamotu beach at New Plymouth (Rogers 1995), and Oriental Bay beach at Wellington (Healy et al. 1990). The most extensively studied case is Pilot Bay beach in Tauranga (Figure 6.14).

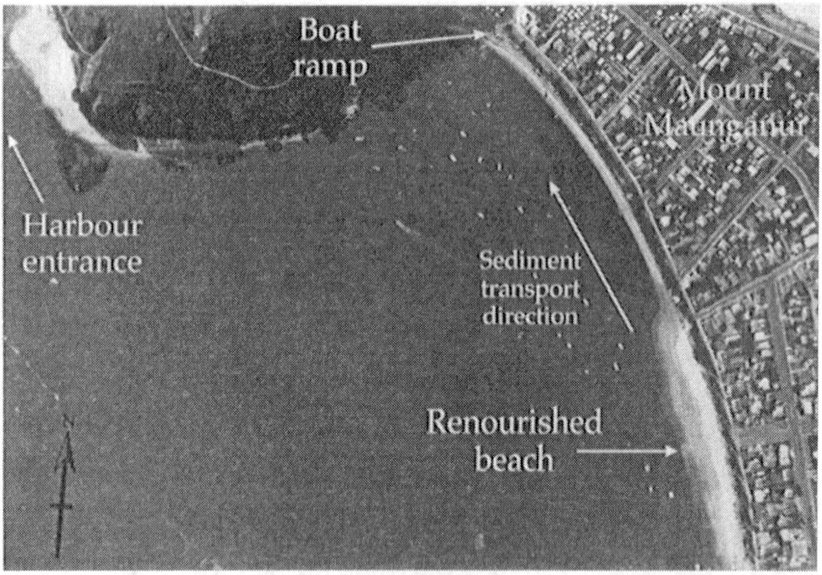

Figure 6.14 Aerial view of Pilot Bay beach shortly after initial renourishment in 1984. The Port of Tauranga is located immediately south of the jetty in the lower right of the photo

Tidal flows within Pilot Bay are strongly ebb-dominated in terms of both flow duration and velocity. This dominance is partly due to the formation of a strong eddy within Tauranga Harbor during flood tide (de Lange and Healy 1990b). Numerical modeling of historical harbor bathymetries indicated that the ebb flow dominance has been a feature of the Pilot Bay for at least 150 years (Black 1984a). Sediment flux measurements indicate that the net transport direction is northwestward, along the Mount Maunganui tombolo towards the harbor entrance.

Since the Second World War the Port of Tauranga has progressively developed the region southeast of Pilot Bay beach. This development has included dredging of shipping channels and construction of wharves and ship-handling facilities. The effect of the developments was to interrupt the supply of littoral sediment to Pilot Bay beach, which began to erode. In 1984, maintenance dredging of an adjacent shipping channel provided sediment suitable for beach renourishment and 21 000 m^3 was emplaced at the south-eastern end of the beach (de Lange 1988; de Lange and Healy 1990b).

Monitoring of the renourished beach demonstrated a continued northwestward sediment flux. It was predicted then that the renourishment material would be completely removed within 13 years (i.e. by 1997), and that renourishment would need to be repeated after 7 years if a reasonable width of beach was to be retained at high tide (de Lange 1988).

Since the completion of the investigation, a prolonged 3.5 year El Niño phase resulted in a higher than normal incidence of south-westerly winds that increased the erosion rate at Pilot Bay beach. However, the renourished beach survived for the minimum 7-year period predicted. The beach was eventually renourished with 10 000 m^3 of sediment in 1993. Further renourishment at 7–13-year intervals is considered necessary to maintain a suitable beach. However, the movement of renourished material and subsequent accretion at the north-western end of the beach caused problems with boat launching facilities. The accumulated material was removed from the boat ramp weekly and transported to another beach for disposal. Eventually in 1995 a system was initiated to recycle $\sim$1,000 m^3 per annum back to the southeastern end of the beach using trucks.

6.7 THREATS TO SHALLOW ESTUARINE SHORELINE ECOLOGY

The most obvious destructive impacts on New Zealand estuaries result from direct constructions such as reclamations and highways smothering the estuarine ecology, or dredging, which may alter the estuarine hydrodynamics (Healy et al. 1988, 1991). More subtle, but likely more pervasive impacts result from changes to the estuarine circulation caused by automobiles and developments and increases in turbidity from catchment earthworks, which induce major changes to the physiognomies of estuaries (Clark 1993).

Direct impacts to the fauna and flora of New Zealand estuarine shorelines have included effects from nearby subdivisions and developments, including siltation, infilling, dredge spoil dumping, and reclamation and highway construction (Creese and Cole 1995). Especially after European colonization and pioneer bush clearing for farming, the catchment erosion regimes were accelerated leading to injection into the estuaries of massive volumes of fine muddy sediment. Under normal estuarine flocculation processes some of this material became deposited over the intertidal sands, producing a covering of mud deposition. A good example of this is Ohiwa Harbor, a barrier-enclosed estuarine lagoon in the Bay of Plenty, which has a very small catchment area and no significant river input. The muddy deposition in the various estuarine valley arms has come from runoff from the adjacent steep hillsides consequent upon clearing of the bush. This change from sandy to muddy intertidal conditions could lead to significant ecological community re-organization from sandy intertidal dominated by the wedge shell *Macomona liliana*, to the present muddy intertidal community dominated by the oval trough shell *Mactra ovata*, mangrove snail *Amphibola crenata*, and the snapping shrimp *Alpheus* sp.

A second example comes from Whangarei Harbor where Millar (1980), from mineralogical analysis, traced the redistribution and relative concentration of cement factory fine sediment washings throughout the harbor embayments, and into the port of Whanagarei, where the port authority had to dredge it from the shipping channel. As a result, the cement factory changed its manufacturing procedures and no longer uses a wet washing system to purify the cement, so the water quality and clarity has improved considerably within the past decade, although no quantitative comparison of the intertidal communities has been undertaken.

Similar impacts have been suffered in the Waitemata Harbor from developments associated with the growth of the city of Auckland (Hume 1983), and in the Manukau Harbor from industrial pollution (Grange 1989) and dredging and spoil dumping (Healy et al. 1989; Ryan 1990).

Impacts from human activities lead to alterations within the estuarine environment which lead to faunal migrations, alterations to the environment which render persistence of populations unlikely by reducing recruitment, and events which directly kill existing populations (e.g. *Amphibola crenata, Helice crassa* – Maxwell 1985, 1987). In the latter study, populations of both species returned to their former densities rapidly, but population dynamics of *A. crenata* were altered; mangroves *Avicennia marina* were little damaged. Several studies have focussed on the effects of human actions in estuaries. Pridmore and colleagues (1991) suggested that juveniles (0.5–2.5 mm) of both *Chione stutchburyi* and *Macomona liliana* drifted in response to application of pesticide. Organisms higher up the food chain could be affected by the interactions of heavy metals and organisms, which have been the subject of several studies (Smith and McVeagh 1991; Stewart et al. 1992). The widespread occurrence of imposex in gastropods is indicative of human impacts; inevitably these are likely to occur in sheltered areas subject to high human use, such as estuaries.

6.8 FUTURE TASKS FOR NEW ZEALAND SHALLOW ESTUARINE RESEARCH

Promising ecological work has already been done on mobility of bivalves and population dynamics in shallow subtidal and intertidal estuarine shores (e.g. Hooker 1995). Information on ways of assessing pollution impacts (Regtien 1992; Hickey 1992) will provide knowledge which is directly applicable to environmental problems. Macroalgae, such as *Ulva lactuca* (Environment BoP 1993) and toxic microalgal blooms (Baldwin 1992), indicate that our understanding of estuarine marine plants and nutrient input from nearby land areas needs to be extended.

Investigation of physical oceanographic properties and hydrodynamics – currents, flushing times, flow dispersion properties, waves, turbidity properties, sediment transport and deposition, and infilling rates of New Zealand estuaries of all types – will be an ongoing future requirement, because here as in most other parts of the world, the estuaries bear the brunt of coastal developments. In particular numerical hydrodynamic and sedimentation model applications will be required for both general and specific applications. Likewise much enhanced understanding of "wave chop" erosion mechanisms and impacts of boat wakes is needed, along with the physics of storm surge heights, which control the building platform levels. Flushing time and physical oceanographic properties are needed for aquaculture (Schiel and Nelson 1990) and for understanding of phenomena such as toxic algal blooms and marina management.

Pridmore and co-authors (1990b, p.94) commented "it is striking that details of the ecology of individual species are lacking and ecological processes controlling such communities are only very poorly understood" with respect to the Manukau Harbour. While the NIWA investigations since then have certainly gone some way to redressing our ignorance, the remainder of our intertidal shores remain largely undocumented. Organisms of estuaries, the hydrology of their environment, and the sediment characteristics comprise a triad of variables that interact with one another in complex ways and require experimental investigations (Underwood and Denley 1984) building on the established and recent work.

ACKNOWLEDGEMENTS

The authors express their sincere thanks to Cynthia Vey for masterly typing of the text across multiple word processing systems. The paper was completed while the senior author was on a Fulbright Fellowship at the Department of Coastal and Oceanographic Engineering, University of Florida, and we gratefully acknowledge Dr Bob Dean for making facilities available there. We thank the independent reviewers, Dr Ashish Mehta, Department of Coastal and Oceanographic Engineering University of Florida, and Dr Bob Creese, Leigh Marine Laboratory University of Auckland for reviewing the manuscript and making helpful suggestions. The views expressed remain those of the authors.

REFERENCES

Adams, N.M. 1983. Checklist of marine algae possibly naturalised in New Zealand. *New Zealand Journal of Botany* 21: 1–12.

Baldwin, R.P. 1992. Cargo vessel ballast water as a vector for the spread of toxic phytoplankton species to New Zealand. *Journal of the Royal Society of New Zealand* 22: 229–242.

Barnett, A.G. 1985. Overview and hydrodynamics. Tauranga Harbor Study Part I & III. A report for the Bay of Plenty Harbor Board.

Batham, E.J. 1956. Ecology of southern New Zealand sheltered rocky shore. *Transactions of the Royal Society of New Zealand* 84: 447–465.

Batham, E.J. 1965. Rocky shore ecology of a southern New Zealand fiord. *Transactions of the Royal Society of New Zealand, Zoology* 6: 215–227.

Batham, E.J. 1969. Benthic ecology of Glory Cove, Stewart Island. *Transactions of the Royal Society of New Zealand, Biological Sciences* 11: 73–81.

Bell, S.S. and G.R.F. Hicks. 1991. Marine landscapes and faunal recruitment: a field test with seagrasses and copepods. *Marine Ecology Progress Series* 73: 61–68.

Bell, S.S., G.R.F. Hicks and K. Walters. 1988. Active swimming in meiobenthic copepods of seagrass beds: geographic comparisons of abundance and reproductive characteristics. *Marine Biology* 98: 351–358.

Black, K.P. 1983. Sediment transport and tidal inlet hydraulics. D.Phil. thesis. Department of Earth Sciences, University of Waikato.

Black, K.P. 1984a. Sediment transport. Tauranga Harbor study part IV. A report for the Bay of Plenty Harbor Board.

Black, K.P. 1984b. Sediment transport, Tauranga Harbour Bridge. A report to the Bay of Plenty Harbour Board.

Black, K.P. and T.R. Healy. 1982. Sediment transport investigations in a New Zealand tidal inlet. *Proceedings 18th Coastal Engineering Conference, ASCE*, pp. 2436–57.

Black, K.P. and T.R. Healy. 1986. The sediment threshold over tidally-induced megaripples. *Marine Geology* 69: 219–234.

Black, K.P. and T.R. Healy. 1988. Formation of ripple bands in a wave convergence zone. *Journal of Sedimentary Petrology* 58: 195–207.

Black, K.P., T.R. Healy and M.G. Hunter. 1989. Sediment dynamics in the lower section of a mixed sand and shell-lagged tidal estuary. *Journal of Coastal Research* 5: 503–521.

Bradstock, M.C. 1989. *Between the Tides: New Zealand Shore and Estuary Life.* David Bateman, Auckland.

Burns, N.M., T.M. Hume, D.S. Roper and R.K. Smith. 1990. Estuaries, p. 81–84. *In*: *Climate Change: Impacts on New Zealand.* Ministry for the Environment, Wellington.

Burton, J.H. 1987. Tidal inlet hydraulics and stability of Maketu Inlet, Bay of Plenty. M.Sc. thesis. Department of Earth Sciences, University of Waikato.

Burton, J.H. and T.R. Healy. 1985. Tidal hydraulics and stability of the Maketu Inlet, Bay of Plenty. *Proceedings 7th Australasian Conference on Coastal and Ocean Engineering*, Christchurch, 139–150.

Cameron, W.M. and D.W. Pritchard. 1963. Estuaries, p. 306–324. *In* M.N. Hill (ed.), *The Sea, 2.* John Wiley, New York.

Cassie, R.M. and A.D. Michael. 1968. Fauna and sediments of an intertidal mud flat: a multivariate analysis. *Journal of Experimental Marine Biology and Ecology* 2: 1–23.

Chappel, J. 1975. Upper quaternary warping and uplift rates in the Bay of Plenty and west coast, North Island, New Zealand. *New Zealand Journal of Geology and Geophysics* 18: 129–155.

Clark, R.B. 1993. *Marine Pollution.* Oxford University Press, Oxford.

Cometti, R.N. and J.E. Morton. 1985. *Margins of the Sea: Exploring New Zealand's Coastline.* Hodder and Stoughton, Auckland.

Cook, S. de C. 1989. Distribution, abundance, and feeding biology of the asteroid *Coscinasterias calamaria.* M.Sc. thesis. Department of Zoology, University of Auckland.

Cotton, C.A. 1942. *Geomorphology.* Whitcomb & Tombs, Wellington.

Creese, R. and R. Cole. 1995. Marine conservation in New Zealand. *Pacific Conservation Biology* 2(1): 55–63.

Crisp, P.A., L. Daniel and P. Tortell. 1990. *Mangroves in New Zealand: Trees in the Tide.* GP Books, Wellington.

Cummings, V.J., R.D. Pridmore, S.F. Thrush and J.E. Hewitt. 1993. Emergence and floating behaviours of post-settlement juveniles of *Macomona liliana* (Bivalvia: Tellinacea). *Marine Behaviour and Physiology* 24: 25–32.

Cummings, V.J., R.D. Pridmore, S.F. Thrush and J.E. Hewitt. 1995. Post-settlement by intertidal benthic macroinvertebrates: do common New Zealand species drift in the water column? *New Zealand Journal of Marine and Freshwater Research* 29: 59–67.

Daniel, L.J. 1984. Mangroves and salt marshes of Ohiwa Harbor. Department of Lands and Survey, Gisborne.

Davidson, R.J. 1989. The bottom fauna from three subtidal locations around Banks

Peninsula, Canterbury, New Zealand. *New Zealand Natural Sciences* 16: 87–95.

Davidson, R.J. 1990. *A Report on the Ecology of Whanganui Inlet, North-west Nelson.* Occasional Publication No. 2, Department of Conservation Nelson/Marlborough Conservancy.

Davidson, R.J. and C.R. Moffat. 1991. *A Report on the Ecology of the Waimea Inlet, Nelson.* Occasional Publication No. 1, Department of Conservation, Nelson/Marlborough Conservancy.

Davies-Colley, R.J. 1976. Sediment dynamics of Tauranga Harbour and the Tauranga inlet. M.Sc. thesis. Department of Earth Sciences, University of Waikato.

Davies-Colley, R.J. and T.R. Healy. 1978. Sediment and hydrodynamics of the Tauranga entrance to Tauranga Harbor. *New Zealand Journal of Marine and Freshwater Research* 12: 225–236.

Davis, R.A. and T.R. Healy. 1992. Holocene coastal depositional sequences on a tectonically active plate margin: southeastern Tauranga Harbor, New Zealand. *Sedimentary Geology* 84: 1–13.

Deely, J. 1992. The last 150 years – the effect of urbanisation, p. 107–121. *In* S.-J. Owen (ed.), *The Estuary – Where Our Rivers Meet the Sea.* Christchurch City Council, Christchurch.

de Lange, W.P. 1988. Wave climate and sediment transport within Tauranga Harbor, in the vicinity of Pilot Bay. D.Phil. thesis. Department of Earth Sciences, University of Waikato.

de Lange, W.P. and T.R. Healy. 1985. Beach nourishment of Pilot Bay Beach, Tauranga Harbor, New Zealand. *Proceedings of 7th Australasian Conference on Coastal and Ocean Engineering*, Christchurch, New Zealand, Vol 1., 501–508.

de Lange, W.P. and T.R. Healy. 1990a. Wave spectra for a shallow meso-tidal estuarine lagoon, Bay of Plenty, New Zealand. *Journal of Coastal Research* 6: 189–199.

de Lange, W.P. and T.R. Healy. 1990b. Renourishment of a flood-tidal delta adjacent beach, Tauranga Harbor, New Zealand. *Journal of Coastal Research* 6: 627–640.

Dickie, B.N. 1984a. Soft shore investigations. Whangarei Harbor Study Technical Report 4. Whangarei Harbor Board, Whangarei.

Dickie, B.N. 1984b. Wading birds: high tide roost surveys. Whangarei Harbor Study Technical Report 5. Whangarei Harbor Board, Whangarei.

Dickie, B.N. 1984c. Rocky shore investigations: Part I. Site descriptions: physical and biological. Whangarei Harbor Study Technical Report 7. Whangarei Harbor Board, Whangarei.

Dinamani, P. 1971. Occurrence of the Japanese oyster, *Crassostrea gigas* (Thunberg) in Northland, New Zealand. *New Zealand Journal of Marine and Freshwater Research* 5: 352–357.

Dinamani, P. 1981. Description of a new species of incubatory oyster from northern New Zealand, with notes on its ecology and reproduction. *New Zealand Journal of Marine and Freshwater Research* 15: 109–119.

Dobbinson, S.J., M.F. Barker and J.B. Jillett. 1989. Experimental shore level transplantation of the New Zealand cockle *Chione stutchburyi. Journal of Shellfish Research* 8: 197–212.

Dromgoole, F.I. and B.A. Foster. 1983. Changes to the marine biota of the Auckland Harbor. *Tane*, 29: 79–96.

Dyer, K.R. 1973. *Estuaries: A Physical Introduction.* John Wiley, London.

Eckman, J.E. 1979. Small-scale patterns and processes in a soft-substratum, intertidal community. *Journal of Marine Research* 37: 437–457.

Eckman, J.E. 1987. Hydrodynamic processes affecting recruitment. *Limnology and Oceanography* 28: 241–257.

Environment Bay of Plenty. 1993. Bay of Plenty Regional Council Natural Environment

Regional Monitoring Network Coastal and Estuarine Ecology Programme 1991/92. Environment BoP Environmental Report No. 42, Whakatane.

Estcourt, I.N. 1967. Distributions and associations of benthic invertebrates in a sheltered water soft-bottom environment (Marlborough Sounds, New Zealand). *New Zealand Journal of Marine and Freshwater Research* 1: 352–370.

Estcourt, I.N. 1975. *Bibliography of Scientific Studies of New Zealand Mainland Estuaries, Inlets, Lagoons, Harbors and Fiords.* Miscellaneous Publication 75, New Zealand Oceanographic Institute, Wellington.

Fincham, A.A. 1974. Periodic swimming behaviour of amphipods in Wellington Harbour. *New Zealand Journal of Marine and Freshwater Research* 8: 505–521.

Forrest, B.M. 1991. Oyster farm impacts on the benthic environment: a study in Mahurangi Harbor. M.Sc. thesis. Department of Zoology, University of Auckland.

Fortier, L. and W.C. Leggett. 1983. Vertical migrations and transport of larval fish in a partially mixed estuary. *Canadian Journal of Fisheries and Aquatic Sciences* 40: 1543–1555.

Gillanders, B.M. and M.J. Kingsford. 1993. Abundance patterns of *Achoerodus viridis* (Pisces: Labridae) on estuarine and exposed rocky reefs: possible linkages, p. 93–98. *In* C.N. Battershill et al. (eds.), *Proceedings of the Second International Temperate Reef Symposium*, 7–10 January 1992, Auckland, New Zealand. NIWA Marine, Wellington.

Gillespie, P.A and A.L. Mackenzie. 1981. Autotrophic and heterotrophic processes on an intertidal mud-sand flat, Delaware Inlet, Nelson, New Zealand. *Bulletin of Marine Science* 31: 648–657.

Grace, R.V. 1966. The bottom communities of the entrance to the Whangateau Harbor. *Tane* 12: 63–70.

Grace, R.V. 1967. An underwater survey of two starfish species in the entrance to the Whangateau Harbor. *Tane* 13: 13–19.

Grace, R.V. 1970. The benthic ecology of the entrance to the Whangateau Harbor, Northland, New Zealand. Ph.D. thesis. Department of Zoology, University of Auckland.

Grange, K.R. 1977. Littoral benthos-sediment relationships in Manukau Harbor, New Zealand. *New Zealand Journal of Marine and Freshwater Research* 11: 111–123.

Grange, K.R. 1979. Soft-bottom macrobenthic communities of Manukau Harbor, New Zealand. *New Zealand Journal of Marine and Freshwater Research* 13: 315–329.

Grange, K.R. 1985. Distribution, standing crop, population structure, and growth rates of black coral in the southern fjords of New Zealand. *New Zealand Journal of Marine and Freshwater Research* 19: 467–475.

Grange, K.R. 1989. *Monitoring of Littoral Macrobenthos and Surface Sediments in Manukau Harbor New Zealand.* Miscellaneous Publication 102, New Zealand Oceanographic Institute, Wellington.

Grange, K.R. and R.J. Singleton. 1988. Population structure of black coral, *Antipathes aperta*, in the southern fjords of New Zealand. *New Zealand Journal of Zoology* 15: 481–489.

Grange, K.R., R.J. Singleton, J.R. Richardson, P.J. Hill and W. de L. Main. 1981. Shallow rock-wall biological associations of some southern fiords of New Zealand. *New Zealand Journal of Zoology* 8: 209–227.

Gregor, K.E. 1995. Grazers of estuarine *Ulva* in Tauranga Harbor. M.Sc. thesis. Department of Zoology, University of Auckland.

Gregory, M.R. and S.A. Thompson. 1973. *Recent Sediments of Waitemata Harbor.* University of Auckland, Auckland.

Greig, M.J. and R. Proctor. 1988. A numerical model of the Hauraki Gulf, New Zealand. *New Zealand Journal of Marine and Freshwater Research* 22: 379–390.

Hay, C.H. and P.A. Luckens. 1987. The Asian kelp *Undaria pinnatifida* (Phaeophyta:

Laminariales) found in a New Zealand harbor. *New Zealand Journal of Botany* 25: 329–333.

Hayward, B.W. and C.J. Hollis. 1993. Ecology of Waimamaku River Estuary, north of Kawerua, North Auckland. *Tane* 34: 69–78.

Healy, T.R. 1985. Field data collection programme and morphological study. Tauranga Harbor Study Parts II and V. A report for the Bay of Plenty Harbor Board.

Healy, T.R. 1994. Harbour shoreline erosion fronting Whareroa Marae; causes and possible remedies. Report to the Tauranga District Council, University of Waikato.

Healy, T.R. and K. Harada. 1991. Definition and physical characteristics of the world's enclosed coastal seas. *Marine Pollution Bulletin* 23: 639–644.

Healy, T.R. and R.M. Kirk. 1992. Coasts, p. 161–183. *In* J.M. Soons and M.J. Selby (eds.), *Landforms of New Zealand*. Longman Paul, Auckland.

Healy, T.R., K.P. Black and W. de Lange. 1987. Field investigations required for numerical model studies of port developments in large tidal inlet harbors, p. 1099–1112. *In* V. Gardiner (ed.), *International Geomorphology*, John Wiley, New York.

Healy, T.R., B. McCabe, R. Grace and C. Harms. 1988. Environmental assessment programme for Tauranga Harbor's dredging and inner shelf spoil dumping. A report for the Bay of Plenty Harbor Board.

Healy, T.R. S. Ryan and C. Harms. 1989. Recent environmental issues concerning dredge spoil dumping, p. 98–101. *In Proceedings of the Ecopolitics III Conference*, University of Waikato, Hamilton, September 1988.

Healy, T.R., R.M. Kirk and W.P. de Lange. 1990. Beach renourishment in New Zealand, p. 77–90. *In* M.L. Schwartz and E.C.F. Bird (eds.), *The Geomorphology of Artificially Renourished Beaches, Journal of Coastal Research Special Issue* 6.

Healy, T.R., B. McCabe and G. Thompson. 1991. Port of Tauranga Ltd environmental impact assessment: channel deepening and widening dredging programme 1991–1992, May 1991.

Healy, W.B. 1980. *Pauatahanui Inlet – An Environmental Study*. New Zealand Department of Scientific and Industrial Research, New Zealand Department of Scientific and Industrial Research Information Series 141, Wellington.

Heath, R.A. 1977. Phase distribution of tidal constituents around New Zealand. *New Zealand Journal of Marine and Freshwater Research* 11: 383–392.

Henriques, P.R. 1980. Faunal community structure of eight soft shore, intertidal habitats in the Manukau Harbour. *New Zealand Journal of Ecology* 3: 97–103.

Hewitt, J.E., G.B. McBride, R.D. Pridmore and S.F. Thrush. 1993. Patchy distributions: optimising sample size. *Environmental Monitoring and Assessment* 27: 95–105.

Hickey, B.A. 1992. *Perna canaliculus* as a biomonitor of water quality: metal accumulation and health indicators. M.Sc. thesis. Department of Zoology, University of Auckland.

Hicks, G.R.F. 1984. Spatio-temporal dynamics of a meiobenthic copepod and the impact of predation-disturbance. *Journal of Experimental Marine Biology and Ecology* 81: 47–72.

Hicks, G.R.F. 1986. Distribution and behaviour of meiofaunal copepods inside and outside seagrass beds. *Marine Ecology Progress Series* 31: 159–170.

Hicks, G.R.F. 1988. Sediment rafting: a novel mechanism for the small-scale dispersal of intertidal estuarine meiofauna. *Marine Ecology Progress Series* 48: 69–80.

Hicks, G.R.F. 1989. Does epibenthic structure negatively affect meiofauna? *Journal of Experimental Marine Biology and Ecology* 133: 39–55.

Hicks, G.R.F. 1992. Tidal and diel in abundance of meiobenthic copepods on an intertidal estuarine sandbank. *Marine Ecology Progress Series* 87: 15–21.

Hooker, S.H. 1995. Biology of the pipi, *Paphies australis*, in the Whangateau Harbor. Ph.D. thesis. Department of Zoology, University of Auckland.

Hubbard, J.C.E. and T.R. Partridge. 1981. Tidal immersion and the growth of *Spartina anglica* marshes in the Waihopai River Estuary, New Zealand. *New Zealand Journal of Botany* 19: 115–121.

Hulberg, L.W. and J.S. Oliver. 1980. Caging manipulations in marine soft-bottom communities: importance of animal interactions or sedimentary habitat modifications. *Canadian Journal of Fisheries and Aquatic Sciences* 37: 1130–1139.

Hume, T.M. 1983. Upper Waitemata Harbour sediments and the inferred impacts of future catchment and estuary use change. Upper Waitemata Harbour Catchment Study. Auckland Regional Authority, NWASCA, Auckland.

Hume, T.M. and J. Dahm. 1991. An investigation of the effects of Polynesian and European land use on sedimentation in Coromandel estuaries. The effect of logging and mining on sediments in Coromandel estuaries. Water Quality Centre Consultancy Report 7106, May 1991.

Hume, T.M. and J.G. Gibb. 1987. The "wooden-floor" marker bed – a new method of determining historical sedimentation rates in some New Zealand estuaries. *Journal of the Royal Society of New Zealand* 17: 1–7.

Hume, T.M. and C.E. Herdendorf. 1988. A geomorphic classification of estuaries and its application to coastal resource management – a New Zealand example. *Ocean and Shoreline Management* 11: 249–274.

Hume, T.M. and C.E. Herdendorf. 1990. Morphologic and hydrologic characteristics of tidal inlets on a headland dominated, low drift coast, northeastern New Zealand. *Journal of Coastal Research, Special Issue* 9: 527–563.

Hume, T.M. and C.E. Herdendorf. 1992. Factors controlling tidal inlet characteristics on low drift coasts. *Journal of Coastal Research* 8: 355–375.

Hume, T.M., M.E. Fox and R.J. Wilcock. 1989. Use of organochlorine contaminants to measure sedimentation rates in estuaries: a case study from the Manukau Harbour, New Zealand. *Journal of the Royal Society of New Zealand* 19: 305–317.

Hume, T.M, R.G. Bell, W.P. de Lange, T.R. Healy, D.M. Hicks and R.M. Kirk. 1992. Coastal oceanography and sedimentology in New Zealand, 1967–91. *New Zealand Journal of Marine and Freshwater Research* 26: 1–36.

Jackson, N.L. and K.F. Nordstrom. 1992. Site specific controls on wind and wave processes and beach mobility on estuarine beaches in New Jersey, USA. *Journal of Coastal Research* 8: 88–98.

Johnson, D.W. 1919. *Shore Processes and Shoreline Development*. Hafner, New York.

Juniper, S.K. 1981. Stimulation of bacterial activity by a deposit feeder in two New Zealand inlets. *Bulletin of Marine Science* 31: 691–701.

KRTA (Kingston, Reynolds, Thom, and Allardice Consultants). 1986. Maketu Estuary study stage 1. Report to the Bay of Plenty Catchment Commission.

Kirk, R.M. 1975. Aspects of surf and runup processes on mixed sand and gravel beach. *Geografiska Annaler* 57A: 117–133.

Knight, G.S. 1974. Benthic community structure in Lyttelton Harbor. *New Zealand Journal of Marine and Freshwater Research* 8: 291–306.

Knox, G.A. 1983. *Upper Waitemata Harbor Catchment Study. Estuarine Ecology*. Auckland Regional Authority, Auckland.

Knox, G.A. and A.R. Kilner. 1973. The ecology of the Avon-Heathcote Estuary. Unpublished report to the Christchurch Drainage Board.

Larcombe, M.F. 1971. Distribution and recognition of intertidal organisms in the Whangateau Harbor, and a classification for sheltered soft shores. Ph.D. thesis. Department of Zoology, University of Auckland.

Larcombe, M.B. 1973. Waitemata Harbor Study. Ecology. Unpublished report to Auckland Harbor Board and Auckland Regional Authority.

Lee, W.G. and T.R. Partridge. 1983. Rates of spread of *Spartina anglica* and sediment

accretion in the New River Estuary, Invercargill, New Zealand. *New Zealand Journal of Botany* 21: 231–236.
Luckens, P.A. 1974. Removal of intertidal algae by herbivores in experimental frames and on shores near Auckland. *New Zealand Journal of Marine and Freshwater Research* 8: 637–654.
Marks, G. 1975. Sedimentoloty of Omaro barrier spit, Whangapoua Harbour, Coromandel Peninsula. M.Sc. thesis. Department of Earth Sciences, University of Waikato.
Mason, R.S. and L.D. Ritchie. 1979. Aspects of the ecology of Whangarei Harbor. Report for Northland Harbor Board and Ministry of Agriculture and Fisheries.
Mathews, R.S. 1982. Physical and chemical characteristics of the Waitara Estuary and coastal embayment with particular reference to effluent disposal. M.Sc. thesis. Department of Earth Sciences, University of Waikato.
Maxwell, G.S. 1985. The chemical flush into Auckland's Tamaki Estuary, following a major fire, on 21 December 1984, at ICI, Mt. Wellington site: An ecological assessment. Unpublished report to ICI.
Maxwell, G.S. 1987. The ecological recovery of the Tamaki Estuary: a report on the post-ICI fire ecological monitoring programme, April 1985 to November 1986. Unpublished report to ICI.
May, J.D. 1984. Mangrove investigations. Whangarei Harbor Study Technical Report 3.
McArdle, B.H. and R.G. Blackwell. 1989. Measurement of density variability in the bivalve *Chione stutchburyi. Marine Ecology Progress Series* 52: 245–252.
McClatchie, S., S.K. Juniper and G.A. Knox. 1982. Structure of a mudflat diatom community in the Avon-Heathcote Estuary, New Zealand. *New Zealand Journal of Marine and Freshwater Research* 16: 299–309.
McDowall, R.M. 1976. The role of estuaries in the life cycles of fishes in New Zealand. *Proceedings of the New Zealand Ecological Society* 23: 27–32.
McKnight, D.G. 1968. Features of the benthic ecology of Chalky and Preservation Inlets. *New Zealand Journal of Marine and Freshwater Research* 2: 716–720.
McKoy, J.L. 1970. Animal-sediment relationships in Wellington Harbor. *Victoria University of Wellington Bulletin of Natural Sciences* 1: 5–20.
McLay, C.L. 1976. An inventory of the status and origin of New Zealand estuarine systems. *Proceedings of the New Zealand Ecological Society* 23: 8–26.
Millar, A.S. 1980. Hydrology and surficial sediments of Whangarei Harbor. M.Sc. thesis. Department of Earth Sciences, University of Waikato.
Morrison, M.A. 1988. Ontogenetic shifts in the ecology of the parore, *Girella tricuspidata*. M.Sc. thesis. Department of Zoology, University of Auckland.
Morton, J.E. and M.C. Miller. 1973. *The New Zealand Sea Shore, 2nd Edition*. Collins, Auckland.
Murray, K.N. 1978. Ecology and geomorphology of Maketu Estuary, Bay of Plenty. M.Sc. thesis. Department of Earth Sciences, University of Waikato.
Naish, T. 1990. Late Holocene mud deposition and diagenesis in the Firth of Thames. M.Sc. thesis. Department of Earth Sciences, University of Waikato.
Owen, S.J. (ed.) 1992. *The Estuary – Where Our Rivers Meet the Sea*. Christchurch City Council, Christchurch.
Partridge, T.R. 1987. *Spartina* in New Zealand. *New Zealand Journal of Botany* 25: 567–575.
Partridge, T.R. and J.B. Wilson. 1987a. Germination in relation to salinity in some plants of salt marshes in Otago, New Zealand. *New Zealand Journal of Botany* 25: 255–261.
Partridge, T.R. and J.B. Wilson. 1987b. Salt tolerance of salt marsh plants of Otago, New Zealand. *New Zealand Journal of Botany* 25: 559–566.
Partridge, T.R. and J.B. Wilson. 1988a. Vegetation patterns in salt marshes of Otago, New

Zealand. *New Zealand Journal of Botany* 26: 497–510.
Partridge, T.R. and J.B. Wilson. 1988b. The use of field transplants in determining environmental tolerance in salt marshes of Otago. *New Zealand Journal of Botany* 26: 183–192.
Pethick, J. 1984. *An Introduction to Coastal Geomorphology*. Arnold, London.
Pickrill, R.A. 1977. Coastal processes, beach morphology and sediments along the northeast coast of the South Island, New Zealand. *New Zealand Journal of Geology and Geophysics* 20: 1–15.
Pickrill, R.A. 1987. Circulation and sedimentation of suspended particulate matter in New Zealand fjords. *Marine Geology* 74: 21–39.
Pillans, B. 1986. A late quaternary uplift map of North Island, New Zealand. *Royal Society of New Zealand Bulletin* 24: 409–417.
Poore, G.C.B. 1968. Succession of a wharf-pile fauna at Lyttelton, New Zealand. *New Zealand Journal of Marine and Freshwater Research* 2: 577–590.
Powell, A.W.B. 1937. Animal communities of the sea bottom in Auckland and Manukau Harbors. *Transactions of the Royal Society of New Zealand* 66: 354–401.
Pridmore, R.D., D.S. Roper, and J.E. Hewitt. 1990a. Variation in composition and condition of the Pacific oyster, *Crassostrea gigas*, along a pollution gradient in Manukau Harbor, New Zealand. *Marine Environmental Research* 30: 163–177.
Pridmore, R.D., S.F. Thrush, J.E. Hewitt and D.S. Roper. 1990b. Macrobenthic community composition of six intertidal sandflats in Manukau Harbor, New Zealand. *New Zealand Journal of Marine and Freshwater Research* 24: 81–96.
Pridmore, R.D., S.F. Thrush, R.J. Wilcock, T.J. Smith, J.E. Hewitt and V.J. Cummings. 1991. Effect of the organochlorine pesticide technical chlordane on the population structure of suspension and deposit feeding bivalves. *Marine Ecology Progress Series* 76: 261–271.
Priestley, S.J. and P.T. Consedine. 1991. Mission Bay Beach replenishment, p. 499–504. *In*: R.G. Bell, T.M. Hume and T.R. Healy (eds.), *Coastal Engineering – Climate for Change: Proceedings of 10th Australasian Conference on Coastal and Ocean Engineering*, Auckland, New Zealand. Water Quality Centre, Publication 21.
Pritchard, D.W. 1967. What is an estuary: physical viewpoint, p. 3–6. *In* G.H. Lauff (ed.), *Estuaries*. American Association for the Advancement of Science Publication 93.
Proctor, R. and M.J.N. Greig. 1989. A numerical model investigation of the residual circulation in Hauraki Gulf, New Zealand. *New Zealand Journal of Marine and Freshwater Research* 23: 421–442.
Putnam, G. 1977. A brief history of New Zealand marine biology. *Tuatara* 22: 189–212.
Rainer, S.F. 1981. Soft-bottom communities in Otago Harbour and Blueskin Bay, New Zealand. *New Zealand Oceanographic Institute Memoir* 80.
Read, G.B. 1984. Persistence of infaunal polychaete zonation patterns on a sheltered, intertidal sand flat. *New Zealand Journal of Marine and Freshwater Research* 18: 399–416.
Regtien, M.H. 1992. Sewage effluent impacts in the Waiwera Estuary: an evaluation of the abundance:biomass comparison (ABC) method. M.Sc. thesis. Department of Zoology, University of Auckland.
Richmond, B. 1977. Geomorphology and modern sediments of Ohiwa Harbour, Bay of Plenty, New Zealand. M.Sc. thesis. Department of Earth Sciences, University of Waikato.
Richmond, B., C.S. Nelson and T.R. Healy. 1984. Sedimentology and evolution of Ohiwa Harbor, a tidally dominated estuary in the Bay of Plenty. *New Zealand Journal of Marine and Freshwater Research* 18: 461–478.
Ritchie, L.D. 1976. Fish and fisheries. Unpublished paper presented to the Nature Conservation Council/Friends of the Earth Symposium, "Why are mangroves

important?". Whangarei, 1976.
Rogers, D.M. 1995. Water quality, surficial sediment quality, and hydrodynamics within the Lee Breakwater Basin, Port Taranaki. M.Sc. thesis. Department of Earth Sciences, University of Waikato.
Roper, D.S. 1986. Occurrence and recruitment of fish larvae in a northern New Zealand estuary. *Estuarine Coastal and Shelf Science* 22: 705–717.
Roper, D.S., S.F. Thrush and D.G. Smith. 1988. The influence of runoff on intertidal mudflat benthic communities. *Marine Environmental Research* 26: 1–18.
Roper, D.S., R.D. Pridmore, V.J. Cummings and J.E. Hewitt. 1991. Pollution related differences in the condition cycles of Pacific oysters *Crassostrea gigas* from Manukau Harbor, New Zealand. *Marine Environmental Research* 31: 197–214.
Roper, D.S., R.D. Pridmore and S.F. Thrush. 1992. Recruitment to the macrobenthos of *Macomona liliana* (Bivalvia: Tellinidae) in Manukau Harbor, New Zealand. *New Zealand Journal of Marine and Freshwater Research* 26: 385–392.
Rutherford, J.C., R.K. Smith, A.B. Cooper and J. Dahm. 1989. Maketu Estuary restoration report. Water Quality Centre Report No. 71312/2.
Ryan, S.G. 1990. Investigation of the sedimentological impact of dredge spoil in the Purakau Channel, Manukau Harbour. M.Sc. thesis. Department of Earth Sciences, University of Waikato.
Schiel, D.R. 1991. Coastal biology and the New Zealand Journal of Marine and Freshwater Research. *New Zealand Journal of Marine and Freshwater Research* 25: 415–427.
Schiel, D.R. and W.A. Nelson. 1990. The harvesting of macroalgae in New Zealand. *Hydrobiologia* 204/205: 25–33.
Sheffield, A., T.R. Healy and M.S. McGlone. 1995. Infilling rates of a steepland catchment estuary, Whangamata, New Zealand. *Journal of Coastal Research* 11: 1294–1308.
Sherwood, A. 1973. Surficial sediments of Raglan Harbor. M.Phil. thesis. Department of Earth Sciences, University of Waikato.
Smith, P.J. and M. McVeagh. 1991. Widespread organotin pollution in New Zealand coastal waters as indicated by imposex in dogwhelks. *Marine Pollution Bulletin* 22: 409–413.
Stanton, D.I., B.B. Bohool and C. Beasley. 1977. Intertidal zone of Delaware Inlet, Nelson, New Zealand. *New Zealand Journal of Marine and Freshwater Research* 11: 577–587.
Stephenson, G. 1986. *Wetlands – Discovering New Zealand's Shy Places.* Government Printing Office Publishing, Wellington.
Stewart, C., S.J. de Mora, M.R.L. Jones and M.C. Miller. 1992. Imposex in New Zealand neogastropods. *Marine Pollution Bulletin* 24: 204–209.
Thrush, S.F. 1988. The comparison of macrobenthic colonization patterns near and away from crab burrows on a sublittoral sand flat. *Journal of Marine Research* 46: 669–681.
Thrush, S.F. and D.S. Roper. 1988. Merits of macrofaunal colonization of intertidal mudflats for pollution monitoring: preliminary study. *Journal of Experimental Marine Biology and Ecology* 116: 219–233.
Thrush, S.F., J.E. Hewitt and R.D. Pridmore. 1989. Patterns in the spatial arrangement of polychaetes and bivalves in intertidal sandflats. *Marine Biology* 102: 529–536.
Thrush, S.F., R.D. Pridmore, J.E. Hewitt and V.J. Cummings. 1991. Impact of ray feeding disturbances on sandflat macrobenthos: do communities dominated by polychaetes or shellfish respond differently? *Marine Ecology Progress Series* 69: 245–252.
Thrush, S.F., R.D. Pridmore, J.E. Hewitt and V.J. Cummings. 1992. Adult infauna as facilitators of colonization on intertidal sandflats. *Journal of Experimental Marine Biology and Ecology* 159: 253–265.
Thrush, S.F., R.D. Pridmore and J.E. Hewitt. 1994a. Impacts on soft-sediment macrofauna: the effects of spatial variation on temporal trends. *Ecological Applications* 4: 31–41.
Thrush, S.F., R.D. Pridmore, J.E. Hewitt and V.J. Cummings. 1994b. The importance of

predators on a sandflat: interplay between seasonal changes in prey densities and predator effects. *Marine Ecology Progress Series* 107: 211–222.

Underwood, A.J. and E.J. Denley. 1984. Paradigms, explanations and generalizations in models for the structure of intertidal communities on rocky shores, p.151–180. *In* D.B. Strong, D. Simberloff, L.G. Abele and A.B. Thistle (eds.) *Ecological Communities: Conceptual Issues and the Evidence.* Princeton University Press, New Jersey.

Van Leeuwe, C.M. 1991. Transport of suspended sediments in the southern Firth of Thames. M.Phil. thesis. Department of Earth Sciences, University of Waikato.

Venus, G.C. 1984. *Paphies australis* (pipis) in Whangarei Harbor. Whangarei Harbor Study Technical Report 6.

Walsby, J. 1990. *Nature Watching at the Beach.* Wilson and Horton, Auckland.

Webb, B.F. 1972. Fish populations of the Avon-Heathcote Estuary. 1. General ecology, distribution, and length frequency. *New Zealand Journal of Marine and Freshwater Research* 6: 570–601.

Webb, B.F. 1973a. Fish populations of the Avon-Heathcote Estuary. 2. Breeding and gonad maturity. *New Zealand Journal of Marine and Freshwater Research* 7: 45–66.

Webb, B.F. 1973b. Fish populations of the Avon-Heathcote Estuary. 3. Gut contents. *New Zealand Journal of Marine and Freshwater Research* 7: 223–234.

Webb, B.F. 1973c. Fish populations of the Avon-Heathcote Estuary. 4. Parasites. *New Zealand Journal of Marine and Freshwater Research* 7: 301–305.

Webb, B.F. 1973d. Fish populations of the Avon-Heathcote Estuary. 5. Records of less common species. *New Zealand Journal of Marine and Freshwater Research* 7: 307–321.

White, J.L. 1979. Recent sediments of Waikareo Estuary, Tauranga, North Island, New Zealand. M.Sc. thesis. Department of Earth Sciences, University of Waikato.

Wood, I.R., R.G. Bell and D.L. Wilkinson. 1994. *Ocean Disposal of Waste Water.* World Scientific, New Jersey.

Woodroffe, C.D. 1982. Litter production and decomposition in the New Zealand mangrove *Avicennia marina* var. *resinifera. New Zealand Journal of Marine and Freshwater Research* 16: 179–188.

Part 2

ENVIRONMENTS AND PROCESSES

The linkage between estuarine environments is examined in a process-response context in the chapters in this part of the book, starting with the relationship between hydrodynamics and geomorphology (Chapters 7 to 9), adding the role of vegetation in geomorphic change, erosion and accretion (Chapters 10 and 11), examining nutrient exchanges (Chapter 12) and then introducing faunal interactions (Chapters 13 to 15). The effect of changes in tidal channels noted in Part 1 is examined in a case study framework by van den Berg et al. (Chapter 7), who identify the relationships between channels, tidal wave assymetry, sediment transport and erosion and deposition. Pethick (Chapter 8) discusses the linkages farther up the cross-shore profile, concentrating on the cyclic morphologic changes on muddy estuarine shores, showing how location in the upper and lower reaches of an estuary affect wave characteristics and morphologic changes. Nordstrom et al. (Chapter 9) identify morphologic changes associated with waves acting on a sand beach, identifying how sediment transfers and landform changes are spatially restricted in estuarine beach environments. Stevenson and Kearny (Chapter 10) elaborate on differences in shoreline environments due to different locations in an estuary (introduced in earlier chapters), and they discuss how shoreline location affects vulnerability to sea level rise on marsh and bluff shorelines. Fonseca (Chapter 11) demonstrates the bi-directional linkage between wave processes, sedimentation and vegetation, underscoring the need for interdisciplinary and inter-environmental research. Subsequent chapters examine factors with high mobility that interact across several estuarine environments, often within a relatively short time frame. Howes et al. (Chapter 12) identify the many pathways that nutrients take in their flow across estuarine shores. Their examination of subsurface exchanges complements the many chapters that examine channel flows. The importance of channel characteristics in relation to adjacent marshes is highlighted by McIvor and Rozas (Chapter 13) in their examination of nekton use of intertidal salt marsh habitat. Their review complements the case study approach used by Able et al. (Chapter 14) who also note the significance of seasonal and life cycle differences and identify a time-interrupted linkage between habitats due to usage of different habitats at different life-history stages. Vargas (Chapter 15) changes the locational focus to the tidal flat in a tropical environment to document linkages to adjacent habitats associated with faunal feeding activities.

Chapter Seven

Hydraulic processes affecting the morphology and evolution of the Westerschelde estuary

JAN H. VAN DEN BERG and CLAIRE J. L. JEUKEN
Institute of Marine and Atmospheric Research, Utrecht

and

AD J. F. VAN DER SPEK
Geological Survey of The Netherlands

7.1 INTRODUCTION

The Westerschelde is part of a system of coastal plain estuaries and tidal basins in the southwestern Netherlands (Figure 7.1). It is a young estuary, that started to evolve only a thousand years ago (Vos and Van Heeringen 1993). At present the Westerschelde is a well mixed estuary, with well developed channels and shoals covering an area of about 300 km^3 (Figure 7.2). The mean amplitude of the semidiurnal tide increases from 3.8 m at the entrance of the estuary to 5.2 m at Antwerpen (Figure 7.3). The total flood discharge of a tidal cycle (flood volume), which amounts to an average of $1100 \times 10^6 m^3$ at Vlissingen, reduces to $70 \times 10^6 m^3$ at Antwerpen. The fresh-water discharge of the Schelde river averages between 50 and 60 $m^3 s^{-1}$ from April to October and 160–180 $m^{3 s-1}$ from December to February. The average chlorinity of the water decreases from 18 $kg m^{-3}$ at Vlissingen to about 3 $kg m^{-3}$ at Antwerpen. At most locations the bed material of the Westerschelde consists of well sorted medium- to fine-grained sand. In deep channels the grainsize is generally coarser than on the shoals. In natural channels and shoal areas admixtures of clay are mostly less

Estuarine Shores: Evolution, Environments and Human Alterations. Edited by K. F. Nordstrom and C. T. Roman.

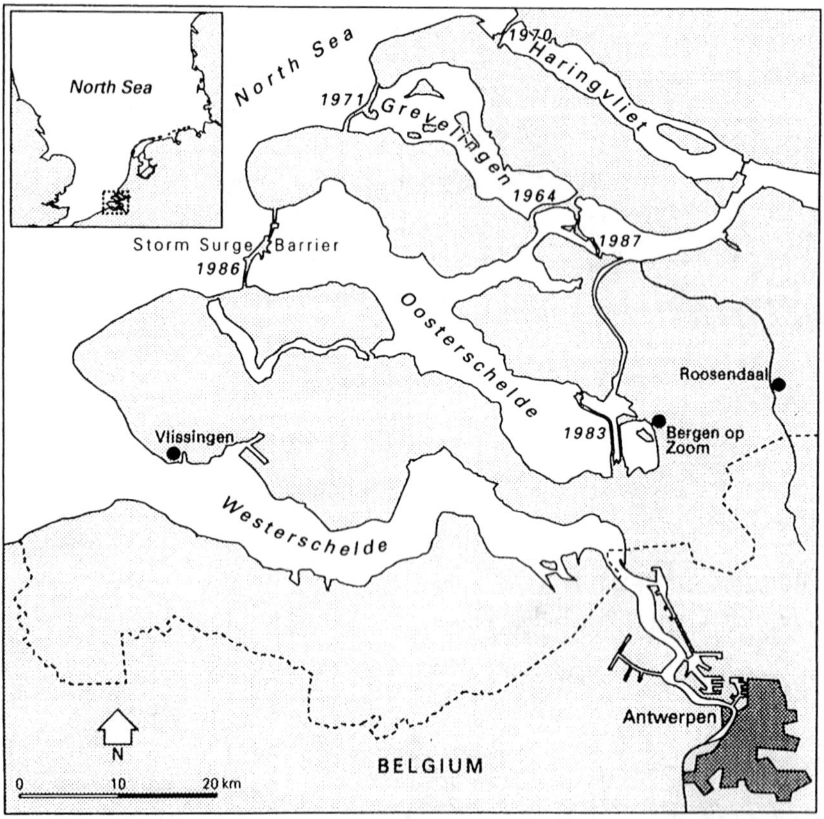

Figure 7.1 Estuaries and inlets of the SW-Netherlands, with dates of closure dams

than 3%, except for locations where older deposits (early Holocene/Tertiary) are exposed.

In the course of the Netherlands Delta Project (Watson and Finkl 1990), a series of coastal constructions were made to protect the SW Netherlands from a repeat of the 1953 storm-surge flooding. The inlets of the northern part of the system of estuaries have been closed by dams. At the entrance of the Oosterschelde a storm-surge barrier was built instead of a closure dam to save the ecology and tidal landscape of the estuary. The Westerschelde is not closed because of its importance as a waterway to Antwerpen, which requires dredging operations to be performed at some relatively shallow areas of the main channel. To keep pace with expanding shipping traffic, maintenance dredging activities have increased in the past decades up to about 12 million $m^3 a^{-1}$ at present. In the near future a further increase of dredging activities may be expected, related to a further deepening of channel bars in the navigation channel.

Human interference in this estuary is not something new. Humans influenced the

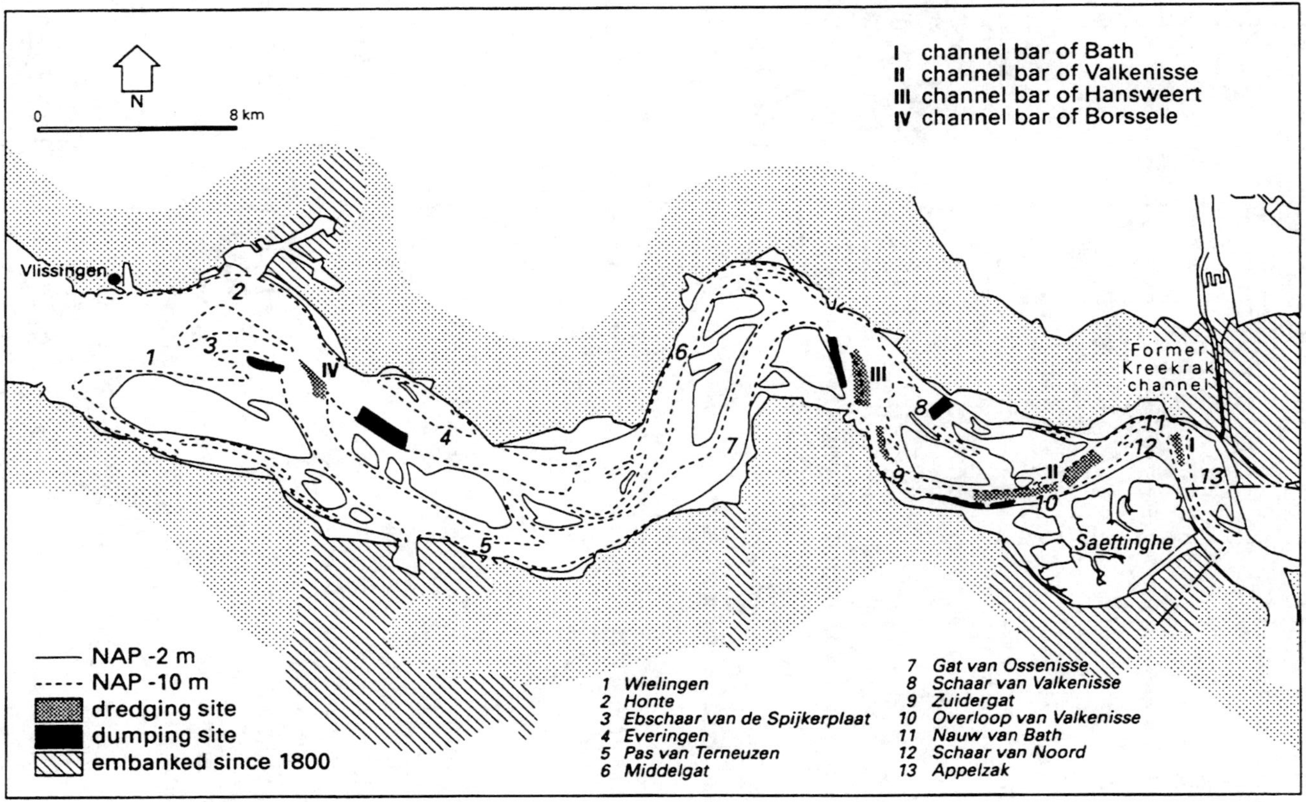

Figure 7.2 The Westerschelde in 1992

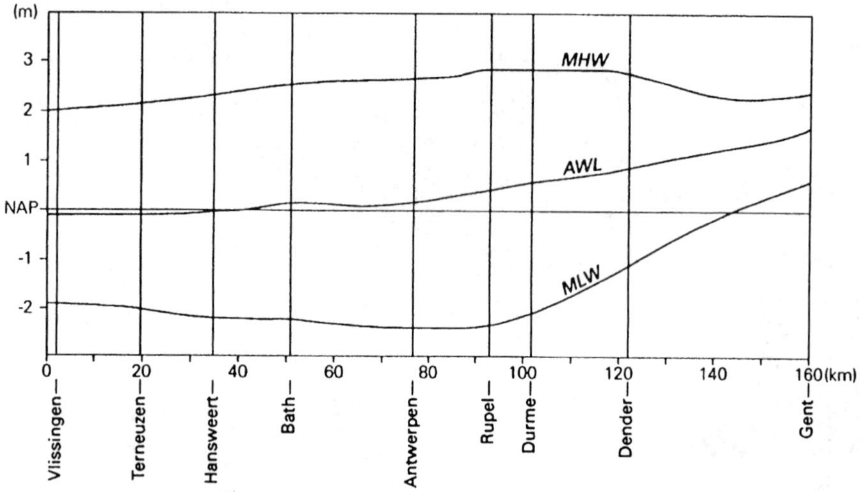

Figure 7.3 Mean High Water (MHW), Mean Low Water (MLW) and Average Water Level (AWL) in the estuary and the lower course of the Schelde river (NAP = Dutch Ordnance Level)

estuary from the moment it started to evolve. Ancient human activities resulted in the lowering of embanked areas, thus creating space for tidal areas. After dike breaches, the lowered polder areas could not be reclaimed and became part of a growing estuary. The purpose of this chapter is to describe and explain the evolution of the Westerschelde estuary by natural processes and increasing human interference.

7.2 PALAEOGEOGRAPHICAL BACKGROUND

The western part of The Netherlands is almost completely covered by a Holocene blanket of sediment, which locally reaches a thickness of 20–40 m. Its development took place under the influence of the rising sea, on a north-to-west dipping surface of Pleistocene deposits. To the west the Holocene sediment wedge is bounded by coastal barrier deposits. The coastal barrier originated during the Atlantic and Subboreal (Zagwijn 1986; Figure 7.4). At the same time, the relatively high elevated Pleistocene surface in the area of the present day Westerschelde became inundated by the rising sea level. Most of the Holocene deposits in this area were laid down by tidal channel systems behind this barrier. During the Atlantic, the barrier coastline retreated and it was still interrupted by many inlets (Figure 7.4). Along the western coast of The Netherlands, the barrier started to prograde by the end of the Atlantic or in early Subboreal times (Beets et al. 1992). Later, a gradual closure of most of the inlet gaps of the barrier took place in these areas. The

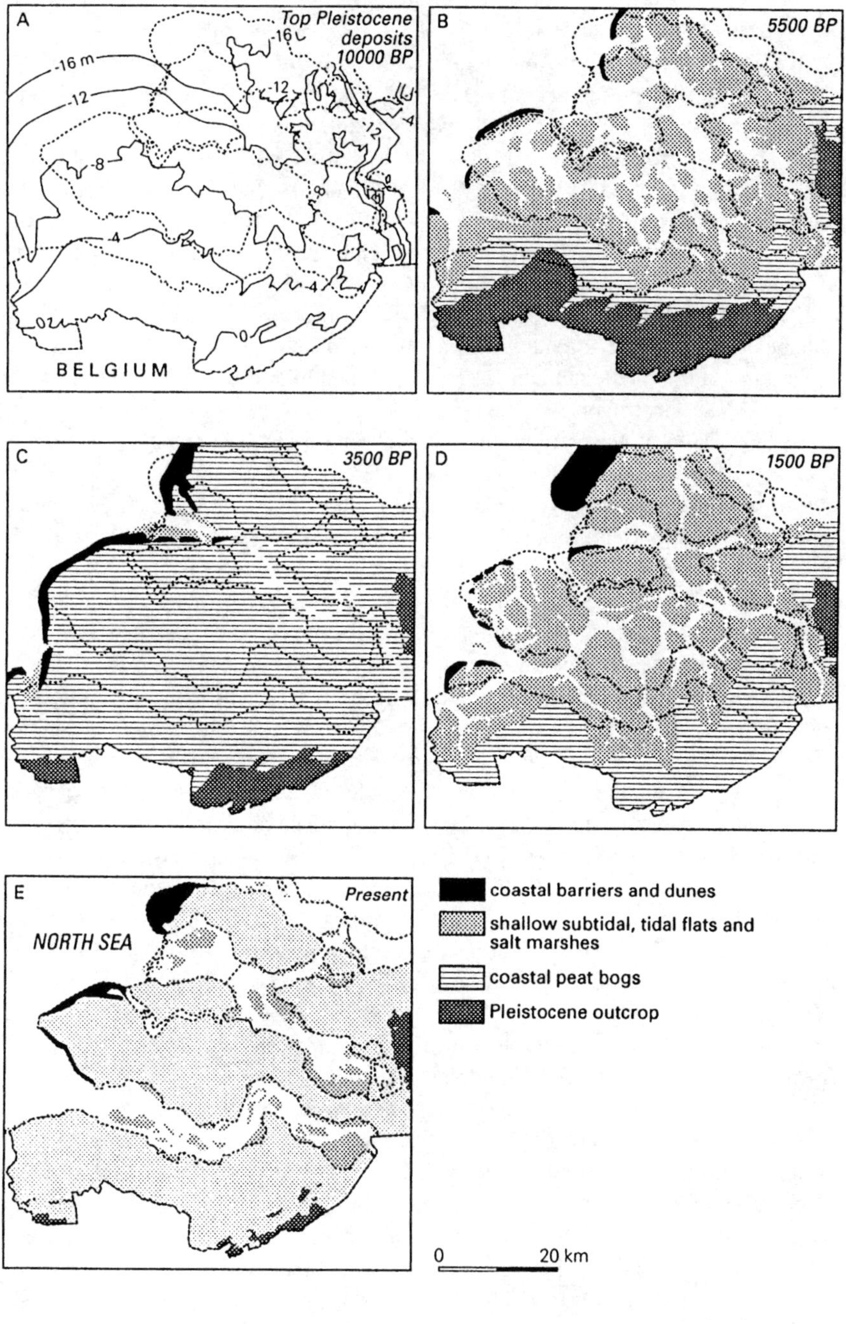

Figure 7.4 Paleogeography of the SW Netherlands (after Vos and Van Heeringen 1993)

closures were caused by a rapid filling of the inshore tidal basins (Calais deposits), made possible by a reduced rate of sea level rise. Barrier progradation is explained partly by sediment supply from degenerating ebb-tidal deltas in front of the inlets during the closure process (Beets et al. 1992). Most of the tidal landscape changed into an extensive fresh water marsh, which resulted in the formation of a thick peat layer (Holland peat). In the SW part of the Netherlands during Roman times, the barrier started to retreat (see Figure 7.4). It became breached at a number of places and new inlet systems evolved. After the second century AD, these systems expanded rapidly. Deep channels were cut and peat layers were removed by tidal currents in many places. Erosion in many areas was soon followed by sedimentation, and during early Mediaeval times many branches of the new channel systems were already completely silted up. However, at the same time, other parts of the tidal channel system were expanding, causing widespread erosion (Vos and Van Heeringen 1993).

Until recently, it was thought that the main ingressions were synchronous events (Zagwijn 1986). Based on this assumption, related deposits are currently classified chronostratigrafically in several Calais and Dunkirk stages. However, evaluation of new and existing archaeological and geological data in the SW-Netherlands (Vos and Van Heeringen 1993) shows that this transgression/regression model does not apply to the Dunkirk stages: in the area of the present day Westerschelde the latest ingression phase started in the ninth century, or even somewhat earlier. More to the north the inlets of the Grevelingen and the Haringvliet started to evolve in the 12th and 13th century respectively (Leenders 1986). The diachronous character of the ingressions excludes external triggers, such as an increase in the rate of sea level rise, to be the principal cause of the initiation of the ingressions as assumed, for example, by Hageman (1969). As far as the later Dunkirk incursions are concerned, human interference is an important factor. During Roman times, ditches were cut to improve the drainage of some marsh areas. Because of the dewatering, surficial peat layers oxidized and the land surface was lowered, which made the area more susceptible to floodings (Vos and Van Heeringen 1993). During the Middle Ages, the influence of humans in creating favourable conditions to the tidal incursions increased. Apart from the effect of artificially improved drainage, a more dramatic lowering of the land occurred in many areas by large-scale stripping of surficial (Holland) peat layers for mining salt and heating purposes; in Biervliet, a small town on the southern bank of the Westerschelde (Figure 7.5), in 1422 more than 300 salt factories were operating (Rottier and Arnoldus 1984).

After the initial start of a tidal incursion, several feed-back mechanisms promoted further expansion of the tidal channel system into the marsh area. The origination of tidal channels caused an improved drainage of nearby marshes, which resulted in a further lowering of these areas; also, owing to an increase in tidal prism, the channels widened at the expense of the original marsh, resulting in a further increase of tidal prism. The deepening of the channels, which accompanied the landward penetration of the channel system, reduces the energy dissipation of the tidal wave. This promoted higher and more devastating storm-surge levels, and lower low water levels, improving drainage and related lowering of peat areas.

7.3 HYDRAULIC EVOLUTION OF THE TIDAL BASIN

The Westerschelde estuary originated quite recently. It started as a new branch of a tidal channel system located in the western part of the present Westerschelde (Figure 7.4) that finally became connected to the Schelde river. During the 14th century the shipping route to Antwerpen started to divert to the Westerschelde (Denucé 1933), indicating the growing importance of the Westerschelde as a distributary of the Schelde river. In the 16th century the channel connecting the Schelde river through the Kreekrak (Figure 7.2) silted up, becoming shallow enough to be waded through at low tide (Van den Berg 1986). The abandonment of the Oosterschelde as a river distributary of the Schelde was finally completed artificially in 1867 and 1871 when two dams were built for a railway connection to Vlissingen.

As a result of warfare in the 14th to 17th century between Holland and Spain (Eighty Years War) and Britain and France (Hundred Years War), many dikes that had been erected along the margins of the Westerschelde during the Middle Ages were breached. This was not only due to poor maintenance; many dikes were breached deliberately, to discourage the enemy. The Westerschelde tidal area reached its maximum size in that century (Figure 7.5). In the past few centuries most of the land losses of the 14th to 17th centuries were reclaimed by embankment of newly formed supratidal marshes. Between 1800 and 1980 15 238 hectares were embanked (Technische Scheldecommissie 1984; Figure 7.2), a third of the tidal area in 1800.

According to an analysis of hydrographic maps drawn between 1878 and 1952 a net sediment import of 1.3 10^6 m^3 per year occurred. This import is related to the siltation of some branches of the Westerschelde and the growth of tidal marshes. The main channel system of the Westerschelde did not show any net loss or gain of sediment (de Looff 1983). At first sight this might look rather strange as a loss in tidal surface area should result in a loss of ebb and flood volume in the channels. This would result in a similar loss of channel cross-sectional area (see Figure 7.6), which would imply a considerable sediment import. This seeming discrepancy is explained by the fact that the reduction in size of the tidal area apparently was compensated by the simultaneous increase in the celerity and amplitude of the tidal wave.

The amplification of the tidal wave is illustrated in Figure 7.7. It is related to the disappearance of the branches and the reduction of intertidal areas along the margins of the Westerschelde, causing the average depth of the estuary to increase and making it more funnel-shaped. As a result, the energy of the tidal wave became less dissipated and more contracted, causing higher amplitudes, especially in the

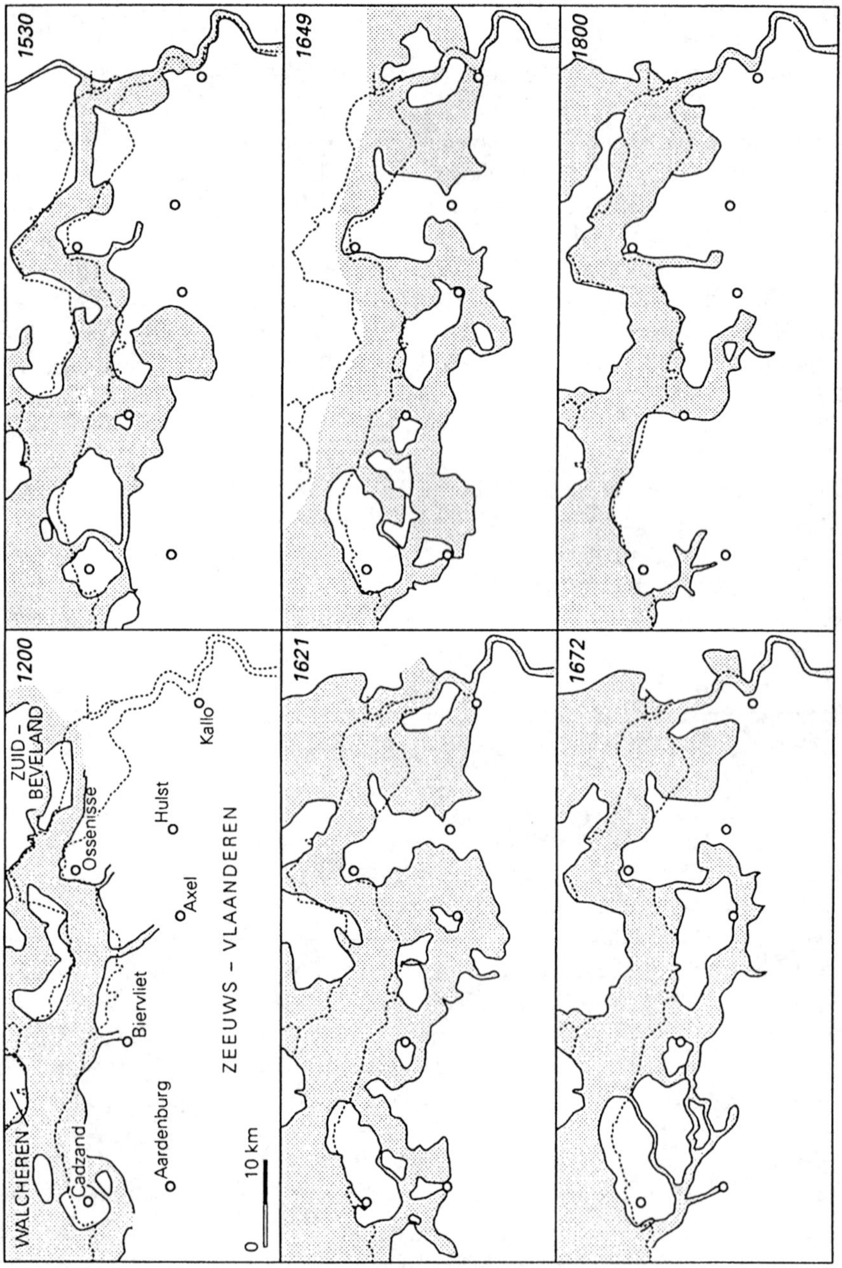

Figure 7.5 Boundaries of the tidal basin of the Westerschelde since AD 1200 (modified after Brand 1985)

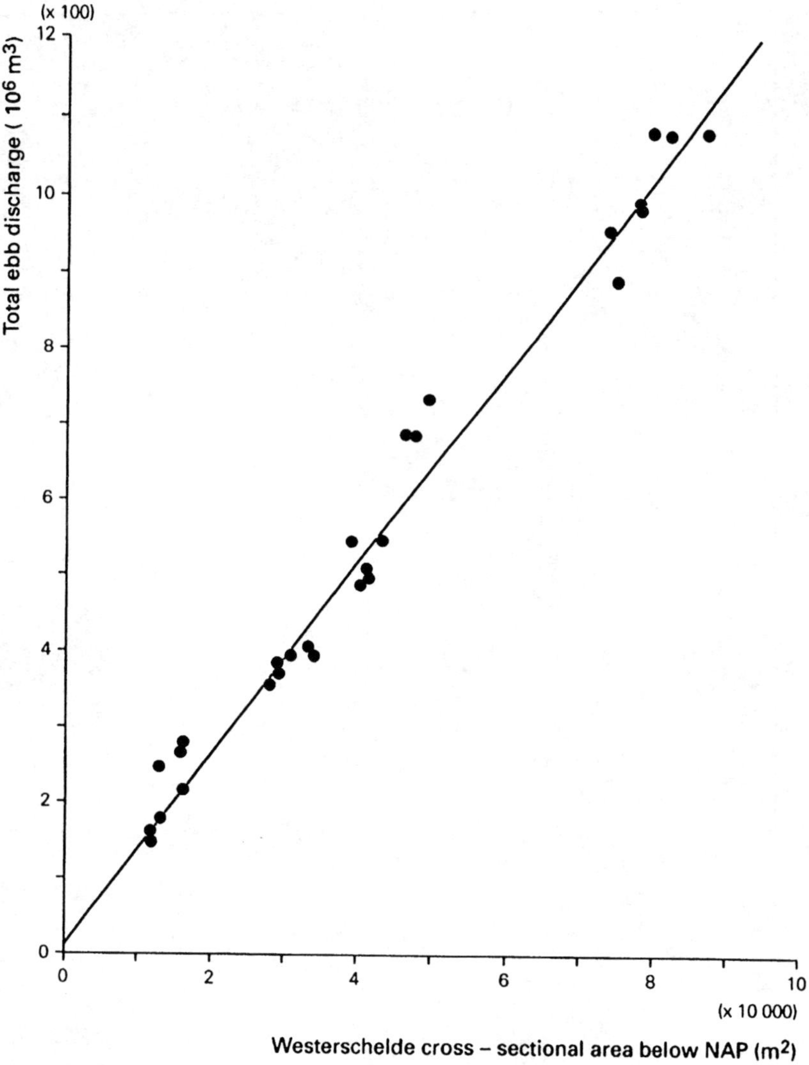

Figure 7.6 Channel cross-sectional area as related to ebb tidal volume (after De Jong and Gerritsen 1984)

inner part of the estuary. The increase of average waterdepth also resulted in a marked increase of the celerity of the tidal wave. In the 16th century, the travelling time of the highest point of the tidal wave from Vlissingen to Antwerpen was about 4.5 hours (Coen 1988). Now the difference in time is only 2 hours. This means that the natural wave period of the estuary at present allows a much better resonance of the tidal wave, which must have contributed to a further amplification of the tide in the basin. The artificial deepening of channel bars in the eastern part of the

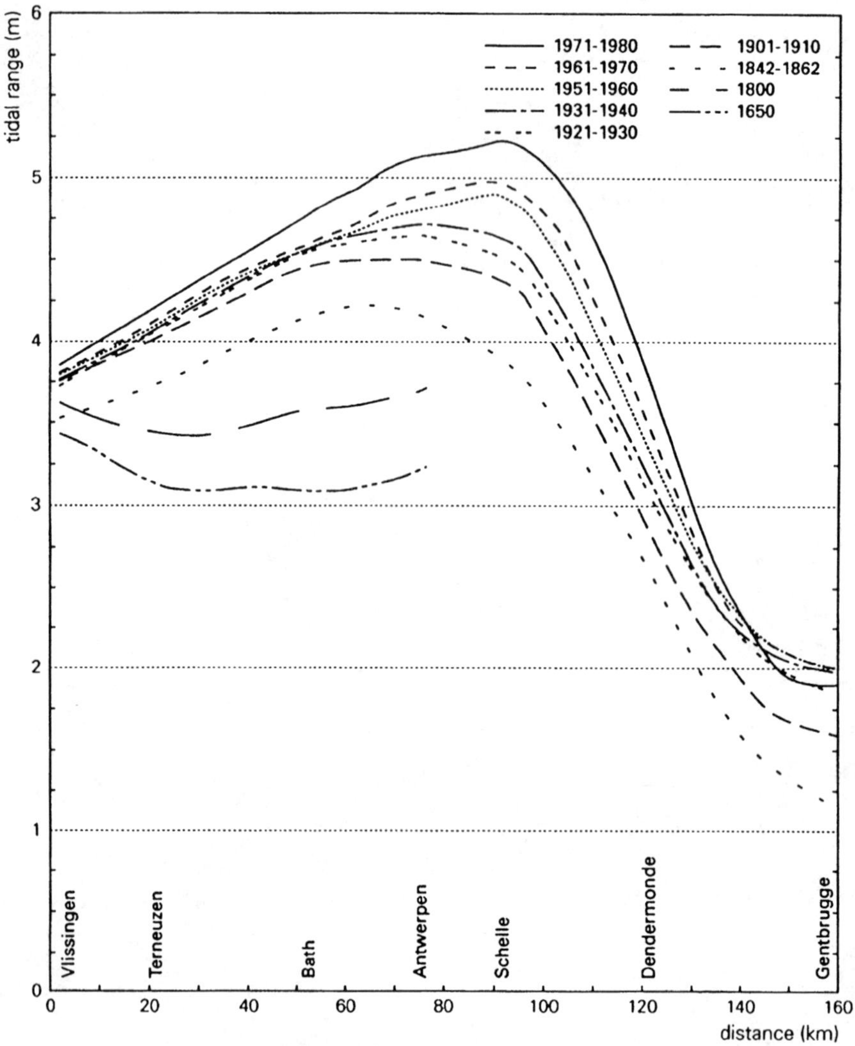

Figure 7.7 Amplitude of the tidal wave in the Westerschelde since AD 1650. Curves for 1650 and 1880 predicted, according to DUFLOW simulations. Remaining curves are derived from 10-yearly reviews of measured water levels along the Schelde estuary

Westerschelde by dredging produced a considerable reduction in the hydraulic resistance of the estuary, which caused a significant lowering of the tidal low water levels in this area, especially after 1970. Finally, some increase in the tidal amplitude resulted from an amplification of the tidal wave in the North Sea recorded during the past century (Figure 7.8). The reason for this amplification is still unknown (De Ronde 1983; Misdorp et al. 1990). The reduction of tidal prism

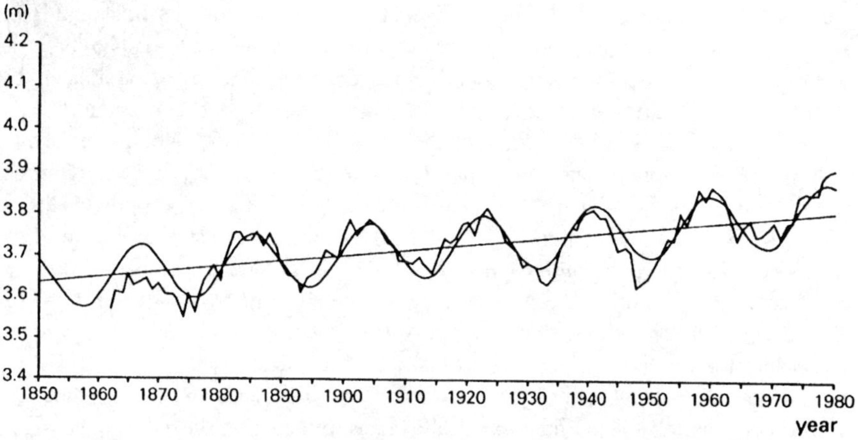

Figure 7.8 Amplitude of the tidal wave at Vlissingen since AD 1862 (after Technische Scheldecommissie 1984)

by loss in tidal area was not only compensated by the increase in amplitude of the tidal wave, but also by an increase in its celerity. In the 16th century, due to the low propagation rate of the tidal wave, high water at Vlissingen coincided with low water in Antwerpen. Now, the water level in Antwerpen is already well above the low water mark when high water is reached in Vlissingen. The increase in propagation rate of the tidal wave has resulted in a better filling of the tidal basin at high tide and thus an increase of tidal prism. Van der Spek (1993) studied the tide characteristics of two former situations of the Westerschelde (1650 and 1800) using the one-dimensional mathematical DUFLOW model (Spaans et al. 1989) and old charts and tide data. Although since 1650 the total tidal area of the Westerschelde reduced by 44%, the computational results suggest that the flood volume diminished by only 13%. From the effect of 44% loss of tidal area, 27% was compensated by increase in tidal range and 4% by increase in tidal wave celerity.

7.4 LONG-TERM SEDIMENT IMPORT AND TIDAL ASYMMETRY

At present, tidal inlets along the Dutch coast are restricted to the SW part and the chain of barrier islands connecting the North Sea with the Wadden Sea. During the Atlantic period a number of inlets were also present along the western part of The Netherlands. In Subboreal times, these inlets were closed as the tidal basin behind then was filled with marine sands and clay (Van der Spek and Beets 1992). The closure of this large basin was achieved in only several thousand years, notwithstanding the fact that sea level during the filling rose at a rate of 0.15 m or more per century. The annual import rate of sediment in the entire basin during this period is estimated at 4–5 million m^3 of sand and 5–7 million m^3 of mud, corresponding to a vertical accretion of 0.30 m per century (Van der Spek and Beets 1992). This illustrates that the basin was an efficient sediment trap. The large net rate of sediment import is not a special feature related to this former tidal basin, but seems to be a common characteristic of many tidal basins and estuaries. For instance, on the basis of hydrographic charts of 1927 and 1967, a net import of 15.2 million $m^3$of sediment into the tidal basin of the Friesche Zeegat (Northern Netherlands) was calculated (Oost and De Haas 1993). This represents a vertical accretion of 0.18 m per century, which is slightly more than the present rate of sea level rise of 0.11 m per century (Misdorp et al. 1990). The net sediment import of the Westerschelde during the past century amounted to almost 1.4 million $m^3 a^{-1}$ on average, corresponding to an average vertical accretion of 0.30 m per century, surpassing the relative sea level rise by about a decimetre.

Two mechanisms account for the sediment trap efficiency of coastal plain tidal basins, such as the basins mentioned above: asymmetry of the tidal wave at the entrance of the basin and settling lag effects. The first mechanism refers to the net import of sand, the second mainly concerns the import of mud. Coastal plain tidal basins are generally found along the margin of shallow seas. Here, the tidal wave crest propagates at a higher velocity than the wave trough. As a result the rising part of the wave surface becomes increasingly steeper, and the falling part steadily flatter. In the case of the Dutch coast, the tidal wave becomes progressively more asymmetric as it propagates from south to north along the coast. In the westernmost part of the Wadden Sea, the asymmetry is temporarily less pronounced, because the tidal wave approaches this area from the relatively deep, central part of the North Sea. At the entrance of tidal basins, the steeper water-level gradient at flood causes higher inshore directed tidal flow velocities during flood as compared to the ebb. As stressed by Dronkers (1986) this is the major cause for the occurrence of greater maximum current velocities during flood than during ebb in the Dutch tidal basins.

In the case of a tidal basin without shoals that emerge at low tide, this tidal asymmetry will result in a net import of sand. Because the transport of sand is proportional to a third (or higher) power of the flow velocity the sand transport during flood is not compensated by the sand transport during ebb. However, if emerging shoals are present in the tidal basin, an important counteracting effect occurs. In the case that large areas of a tidal basin consist of shoals, the maximum of the flood discharge will shift to a later stage in the flood period at a relatively high water level, during which the shoals become flooded. Such a shift means an increase in cross-sectional area of the main channels at the moment of maximum discharge, which implies a reduction in flow velocities. On the other hand the presence of large tidal flats may enhance the maximum ebb current in these channels, as noted by Speer and Aubrey (1985). The decrease in water level on the shoals lags behind the fall in the large channels (Dronkers 1986). As a result, the maximum of the ebb discharge shifts to a later stage of the tidal cycle, to a lower water level. The effect is, as shown by model computations by Friedrichs et al.

(1990), that the ratio of flood to ebb sand transport decreases with increasing amount of shoal area.

In general terms, the deformation of the tide by the presence of shoals means a decrease of flood current velocities and an increase in ebb velocities, which counteracts the net import effect by tidal wave asymmetry at the seaward boundary of the estuary. One could imagine a kind of equilibrium condition of the amount of shoal area in a tidal basin, in which the morphological effect of the shoals on the tide would be just enough to eliminate the import effect due to tidal wave asymmetry. This would enable the prediction of equilibrium hypsometry as a function of basin geometry and tidal wave characteristics at the entrance of the basin.

Such an equilibrium is not expected because of the second important mechanism of sediment import, the settling-lag and scour-lag effects as described by Postma (1967) and Van Straaten and Kuenen (1957). Both effects favour a net transport of mud particles into a tidal basin and into tidal marshes. The lag effects refer to the import of mud into estuaries, whereas the tidal wave asymmetry effect refers to sand import. This means that theoretically two morphology-related stages of infilling of a coastal plain tidal basin can be distinguished, which can be presented in a schedule as shown in Figure 7.9. The rate of infilling of basins in their first stage of evolution increases with increasing asymmetry of the tidal wave. This might explain the rather extreme rate of filling of the Holland Basin in Subboreal times. The model suggests that the rate of filling of a tidal basin in its second stage is related to the availability of (suspended) mud. Some supporting evidence of this is found in the SW Netherlands. The Oosterschelde and the Westerschelde – both mature stage tidal basins in the sense of the model – show a remarkable difference in suspended mud concentration. During 1972–1975 the mean mud concentration

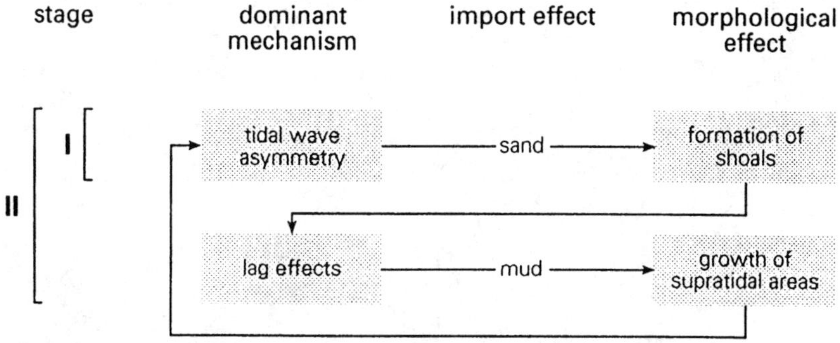

Figure 7.9 Fill processes in small tidal basins: (I) initial stage: infilling dominated by the tidal wave asymmetry effect (formation of shoals emerging at low tide); (II) mature stage: channel and shoal morphology in 'dynamic equilibrium': infilling caused by lag effects (growth of supratidal areas, resulting in a reduction of intertidal shoal area and thus restoring some of the tidal wave asymmetry effect)

measured during calm sea in the Oosterschelde was about 30 mg l^{-1}; in the Westerschelde the concentrations were about twice as much (Terwindt 1977). This occurs because the Westerschelde is a real estuary with a characteristic turbidity maximum due to (estuarine) density circulation, whereas the Oosterschelde has no significant fresh-water inflow and therefore lacks a turbidity maximum. The difference in the suspended mud concentration is in accordance with the rate of reduction of the size of the two basins as a result of the embankment of marshes during the past century; the tidal basin of the Westerschelde reduced at an annual rate of 0.18%, the rate of reduction of the Oosterschelde basin was six times smaller.

The amount of sediment deposited between 1880 and 1950 in shallow areas along the margins of the Westerschelde is estimated at $100 \times 10^6 \, m^3$. This rapid sedimentation and related expansion of tidal marshes in the Westerschelde is

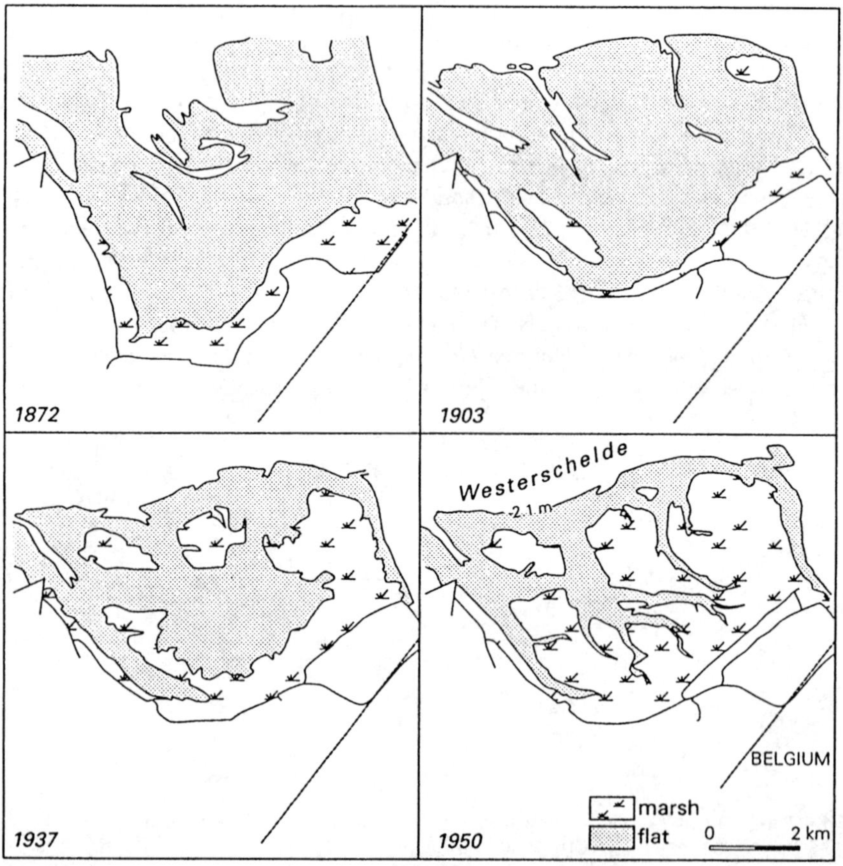

Figure 7.10 The Drowned Land of Saeftinghe since 1872 (modified after Sponselee and Buise 1979)

illustrated by the Saeftinghe salt marshes, located in the eastern part of the estuary (Figure 7.2). The area was embanked during the Middle Ages (Figure 7.5). In 1584, during the war against Spain, the dikes of the polder were deliberately breached by Dutch troops in an effort to prevent the siege of Antwerpen, but Antwerpen was taken and a large polder area became part of the Westerschelde tidal system. In the following centuries, salt marshes developed and part of the lost land was reclaimed (Figures 7. 2 and 7.10). The process of sedimentation and expansion of salt marshes is still going on. According to detailed topographical measurements, a net deposition of about $25 \times 10^6\,m^3$ occurred in the period 1949–1992; during the past 30 years the accretion was $0.013\,m\,a^{-1}$ on average (Krijger 1993). Spatial differences in the rate of sedimentation are indicated in Figure 7.11. Accumulation in channels was generally faster than in marsh areas. The extreme sedimentation in the westernmost channel is possibly related to the proximity of a dumping site of dredged spoil in the adjacent main channel area. In the period 1965–1990 more than $12 \times 10^6\,m^3$ of sand dredged from channel bars in the shipping route to Antwerpen was dumped at this location (Van den Berg et al. 1991).

The quasi-equilibrium condition at which the sand import effect of the tidal wave asymmetry at the entrance of the basins is eliminated by the morphological effect of shoals is not just a function of the available shoal area but the amount of shoal area that is flooded or drained at the same time. In the case of large basins, such as the Westerschelde, a considerable time lag exists between the flooding of the shoals near the entrance and intertidal areas at the landward end of the basin. In

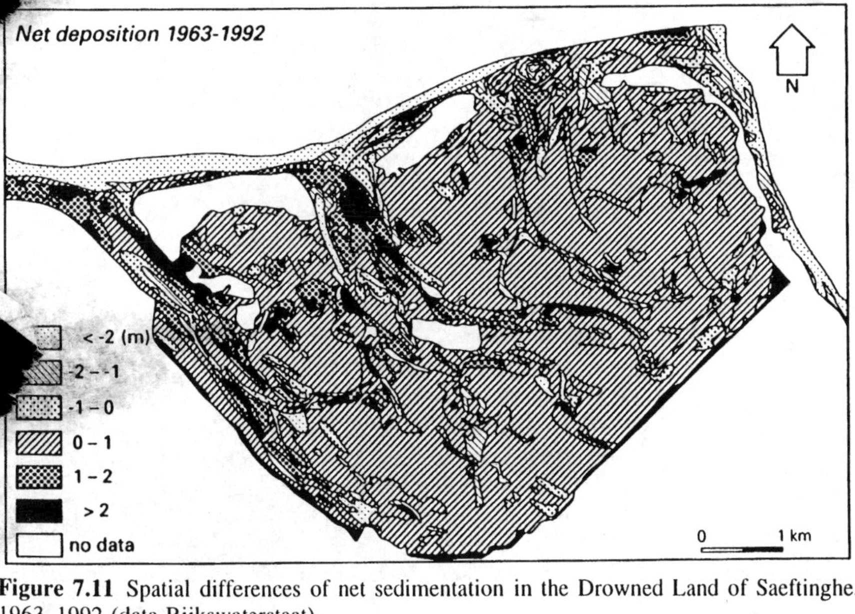

Figure 7.11 Spatial differences of net sedimentation in the Drowned Land of Saeftinghe, 1963–1992 (data Rijkswaterstaat)

such a basin an increase in the total amount of shoal area generally will result in a decrease of the average water depth of the basin and therefore a slowing down of the propagation of the tidal wave. An increase in the total amount of shoal area may not always result in an increase of area that is flooded at the same time. In the Westerschelde during the past centuries, the opposite seems to have occurred. According to Van der Spek (1993) the total amount of shoal area (between NAP −2 m and NAP + 2 m) reduced from 340 million m^2 in 1650 to 200 million m^2 in 1800 and 90 million m^2 in 1968. However, the model simulations of the three situations show that the effect of flooding of shoals as expressed in a peak of flo discharge near high water level is most obvious in 1968 (Figure 7.12). The to amount of shoal area decreased, but the total area that flooded simultaneousl increased. This is explained by the dramatic increase of the propagation rate of th tidal wave in the Westerschelde discussed before. However, the larger extension the shoals in 1650 caused a more retarding ebb in 1650. It seems that both flood ar ebb currents reached higher maximum values in 1650 than at present. However, th ratio of maximum flood to maximum ebb current reduced from 1.30 to 1.21, whic suggests that the Westerschelde in 1650 was a more sand-importing syster is now.

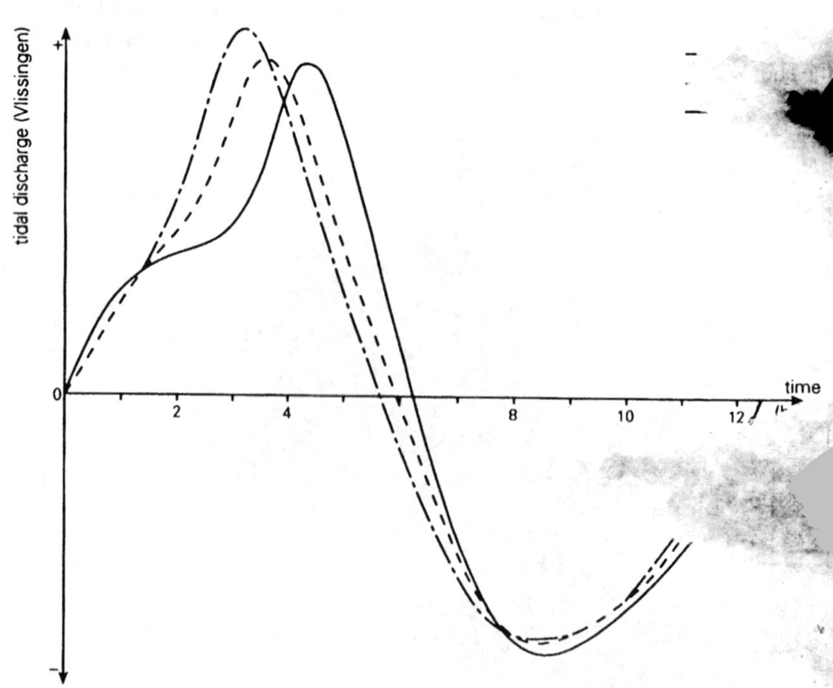

Figure 7.12 Tidal discharge during the average tidal cycle at Vlissingen according to DUFLOW simulations. Based on data of Van der Spek (1993)

7.5 RECENT MORPHOLOGICAL BEHAVIOUR OF THE CHANNELS AND SHOALS IN THE WESTERSCHELDE

The present morphology of the Westerschelde is representative of a mature tidal basin in its second stage of evolution, according to the model described earlier. The overall morphology consists of a system of mutually evasive ebb and flood channels separated by (inter)tidal shoals (Van Veen 1950). The main ebb and flood channels show great differences in morphology. The ebb channels exhibit a meandering pattern, with relatively deep bars. The flood channels are relatively straight channels more or less parallel to the general alignment of the estuary. Generally the bars at the end of the flood channels are shallower than the bars in the ebb channels. These differences in morphology are probably related to an asymmetry in the ebb and flood discharge cycle. The draining ebb flow reaches its maximum discharge near average water level (NAP-level). The flood reaches its maximum discharge at the time that the shoals are inundated. Owing to this asymmetry, the ebb is a less diverging flow than the flood flow, resulting in the relatively deep bars in the ebb channel. As a second result the ebb current causes most of the bend migration of the larger channels, whereas the flood flow tends to cross-cut these channel bends.

These different characteristics are illustrated in Figure 7.13 by the location and shifts of the thalwegs of the main channels since 1800. Until approximately 1930 the morphological evolution of the channel and shoal system was dominated by the ... ebb channels due to meandering process. The annual migration rates in ... bends were in the order of tens of metres. Around 1930 most ebb channels had reached the embankments along the Westerschelde. Further migration of the channel thalwegs caused a steepening of the channel banks along the estuary, resulting in an increasing risk of dike breaches by bank failures. Failure of the channel banks, i.e. slides with more or less circular failure surfaces, sometimes involving more than a million cubic metres of sediment, has been observed at several locations along the Westerschelde. Along some stretches of the coastline additional protections of gravel or slag are made, down to the toe of steep channel bank slopes, to eliminate the threat of dike breaches by slope failures. In the near future these measures will also be extended to areas without a direct thread to dikes, in order to safeguard valuable saltmarsh areas. Bank failures also occur along steep ... the margin of intertidal shoals (Figure 7.14).

... movement is obstructed by coastal defence structures, ebb ... migrate in a seaward direction. This meandering process is often accompanied by an expansion of the bend. Because of the resulting increase in channel length, small connecting channels in the inner bend of the meander develop. These connecting channels often originate in the bar-area of the flood channel. At one location such a small channel developed into a new major channel (indicated by A in Figure 7.13). This new channel gradually took over part of the tidal discharges of the expanded ebb channel, resulting in large sedimentation in the

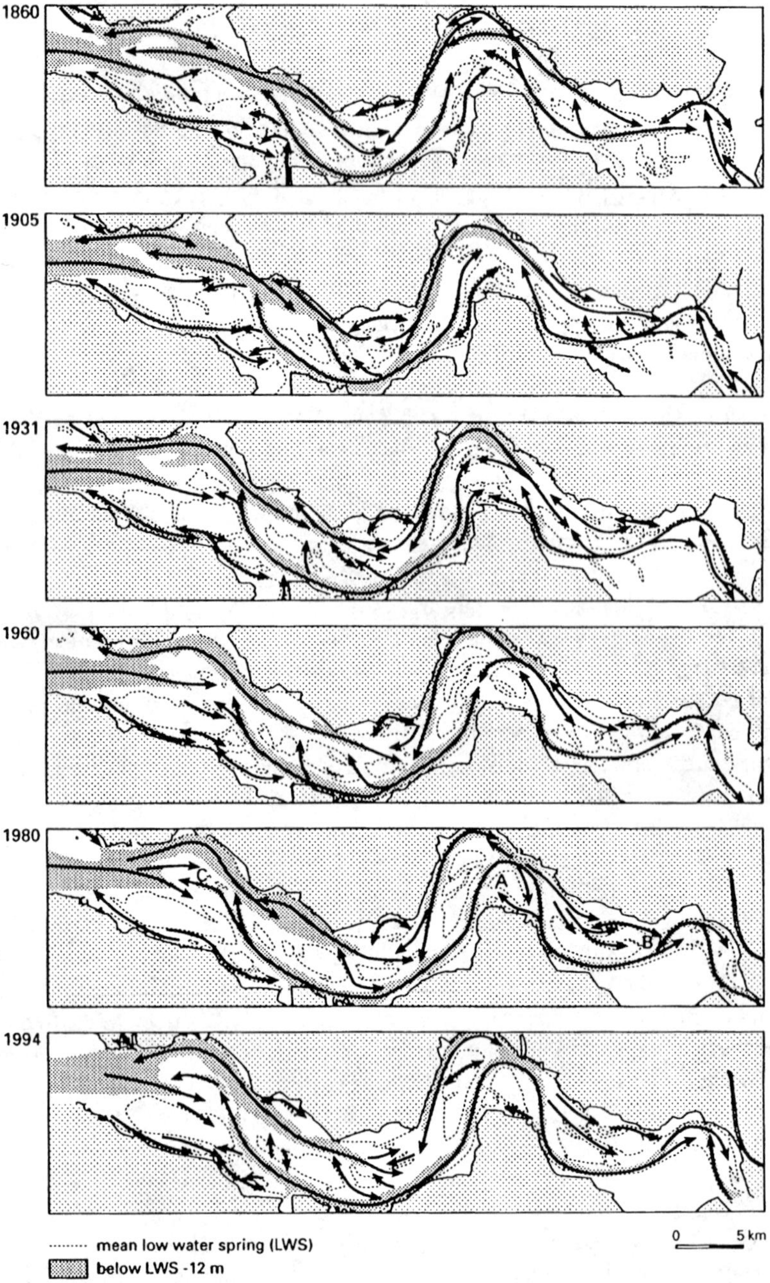

Figure 7.13 Location of channel thalwegs in the Westerschelde since 1860

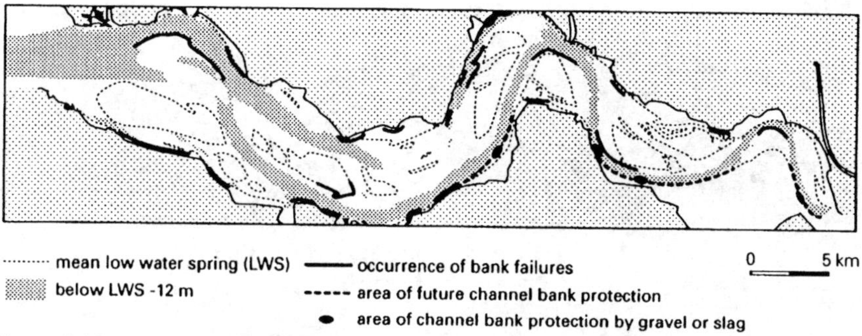

Figure 7.14 Occurrence of coastal defence structures reaching more than 10 m below mean sea level and locations of regular bank failures

older channel. This entire cycle of bend expansion and final bend cut-off lasted more than 100 years. At two other locations (B and C in Figure 7.13) similar connecting channels developed, although these channels did not develop into new major channels. At location B migrating connecting channels with a lifetime of less than a decade occurred (Kleinjan 1938; Sterling and Roovers 1967). Since approximately 1930 these migration cycles have been obstructed by dredging. The connecting channel at location C originated in the early 1960s and rapidly expanded and migrated in a northerly direction. The development of connecting channels at locations B and C noticeably influences the dimensions of the bars in the main ebb channel (Kleinjan 1938; Blick and Ruijter 1994).

The development of a new major channel from a small connecting channel is rather exceptional. Generally, the connecting channels between the main channels remain small, but they are the most dynamic elements of the channel system and are found at many places. Connecting channels originate due to water level differences between the main channels. These water level differences are primarily due to the differences in channel dimensions (e.g. channel length, cross-sectional area) between the main ebb and flood channels. In addition, water level differences between the main channels may exist due to the Coriolis force, loss of momentum of the tidal flow or centrifugal forces (Figure 7.15). Probably, in most cases, more than one process is operating. Most connecting channels are marked by rapid lateral migration (up to $100\,\mathrm{m\,a^{-1}}$). These rapid migration rates are due to the sharp bend at the places where the connecting channel splits from the main channel. Owing to the migration, the channel shifts from the position with a maximum water level gradient to a less optimal location. This will reduce the size of the channel and eventually lead to its disappearance. In the meantime, a new connecting channel is formed at the place of maximum water level gradient. These cyclic processes of origination, migration and degeneration of connecting channels are found at various places in the Westerschelde. An example of the cyclicity of connecting channels along the margin of the Nauw van Bath is shown in Figure 7.16. Besides differences in dimensions of the main channels, the water surface slopes here are

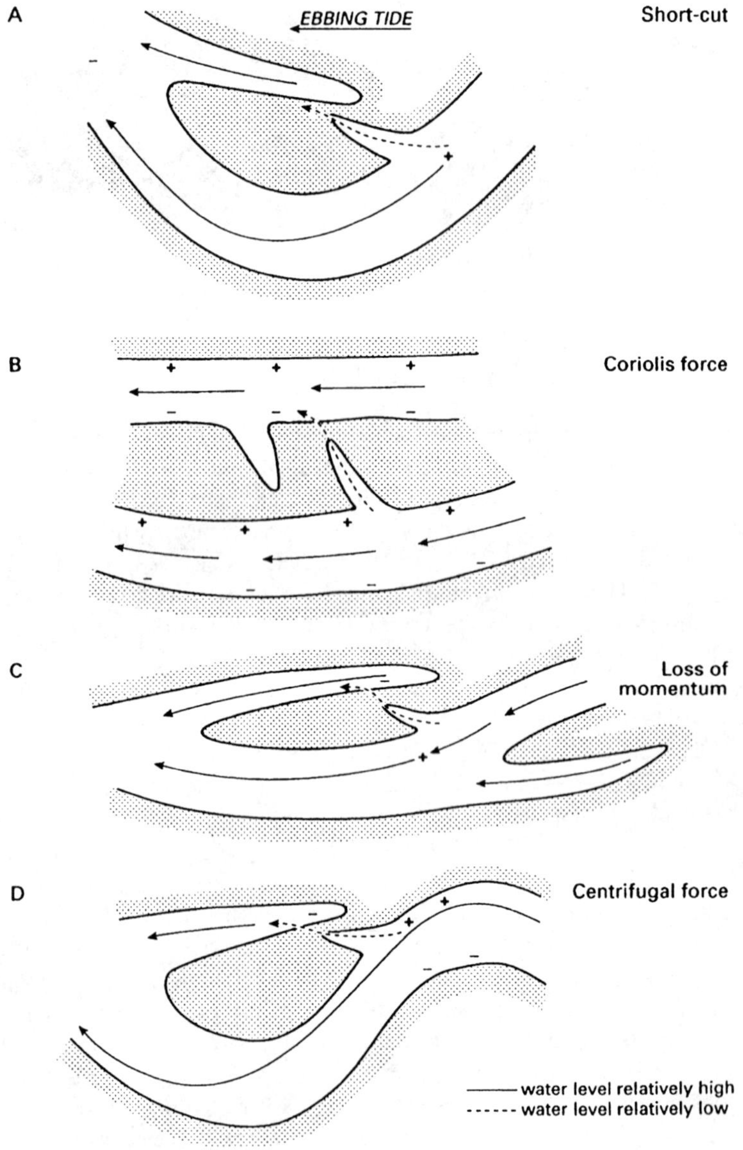

Figure 7.15 Cause of cross-shoal water level difference and the existence cross-connecting channels

possibly enhanced by a combination of centrifugal forces, Coriolis force and loss of momentum. Generally, the influence of connecting channels on the banks of the estuary is negligible.

The cross-sectional surface of ebb and flood channels is proportional to the

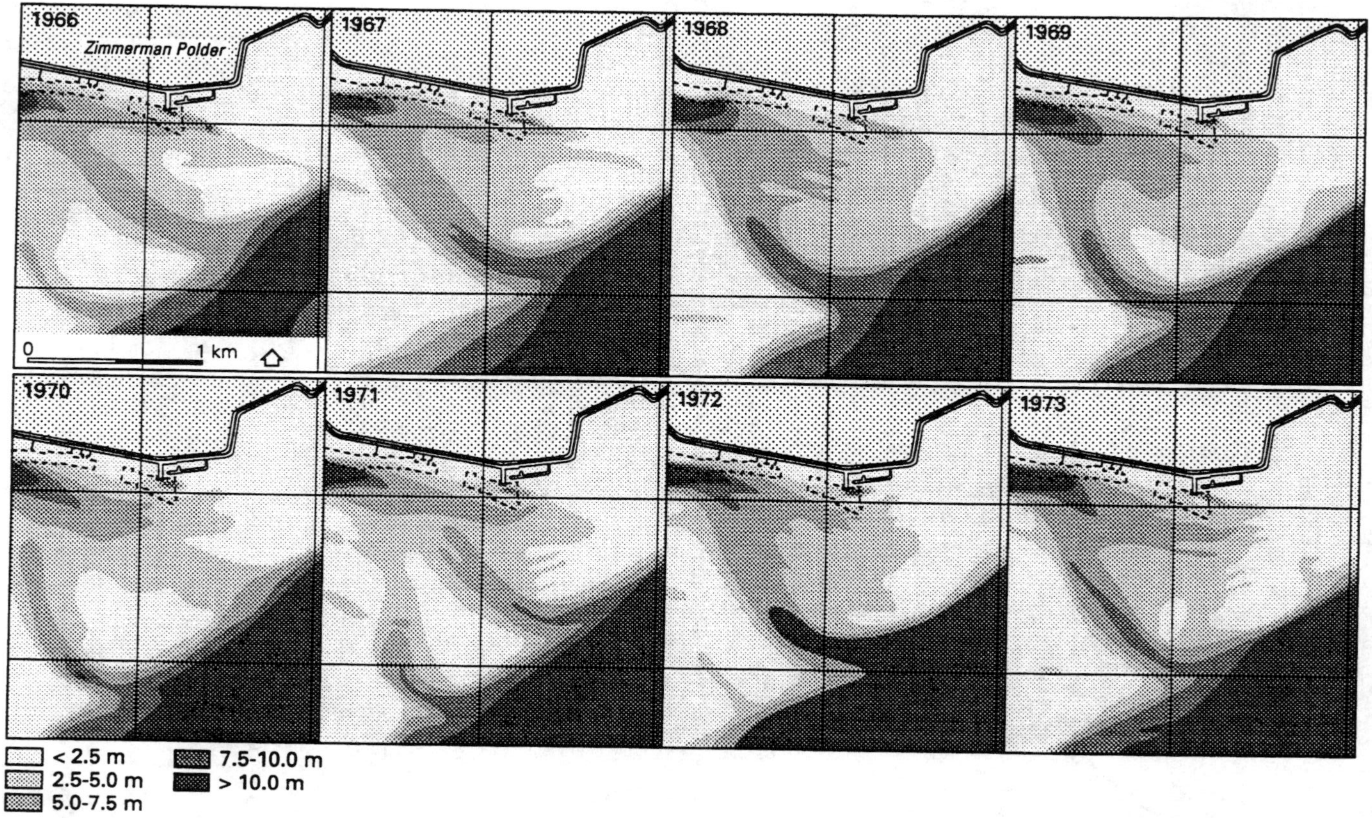

Figure 7.16 Cyclicity of cross-connecting channels along the margin of the Nauw van Bath main ebb channel

volume of the dominant tide (Figure 7.6). However, the cross-sectional area is larger at places where an ebb channel in downcurrent direction reunites with a flood channel or where connecting channels are present. Here the ebb flow expands and the channel becomes relatively shallow. Especially in the eastern part of the Westerschelde the natural depths at these bars in the main ebb channel is insufficient for navigational needs. Because of the increase in the size of ships, the depth of the bars was increased artificially from NAP −10 m after world war II, to NAP −14.5 m at present (Figure 7.17). A second artificial deepening to NAP −16 m is planned for the near future. In order to maintain the bars at the desired depth, the magnitude of the dredging also increased (Figure 7.17). Most of the dredged spoil is disposed at a number of dumping sites inside the tidal system (see Figure 7.2).

The system of tidal channels and flats in the Westerschelde can be subdivided into five bend groups. Each bend group consists of a meandering ebb channel and a straight flood channel, separated by (inter)tidal flats and connected by one or more connecting channels. Each of these bend groups is marked by specific morphological changes. In the remaining part of this chapter the major morphological changes and human interferences are summarized. The analysis of morphological behaviour of the channels and shoals is based on detailed five-yearly soundings of the Westerschelde from 1965 onwards.

The most easterly located bend group is formed by the ebb channel Nauw van Bath and the flood channels Schaar van Noord and Appelzak (Figure 7.2). In this area, the first dredging activities started around 1930. These dredging activities obstructed the formation of large migrating connecting channels between the ebb channel and flood channel. Until 1960 migrating connecting channels also occurred between the Nauw van Bath and the flood channel Appelzak. A guide wall that was built in the early seventies to conduct the ebb flow, resulted in the final disappearance of the connecting channels and slight sedimentation in the flood channel Appelzak. The artificial deepening of the bars in the main channels resulted in an enlargement of the adjacent part of the ebb channel. Between 1965 and 1980 the channel volume below NAP −2 m increased by more than 25%. The average channel depth increased by approximately 1.5 m (Jeuken 1993).

The second bend group encloses the ebb channels Overloop van Valkenisse and the Zuidergat, and the main flood channel Schaar van Valkenisse. In this area approximately 5 million $m^3 a^{-1}$ is dredged. Most of the dredged spoil has been disposed at the eroding outer channel bend of the Zuidergat and Overloop van Valkenisse. Despite the dumping, the erosion rate is approximately $5 m a^{-1}$. Most of the disposed sediment is rapidly removed by the tidal currents (Jeuken 1994). The artificial deepening of the ebb channel bars caused a similar enlargement of the adjacent parts of the ebb channel as in the first bend group. The disposal of large volumes of dredged material at the entrance of the flood channel Schaar van Waarde induced an increase of the local tidal flat volume (above NAP −2 m) of more than 30% between 1975 and 1985. Since 1955 the flood channel Schaar van

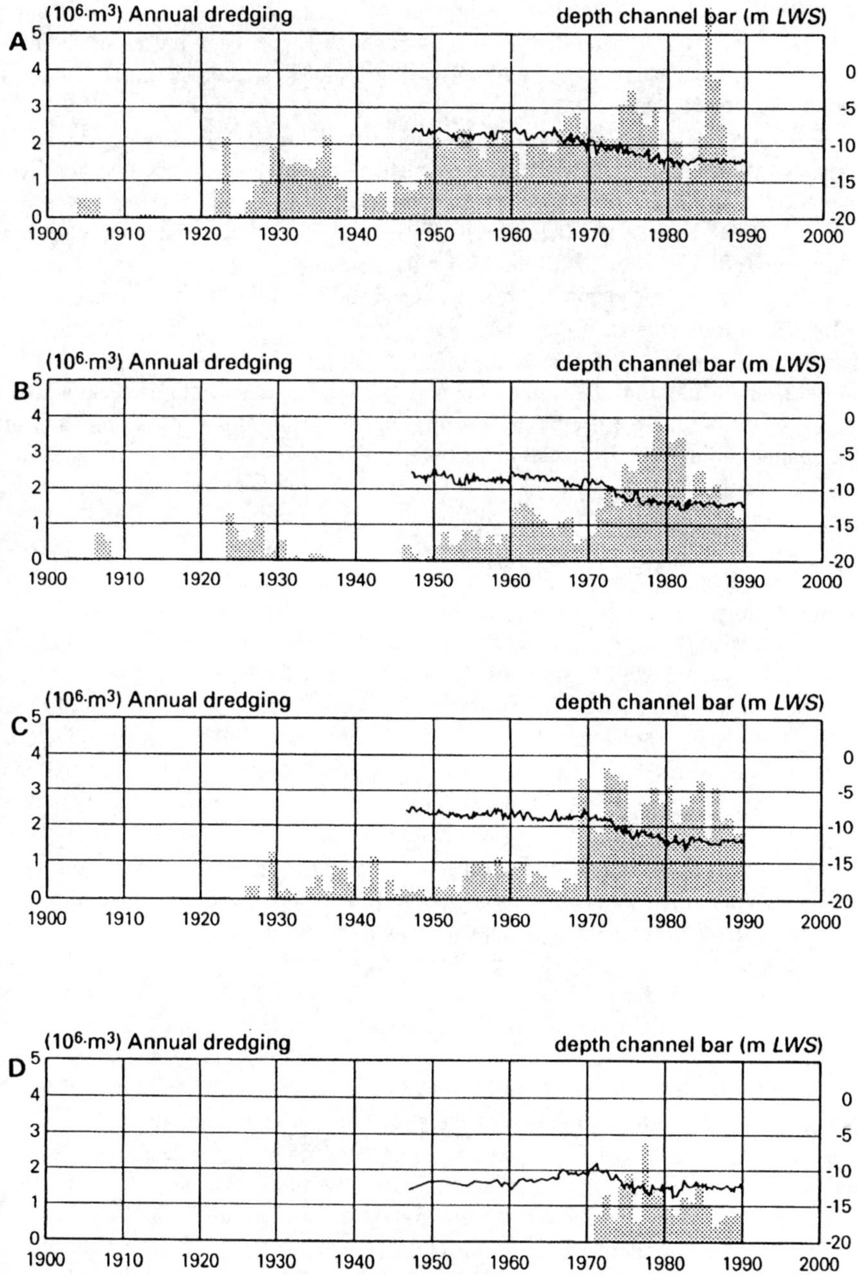

Figure 7.17 Annual dredging and the minimal depth of the channel bars at (A) Bath, (B) Valkenisse, (C) Hansweert and (D) Borssele. Depths in metres with respect to mean low water at spring tide. For location of channel bars, see Figure 7.2

Valkenisse gradually rotated in a southwesterly direction over a distance of approximately 900 m. Migrating connecting channels have been present in the bar-area of the flood channel since 1985. The lifetime of these connecting channels is approximately 7 years.

The morphological processes in the third bend group, Middelgat – Gat van Ossenisse, are dominated by the process of bend cut-off as described earlier. The meandering and expansion of the Middelgat has been going on since at least 1800. The actual bend cut-off, initiated by the formation of a small connecting channel, started in 1951. Between 1951 and 1970 this connecting channel rapidly expanded forming a new major channel. As a result, the tidal volume through this channel increased at the expense of the tidal volume through the ebb channel Middelgat. A second consequence of this bend cut-off was that large volumes of sand could be stored in the Middelgat. Between 1965 and 1990 the channel volume (below NAP −2 m) of the Middelgat decreased by approximately 35 million m^3, a reduction of the channel volume of 20 % (Jeuken 1993). The intertidal flat volume between the main channels increased by approximately 40 %. The channel axis of the northern part of the new channel is slightly rotating. The migration tendency is small due to the limited freedom of movement (imposed by the embankments along the estuary).

The fourth bend group (Everingen–Pas van Terneuzen) is probably the most stable channel system when considering the major morphological changes. The volumetric changes of the main ebb and flood channel between 1965 and 1990 are below 6% (Jeuken 1993). Also the ebb and flood tidal volumes remained stable. The tidal flat volume increased between 1965 and 1985 by 15%. Since 1955 the outlet of the ebb channel Pas van Terneuzen has slightly expanded in a southwesterly direction over a distance of approximately 500 m. Although the net changes in channel volume are small, the reworking of sediment is probably large due to the presence of five connecting channels. The connecting channels in the bar-area of the flood channel are particularly dynamic. Annual migration rates of the connecting ebb channel amount to 100 m and more. Three connecting channels are cross-connecting channels between the main channels. They owe their existence to cross-shoal water level differences (Figure 7.15A and C). The migration rate of these connecting channels, about $40\,m^3\,a^{-1}$, is considerably smaller than the migration rate of the connecting channels in the bar area of the flood channel.

The fifth bend group consists of the ebb channel Honte and the flood channel Wielingen. The gradual migration of the ebb channels Honte and the outlet of the Pas van Terneuzen (Figure 7.2) implied an increase of water surface slopes across the bar-area of the flood channel. This caused the formation of connecting channels in around 1960. The morphological evolution of the bend group is dominated by the expansion and migration of this connecting channel since 1960. The connecting channel started to degenerate in 1988. New small connecting channels have formed in the bar area of the flood channel since 1990.

7.6 FURTHER DEEPENING OF CHANNEL BARS IN THE NEAR FUTURE

In order to improve further the navigation possibilities of large vessels, the minimal depth of the main shipping route to Antwerpen will be increased from 14.5 to 16.5 m in the near future. If the present strategy of disposal of dredged spoil at nearby shallows away from the main channel is followed, the present amount of 12 million $m^3 a^{-1}$ of maintenance dredging could increase to 15 or 20 million $m^3 a^{-1}$. Initially, much more sediment must be dredged to reach the desired design of the navigation channel. This volume of sediment is far too large to be stored at the present dumping sites. Also, it is questionable whether natural processes are able to remove the increased amount of dredged material off the dumping sites. These areas may therefore gradually become choked, eventually causing dramatic and rather unpredictable changes to the remaining channels and shoals and the sediment dynamics of the estuary. Therefore several possible schemes for the maintenance of the navigation channel are being studied by the Board of Ways, Waterways and Harbours of Zeeland province, together with the Institute of Nature Conservation. The main objective of these schemes is to improve the navigation on the Westerschelde with minimum dredging effort in such a way that the characteristic system of flood and ebb channels in the long term is safeguarded. One of these ideas is to increase the tidal prism of the estuary. This would increase the natural depth of the channel bars. A reduction of the dredging may also be reached by a more sophisticated way of morphologically oriented removal and disposal of material.

An increase the tidal prism can be achieved in two ways: (1) by giving some polders back to the tidal system after lowering their surface down to the low water line and (2) by mining some of the dredged material instead of dumping it back into the system. The dredging activities are largely concentrated in the eastern part of the Westerschelde.

Most of the material dredged is dumped at nearby locations. One of the envisaged strategies of morphologically oriented dredging is to transport and dump the sediment at more remote locations, in the western part of the Westerschelde. The resulting increase of travelling time of the sediment to compensate the loss by dredging might reduce the sedimentation rate at the channel bars, thus reducing the dredging effort. Other possible reductions of dredging may be found by employing a dredging that follows better the natural alignment of the channel system. In addition to this the morphology adjacent to the channel bar could be adapted by the dumping of dredged spoil, in such a way that the tidal flow becomes more concentrated over the channel bar.

At present the sediment import of the Westerschelde is very small compared to the sediment volumes that circulate in its five channel bend groups. This fact may indicate a delicate equilibrium, that might be disturbed by side-effects of future navigation channel maintenance strategies. Therefore the possible consequences of these plans, or a combination of these, to the sediment system as a whole must be known before implementation. The possibilities of predicting the consequences of a dramatic change in dredging activities and maintenance strategy to the sediment dynamics of the estuary are still very limited, as no reliable mathematical morphodynamic simulation models are available. Therefore any change in strategy will be made stepwise. In this way it is hoped that possible undesired developments can be identified early, to be able to minimize these effects by adapting the maintenance strategy accordingly.

7.7 CONCLUSIONS

From its origins, less than a millennium ago, the evolution of the Westerschelde estuary has been influenced by man. The expansion of the estuary in the 14th to 17th century was mainly an inadvertent effect of land use. On the other hand the large reclamation of the former land losses in the past centuries was made possible by the natural process of rapid accretion of marginal flats and saltmarshes. In the past decades the increasing dredging and dumping activities mark a change to deliberate man-made modifications. As a reaction to this, significant changes in the overall morphology have occurred, such as the increase in height of some large shoals in the eastern part of the estuary. In the near future the minimal depth of the channel bars will be increased from 14.5 to 16.5 m. In order to reduce the effects of a further deterioration of the natural environment by this interference, several alternative dredging strategies are being considered. Since the middle of the 17th century the size of the tidal basin of the Westerschelde has reduced by more than 40%, but the tidal prism has not reduced, because the loss of tidal basin storage was compensated by an amplification and acceleration of the tidal wave in the estuary.

REFERENCES

Beets, D.J., L. Van der Valk and M.J.F. Stive. 1992. Holocene evolution of the coast of Holland. *Marine Geology* 103: 423–443.

Bliek, A.J. and M.N. de Ruijter. 1994 Verklarend onderzoek drempels Westerschelde. Report Svasek engineering consultants, project 940, Rotterdam: 82 pp.

Brand, K.J.J. 1985. Zeeuws-Vlaanderen, een gebied met een lange en rijke bedijkingsgeschiedenis. *Waterschapsbelangen* 70: 382–393.

Coen, I. 1988. Ontstaan en ontwikkeling van de Westerschelde. *Water* 43: 156–162.

De Jong, H. and F. Gerritsen. 1984. Stability parameters of Western Scheldt estuary, p. 3078–3093. *In Proceedings Engineering Conference*, American Society of Civil Engineers, Houston.

De Looff, D. 1983. Inhoudsveranderingen en zandbalans Westerschelde. Report WWKZ-83.V003, Rijkswaterstaat, Middelburg: 26 pp.

De Ronde, J.G. 1983. Changes of relative mean mea level and of mean tidal amplitude along the Dutch Coast, p. 131–142. *In*: A.R. Ritsema and A. Gürpinar (eds.), *Seismicity and Seismic Risk in the Offshore North-Sea Area. Reidel, Dordrecht.*

Denucé, J. 1933. *De loop van de Schelde van de zee tot Rupelmonde in de XVe eeuw*. Veritas, Antwerpen.

Dronkers, J. 1986. Tidal asymmetry and estuarine morphology. Netherlands. *Journal of Sea Research* 20: 117–131.

Friedrichs, C.T., D.G. Aubrey and P.E. Speer. 1990. Impacts of relative sea-level rise on evolution of shallow estuaries, p. 105–122. *In*: R.T. Cheng (ed.), *Coastal and Estuarine Studies* 38. Springer, New York.

Hageman, B.P. 1969. Development of the Western Part of the Netherlands during the Holocene. *Geologie en Mijnbouw* 48: 373–388.

Jeuken, M.C.J.L. 1993. Morfologische veranderingen van de platen en geulen in de Westerschelde gedurende de periode 1965–1990. Report NWL-93.13. Rijkswaterstaat, Middelburg.

Jeuken, M.C.J.L. 1994. Ontwikkelingen van de geulen en platen in het Valkenissegebied tussen 1986 en 1993. Report NWL-94.15. Rijkswaterstaat, Middelburg.

Krijger, G.M. 1993. Het verdronken land van Saeftinghe komt weer boven water. Report GWWS-93.838. Rijkswaterstaat, Middelburg.

Kleinjan, I. 1938. Het gebied van de Westerschelde nabij Bath. Rapport R88a, 88b, 88c. Rijkswaterstaat, Dordrecht.

Leenders, K.A.H.W. 1986. 2000 Jaar kustontwikkeling van Cap Gris Nez tot Hoek van Holland. Report NZ-N-86.19 Rijkswaterstaat, Den Haag.

Misdorp, R., F. Steyaart, F. Hallie and J. De Ronde. 1990. Climate change, sea level rise and morphological developments in the Dutch Wadden Sea, a marine wetland, p. 123–131. *In*: J.J. Beukema, W.J. Wolff and J.J.W.M. Brouns (eds.), *Expected Effects of Climatic Change on Marine Coastal Ecosystems*. Kluwer, Dordrecht.

Oost, A.P. and H. De Haas. 1993. Het Friesche Zeegat: Morfologisch-Sedimentologische Veranderingen in de Periode 1927–1970. Report, Department of Earth Sciences, Utrecht University, Utrecht.

Postma, H. 1967. Sediment transport and sedimentation in the estuarine environment, p. 158–179. *In*: G.H. Lauff (ed.), *Estuaries*. American Association for the Advancement of Science, Washington.

Rottier, H. and H. Arnoldus. 1984. *De Vlaamse kustvlakte van Calais tot Saeftinge*. Den Boer, Middelburg.

Spaans, W., N. Booij, N. Praagman, R. Noorman and J. Lander. 1989. DUFLOW, a micro-computer package for the simulation of one-dimensional unsteady flow in open channel systems. SAMWAT, Delft.

Speer, P.E. and D.G. Aubrey. 1985. A study of non-linear tidal propagation in shallow inlet/estuarine systems, part II, theory. *Estuarine Coastal and Shelf Science* 21: 207–224.

Sponselee, G.M.P. and M.A. Buise. 1979. *Het Verdronken Land van Saeftinghe*. Duerlink-Krachten, Kloosterzande.

Sterling, A. and P. Roovers. 1967. Modelstudies aangaande de verbetering van de bevaarbaarheid van de Westerschelde. *De Ingenieur* 79: 41–68.

Technische Scheldecommissie. 1984. Verdieping Westerschelde: studierapport programma 48' 43'. 1. Rijkswaterstaat, Middelburg.

Terwindt, J.H.J. 1977. Mud in the Dutch delta area. *Geologie en Mijnbouw* 56: 203–210.

Van den Berg, J.H. 1986. Aspects of sediment- and morphodynamics of subtidal deposits of the Oosterschelde (the Netherlands). Thesis, Utrecht; also in: *Rijkswaterstaat Communications 43*. Rijkswaterstaat, The Hague.

Van den Berg, J.H., D. Schouten and C.J. Van Westenbrugge. 1991. Zandbalans Westerschelde, 1965, 1970, 1975, 1980, 1985. Report NWL -91.36. Rijkswaterstaat, Middelburg.

Van der Spek, A.J.F. 1993. Getijsimulatie voormalige getijdebekkens, deel I: de

Westerschelde in 1650 en 1800. Report R 93-10. Institute for Marine and Atmospheric Research Utrecht, Utrecht University, Utrecht.

Van der Spek, A.F. and D.J. Beets. 1992. Mid-Holocene evolution of a tidal basin in the Netherlands: a model for future changes in the northern Netherlands under conditions of accelerated sea-level rise? *Sedimentary Geology* 80: 185–197.

Van Straaten, L.M.J.U. and P.H. Kuenen. 1957. Accumulation of fine-grained sediments in the Dutch Wadden Sea. *Geologie en Mijnbouw* 19: 419–432.

Van Veen, J. 1950. Eb- en vloedscharen in de Nederlandse getijwateren. *Tijdschrift Koninklijk Nederlands Aardrijkskundig Genootschap* 67: 303–335.

Vos, P.C. and R.M. Van Heeringen. 1993. The impact of environmental changes on the occupation history in the south-western coastal area of the Netherlands (province of Zeeland). Report 1459. Rijks Geologische Dienst, Haarlem.

Watson, I. and W. Finkl. 1990. State of the art in storm-surge protection: the Netherlands Delta Project. *Journal of Coastal Research* 6: 739–764.

Zagwijn, W.H. 1986. *Nederland in het Holoceen. Geologie van Nederland, 1.* Rijks Geologische Dienst, Haarlem.

Chapter Eight

The geomorphology of mudflats

JOHN S. PETHICK
University of Cambridge

8.1 INTRODUCTION

The geomorphology of intertidal mudflats is both complex and imperfectly understood. There is an extensive literature on the sedimentology and tidal processes acting on mudflats and on their invertebrate ecology and conservation importance, but very little attention has been given to the morphology of these important areas. This importance lies not only in the intrinsic value of these intertidal areas but also in their interaction with other tidal landforms: on the one hand, between the mudflat and the larger estuary or coastal bay in which they lie and for which they act as the tidal boundary, and, at a smaller scale, in their relationship with intertidal wetlands – saltmarshes or mangroves – which depend upon an efficient mudflat morphology for their maintenance.

This chapter reviews some of the existing studies of mudflat geomorphology and places these within the larger estuarine framework. The review highlights the urgent need for further research on several crucial aspects, including the morphological relationship between saltmarshes and mudflats, the importance of wind waves to mudflat morphology, and the temporal response of the intertidal geomorphology to wave and tide return periods. Above all it emphasises the need for a more detailed investigation of the relationship between large-scale estuarine form and process and the smaller-scale mudflat morphodynamics. Any attempts to describe either the mudflat or the marsh without reference to the other must be regarded as seriously flawed but to attempt to describe the intertidal unit without reference to the large-scale estuarine dynamics would be equivalent to trying to describe the functioning of a carburettor without mentioning the engine.

Estuarine Shores: Evolution, Environments and Human Alterations. Edited by K. F. Nordstrom and C. T. Roman.

8.2 ESTUARIES AND MUDFLATS

Estuaries are coastal landforms that respond primarily to tidal energy inputs, a response which involves the development of a morphology that can dissipate energy without suffering long-term change. The processes involved in such a response are diverse and complex, but the basic principle is straightforward: failure to dissipate energy inputs exceeding a critical stress–strain threshold leads to morphological change and increases the probability of achieving a form which experiences subcritical stresses. An estuary thus achieves an overall form in which tidal energy per unit area, over a significant time interval, is reduced to a level at or below the strength of the materials making up its bed and banks – principally those of the intertidal zone. The facts that these bed and bank materials are often deposits of cohesive clays and silts and that the steady state referred to is a probabilistic one, achieved over a long time period, provide the complexity of the intertidal geomorphology. Viewed in these terms the intertidal mudflats and saltmarshes are the key to estuarine morphology since they define the shape of the estuary: mudflats are the banks of the estuary channel.

Mudflats, however, not only represent the channel banks of an estuary, responding to tidal processes, they are also an end-member of a spectrum of beaches which respond to inputs of wind wave energy. Beaches, whether they are composed of cohesive or non-cohesive sediment, respond to rapid variations in incident wave energy by short-term adjustments of their plan-form and profile, adjustments which are achieved by both depositional and erosional processes. Intertidal mud beaches thus experience temporally periodic processes of erosion and deposition – tidal energy drives the morphologically-enabling depositional processes while subsequent morphological adjustments are carried out in response to wind wave energy.

8.2.1 Mudflat location and extent

One of the most complex issues facing an understanding of mudflat geomorphology is that neither of the twin energy inputs of tide and wind wave is independent of the morphology. Both tidal processes and wave activity are dependent on the size and form of the estuary channel as defined by the location and extent of the mudflats – so that geomorphological explanation must involve an apparently circuitous argument which is characteristic of many estuarine problems. In their seminal paper on estuarine channel development, Wright and co-authors (1973) emphasise this difficulty:

Neither the channel morphology nor the tidal properties can be explained solely in terms of each other, though the two are mutually dependent. Simultaneous co-adjustment of both process and form has yielded an equilibrium situation in which further adjustment is nonadvantageous.

It is this simultaneous co-adjustment between process and form in the tidal environment that has prevented the development of existing hydrodynamic models to allow prediction of long-term morphological responses. These existing 'bottom up' models, which attempt to define mudflat morphology by the integration of small-scale depositional and erosional process over time and thus to characterise the entire estuary form, fail to appreciate this complex interaction. There is, consequently, an urgent need for the development of 'top-down' models (for example, Capobianco et al. 1993) which begin by defining the steady-state response of an estuary to tidal inputs and then examine the smaller-scale interactions between mudflat process and morphology. Two approaches to this problem are described here, first a 'top-down' model based upon tidal resonance, second an approach based upon tidal prism theory. In each case it is the location and extent of the mudflats within the estuary that form the focus for model predictions.

8.2.2 Tidal resonance models

Estuaries maximise tidal energy dissipation by evolving towards a trumpet-shaped plan-form and a wide, shallow cross-section. Considerable variation may be shown in the exact nature of these morphological adjustments, depending on such factors as tidal range, sediment supply and geological controls. Tidal range, in particular, appears to be a major control of estuary plan-form. The rate at which the estuary narrows inland decreases with the tidal range, so that mesotidal estuaries tend to be short and stubby in plan while macrotidal estuaries are long and more funnel-shaped. (e.g. Hayes 1975; Pethick 1984) The landward exponential decrease in width, so characteristic of estuaries throughout the world, implies that tidal energy is forced into a channel which narrows rapidly landward, thus decreasing its hydraulic efficiency and maximising frictional drag – so that tidal range is reduced to zero at the head of the estuary. Most of this frictional drag is exerted within the intertidal zone of the estuary, which consequently adjusts to the imposed energy inputs from the tide. Wright and colleagues (1973) imply that, since the form of the tidal wave in an estuary depends, critically, on the mean water depth, intertidal deposition resulting in extensive mudflats would decrease the mean channel depth and at the same time reduce the tidal wave length. Continued deposition means that the tidal wave length approaches a multiple of four times the estuary tidal length, at which stage tidal resonance is set up and energy is trapped within the estuary channel. The resultant resonant tide is a standing wave characterised by a nodal point at the estuary mouth associated with maximum flow velocities, and by an anti-node and zero velocities at the estuary head (McDowell and O'Connor 1977). This initial energy gradient results in an extension of the mudflats towards the head of the estuary, confining the channel and giving the characteristic flared channel plan-form in which overall tidal energy dissipation per unit bed area is uniform. The morphology of the

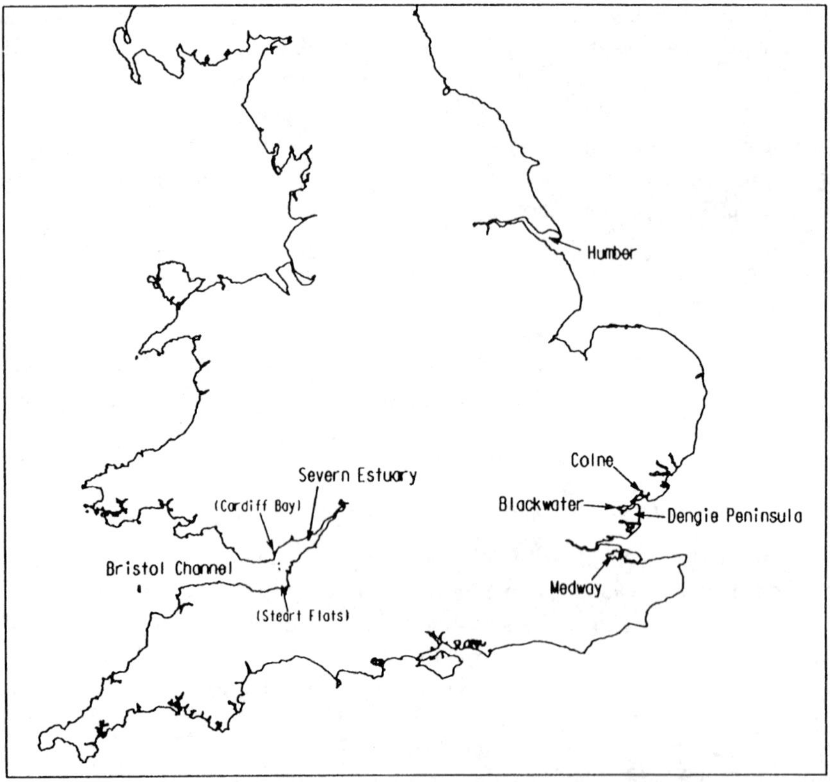

Figure 8.1 Location of places mentioned in the text

alluvial estuaries of the eastern seaboard of Britain (Figure 8.1), which, due to a high level of sediment supply, have developed relatively rapidly over the Holocene, conform closely to this relationship between estuary length and tidal wave length (Figure 8.2A), suggesting that such a tidal resonance model may be applicable in many cases (Pethick 1991; 1994).

The distribution and extent of the mudflats within a tidally resonant estuary may therefore be seen as a self-regulating morphological response to an energy gradient which itself is dependent on the morphology. Thus changes imposed on the system from outside, such as sea level rise, can increase the channel mean depth, forcing the system out of equilibrium. However, the resultant seaward migration of the nodal point reduces velocities in the estuary channel allowing increased intertidal deposition and restoration of the equilibrium. It is interesting to note that reclamation of intertidal mudflats and saltmarshes has a similar effect to that of sea level rise. Reclamation means that extensive shallow water areas are removed from the active estuary channel, resulting in an increase in mean depth and forcing the nodal point seaward. In eastern Britain, however, such reclamation is achieved by

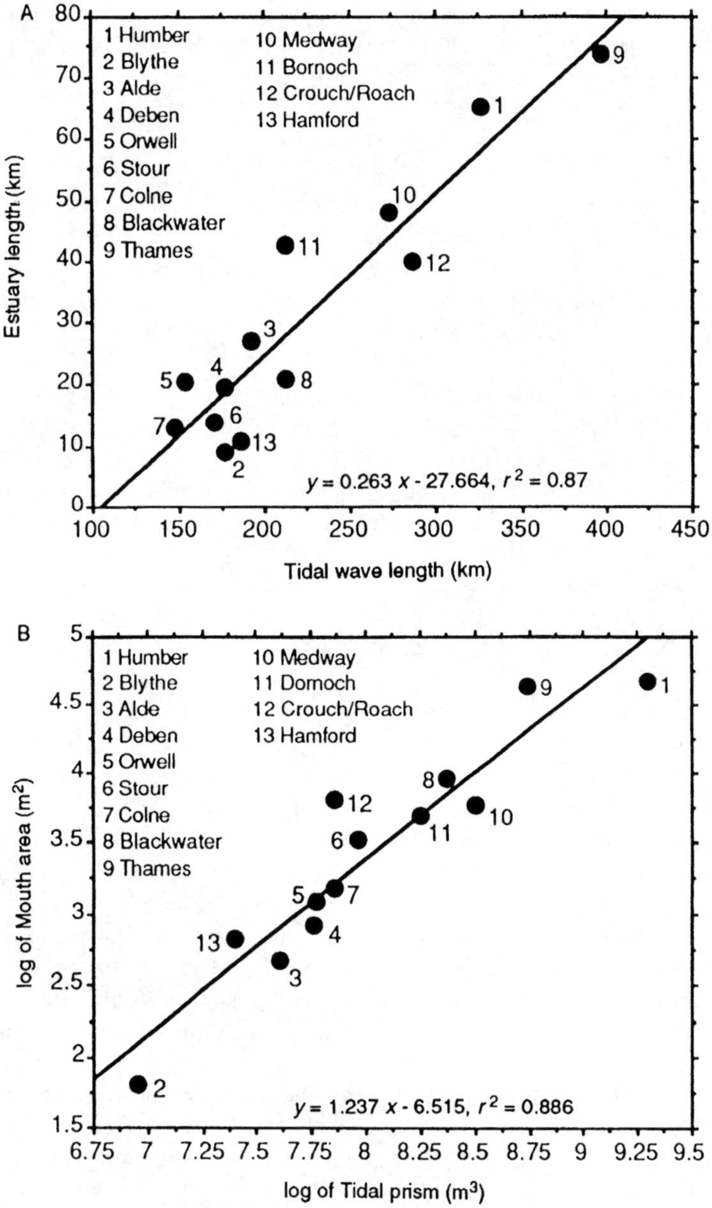

Figure 8.2 The morphology of the estuaries of the east coast of the UK indicates the importance of mudflat location and extent to the overall system. Tidal lengths approximate 1/4 tidal wave length (A) suggesting a resonant system dependent on shallow water intertidal areas, while cross-sectional area at the mouth (B), again largely determined by intertidal deposits, is related to tidal prism

the construction of embankments which confine the estuary channel so that subsequent intertidal deposition is restricted and the channel fails to return to its equilibrium state.

8.2.3 Tidal prism models

The increasing constriction of the estuary channel in the landward direction, increases frictional drag on the tidal wave and results in an exponential landward decrease in the tidal prism of the estuary. The tidal prism, or total volume of water exchanged across a cross-section on a single tide, determines the discharge, and therefore the velocities, of flow across any cross-section (O'Brien 1931; O'Brien and Dean 1972).

This relationship has been shown to be ubiquitous in its applicability. Gao and Collins (1994), for example, provide support from an extensive world-wide data set while Pethick (1994) has shown that the estuaries of the eastern seaboard of Britain again conform to the basic relationship (Figure 8.2B). One criterion on which this relationship could rest centres around the attainment of a critical shear stress for a stable inlet which, according to many authors, is around $1.0\,\mathrm{m\,s^{-1}}$ (Bruun 1978; Bruun and Gerritson 1960; Escoffier 1940; O'Brien 1931; O'Brien and Dean 1972). Thus the inlet cross-section will attain an area which will provide velocities and bed shear stress levels at, or immediately below, the critical erosion threshold.

In general terms, this relationship appears to apply to the inlet or mouth of an estuary and to any cross-section within the tidal basin and has been used as the basis for a one-dimensional model of estuarine morphology by van Dongeren and de Vriend (1994). The model simulates the morphological response of tidal channel and mudflats to externally imposed changes in tidal prism (such as those produced by sea level rise or partial closure of the tidal basin inlet). This model represents an important step forward in that it defines the equilibrium morphological framework within which small-scale adjustments to the intertidal zone are referred. Nevertheless, it still fails to provide the simultaneous feedback between morphology and energy inputs which was described by Wright and co-authors (1973) and which must form the basis of any long-term tidal geomorphological model.

The two approaches to large-scale estuarine modelling described here are not mutually exclusive and future work should attempt to incorporate them, and other approaches, into a unified 'top-down' model in which, for example, morphology and tidal energy inputs are seen as co-adjusting according to the resonance model, thus attaining the equilibrium relationship between tidal prism and tidal flats described by van Dongeren and de Vriend (1994). Once this is achieved and the location and extent of tidal mudflats can be placed into the overall framework, the smaller-scale interactions between sediment processes and mudflat morphology may be examined.

8.2.4 Cross-sectional form

The developmental changes in the plan-form of the estuary are paralleled by changes in the cross-sectional area and depth of the estuarine channel. Evidence from Holocene cores, such as that presented by Brew (1990), suggests that in the early stages of development, as sea level rose in the early Holocene, intertidal mudflats were poorly developed and channel cross-sections exhibited much smaller width/depth ratios than they do today. In these cross-sections, characterised by low-elevation intertidal mudflats and shallow subtidal channels, the speed of the tidal wave is determined mainly by the tidal stage; thus, at low tide, the trough of the tidal wave progresses slowly in the shallow water compared with the crest of the wave, which progresses more rapidly due to the greater mean depths in the channel at high tide. This leads to an increasing asymmetry in the tidal wave so that the flood limb becomes steep and of short duration while the ebb duration increases. Such a tidal wave is known as 'flood-dominant' (Dronkers 1986) since current velocities on the flood are greater than those of the ebb. Flood-dominant estuaries tend to input greater amounts of sediment than they export, so deposition rates are high and the intertidal mudflats thus accumulate sediment and increase their elevation.

In addition to the flood-dominant flow in such developing estuaries, the attainment of a partial standing tidal wave means that velocities are maximised at mid-tide and decrease to zero at high and low water. Consequently, most of the intertidal deposition takes place at elevations above mid-tide so that a convex mudflat profile is attained with a near-horizontal upper surface and a marked break in slope at mid-tide (e.g. Evans 1965). As the upper surfaces of the intertidal mudflats increase in elevation, the convex profiles begin to define an estuary cross-section in which the mean water depth at high tide is actually less than that at low tide. This apparent paradox is due to the fact that the high tide cross-section is dominated by the presence of the shallow water over the high intertidal flat areas, while the low tide cross-section is confined to the deeper, subtidal section of the channel. Consequently, as the estuary develops over time, the progress of the tidal wave crest becomes slower than that of the wave trough and an ebb-tide dominance is set up in which more sediment is exported from the system than imported (Fitzgerald et al. 1976; Boon and Byrne 1981; Speer and Aubrey 1985). The result is that, in the mature estuary, sedimentation rates within the intertidal zone are reduced or even reversed, leading to erosion of the intertidal zone. Erosion of the intertidal mudflats is more likely to be the result of wind wave action than of tidal flows, a subject discussed later in this chapter, but the resultant suspended sediment is transported out of the channel by the dominant ebb flows, reducing mudflat elevation and returning the estuary to a morphology more typical of a flood-dominated system.

At this stage in the development of the estuary, intertidal deposition on the upper mudflats may be expected to increase once more so that the overall estuary channel

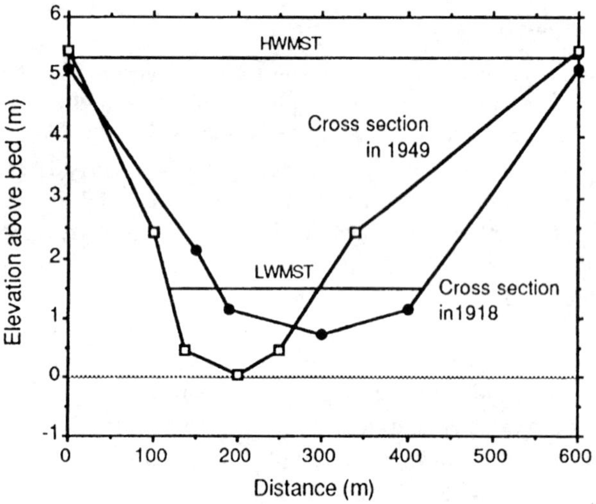

Figure 8.3 Cross-sections of the Colne Estuary in eastern England show development from a flood-dominant form in 1918 to an ebb-dominant form by 1949

morphology begins to oscillate around a quasi-steady state – a form of homeostasis which may characterise many of the estuaries of eastern Britain today. For example, the cross-section of the estuary of the River Colne in south-east England (Figure 8.3) appears to have changed from an ebb-dominant to a flood-dominant form in the first half of the present century, without any obvious external forcing factors; no anthropogenic changes occurred in this estuary over the relevant period, and sea level rise here is identical to that in adjacent estuaries which have experienced no such cross-section modification. If mudflat morphology does experience internally induced periodic change, then it is of crucial importance for coastal managers and conservationists, who may otherwise mistake erosion for long-term deterioration rather than natural homeostasis and take inappropriate mitigating action.

8.2.5 Sedimentary processes

The details of sediment transport, deposition or erosion processes within the intertidal environment need not be discussed here; an extensive literature exists on this subject (for example, Postma 1961; Anderson 1973; Amos and Mosher 1985; Dyer 1986; Mehta 1988; Bishop et al. 1992; Kirby 1992). The discussion here centres on the morphology of mudflats rather than their processes, which are far more complex than envisaged in the brief outline given above. The effect of flood and ebb dominance on suspended sediment loads in an estuary, for example, is of far less importance than for bed load material; thus deposition rates on intertidal mudflats are not directly affected by this aspect of tidal asymmetry. However,

Dronkers (1986) points out that flood-dominant estuaries are characterised by a more protracted slack water period at high tide than at low tide, leading to sedimentation on the upper intertidal zone rather than in the subtidal channels and resulting in a net landward movement of suspended sediment into the estuary. In contrast, ebb-dominant estuaries tend to exhibit longer low water slack periods which encourage seaward movement of sediment, so that these estuaries are sediment sources rather than sinks.

Another complicating factor is noted by several authors (e.g. Postma 1961; Van Straaten 1965; Speer and Aubrey 1985), who point out that the fall in water elevation during the ebb tide in ebb-dominant estuaries is more rapid within the deep central channels than over the extensive, high-elevation, intertidal mudflats. This leads to the development of a marked hydraulic gradient across the mudflats which accentuates the high ebb velocities. The development of wind waves in the shallow water covering the mudflats during the ebbing tide results in resuspension of sediment which is then carried seawards by these ebb currents – leading to low sedimentation rates or even net erosion as noted above; this critical role of wind waves in the resuspension of sediments on intertidal mudflats is reviewed below.

8.3 WAVES ON MUDFLATS

The tidal dynamics of an estuary, described above, can account for the extent, location and depositional morphology of intertidal mudflats. The erosional profile of the intertidal zone, however, and, crucially, the location of the saltmarsh/mudflat boundary, are probably more the product of wind wave processes than of tidal action. Despite the differences in the sedimentary processes between cohesive and non-cohesive environments, the morphological response of cohesive sediment mudflats to wave energy is essentially similar to that of non-cohesive sand beaches. Within this overall uniformity, details of the cohesive sediment equilibrium profile differ from those for sandy beaches principally because of the greater wave attenuation properties of the mudflat surface (Maa and Mehta 1987).

8.3.1 Waves and intertidal profiles

The depositional process on an intertidal mudflat is usually typified by an expression such as that proposed by Krone (1962) in which the depositional rate is dependent on the suspended sediment concentration, the sediment fall velocity and the excess bed shear. Similarly, the erosion rate can be expressed following several authors (e.g. Krone 1962; Odd and Owen 1972; Owen 1976; Dyer 1986) as dependent on bed sediment density and excess bed shear. In both cases excess bed shear is the difference between the actual applied stress and the critical threshold values. Bed shear stresses are partly due to tidal flows in the estuarine channel and partly to wind-wave action. The rapidly oscillating flows under wind waves, however, result in much higher shear stresses than under the longer flood and ebb limbs of the tidal wave. This means that wind waves have a greater potential for the erosion of intertidal deposits than do tidal flows, although in the sheltered estuarine environment they are both temporally intermittent and less energetic than at the open coast; moreover, since bed shear stress under waves is inversely related to water depth (e.g. Madsen and Grant 1976) it is only in the shallow water over the upper mudflats that they have any major effects. The inverse relationship between bed shear and water depth means that, as the mudflat surface elevation rises and mean water depths decrease, so the bed shear stress increases until the critical erosion threshold is approached, thus defining a stable mudflat surface elevation for a specific wave height. Waves weaken the mudflat surface by the oscillatory motion of water and thus allow resuspension of consolidated sediment (Tubman and Suhaya 1976; Maa and Mehta 1987), they do not provide a mechanism for the removal of the eroded mud which is achieved by tidal currents. Since, within mature estuaries, the major part of the mudflat is covered only at high water, wind-wave resuspension of sediment is affected by ebb-dominant currents resulting in a net loss of material from the estuary and causing morphological changes which, as outlined above, can eventually lead to the inception of flood dominance and restoration of the mudflat profile.

One of the crucial factors for the long-term development of mudflats is that, for the same sediment, the critical erosion stress is significantly greater than that for deposition. This increase is partly explained by the settling-lag model of Postma (1961) and partly by the biological and physical changes that occur in a sediment once deposited (e.g. Paterson et al. 1990). These changes continue for some time after deposition and mean that partial re-erosion of intertidal sediment by tidal currents is normally restricted to the same tidal period in which it was deposited. Mudflat accretion thus results in a slow but continuous increase in surface elevation up to a critical limit, set by two factors. First, as the channel cross-sectional area decreases, the critical erosion threshold under tidal flows is approached. Second, as surface elevation rises so water depths decrease and the bed shear stresses generated by wind waves increase in the tidal frame until a critical limit is reached.

8.3.2 Morphological response to wave return intervals

Since bed shear stress under waves is inversely related to water depth it follows that, at high tide, waves of a given height at a single point on a mudflat profile will set up bed shear stresses which will define a critical elevation for the mudflat surface. If the surface falls below this elevation, water depths will increase and deposition will occur, and conversely, increases above the critical elevation will result in erosion. The complexity of the mudflat morphology arises due to the fact that waves of widely different height occur at periodic intervals and that, as these waves propagate over the mudflat, changes in water depth occur, due to profile slope and tidal stage, which determine the actual bed shear.

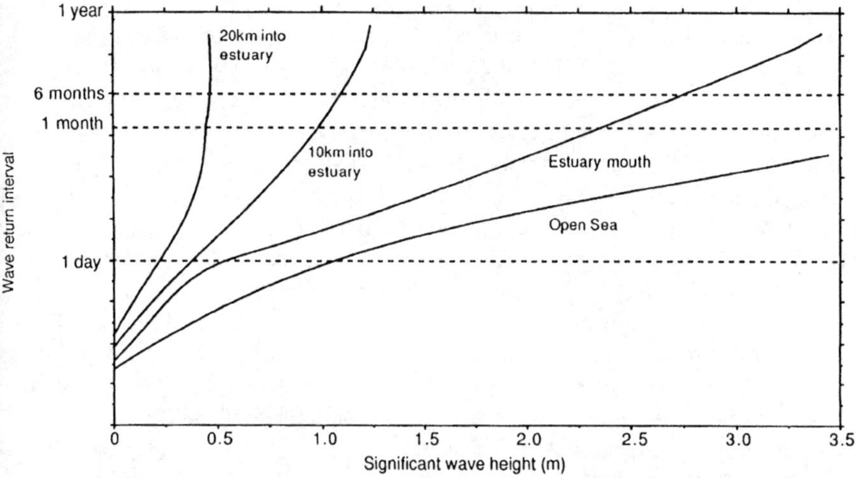

Figure 8.4 Wave return intervals from the Blackwater Estuary, eastern England show distinction between open sea waves in the outer estuary and the truncated, fetch-limited waves in the inner estuary which are locally generated

The process of wind-wave erosion, unlike that of tidal deposition or erosion, takes place intermittently depending on meteorological conditions. Work by Pethick (1992) on the estuaries in eastern England has identified two types of wave activity in the seaward portion of the estuary and in the inner portion of the estuaries, resulting in distinct morphological responses.

In the seaward section of the estuaries, open sea waves are able to propagate into the channel, depending upon the wave characteristic and estuary dimensions, for a distance which can vary between 10 km (e.g. in the Blackwater Estuary) and 25 km (e.g. in the Humber Estuary). The return-interval distribution for these sea waves is open-ended – that is, larger waves are associated with longer return intervals, so that the intertidal zone is subjected to increasingly high-magnitude events although at rarer intervals (Figure 8.4). These rare wave events cause erosion of the upper mudflats and transfer sediment to the lower intertidal zone, resulting in a flatter, wider profile. Profile recovery between such erosive events is achieved by rapid deposition on the upper mudflats during intervening periods of low wave energy.

As a consequence, mudflat morphology in these outer estuaries appears to undergo periodic variations but with a long-term average form which represents a balance between wave return interval and deposition rates. If deposition rates are lower than necessary for morphological recovery between erosion events, due to sediment inhibition for example, then the mudflat surface becomes progressively lower and wider. This in turn decreases maximum bed shear stress, since this is related to water depth and to the distance over which wave energy is attenuated, and a new quasi-steady state morphology is established. Similarly, if the return interval between erosive wave events affecting the upper mudflat increases, as occurs during periods of sea level rise, then again a wider and lower profile results.

In the inner estuaries a quite distinct set of process elements and morphological responses are observed. In the first place open sea waves do not affect these areas so that all waves are generated within the estuary itself and are fetch-limited – that is, the maximum wave height is determined, uniquely, by the width of the water surface over which a given wind is blowing. This produces a distinctive wave height distribution (Figure 8.4), with a marked upper limit determined by the maximum fetch length, and means that the mudflat in these inner areas experiences a limited wave energy input to which it responds. Thus the mudflat profile develops a slope and width which are capable of fully attenuating the maximum wave height.

In general, the landward decrease in wave energy in an estuary means that mudflat profiles are likely to become steeper, narrower and higher in the tidal frame. A survey of intertidal profiles in the Humber estuary supports this hypothesis (Figure 8.5). Profiles at the estuary mouth, that is between 0 and 10 km inland, are long (1 km) and flat (1:500), while their average height in the tidal frame is 0 m MTL. Further into the estuary, at 45–60 km, profiles are narrow (10–300 m), steep (1:50) and relatively high in the tidal frame (+1.0 m MTL).

The relationship between waves and intertidal morphology outlined here is governed by spatial and temporal variations in the wave climate of shallow water areas and the attenuation of waves over cohesive sediment profiles with varying rheology. Both these aspects need to be more thoroughly investigated before any progress can be made into these smaller-scale aspects of mudflat geomorphology.

8.4 TEMPORAL CHANGES

The discussion above suggests that two major components of variation in mudflat morphology are present. First, spatial variations in the cross-sectional profiles of the intertidal zone can be expected along the longitudinal axis of an estuary. These changes reflect the variations in tidal and wave energy with increasing distance inland. Second, temporal variations exist on a number of scales: short-term responses to meteorological conditions, medium-term adjustments to environmental changes, such as those in the rate of sea level rise or sediment availability, and long-term change, principally the adjustment to Holocene sea level rise within the estuary.

8.4.1 Short-term periodic changes

The sedimentological differences between mudflats and sand beaches provide many important points of contrast between the two systems, but perhaps the major difference lies in the response time to short-term changes in wave energy. The rapid changes which take place in the sedimentary properties of silts and clays once they have been deposited, due to physical and biological mechanisms, mean that re-

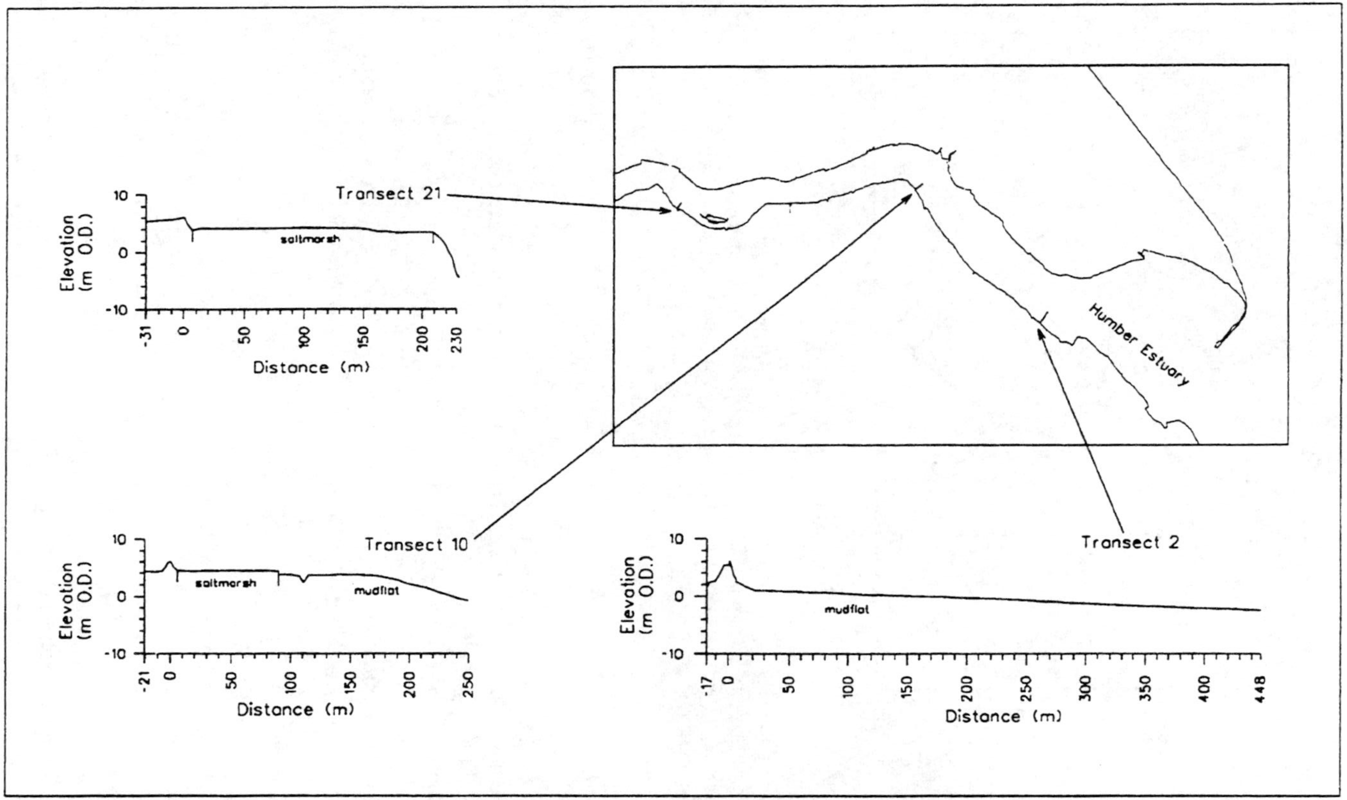

Figure 8.5 Intertidal cross-sections from the Humber Estuary, eastern England. The convex profiles become narrower, steeper and higher in the tidal frame further landwards

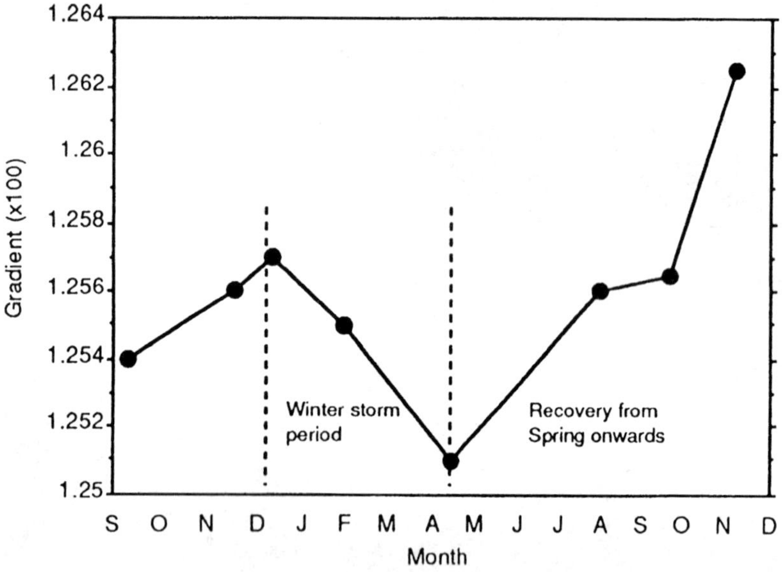

Figure 8.6 Response of the mudflat gradient (tangent) to wave action in the Dengie Peninsula, eastern England. The decrease in gradient as a response to individual storm events is cumulative over the winter period, while recovery occurs over an eight-month period in spring and summer

erosion thresholds are much higher than deposition thresholds. This contrasts with the thresholds for sand beaches where major rheological changes do not occur immediately after deposition and, consequently, where the differences between erosion and deposition thresholds are less marked. These differences in erosion and deposition thresholds mean that sand beaches are much more responsive to slight changes in wave energy conditions than are mudflats, which change from depositional to erosional regime only in response to a major increase in wave energy. Thus mudflat morphology responds to periodic wave events with a much longer return interval than do sand beaches – giving them the appearance of long-term stability which, in fact, they do not possess.

Observation of mudflat response to wave events on the Dengie Peninsula, on the east coast of England, shows that their response to, and recovery from, a series of major storm events over a period of three months was so slow as to allow the morphological changes to each high-energy event to accumulate over this period, while recovery took place over the ensuing spring and summer period and lasted some eight months. These changes involved the widening and flattening of profiles during the sequence of storm events and subsequent deposition and steepening in the subsequent months (Figure 8.6). Recovery, that is deposition on the upper tidal flats leading to an overall steepening of the profile, was brought about partly by the lower wave energy during the spring and summer period but was also due to an increase in biological activity on the mudflat surface – a seasonal response noted by many authors (e.g. Frostick and McCave 1979; Paterson 1989).

Thus short-term periodic changes in mudflat morphology may be defined as those occurring over annual seasons (winter/summer) rather than as responses to individual storms, although this definition should not ignore the fact that individual storm events do provide the intermittent energy inputs necessary to give the overall cumulative seasonal effect.

8.4.2 Medium-term changes

Medium-term response in mudflat morphology may be thought of as occurring over periods measured in decades; such changes are mainly secular – that is unidirectional – although periodic changes do occur over decadal time intervals.

Medium-term secular responses occur as a response to changes in a number of environmental factors including sediment supply, wave climate, wave return intervals or sea level change. The last three factors are interrelated so that it is difficult to distinguish between them insofar as they cause morphological changes in intertidal mudflats. An increase in sea level in an estuary results in an increase in mean water level, which in turn decreases wave attenuation over intertidal areas and increases the joint probability of wave and tide events at any given elevation on the intertidal profile. Thus sea level rise can be expected to result in a similar intertidal morphological response to that for short-term storm events – a widening, flattening and lowering of the profiles.

Sea level rise also means that bed shear stress decreases in the deeper water and thus the attenuation of tidal energy is reduced while, since estuarine channels are trapezoidal, tidal prism increases. The morphological response to these changes is for the estuary to widen and become more shallow so that the increased tidal prism is accommodated in a channel whose high levels of frictional drag are maintained. This morphological response is identical to that for an increase in wave energy and an increase in wave return intervals, so sea level rise sets up a number of reinforcing morphological effects. Reference to the tidal resonance model discussed above also indicates a similar response to sea level rise in which the nodal point is moved seaward thus reducing velocities at the inlet and causing increased deposition in a wider channel. The effect of these combined responses can be seen in the Blackwater estuary where the rate of sea level rise due to land subsidence is at present 4 mm per year. Here the outer estuarine channel widened by over 1000 m between 1820 and 1992 (Figure 8.7) and the intertidal areas flattened and lowered (Figure 8.8). The sediment released by the erosion of the intertidal areas was not lost to the system but appears to have been deposited in the subtidal channel of the estuary, resulting in a general increase in width/depth ratio and attenuation of the tidal wave. The Blackwater, in common with many east coast UK estuaries, is flanked by an almost continuous line of flood

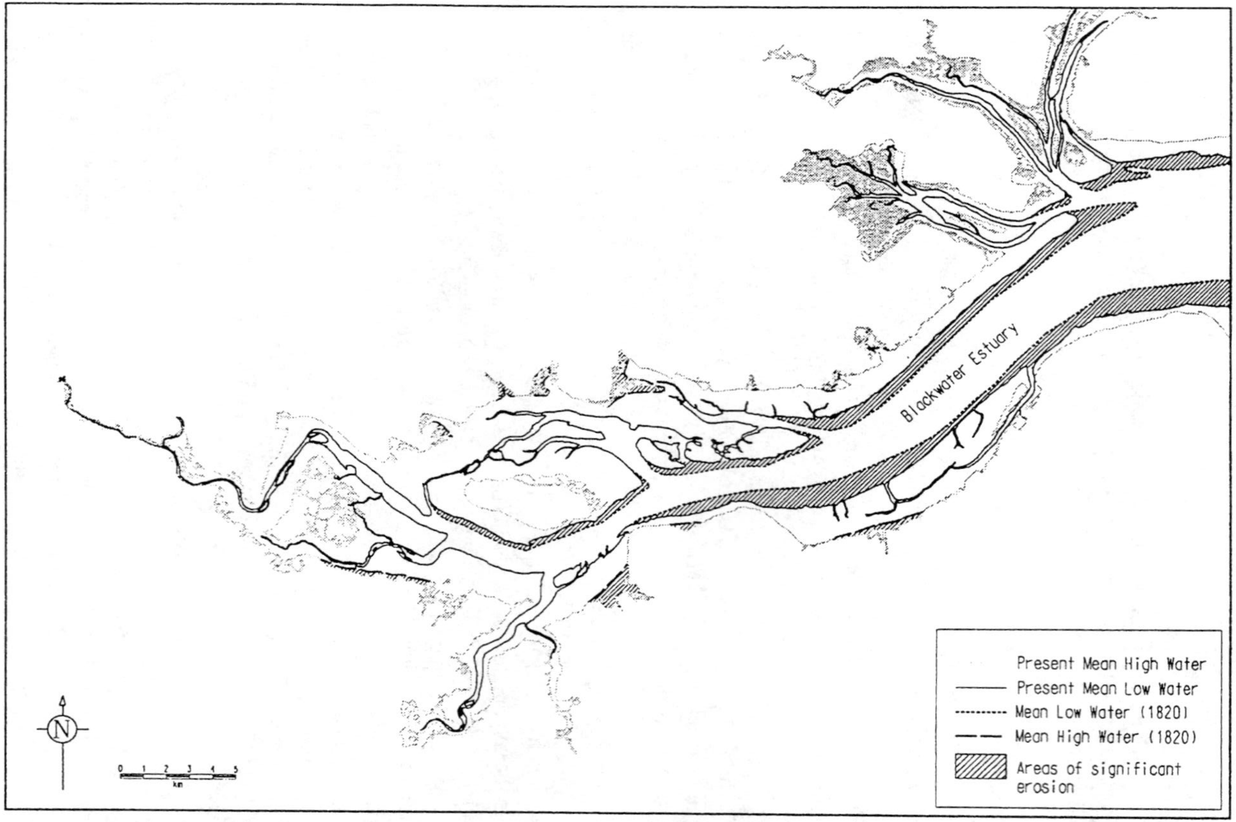

Figure 8.7 Intertidal changes in the Blackwater estuary, eastern England, compiled from maps of the area surveyed from 1820 to the present day. The widening of the estuary channel at low water is not reflected to the same degree at high water due to the presence of flood embankments

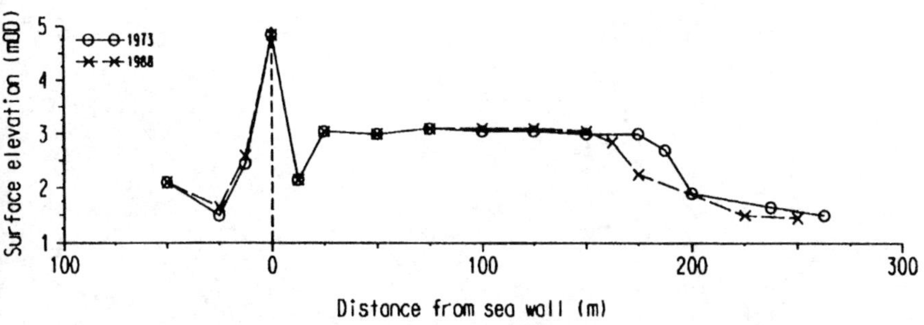

Figure 8.8 Changes in the intertidal profile on the Blackwater estuary between 1973 and 1988. The saltmarsh extends between 25 m and 150 m from the sea wall; its vegetated surface experienced 0.10 m of vertical accretion in the 15-year period while the saltmarsh–mudflat boundary eroded by 25 m in the same period. The intertidal mudflat extends from 150 m to 275 m from the sea wall; its cross-profile widened and flattened over the surveyed period

embankments which prevent onshore migration of the intertidal profile as sea level rises so that the upper intertidal area is progressively truncated as sea level rise occurs. At first this involves only the upper intertidal areas, mainly saltmarsh, but as these are eroded so progressive lowering of the mudflat profiles is initiated. This leads to a reduction in wave and tidal amplitude attenuation, increasing erosion rates and flood risks for low-lying agricultural and urban land. Moreover, the loss of intertidal habitat in this and other east coast estuaries is regarded as of major environmental concern.

Medium-term periodic changes, such as those set up by oscillations between flood- and ebb-dominant mudflat morphology as discussed above, are more difficult to identify in view of the paucity of long-term morphological data, which must cover periods of hundreds of years in order to identify decadal periodicities. Nevertheless, some evidence is available to suggest that swings from accretional to depositional regimes have taken place in the UK; this suggests rather than confirms such periodicity. In addition to the case of the River Colne, discussed above (Figure 8.3), Kirby (1992) notes that the Cardiff Bay area within the Severn Estuary has experienced a change from erosion to accretion over the past 90 years, while the Medway estuary in eastern England has experienced a change in the reverse direction, from accretion to erosion over the past 150 years (Kirby 1990), although over the past decade some recolonisation of the eroded mudflats by saltmarsh vegetation suggests that a further reversal back to accretion is imminent in this estuary (Figure 8.9). The morphological feed-back mechanism involved, as outlined above, relies on wave erosion and subsequent tidal sediment transfers within the intertidal zone. Given the slow response time of the intertidal morphology in an estuary, a complete sequence may occur over periods measured in decades or centuries, rather than years or decades.

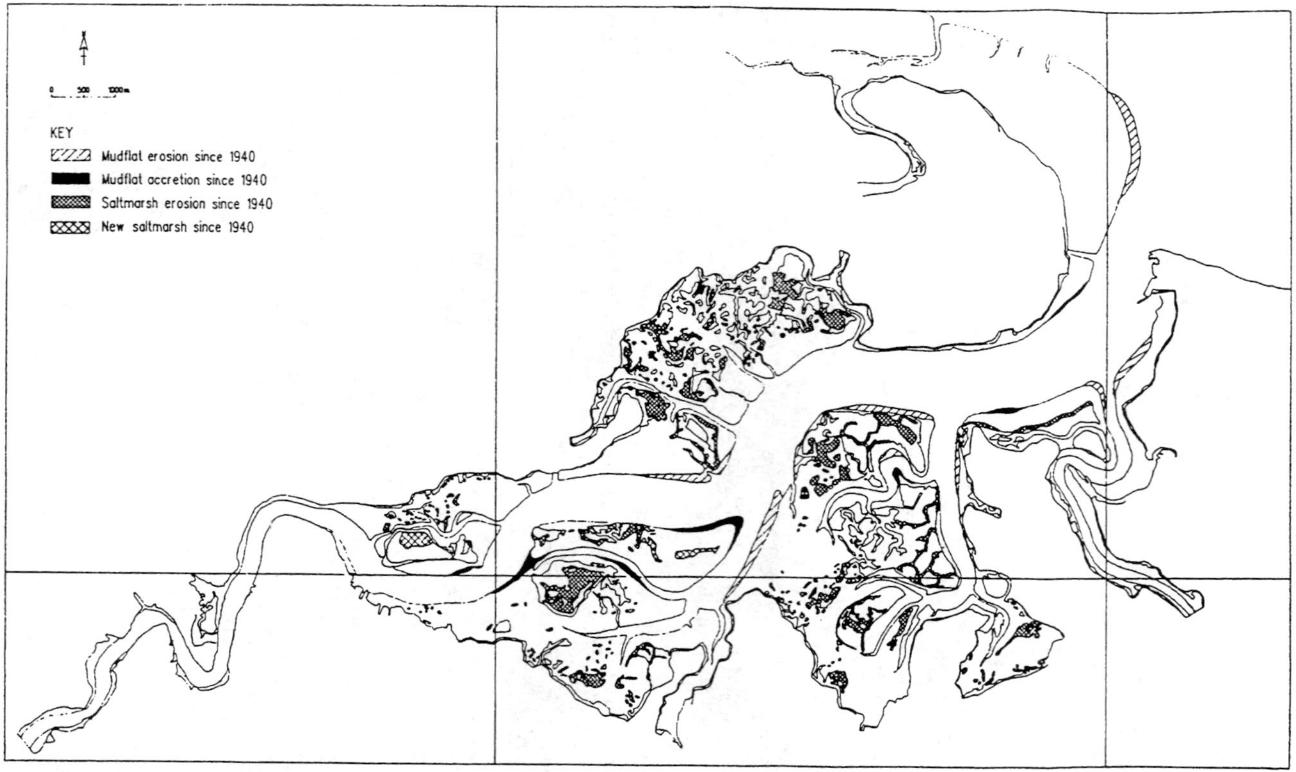

Figure 8.9 Changes in the intertidal morphology of the Medway estuary since 1940. Although the general tendency is for erosion of both mudflats and saltmarshes, continuing the trend of the past 100 years, there are signs of intertidal accretion suggesting that a critical limit to erosion is imminent

8.4.3 Long-term changes

The medium-term periodic changes outlined in the previous section, form part of a long-term secular response to the Holocene sea level rise which initiated, or in many cases reinstated, estuarine processes some 6000 years ago. The development of these estuarine processes has been outlined above and can be seen to give rise to a sequence of morphological changes, especially in the intertidal zone. Thus the initial low intertidal flat morphology with shallow subtidal channels is replaced by a more mature stage of high tidal flats and deeper subtidal channels. This sequence may then be repeated to give the periodicity discussed above, which could be seen as representing a series of oscillations around a quasi-steady state morphology.

These changes involve progressive modification to the cross-estuary shape of the intertidal profiles. However, characterising mudflat morphology using two-dimensional cross-sections is often misleading; these fail to give an indication of the general three-dimensional shape of the intertidal zone of an estuary. Several authors have suggested that this three-dimensional intertidal morphology may be represented using the hypsometric interval method originally developed by Strahler (1952) in which intertidal surface area is plotted against tidal stage. Boon and Byrne (1981), for example, looked at the hypsometry of tidal inlet systems in the US; Dieckmann and Partensky (1985) used this approach to characterise the estuaries of the German Bight, while Kirby (1992) defined hypsometric (or hypsographic) curves for four British estuaries. Dronkers (1986) suggested that the intertidal zone of an accreting area such as the Wadden Sea in the Netherlands was typified by a convexo-concave hypsometric curve in which more than 50% of the mudflat area lies below mean tide level. On the other hand, he showed that the hypsometric curve for an eroding estuary such as the Oosterschelde, again in the Netherlands, although of a similar general convexo-concave shape, lay higher in the tidal frame – more than 50% above mean tide level – and had a more pronounced lower convexity and less obvious upper concavity (Figure 8.10). Kirby (1992) provides a composite hypsometric curve (Figure 8.11) for his sample of British estuaries together with those from the German Bight as recorded by Dieckmann and Partensky (1985). Again, all curves are convexo-concave within the intertidal zone, with the exception of that for Stert Flats, in the Bristol Channel, a composite sand and mudflat beach whose location may be considered more open coast than estuarine. The complete series may be interpreted as representing a sequence in which the lower, flatter profiles represent either the initial stages in the long-term secular development of estuaries or a phase in the periodic oscillation around a steady-state morphology. It is worth noting here that, if the latter explanation obtains, the lower, flatter profile may represent either the end of an erosional phase or the inception of an accretionary phase, so that attempting to infer process from form is not possible.

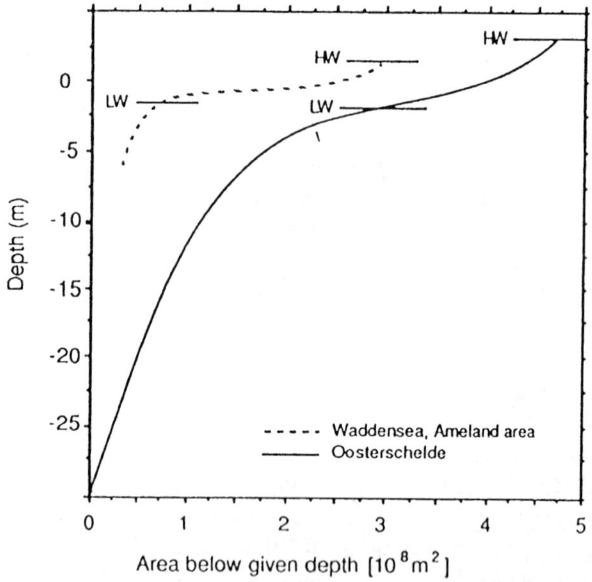

Figure 8.10 Hypsometric curves for an eroding estuary (the Oosterschelde) and an accreting estuary (Ameland, Waddensea). Note the eroding estuary mudflats lie higher in the tidal frame with a more pronounced lower convexity. Compare with the Colne cross-sections shown in Figure 8.2. (After Dronkers 1986)

8.5 MUDFLATS AND SALTMARSHES

While this chapter is primarily concerned with the morphology of mudflats it would be inappropriate, indeed inexcusable, to exclude some discussion of saltmarshes, since the morphodynamics of these upper vegetated surfaces of the intertidal zone are so closely related to that of the unvegetated mudflats that the two zonal types may be regarded as part of the same geomorphological unit. Saltmarsh morphology has been studied in much greater detail than that of mudflats and only a brief discussion is possible here.

The saltmarsh/mudflat intertidal system can be regarded as a single integrated whole whose physical processes are interdependent. Saltmarshes develop on the upper intertidal surface where wave and tidal attenuation by the outer unvegetated mudflat provides low-energy conditions – suitable for vegetation colonisation. Since critical erosion thresholds are predominantly exceeded by wind waves, the exact location of the saltmarsh/mudflat boundary appears to reflect the average wave conditions in the estuary, but to respond to above-average wave events by regression (that is, landward recession) and below-average events by transgression (that is seaward advance). Pethick (1992) pointed out that, for saltmarshes in the outer part of an estuary, the possibility of increasingly high magnitude, but rarer, wave events meant that the position of the saltmarsh/mudflat boundary must be

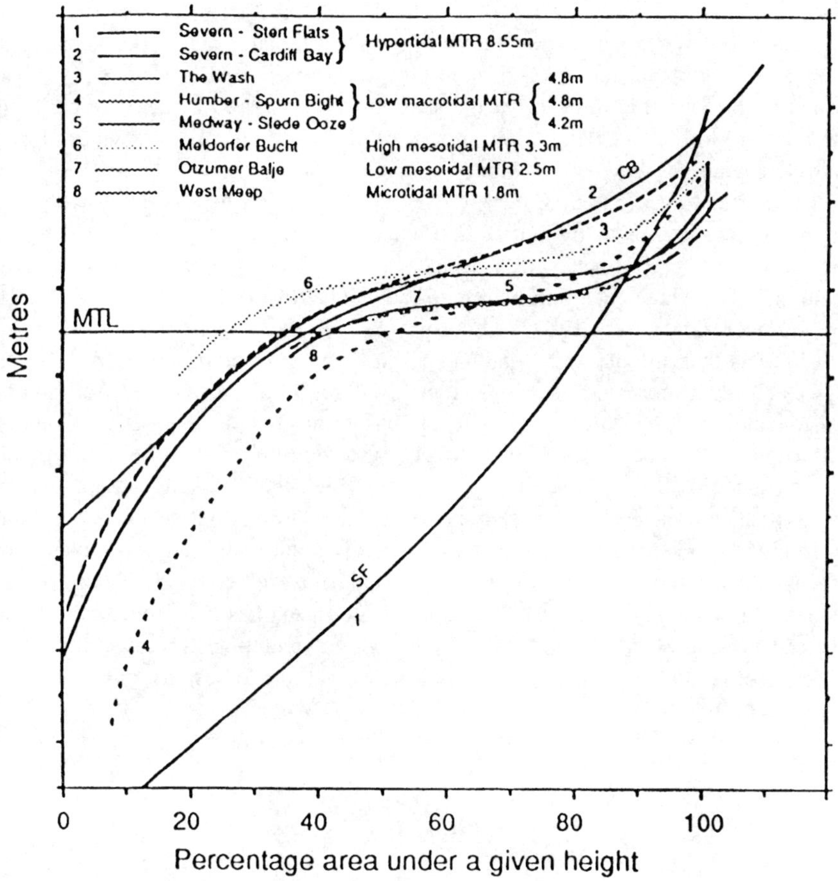

Figure 8.11 Hypsometric curves for a sample of British and German estuaries. The variations to the overall convexo-concave shape may represent a developmental sequence. Note that Stert Flats in the outer Severn may be more representative of an open coast than an estuary. (After Kirby 1992)

constantly shifting: eroding landwards during such rare events and then affecting a depositional recovery during intervening calm conditions.

Once developed in the hydraulic shelter of the mudflat, the saltmarsh vegetation itself provides a hydraulically rough surface that is capable of attenuating excess wave energy which may propagate over the mudflat; thus the saltmarsh and mudflat together provide a more stable environment than either surface could by itself. Moreover, the drag provided by the saltmarsh vegetation on tidal and wave-driven currents results in a velocity profile which prevents re-erosion of sediment except under the most severe wave conditions (Pethick et al. 1990). This means that sediment accretion rates are maximised so that the saltmarsh acts as a sediment store. During high-energy wave events in the outer estuary, the erosion of the upper mudflat can result in the recession of the saltmarsh edge, releasing large volumes of sediment that are deposited on the lower mudflat, so lengthening and flattening the intertidal profile. At the same time the erosion of the marsh allows the mudflat profile to extend landwards so again providing a flatter, wider profile and increasing the wave attenuation. This means that the morphological response of the intertidal profile to high-energy wave events is considerably enhanced when the profile consists of both mudflat and marsh – a relationship that can be likened to that which exists between sand beaches and sand dunes – a symbiotic relationship in which each unit both gains advantage from, and supplies advantage to, the other.

Mudflats that do not exhibit saltmarshes at their upper margins are likely to be less stable than those with an upper marsh, taking longer to recover from storm erosion and in some cases experiencing long-term erosion. This is particularly noticeable in those east coast UK estuaries suffering the effects of sea level rise. The presence of flood embankments along the high tide mark of these estuaries means that, as sea level rises, saltmarshes are unable to migrate landward. The erosion of the outer marsh edge thus progressively reduces the saltmarsh extent – a process now referred to in Britain as 'coastal squeeze' (English Nature 1992; Doody, this volume Chapter Eighteen). Once the marsh has been removed by this process, erosion of the upper mudflat is accelerated resulting in a lowering of the upper profile and intensification of the erosional processes, as outlined in a previous section.

8.6 FURTHER RESEARCH

More questions are raised in this chapter than have been answered – and it is clear that mudflats remain a complex interdisciplinary area of research. Many of the major issues requiring investigation will be evident from the discussion given above but it may be useful to summarise some of the main topics here. Four basic problems can be identified:

1. The relationship of mudflats to estuarine morphodynamics.
2. The depositional and erosional processes on intertidal profiles, including both mudflats and saltmarshes.
3. The temporal development of estuarine morphology over a range of periods up to and including the Holocene.
4. The need to integrate estuarine process–form responses within a predictive model framework.

Each of these general topic areas can be used to generate a series of specific hypotheses by the individual investigator, as given in the following examples.

8.6.1. Mudflats and estuarine morphodynamics

The tidal inputs into estuaries depend to a large extent on the nature of the adjoining sea bed morphology so that, for example, asymmetric tides characterise shelf sea estuaries. Does intertidal morphology reflect these variations in inputs to the system?

Can the relationship between estuary mouth cross-section and tidal prism (e.g. O'Brien 1931) be used to generate a three-dimensional model of the whole estuarine system or does the interdependency of form and tidal properties (Wright et al. 1973) prevent the development of such a simple unidirectional model?

The development of partial or complete standing wave tides in an estuary is reflected in major changes in, for example, the tidal stage – velocity distribution and tidal amplitude. Do such differences result in variations in intertidal morphology, so that, for example, estuaries which experience a progressive tidal wave and high tide velocity maxima exhibit concave intertidal profiles?

8.6.2 Process studies

Intertidal mudflats are beaches responding to wind wave energy and, at the same time, the banks of estuarine channels responding to tidal current velocities. What is the relative importance of these two process–response systems? Can spatial difference in their relative importance be recognised either within or between estuaries?

Does the profile shape of intertidal mudflats represent a response to variations in erosion and deposition? For example, do concave profiles represent erosional forms and convex profiles a response to deposition?

What are the mechanisms by which intertidal mudflats attenuate wave energy? Can such attenuation be modelled?

8.6.3 Temporal development

What are the periodicities involved in the erosional/depositional phases on intertidal mudflats? Can these be related to short-term variations in tidal inputs (semi-diurnal or spring neap changes in velocity); to medium-term process changes such as annual changes in surface flora or wave energy; or to long-term homeostatic mechanisms?

There is an urgent need for the reconstruction of the palaeogeographies of estuaries during the Holocene, using seismic or coring techniques supported by rigorous dating methods of recognisable intertidal palaeo-surfaces. Such work would allow the long-term development of estuarine intertidal morphological response to sea level change to be identified and predictive models constructed.

8.6.4 Predictive models

Prediction of intertidal mudflat behaviour over medium to long time periods remains one of the central challenges to coastal scientists. The range of models, from empirical extrapolations using Holocene reconstructions, through morphodynamic models using iterative process response relations with morphological feedback, to top-down models utilising regime or energy system approaches, requires urgent investigation.

8.7 CONCLUSION

The morphology of intertidal mudflats represents a response to two distinct energy sources and varies both spatially within the estuary and periodically over time. From the discussion given in this chapter, some broad generalisations emerge. Mudflats are the banks of tidal estuaries, so that their geomorphology can only be assessed in the context of the wider response of the estuary as a whole. The complex form ratios of estuaries, linking gross morphological properties such as mean depth, width and tidal length, develop so that tidal forces are balanced by morphological damping to produce a channel system under tension. The system is, however, characterised by the simultaneous adjustment of tidal inputs and estuarine morphology so that a simple process–form explanation is precluded. Modelling this system using conventional hydrodynamic models may therefore be inappropriate and an approach using some form of top-down modelling is advocated. Such models must define the equilibrium state of an estuary before details of process–form interaction may be defined.

Short-term responses of the intertidal mudflat system are mainly driven by wind-wave energy; morphological responses are essentially similar to those of sand beaches although differing in sediment processes and in response time. A distinction is made between the mudflat morphology of the outer estuary, where open sea waves are able to propagate, and that of inner estuary, where only internally generated waves occur.

Medium-term periodic changes to mudflat morphology may take place over time-scales measured in decades to centuries and represent oscillations around a quasi-steady state morphology. The sequence involves the development of high-elevation, convex mudflats during flood-dominant phases which produce ebb-dominant conditions and lead to the erosion of the upper mudflats to give low-elevation concave profiles. This morphology is associated with flood-dominant conditions, accretion is resumed giving high-elevation convex profiles once more. If these periodic changes do take place, then care must be exercised in the interpretation of erosional conditions on mudflats as indicative of long-term deterioration.

Medium- to long-term secular changes in the morphological response of the intertidal system, occurring over periods of decades, include those brought about by sea level rise, in which estuary cross-sections are observed to widen and shallow thus allowing increased attenuation for the increased tidal energy forced into the system. This response is itself reinforced by minor changes to the intertidal profiles brought about by the increased wave energy at the shore due to the deeper water conditions. The inability of many estuaries to respond to such changes due to the presence of flood embankments causes the loss of mudflat and marsh rather than their landward transgression and is seen as a major management issue in the immediate future.

Finally the importance is emphasised of the process relationship between unvegetated mudflat and vegetated saltmarsh which enables short-term morphological responses to storm events. Saltmarshes can only occur in the wave shadow of mudflats but themselves provide additional wave attenuation and a supply of sediment to the mudflat in severe storm events. Thus a stable intertidal system consists of both mudflat and saltmarsh (or mangrove) in an integrated landform.

The increasing interference with natural intertidal systems over the past few centuries and the increasing importance of the remaining areas both economically and ecologically, provide an urgent need for further research into mudflat form and function. Management of estuaries for navigation may involve channel dredging, which can have serious impacts on the morphology of the intertidal mudflats and saltmarshes. Similarly reclamation of intertidal areas for urban, industrial or port development can cause major changes in the morphodynamic balance of the whole estuary. Management urgently needs research into these negative impacts so that long-term strategic plans may be initiated to alleviate and even reverse the deterioration of these important systems.

REFERENCES

Anderson, F.E. 1973. Observations of some sedimentary processes acting on a tidal flat. *Marine Geology* 14: 101–116.

Amos, C.L. and Moster, D.C. 1985. Erosion and deposition of fine grained sediments from the Bay of Fundy. *Sedimentology*, 32: 815–832.

Bishop, C.T., M.G. Skafel and R.B. Nairn. 1992. Cohesive profile erosion by waves. *Proceedings of 23rd Coastal Engineering Conference., ASCE, New York* 3: 2976–2989.

Boon, J.D.I. and R.J. Byrne. 1981. On basin hypsometry and the morphodynamic response of coastal inlet systems. *Marine Geology* 40: 27–48.

Brew, D.S. 1990. *Sedimentary environments and Holocene evolution of the Suffolk estuaries.* Ph.D. dissertation, University of East Anglia, Norwich, UK.

Bruun, P. 1978. *Stability of Tidal Inlets*. Elsevier, Amsterdam.

Bruun, P. and Gerritson, F. 1960. *Stability of Tidal Inlets.* N. Holland Publ., Amsterdam.

Capobianco, M., H.J. de Vriend, R.J. Nicholls and M.J.F. Stive. 1993. Long term evolution of coastal morphology and its effects on the coastal environment, p. 889–903. *In MEDCOAST 93 (1st Int. Conf. Mediterranean Coastal Env.)*, Turkey.

Dieckmann, R. and H.W. Partensky. 1985. Quantitative geomorphology of tidal flat areas, p. 53–62. *In* Sigbjarnarson, J. (ed.), *Iceland Coastal and River Symposium.* Library of the

National Energy Authority, Reykjavik, Iceland.

Dronkers, J. 1986. Tidal asymmetry and estuarine morphology. *Netherlands Journal of Se Research* 20: 117–131.

Dyer, K.R. 1986. *Coastal and Estuarine Sediment Dynamics*. John Wiley, Chichester.

English Nature. 1992. *Coastal Zone Conservation. English Nature's Objectives, Rational and Recommendations*. English Nature, Northminster House, Peterborough.

Escoffier, E.F. 1940. The stability of tidal inlets. *Shore and Beach* 8: 114–115.

Evans, G. 1965. Intertidal flat sediments and their environment of deposition in the Wash *Quaternary Journal of Geological Society of London* 121: 209–245.

Fitzgerald, D.M., D. Nummedal and T.W. Kana. 1976. Sand circulation patterns at Pric Inlet, SC, p. 1868–1880. *Proceedings of 15th Coastal Engineering Conference, Honolulu* vol. II.

Frostick, L.E., and I.N. McCave. 1979. Seasonal shifts of sediment within an estuar mediated by algal growth. *Estuarine, Coastal and Marine Science* 9: 569–576.

Gao, S. and M. Collins. 1994. Tidal inlet equilibrium in relation to cross-sectional area an sediment transport patterns. *Estuarine, Coastal and Shelf Science* 38: 157–172.

Hayes, M.O. 1975. Morphology of sand accumulation in estuaries. *In* L. Cronin (ed.) *Estuarine Research: II*. Academic Press, New York.

Kirby, R. 1990. The sediment budget of the erosional intertidal zone of the Medway estuary Kent. *Proceedings of Geological Association* 101(1): 63–77.

Kirby, R. 1992. Effects of sea-level rise on muddy coastal margins. *In* D. Prandle (ed.) *Dynamics and Exchanges in Estuaries and the Coastal Zone*. American Geophysica Union, Washington, DC.

Krone, R.B. 1962. *Flume Studies of the Transport of Sediment in Estuarial Shoaling Processes*. Hydraulic Engineering Research Laboratory. University of California Berkeley.

Maa, P.-Y. and A.J. Mehta. 1987. Mud erosion by waves: a laboratory study. *Continenta Shelf Research* 7: 1269–1284.

Madsen, O.S. and W.D. Grant. 1976. Quantitative description of sediment transport by waves, p. 1093–1112. *In 15th Coastal Engineering Conference*. ASCE, New York.

McDowell, D.M. and B.A. O'Connor. 1977. *Hydraulic Behaviour of Estuaries*. Macmillan London.

Mehta, A.J. 1988. Laboratory studies on cohesive sediment deposition and erosion p. 427-445. *In* J. Dronkers and W. van Luessen (eds.), *Physical processes in estuaries*. Springer Verlag, Berlin.

O'Brien, M.P. 1931. Estuary tidal prism related to entrance areas. *Civil Engineering* 1(8) 738–739.

O'Brien, M.P. and R.G. Dean. 1972. Hydraulic and sedimentary stability of tidal inlets, p 761–780. *In 13th Coastal Engineering Conference*, ASCE, New York.

Odd, N.V.M., and Owen, M.W. 1972. A two layer model of mud transport in the Thame estuary. *Proceedings of the Institute of Civil Engineers Supplementary Vol. 1972* 175-205.

Owen, M.W. 1976. *Determination of settling velocities of cohesive mud*. Hydraulics Research Report No IT 161.

Paterson, D.M. 1989. Short term changes in the erodibility of intertidal cohesive sediments related to the migratory behaviour of epipelic diatoms. *Limnology and Oceanography* 34(1): 223–234.

Paterson, D.M., R.M. Crawford and C. Little. 1990. Subaerial exposure and changes in the stability of intertidal estuarine sediments. *Estuarine, Coastal and Shelf Science* 30: 541-556.

Pethick, J.S. 1984. *An Introduction to Coastal Geomorphology*. Edward Arnold, London.

Pethick, J.S. 1991. Tidal dynamics and estuarine management in the Humber Estuary, England, p. 1456–1468. *In* O. Magoon (ed.), *Coastal Zone 91* Vol. 2. ASCE, Long Beach, CA.

Pethick, J.S. 1992. Saltmarsh geomorphology, p. 41–62. In J.R.L. Allen and K. Pye (eds.), *Saltmarshes*. Cambridge University Press, Cambridge.

Pethick, J.S. 1994. Estuaries and wetlands: form and function. *In* R.A. Falconer (ed.). *Wetland Management*. Institution of Civil Engineers, London.

Pethick, J.S., D. Leggett and L. Husain. 1990. Boundary layers under saltmarsh vegetation developed in tidal currents, p. 113–124. *In* J.B. Thornes (ed.), *Vegetation and Erosion*. John Wiley, Chichester.

Postma, H. 1961. Transport and accumulation of suspended matter in the Dutch Wadden Sea. *Netherlands Journal of Sea Research* 1: 148–190.

Speer, P.E. and D.G. Aubrey. 1985. A study of non-linear tidal propagation in shallow inlet/estuarine systems. *Estuarine, Coastal, and Shelf Science* 21: 207–224.

Strahler, A.N. 1952. Hypsometric (area–altitude) analysis of erosional topography. *Geological Society American Bulletin* 63: 1117–1141.

Tubman, M.W. and J.N. Suhaya. 1976. Wave action and bottom movement in fine sediments, p. 1168–1183. *Proc. 15th Conf. Coastal Eng.*, Vol. 2. ASCE, New York.

van Dongeren, A.R. and H.J. de Vriend. 1994. A model of morphological behaviour of tidal basins. *Coastal Engineering* 22: 287–310.

Van Straaten, L.M.J.U. 1965. Coastal barrier deposits in south and north Holland. *Mededelingen Rijks Geologische Dienst* 17: 41–75.

Wright, L.D., J.M. Coleman and B.G. Thom. 1973. Processes of channel development in a high-tide range environment: Cambridge Gulf–Ord River Delta, W Australia. *Journal of Geology* 81: 15–41.

Chapter Nine

Wave and current processes and beach changes on a microtidal lagoonal beach at Fire Island, New York, USA

KARL F. NORDSTROM
Rutgers University
NANCY L. JACKSON
New Jersey Institute of Technology
JAMES R. ALLEN
National Biological Service

and

DOUGLAS J. SHERMAN
University of Southern California

9.1 INTRODUCTION

Microtidal beaches occur in estuaries adjacent to open coasts where regional tidal ranges are low, such as on the Gulf Coast of the USA, but they also can occur in regions where higher tidal ranges occur on the open coast. Examples of the latter type of beach are found in Albemarle and Pamlico Sounds, North Carolina (where tidal range in the estuaries is 0.15 m and tidal range on the ocean side is 1.0 m) and many of the narrow, shallow lagoons behind barrier islands and spits on the northeast coast of the USA, such as Fire Island and Island Beach in New York and New Jersey (Figure 9.1). Tidal ranges in the lagoons near the middle of these two barriers are 0.15 m to 0.21 m, whereas tidal ranges on the adjacent ocean coast range from 1.25 m to 1.43 m.

Estuarine Shores: Evolution, Environments and Human Alterations. Edited by K. F. Nordstrom and C. T. Roman.

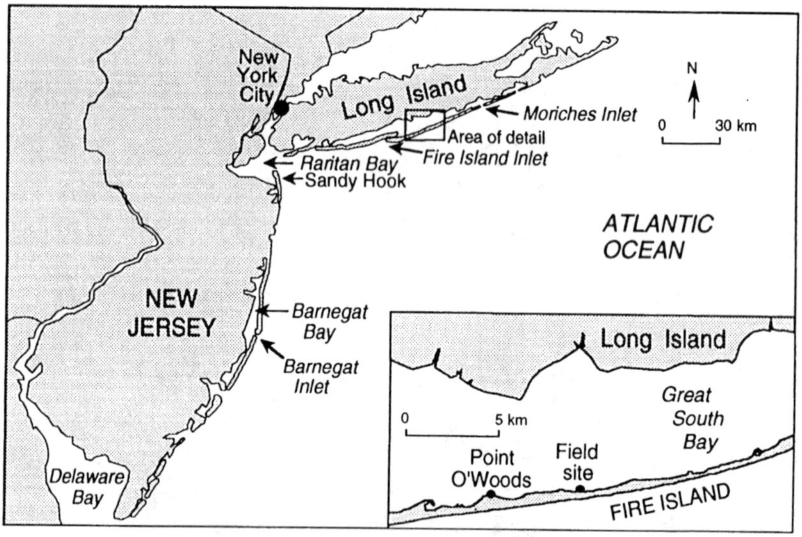

Figure 9.1 Location of field site at Fire Island, New York

Microtidal estuarine beaches have been examined in the past (Rosen 1977, 1980; Wang and Chang 1978; US Army Corps of Engineers 1981; Love et al. 1985; Nordstrom et al. 1989, 1991; Armbruster et al. 1995), but little is known about the wave regimes on these beaches or the relationship between wave and current energies and beach change. Inferences have been made about microtidal estuarine beaches based on qualitative analysis of topographic features (Rosen 1977, 1980; Nordstrom and Jackson 1992; Nordstrom 1992). These studies imply that: (1) changes in water level associated with storm surge play a relatively greater role in beach morphodynamics on microtidal beaches (tidal range <1.0 m according to Hayes 1979) than on mesotidal beaches; (2) the resonant growth of long-period waves will be more conspicuous on microtidal beaches (Nordstrom and Jackson 1992); and (3) effects of tidal currents are reduced, resulting in more conspicuous shore-perpendicular forms (Hayes 1975; Rosen 1975).

Changes in estuarine water levels due to atmospheric forcing can be locally induced (e.g. by winds blowing across an estuary) or they can be induced by flow of water through inlets as a result of raised water levels generated on the open ocean coast (Dupre 1985). Water-level changes due to wind can greatly exceed tidal ranges on microtidal shorelines (Wong and Wilson 1984; Nordstrom 1992). In general, an onshore wind raises water levels, whereas an offshore wind lowers water levels, and winds can pile water up on one side of the bay while lowering water levels on the other side. Some studies conducted on estuarine beaches (Jackson 1992, 1995) show that an opening to the sea on the downwind side of an estuary can result in lower water levels on the downwind estuarine beach despite onshore winds at that site. The applicability of this concept has been examined for beaches close to inlets on microtidal shorelines (Armbruster et al. 1995) but not for beaches far from inlets in shallow lagoons, where hydraulic inefficiency may damper inlet effects.

Differences in tidal range cause differences in the amount of time wave energy is concentrated in a given location on a beach (Wright et al. 1982; Jago and Hardisty 1984). High water levels can be sustained for longer periods on beaches with low tidal ranges because the elevation is determined by wind and wave characteristics, rather than tidal effects. The surf and swash zones of microtidal beaches thus remain in the same location for long periods of time (Rosen 1977). A more fixed surf zone position increases the likelihood of resonant growth of long-period energy (Wright et al. 1982).

Low tidal ranges tend to favor bars on the low tide terraces and transverse bar and rip morphology (Wright et al. 1986; Short 1991), and field observations of estuarine beaches in microtidal environments reveal features resembling these bars (Nordstrom 1992). The relatively large size of these bars and their attachment to the beach foreshore imply that they affect change on the upper beach and that sediment may be exchanged between them (Bruner and Smosna 1989), but sediment transfers between the upper foreshore and these bars have not been monitored. The bars should be more dynamic than on mesotidal shorelines because they occupy a larger portion of the intertidal frame.

The purpose of this chapter is to evaluate the relationship between winds, waves, currents and beach change on a microtidal beach in a shallow lagoon-type estuary using field data gathered during a 22-day experiment on the bay side of Fire Island, New York. Research issues addressed include: (1) how much effect storm surge would have during moderate winds; (2) whether the inverse relationship between wind direction and water level due to ocean influences could occur at distances greater than 15 km from an inlet; (3) whether there is a conspicuous buildup of long-period energy due to relative stationarity of the breaker zone; (4) how the mobility of the transverse bars compares to the mobility of the upper foreshore and low tide terrace; and (5) whether the upper foreshore and bars are linked through sediment exchanges.

Previous papers resulting from this field deployment include an analysis of sediment mixing by waves and swash (Sherman et al. 1994) and a preliminary description of beach profile changes and sand transport by wind (Jackson et al. 1993).

9.2 FIELD SITES AND METHODS

9.2.1 Field sites

The Fire Island field site is a sand beach on the bay side of a barrier island fronting Great South Bay (Figures 9.1 and 9.2). The site is 17 km east of Fire Island Inlet and 32 km west of Moriches Inlet (Figure 9.1). Mean tidal range at Point O'Woods,

Figure 9.2 Fire Island, looking southwest. The study site is in the lower left of the photo

2.5 km to the west of the field site, is 0.21 m; spring tidal range is 0.24 m (NOAA 1995). The shoreline along this portion of the bay is characterized by a low, narrow sand beach fronting portions of eroding salt marsh, an overwash platform formed by oceanic overwash and isolated dunes that formed in the past as a result of transport by processes initiated on the oceanside. Small cuspate forelands that have accreted as a result of bay processes occur in places, but most of the shoreline shows evidence of erosion, including the outcrops of peat substrate, exhumed tree roots and scarped dunes that are commonly found on eroding microtidal lagoon beaches (Nordstrom 1992; Armbruster et al. 1995).

The strongest and most frequent winds with an onshore component on the bay shoreline blow from the northwest, but strong northeast winds occur during passage of low-pressure centers. Fetch distances are 12 km to the northwest and 15 km to the northeast. The upper foreshore and the low tide terrace (Figure 9.3) are separated by a conspicuous break in slope, as is common on estuarine beaches (Nordstrom 1992; Jackson and Nordstrom 1992). The slope of the upper foreshore averages 8.0°; the slope of the low tide terrace is <0.5°. The transverse bars found on the inner low tide terrace have a local relief between 0.1 and 0.2 m. Their stoss slopes face in the direction of the northwesterly winds.

The upper foreshore of the study beach is on the bayward portion of an ocean overwash lobe that was, in turn, reworked by bay overwash, forming a beach ridge of bay origin. There is a 2.0° difference in beach orientation in the eastern (azimuth

Figure 9.3 Study site 29 March, looking west. The small rods extending across the foreshore and onto the low tide terrace are the profile rods spaced at 1 m intervals along profile line E (identified in Figure 9.4). The instrument stand for the current meter and pressure transducer located 11 m from the baseline for topographic surveys (Figure 9.4) is the conspicuous feature at the break in slope. The photo was taken after two days of strong northwesterly winds, and the upper foreshore shows deflation pedestals resulting from aeolian action in the previous 24 hours (reported in Jackson et al. 1993). Water levels are lower than normal, revealing quasi-bar along the profile line W (extending bayward from meteorological mast) and Bar W beyond it (identified in Figure 9.5)

289) and western (azimuth 287) portions of the study site where beach profile changes were monitored (Figure 9.3). The foreshore in the western portion of the study area is truncated by a 0.24 m high erosion scarp cut into the overwash platform. This scarp is inactive under low and moderate wave-energy conditions because wrack from vegetative litter (largely eel grass, *Zostera marina*) prevents wave uprush from removing sediments. Runup on the beach to the west is not restricted by a scarp, and the beach is wider (Figure 9.3). Beach sediments are predominantly medium-sized quartz and feldspar sands. Heavy minerals include ilmenite, garnet and magnetite (Williams and Leatherman 1985).

9.2.2 Data collection and analysis

The field deployment occurred from 11 March to 7 April 1992; data were collected in daily time series 15 March to 5 April. This time period was selected because mid-latitude cyclonic storms are common then and because the beach is clear of much of the vegetative litter that accumulates in the fall and early winter. Wind, wave and current data were gathered daily at high tide. Data were collected throughout selected high-wind events to characterize spatial and temporal changes in winds, water levels, waves and currents. Data were gathered at 0.5 s intervals for 17.1 min on a Sea Data model 1255B-27 data logger.

Wind speed was monitored using four Gill 3-cup anemometers. Wind direction was measured using a Gill microvane. These instruments were mounted on a 6 m high meteorological tower placed on the overwash platform 3.5 m landward of the scarp on the western cross-shore profile line (Profile W, Figures 9.3 and 9.4). The placement of the meteorological tower near the water on the bayside provides an estimate of the velocity of onshore winds that can be used to assess the potential for wave generation in the bay and the potential for generation of surge (including the negative surge resulting when water is blown from the coast on the oceanside). Winds that are offshore on the bayside are affected by the topography and vegetation of the barrier island. The offshore winds monitored at the site are thus considerably lower than the regional wind conditions; they may differ from those winds in direction; and they do not provide a meaningful picture of the potential for creation of waves in the bay or oceanside storm surge that could enter the inlets.

Data on flow velocities were measured approximately 0.12 m above the bed using three Marsh-McBirney model 511 bidirectional electromagnetic current meters. The current meters were placed in a cross-shore array on the eastern line (Profile E) but were offset to the west (Figure 9.4) so the instrument stands would

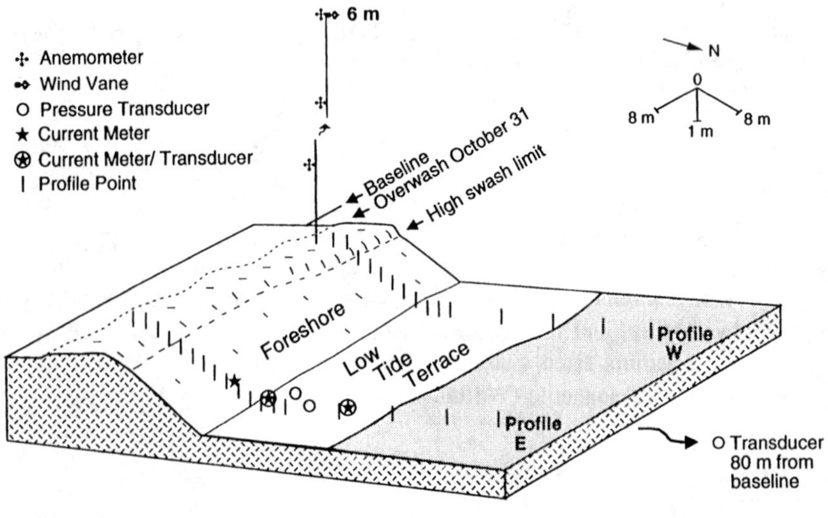

Figure 9.4 Location of instruments and rods used as profile sampling points along Profile E and Profile W. The southern limit of the overwash platform represents effects of a storm that occurred 31 October 1991

not affect processes at the rods used to identify changes in sediment level. One current meter was at the break in slope on the low tide terrace; one current meter was 4 m farther bayward on the inner portion of the low tide terrace. The third current meter was used on the upper foreshore to obtain perspective on swash processes. The sensor of this instrument had to be relocated frequently to keep it within the migrating swash zone, and the current meter was deployed only on days when tracer sediments were injected, as described below. The sensor came out of the water periodically during data runs because of shallow water depths, so data are only examined for portions of each record. Swash zone data reported in this chapter are for 31 March and 3 April because of the completeness of records for those days.

Water levels were measured using five pressure transducers. Two of these were co-located with current meters (Figure 9.4). Two other transducers were placed to form a triangular array to identify wave angle. The fifth transducer was placed 80 m offshore of the foreshore to obtain data on conditions representative of deeper water. Significant wave heights were calculated as 4 $\times$ standard deviation of the wave record (CERC 1984). Wave periods represent the peak energy variance derived from the spectral estimates of pressure transducer data. Modal wave and current characteristics were derived from the daily high tide records.

Daily changes in beach-surface elevation were monitored along two shore-normal profile lines that were selected to represent the different beach forms found along this shoreline reach. Runup on Profile W was restricted by the scarp. This profile had an accumulation of sediment on the low tide terrace just bayward of the break in slope. We refer to this feature as a quasi-bar to distinguish it from the well formed transverse bars to the east and west (Figure 9.5) and the more planar low tide terrace on which it formed. Profile E had had a wider foreshore, and swash uprush was unrestricted; the low tide terrace was lower in elevation along this profile.

Additional elevation changes were monitored on the two transverse bars located closest to the cross-shore profile lines (Figure 9.5). West Bar is about 0.07 m higher in elevation and is longer and wider than East Bar and it thus has a greater effect on, and responsivity to incident waves.

Changes in beach elevation were measured at successive low tides using 6.4 mm diameter profile rods as local subdatums (Figures 9.3 and 9.4). Profiles E and W were monitored at 1.0 m intervals on the upper foreshore and inner low tide terrace and at 5.0 m intervals farther out on the less active low tide terrace. A loose-fitting washer was placed over the rods on Profile E to enable identification of depth of activity of sediments according to the procedure identified in Greenwood and Hale (1980). The value used as maximum depth of activity is the greater of the two values of washer depth below the initial and post-event surface elevations.

The profile rods on the transverse bars (Figure 9.5) were oriented perpendicular to their axes. The bars are asymmetric, with steeper slopes on the southeast (lee) side. Spacing of rods was designed to obtain detail on movement of the lee slope. Rods were placed at 3 m intervals (3 and 6 m northwest of the crest, on the crest, and 3 m southeast of the crest). A rod was also placed on the lee slopes of the bars

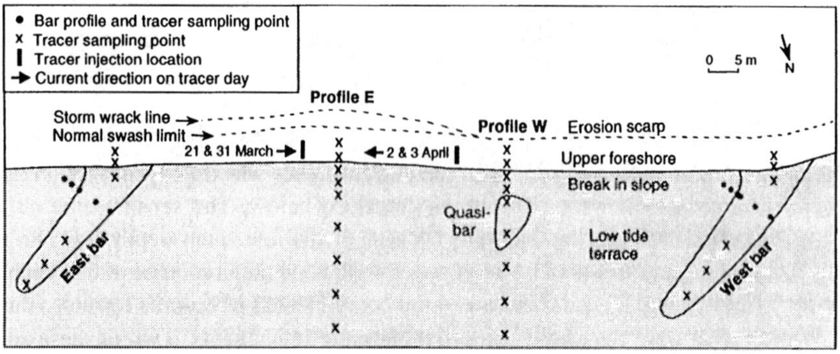

Figure 9.5 Tracer injection locations and sampling points on bars and cross-shore profiles

1.5 m from the crest. (Supplemental stakes were added on the stoss and lee side of the west bar 0.1 m from the crest to identify microtopographic changes.)

Dyed sand was injected into the swash zone on four different days to provide insight to the relative proportion of sediment that is transfered from the foreshore to the bars. Batches of fluorescent tracer sands were made in four different colors using the polyvinyl chloride acetate resin recipe described in McArthur (1980). Tracer was injected in a line across the swash zone on 21 March and 31 March, when current directions were toward the west (Figure 9.5), and on 2 April and 3 April, when current directions were toward the east. Sand samples were taken using 100 mm long, 50 mm diameter clear plastic tube cores, yielding an approximately 245 g sample when dry. The samples all were taken at low water on 4 March. Sampling points (Figure 9.5) included: (1) the locations transverse to the bar axes that were used to sample elevation changes; (2) points along the bar axes extending 8, 13, and 21 m bayward of the rod located at the crest of the bar; (3) 5 m landward of the rod at the crest (just bayward of the break in slope); and (4) two points low on the foreshore at 1.5 m intervals along a line normal to the foreshore. Profiles E and W were sampled at 2 m intervals on the upper foreshore and inner low tide terrace and at the 5 m profile points farther bayward. Grain size statistics were run on selected cores to compare sediments on the bar with sediments gathered on the upper foreshore and presented in previous studies (Jackson et al. 1993; Sherman et al. 1994). These sediments were washed, dried and sieved at 0.5 phi intervals. Inclusive graphic statistics were calculated according to the procedure in Folk (1974).

9.3 RESULTS

9.3.1 Wind, wave and current characteristics

Table 9.1 provides wind and wave data for records that were taken closest to the time of highest water on each day. This time was selected because wave energies

Table 9.1 Wind and wave data taken near times of high water

Day	Time (h min)	Wind direction (°)	Relative wind direction (°)*	Wind speed ($m\,s^{-1}$)	Water depth (m)	Sig. wave height (m)	Wave period (s)
March							
15	−1 52	308	72	10.2	ND	0.09	2.3
16	−1 26	318	62	4.5	ND	0.13	2.1
17	+1 04	234	−34	4.7	ND	0.05	1.5
18	+0 47	016	4	3.8	ND	ND	ND
19	+0 45	064	44	10.2	0.47	0.12	2.3
20	+2 31	220	−20	2.6	0.65	0.09	2.0
21	+2 13	348	32	7.6	0.70	0.13	2.4
22	−0 32	186	−14	1.7	0.66	0.06	ND
23	−0 43	011	9	3.1	0.75	0.10	1.8
24	−1 53	245	−45	4.4	0.72	0.04	1.3–2.5
25	0 54	177	−23	2.1	0.41	0.03	2.1
26	+0 43	121	−79	4.5	0.27	0.08	2.1
27	0 00	203	−3	4.5	0.50	0.07	ND
28	−0 20	303	77	12.9	0.39	0.10	2.6
29	−1 56	311	69	10.3	0.35	0.07	2.3
30	+1 00	234	−34	2.2	0.42	0.05	2.0
31	+0 05	004	16	7.6	0.39	0.11	2.3
April							
1	+2 58	194	−6	2.0	0.48	0.06	2.5
2	+0 06	338	42	5.3	0.49	0.09	2.1
3	−0 01	302	78	10.4	0.41	0.11	2.3
4	+0 22	238	−38	4.4	0.26	0.10	1.3–1.8
5	−0 45	310	70	8.6	0.37	0.13	2.2

* Degrees relative to shore–normal; negative sign indicates offshore wind.
Data for wind speed are taken from the anemometer at 6 m elevation. Wave data are taken from the pressure transducer located on the low tide terrace closest to the upper foreshore (Figure 9.4). Water depths are relative to the elevation of the bed on 18 March. ND means no data collected. Pressure transducers were placed in temporary positions 15 to 17 March, and water depths were not tied into the local datum.

are often highest during high water when energy loss prior to breaking is minimized. The time of each record relative to predicted high tide is highly variable because the time of highest water is determined by both atmospheric forcing and tidal characteristics. Significant wave heights during onshore winds 10.0 m s^{-1} are low and not much greater than during low-speed onshore winds (Table 9.1). As a result, the correlation between wind speed and wave height is low ($r^2 = 0.1$ for the data portrayed in Table 9.1). Breaker periods are low (2.6 s) as expected for waves that are generated locally within a small basin.

Water depths ranged from 0 m to 0.75 m above the break in slope at the low tide terrace. This range is greater than the 0.21 m mean and 0.24 m spring range. Water levels in Table 9.1 are not always correlated with wind direction at the site because surge levels are a function of the regional shoreline orientation, basin configuration, and ocean surge that passes through the inlets into the bay, and there is a time lag between changes in wind and changes in water level. There is no conclusive evidence for a direct or indirect relationship between wind direction and water level at the site; high water levels can be associated with both onshore and offshore winds.

Energy spectra for wave and current records taken at high water during relatively high wind speeds that blew nearly directly onshore are broad-banded (Figure 9.6). The spectra lack significant energy at periodicities longer than the incident waves, as is common where waves are locally generated in short-fetch

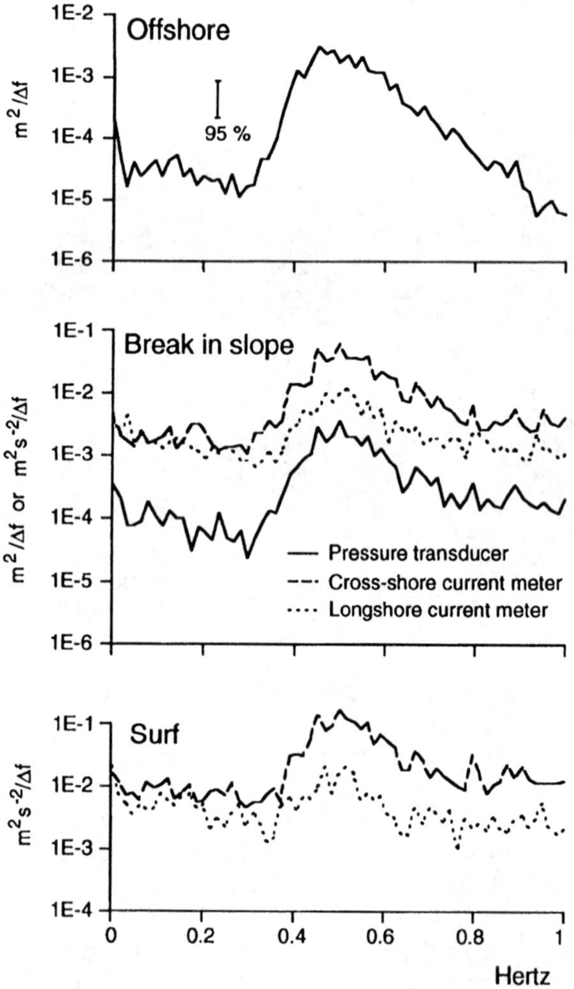

Figure 9.6 Energy spectra for record taken at 08:30 31 March, showing broad-banded character and lack of long-period energy. Peak period is about 2.3 s

environments (Nordstrom and Jackson 1992). Data from the pressure transducer 80 m from the baseline and the pressure transducer at the break in slope show similar levels of energy at incident wave periods, but there are higher levels of energy at greater and lower frequencies at the break in slope. The energy in the cross-shore and longshore flows is greater in the surf zone than on the low tide terrace at all frequencies.

The steep foreshore and low wave energies result in a narrow surf/swash zone. The width of this zone was no greater than 3.0 m during the events monitored during the field deployment. The magnitude of the cross-shore flows in the swash zone (indicated by the standard deviation of the current meter) was generally low. They did not exceed 0.21 m s^{-1} during the 10.0 m s^{-1} wind that occurred on 3 April (Table 9.1). Mean cross-shore flows in the swash on 3 April were at a maximum of 0.04 m s^{-1} when wave height at the break in slope was 0.10 m, but means of other records during that day were all below 0.03 m s^{-1}. The greatest longshore current velocities monitored were 0.24 m s^{-1} on 3 April, when 11.7 m s^{-1} winds blew at an angle of 71° from shore normal and the height of significant waves was 0.13 m.

9.3.2 Beach characteristics

Grain size analysis of sediments collected 26 and 29 March indicates that foreshore sediments have a mean (m_z) of about 1.30 phi (0.41 mm) and a sorting value (s_I) of about 0.30 phi. Sediments on the low tide terrace are only slightly finer and more poorly sorted.

Sweep zones of daily beach profile changes for the two cross-shore lines (Figure 9.7) indicate that the width of the active foreshore is about 10.0 m at Profile E and about 5.0 m at Profile W, where swash uprush is restricted by the litter at the scarp. The slope of the foreshore during the study period did not vary by more than 2°. There is greater mobility on the low tide terrace at Profile W than at Profile E (Figure 9.7). This greater mobility occurs on the low-relief quasi-bar. Daily observations of the location of the upper limit of swash indicate that water levels reached the scarp on Profile W on 9 days. Conspicuous surface elevation changes occurred at the base of the scarp, but the wrack was sufficiently thick to prevent landward displacement of the scarp. There is greater change high on the beach at Profile E where there is no scarp or thick vegetative litter to restrict swash uprush.

The sweep zone at Profile E (Figure 9.7) reveals a horizontal displacement of the beach of 0.77 m at mid-foreshore, although net change over the 20 days was only 0.14 m. The greatest net horizontal change on the mid-foreshore on Profile W over the 20 days was 0.75 m, but the bulk of the upper foreshore was less mobile on a daily basis than on Profile E.

The greatest short-term erosional episodes on Profile E occurred 19–20 March and 1–2 April (Figure 9.8). There was considerable mobility low on the upper foreshore on Profile W 19–20 March (Figure 9.8), but little net change occurred

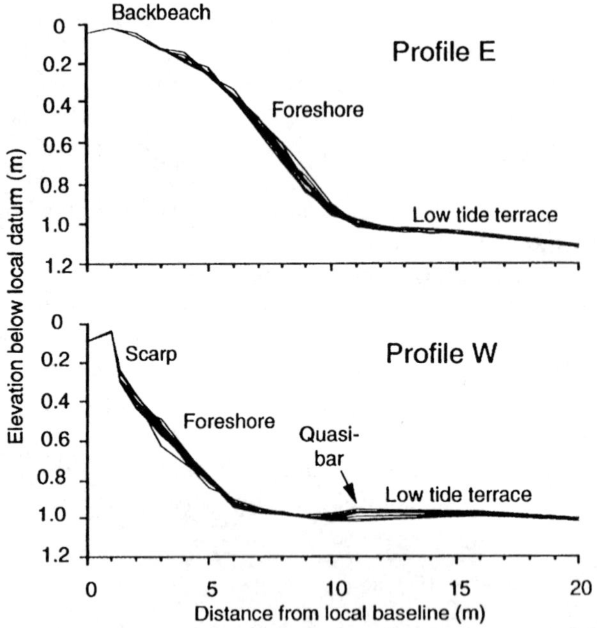

Figure 9.7 Sweep zone profiles on shore-normal sampling lines for the period 17 March to 5 April

along the entire profile. The greatest erosion on Profile W occurred 27–28 March when onshore winds were the strongest of the monitoring period (Table 9.1). The maximum amount of linear retreat on the upper foreshore was 0.43 m at this time. The beach at Profile E showed slight accretion during this period. A relatively great amount of accretion occurred on Profile E between 18 and 19 March, but otherwise profile changes are relatively small throughout the monitoring period. The form of profile change indicates that linear retreat (with little change in slope) and onshore–offshore transport (with local changes in slope) occur, although linear retreat is more conspicuous. Changes on the foreshores of the two profiles were inversely related at times (erosion on one and deposition on the other), suggesting that one beach may function periodically as a feeder beach for the other.

Comparison of daily changes in beach profiles indicates that the volume of change on the low quasi-bar can be greater at times than the volume of change on the foreshore. The changes on these two features can be inversely related (erosion on the foreshore with deposition on the quasi-bar or vice versa), directly related (e.g. erosion on both) or unrelated (erosion on the foreshore with no net change on the quasi-bar).

The greatest depths of activity across the beach were during the strong onshore winds that occurred 28–29 March (Figure 9.9). The greatest depths extend over a short horizontal distance low on the upper foreshore, with the maximum depth in the region of wave breaking near time of high water. Depths of activity on the low tide terrace were 20 mm during this event.

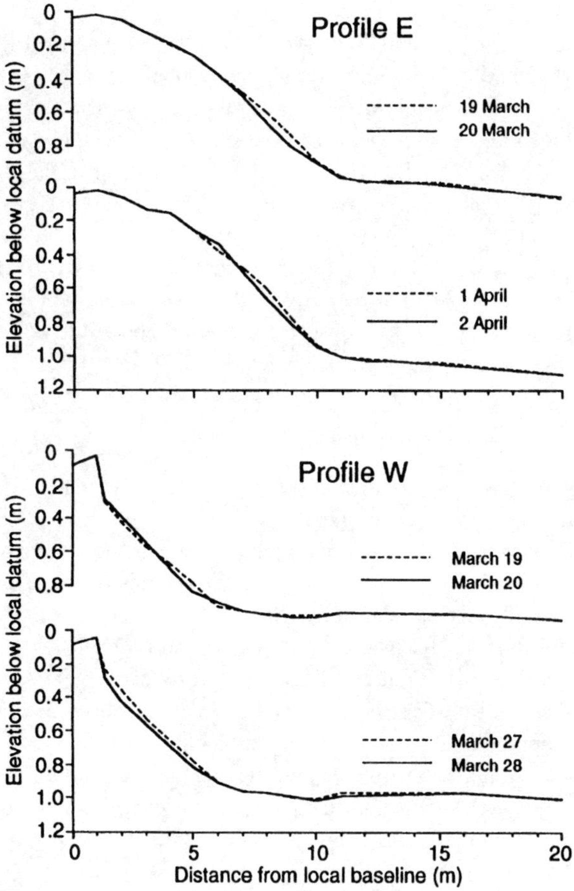

Figure 9.8 Depths of sediment activity across Profile E March 28 to 29

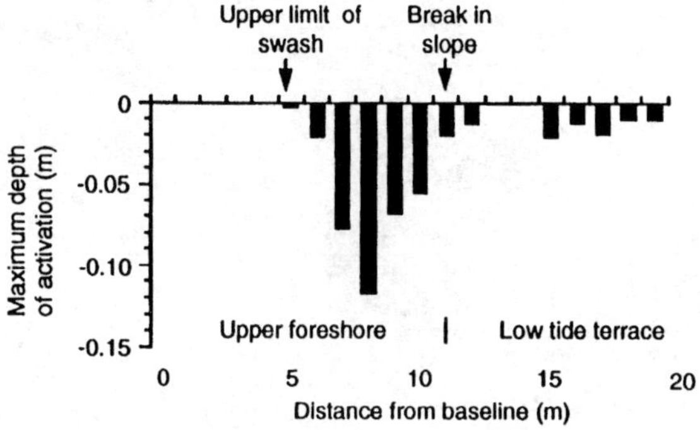

Figure 9.9 Depths of sediment activity across Profile E March 28 to 29

The sweep zones of daily changes at the rods placed at the initial position of th crests of on the transverse bars (Figure 9.10) varied over a vertical range of 36 mr at East Bar and 43 mm at West Bar. The vertical range was nearly 0.1 m 7.5 m fror the first stake on the stoss side of West Bar, where the crest was displaced abou 4.5 m during the study period. Displacement of the stoss slope was about 1.7 m a East Bar and 3.3 m at West Bar (Figure 9.10). Fluctuation in the direction of ba movement indicates that they are driven eastward by northwesterly wind events an westward by northeasterly wind events. Net movement indicates that th northwesterly winds were dominant during the study period. Surface elevatio changes on West Bar are greater than on East Bar, and the variability is greater tha on East Bar, presumably due to its greater elevation above the bed and its greate exposure to wave action during higher water levels.

9.3.3 Sediment transfers from foreshore to bars

Figure 9.11 shows recovery of tracer grains the day after the last injection. Th tracer injected on 21 March (Figure 9.11A) had 14 days to disperse, and th distribution of tracer particles from that injection reflects the effects of both easterl and westerly wind conditions. There are more recoveries of tracer particles on th bars from that batch than there are of particles from other batches. The only othe tracer recovered on either bar was three grains found on East Bar two days after i was injected on 2 April (Figure 9.11C). The results of the tracer experiment indicate that there is transfer of sediment from the foreshore to the bars, but th quantities are appreciable only near the attachment of the bars to the uppe foreshore. The large number of tracer particles remaining on the foreshore 14 day after injection (Figure 9.10A) indicates that wave energies are not sufficient t disperse foreshore sediments fully, and longshore transport in one direction durin the day of injection is reversed on other days.

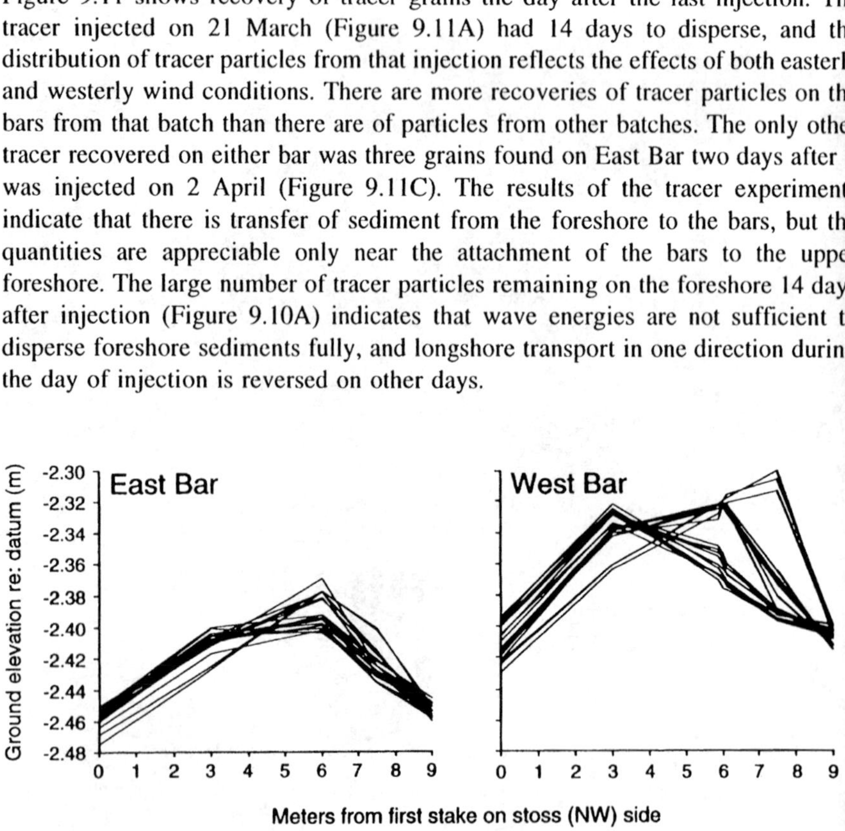

Figure 9.10 Sweep zone profiles on transverse bars for the period 17 March to 5 April

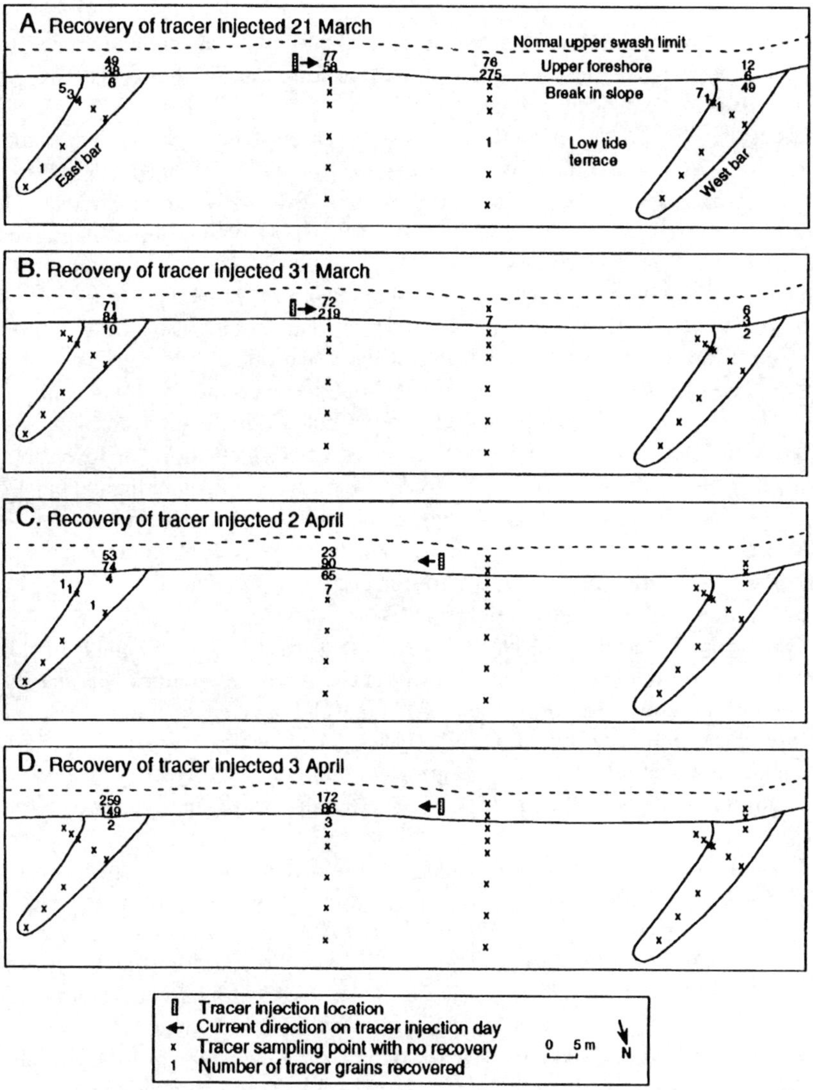

Figure 9.11 Tracer recovery on 4 April. Amounts of tracer deployed were 5036 g (21 March), 4813 g (31 March), 5000 g (2 April) and 4680 g (3 April). The figure is divided into four parts to distinguish results of the four different injections. The total number of tracer grains at a sampling location would be represented by the sum of the numbers portrayed on panels A to D

9.4 DISCUSSION

Changes in water level caused by atmospheric forcing can greatly exceed tidal ranges in microtidal estuaries. Hurricane surge levels in the microtidal Galveston Bay, Texas have reached about 2.4 m higher than mean high water elevation (Garcia and Flor 1984). 'Wind tides' of up to 0.9 m have been observed in estuaries that have astronomical tides of 0.15 m (Love et al. 1985). Water level may reach 1.8 m above average once in about 10 years in a location where tidal range is less than 0.15 m at Roanoake Island, North Carolina (US Army Corps of Engineers 1981). Armbruster and co-authors (1995) reported an increase in water level of 0.89 m over high tide in a location where the mean tidal range is only 0.43 m. This condition occurred during an offshore (bayside) wind, when water flowed into the bay due to an increase in water level on the oceanside. The range between the highest and lowest water levels in our study did not exceed 0.75 m, indicating that surge levels were high relative to tidal range but not high in an absolute sense. Wong and Wilson (1984) indicate that water level fluctuations at low subtidal frequencies in Great South Bay are forced primarily by water entering the bay from the ocean, but direct local wind forcing within the bay is noticeable and can help counteract the effect of ocean forcing. The greater distance from the inlet at our site (relative to the site studied by Armbruster et al.) may add to the effect of local winds in diminishing surge levels.

Nordstrom and Jackson (1992) speculated that the wave energy field should be complex on the upper foreshores of microtidal estuarine shorelines and that long-period energy should be conspicuous. Although the breaker, surf and swash zones at Fire Island do not migrate far on the beach (horizontally or vertically) with the rise and fall of the tide, there is no significant buildup of infragravity energy, even with onshore wind velocities 13 m s^{-1}, indicating that wave energy may be too low for this to occur.

Wave heights monitored in this study are lower than would be expected for the fetch distances and wind speeds. Significant wave heights monitored by Armbruster and colleagues (1995) were up to 0.18 m for wind speeds 10 m s^{-1} and fetch distances little greater than 5 km. The lower energy in the waves at the Fire Island site is believed to be due to shoaling resulting from shallow water depths on the low tide terrace. Inspection of navigation charts reveals that bay depths 1.0 km from the shoreline at Fire Island are 2.7 m and the deepest part of the bay toward the northwest is about 4.6 m; bay depths 1.0 km from the site monitored by Armbruster and colleagues are about 6.1 m, with depths 9.0 m in the deepest part of their bay. The maximum water level they reported was 1.67 m above MLW, indicating far greater water depths at their sampling site.

Armbruster and co-authors (1995) monitored 1.75 m of erosion associated with peak wind speeds of 10 m s^{-1}. Geomorphic changes monitored at Fire Island indicate that storms with onshore wind speeds 13 m s^{-1} are not effective erosional agents. The vegetation wrack at Profile W was sufficient to prevent beach reworking during onshore winds of this speed. The small amount of shoreline retreat for weather systems monitored at Fire Island in this study may be due to the greater availability of sediment in the longshore transport system or to the lower breaking wave energies for a given wind speed. The retention of tracer in the system and the lack of significant change in foreshore position throughout the study indicate that there is little cross-shore sediment exchange and that gross rates of longshore transport may greatly exceed net rates. The form of the overwash deposits behind the active beach indicate that surges in this microtidal estuary can create water levels well above the elevation of the low tide terrace and cause considerable geomorphic change, but these conditions were not monitored here.

Transverse bars occur on estuarine beaches with greater tidal ranges than at Fire Island, but their dimensions are small relative to the dimensions of the foreshore. The transverse bars at Lough Foyle, Ireland, where spring tidal range is 2.3 m, are 0.05 m high (R.W.G. Carter, personal communication). The transverse bars at Assateague Point, Virginia, USA, where spring tidal range is 1.3 m, are 0.10 to 0.15 m high (Bruner and Smosna 1989). Transverse bars on mesotidal estuarine shorelines are rarely of sufficient amplitude to cause waves to break at raised water level, and they may not normally be an effective energy filter for storm waves (Nordstrom 1992). The transverse bars at Fire Island, in contrast, are large relative to the small dimensions of the foreshore and the intertidal area (Figure 9.3), resulting in relatively great amounts of morphologic change on the bars compared to the foreshore (Figures 9.7 and 9.10). The variability of West Bar is greater than East Bar, presumably due to its greater elevation above the bed and its greater exposure to wave action during higher water levels.

Bruner and Smosna (1989) speculate the bars in their study form as a result of the removal of sediment from the upper foreshore during storms. We observed evidence of transfer of sediment to the bars, but we did not observe major form changes associated with the weather systems monitored here.

9.5 SUMMARY AND IMPLICATIONS

Our investigation of storms with mean wind speeds 13.0 m s^{-1} in a shallow microtidal estuary indicate that: (1) storm surge can greatly exceed tidally induced water level changes; (2) raised water level can occur during offshore (bayside) winds, because of ocean influences, even at distances 15 km from the nearest inlet; (3) long-period energy is not conspicuous on the upper foreshore; (4) wave heights can be low (0.13 m) for fetch distances 12 km; (5) beach change occurs across a limited horizontal (5.0 m) and vertical (1.0 m) distance; (6) there is little volumetric change on the upper foreshore; (7) the transverse bars can undergo great amounts of morphologic change relative to the foreshore; and (8) there is evidence of transfer of at least some sediment from the foreshore to the bars.

The mobility of the transverse bars during light-intensity storms and th exchanges of sediment between them imply that they play a more fundamental rol in the morphodynamics of microtidal estuarine beaches than they do on mesotida estuarine beaches. Future study should be directed toward more intensiv monitoring of these bars to specify their relationship to the foreshore. The lac of significant geomorphic changes under light-intensity storms implies that longe field deployments may be required to assure process–response monitoring of th intensive weather systems that are responsible for major topographic features an shoreline erosion.

ACKNOWLEDGMENTS

Financial support for this project was provided by the National Science Foundation Geography and Regional Science Branch and the National Park Service, Nort Atlantic Regional Office. We are grateful to Jim Ebert for local arrangements an Jack Hauptmann, Superintendent of Fire Island National Seashore. This i Contribution Number 95-26 of the Rutgers University Institute of Marine an Coastal Sciences.

REFERENCES

Armbruster, C.K, G.W. Stone and J.P. Xu. 1995. Episodic atmospheric forcing and baysid foreshore erosion: Santa Rosa Island, Florida. *Gulf Coast Association of Geologica Societies Transactions* 45: 31–37.

Bruner, K.R. and R.A. Smosna. 1989. The movement and stabilization of beach sand o transverse bars, Assateague Island, Virginia. *Journal of Coastal Research* 5: 593–601.

Coastal Engineering Research Center (CERC). 1984. *Shore Protection Manual.* U.S. Arm Corps of Engineers, Vicksburg, MI.

Dupre, W.R. 1985. Geological effects of Hurricane Alicia (August 18, 1983) on the uppe Texas coast. *Transactions, Gulf Coast Association of Geological Societies* 35: 353–359

Folk, R.L. 1974. *The Petrology of Sedimentary Rocks*. Hemphill, Austin, TX.

Garcia, A.W. and T.H. Flor. 1984. Hurricane Alicia storm surge and wave data. US Arm Corps of Engineers CERC Technical Report 84-6.

Greenwood, B. and P.B. Hale. 1980. Depth of activity, sediment flux and morphologica changes in a barred nearshore environment, p. 89–109. *In* S.B. McCann (ed.), *Th Coastline of Canada.* Paper 80-10. Geological Survey of Canada, Ottawa.

Hayes, M.O. 1975. Morphology of sand accumulation in estuaries: an introduction to the symposium, p. 3–22. *In* L.E. Cronin (ed.), *Estuarine Research*, Volume II. Academic Press, New York.

Hayes, M.O. 1979. Barrier island morphology as a function of tidal and wave regime, p. 1-27. *In* S.P. Leatherman (ed.), *Barrier Islands.* Academic Press, New York.

Jackson, N.L. 1992. The effect of wind speed and direction on nearshore processes and cross-shore sediment mobility on an estuarine beach, Delaware Bay, New Jersey. Unpublishe

Ph.D. dissertation, Rutgers University, New Brunswick, NJ.
Jackson, N.L. 1995. Wind and waves: influence of local and non-local waves on mesoscale beach behavior in estuarine environments. *Annals of the Association of American Geographers* 85: 21–37.
Jackson, N.L. and K.F. Nordstrom. 1992. Site-specific controls on wind and wave processes and beach mobility on estuarine beaches. *Journal of Coastal Research* 8: 88–98.
Jackson, N.L., K.F. Nordstrom, D.J. Sherman, J.R. Allen, S.L. Namikas and D. He. 1993. Shoreline processes and beach changes on the lagoon shoreline at Fire Island National Seashore, NY, USA, p. 2068–2082. *In Coastal Zone 93*. American Society of Civil Engineers, New York.
Jago, C.F., and J. Hardisty. 1984. Sedimentology and morphodynamics of a macro-tidal beach, Pendine Sands, SW Wales, p. 123–154 *In* B. Greenwood and R.A. Davis, Jr. (eds.), *Hydrodynamics and Sedimentation in Wave-Dominated Coastal Environments*. Elsevier, New York.
Love, K.M., H.S. Chafetz, L. Bruno and C. Buczynski. 1985. Bay margin sand distribution, north Texas coast: a model for sediment distribution in microtidal bays and lagoons. *Transactions, Gulf Coast Association of Geological Societies* 35: 439–443.
McArthur, D.S. 1980. Fluorescent sand tracer methodology for coastal research. *Melanges* Louisiana State University, Baton Rouge, LA.
National Oceanic and Atmospheric Administration (NOAA) 1995. *Tide Tables 1995, East Coast of North and South America*. US Department of Commerce, Washington.
Nordstrom, K.F. 1992. *Estuarine Beaches*. Elsevier Science Publishers, London.
Nordstrom, K.F. and N.L. Jackson. 1992. Two-dimensional change on sandy beaches in meso-tidal estuaries. *Zeitschrift für Geomorphologie* 36: 465–478.
Nordstrom, K.F., N.L. Jackson, K.A. Jagger and J.R. Allen. 1989. Backbarrier beaches on developed shoreline, p. 23–32. *In* D.K. Stauble (ed.), *Barrier Islands: Processes and Management*. American Society of Civil Engineers, New York.
Nordstrom, K.F., H.M. Sterr and N.L. Jackson. 1991. Comparison of estuarine beaches in Germany and USA, p. 2409–2420. *In Coastal Zone 91*. American Society of Civil Engineers, New York.
Rosen, P.S. 1975. Origin and Process of Cuspate Spit Shorelines, p. 77–92. *In* L.R. Cronin (ed.), *Estuarine Research*. Academic Press, New York.
Rosen, P.S. 1977. Increasing shoreline erosion rates with decreasing tidal range in the Virginia Chesapeake Bay. *Chesapeake Science* 18: 383–386.
Rosen, P.S. 1980. Erosion susceptibility of the Virginia Chesapeake Bay shoreline. *Marine Geology* 34: 45–59.
Sherman, D.J., K.F. Nordstrom, N.L. Jackson and J.R. Allen. 1994. Sediment mixing-depths on a low-energy reflective beach. *Journal of Coastal Research* 10: 297–305.
Short, A.D. 1991. Macro-meso tidal beach morphodynamics – an overview. *Journal of Coastal Research* 7: 417–436.
US Army Corps of Engineers. 1981. *Low Cost Shore Protection*. Office of the Chief of Engineers, Washington, DC.
Wang, Y.H. and T.H. Chang. 1978. Littoral drift along bayshore of a barrier island. Proceedings, *16th Coastal Engineering Conference*, 1614–1625.
Williams, A.T. and S.P. Leatherman. 1985. Sedimentologic analysis, p. 177–203. *In* S.P. Leatherman and J.R. Allen (eds.), *Geomorphic Analysis: Fire Island Inlet to Montauk Point Long Island, New York, Reformulation Study*. National Park Service, NARO, Boston, MA.
Wong, K-C. and R.E. Wilson. 1984. Observations of low-frequency variability in Great South Bay and relations to atmospheric forcing. *Journal of Physical Oceanography* 14: 1893–1900.

Wright, L.D., P. Nielsen, A.D. Short and M.O. Green. 1982. Morphodynamics of a macrotidal beach. *Marine Geology* 50: 97–128.

Wright, L.D., P. Nielsen, N.C. Shi and J.H. List. 1986. Morphodynamics of a bar-trough surf zone. *Marine Geology* 70: 251–285.

Chapter Ten

Shoreline dynamics on the windward and leeward shores of a large temperate estuary

J. COURT STEVENSON and MICHAEL S. KEARNEY
University of Maryland

10.1 INTRODUCTION

The Chesapeake Bay, with a drainage basin of 191 500 km^2 (Ellison and Nichols, 1976), is the largest estuary in the United States, and is the focus of several widely publicized environmental problems. Most of these problems are related to eutrophication, such as loss of submersed aquatic vegetation in the shallows, and anoxia in the mainstem (Stevenson and Confer 1978, Malone et al. 1993); however, concerns have broadened to include other issues. Apart from the decline in harvestable resources (oysters, striped bass, and crabs) which traditionally captured headlines, the relationships between shore erosion, sediment inputs and sea level rise are receiving increasing attention. The present Chesapeake Bay was formed over the past 10 000 years by rising sea levels from melting late Wisconsinan ice sheets, drowning the old Susquehanna River channel that had been cut as far seaward as the continental shelf-break. During the Holocene, the overall rise in sea levels shifted shorelines inland, resulting in an ever-enlarging Bay. Nevertheless, although sea level rise underlies much of the past (and recent) history of the Chesapeake, its manifestation in landforms and processes varies considerably depending on which side of the Bay is considered (as well as other factors).

The western coastline has a significantly higher relief, with bluffs rising over 50 m at Calvert Cliffs at mid-Bay. The large rivers on the western side of the Bay have extensive watersheds draining not only the upper Coastal Plain and the

Estuarine Shores: Evolution, Environments and Human Alterations. Edited by K. F. Nordstrom and C. T. Roman.

Figure 10.1 Satellite view of the middle and upper part of Chesapeake Bay, with major localities and sites mentioned in the text indicated. The image shows band 7 of a Thematic Mapper composite acquired in August 1993

Piedmont Plateau, but also the Appalachian Mountain regions. In comparison, most cliffs on the Eastern Shore are classified as low bluffs (below 6 m), and watersheds here are of a more modest size with much less overall relief. The western shore has a tendency for more smoothed shorelines (Figure 10.1), indicating ample sediment supplies and longshore transport. Despite an estimated terrestrial input of 6.9×10^6 tons of sediment per year debouching from riverine sources into the Chesapeake system (USGS data, in USACE 1990), the Eastern Shore has a comparatively jagged perimeter with many coves, indicative of sedimentary deficits. Nixon (1980) has emphasized that the Chesapeake Bay historically has had a much higher biological productivity than other temperate estuaries. We suggest that one of the keys to this productivity lies in the abundance of sheltered shallow coves (left unfilled by rates of sedimentation that were historically slow) that allowed once luxuriant grass beds to nourish and protect from predators a variety of juvenile fish and crabs (Orth et al. 1984; Stevenson 1988; Lubbers et al. 1990).

Sea level rise, and its manifestation in the erosion of the Bay's shores, is hardly an environmental newcomer to the Chesapeake [though the scientific community has sometimes been slow to acknowledge it – e.g. Hunter (1914) incorrectly surmised that the marshes were rapidly building south of the Choptank River on the Eastern Shore]. Its effects on the Bay's shores have been noted since colonial times (Harrison 1915), as the encroachment of rising waters destroyed early sites and forced a gradual retreat to higher ground. By the beginning of the 19th century, shore erosion, driven by rising sea levels, had erased the last vestiges of the original fort at the Jamestown settlement (Hobbs et al. 1994). Old plantations that once figured prominently in the early colonial history of the lower James River and other estuaries have been drastically reduced in size, and several 'dwelling houses' along with the family graveyards have since fallen into the Bay's waters as shorelines receded. One reason for this was the 17th century custom of building houses less than 100–150 feet (30–45 m) from the shoreline of tidal waters (Durand 1686), which served as the main transportation links in a trackless landscape.

In stark contrast to this picture of historical shore erosion is the general portrayal of a rapidly 'geological-aging' Chesapeake Bay, resulting from accelerated infilling (Wolman 1968; Schubel and Carter 1976, Schubel 1981). Sedimentation in the upper parts of Bay tributaries has been the theme of several studies on the western shore of Chesapeake Bay (Gottschalk 1945; Froomer 1980; Scatena 1987; Dugger 1990; Marcus and Kearney 1991). Shoaling of estuaries has been cited as the prime reason for the decline of 18th century towns such as Port Tobacco, London Town, Bladensburg and Joppa – all western shore ports. However, despite small-scale infilling of tidal creeks near the Maryland towns of Princess Anne in Somerset Co. and Easton in Talbot Co., comparable examples are difficult to find on the Eastern Shore. This difference has been simply explained by the large-scale erosion of highly erodible agricultural soils on the western shore, which produced higher supplies of sediment than the flat terrain of the Eastern Shore (Scatena 1987). More detailed analysis suggests that shore erosion actually accounted for more sediment deposition in the Port Tobacco tributary than fluvial inputs (Dugger 1990), and this appears to be the case in the South River also (Marcus and Kearney 1991).

In this review, we discuss the impacts of sea level rise on marshes and shorelines of both sides of Chesapeake Bay. Relative sea level rise along any coast manifests itself in different ways, depending on the rate of apparent sea level rise (including land submergence), availability of upland sediment supplies, elevation of the eroding shoreline, tidal range as well as exposure to prevailing winds and wave attack (not to mention anthropogenic features). In a large estuarine system like the Chesapeake, shoreline processes also are a function of the diversity of shoreline types, ranging from the high cliff shorelines of western shore tributaries (like the Severn and Rappahannock) to low marshy shorelines that characterize much of the lower part of the Maryland Eastern Shore. We conclude with a bay-wide perspective on the sediment budget of the Chesapeake, suggesting that the influx of particulates is not high enough to keep pace with relative sea level rise (including submergence).

10.2 SHORELINE CHANGE AND SEA LEVEL RISE

The first quantitative comparisons (cf. Singewald and Slaughter 1949) of modern maps with the first accurate charts (ca. AD 1850) revealed considerable shoreline retreat in many areas of the Bay. More recent work (Kearney and Stevenson 1991), using a variety of old records, shows that wholesale abandonment of many old and stable island communities occurred within a generation in the early 1900s. In the decades since World War II, the quickening tempo of sea level rise (cf. Kearney and Stevenson 1991) has generated a growing number of shore erosion studies, a phenomenon that amply reinforces the appreciation of coastal communities in the Chesapeake of the increasing shore erosion risk (e.g. Byrne and Anderson 1978). Along with shore erosion, the disappearance of marshlands is occurring at an alarming rate on the Eastern Shore (Stevenson et al. 1985a; Kearney et al. 1988).

Present rates of sea level rise in Chesapeake Bay, ranging since 1940 from 2.5 mm a^{-1} at Baltimore to 3.6 mm a^{-1} near the Bay mouth, are among the highest along the US Atlantic Coast (Stevenson et al. 1986). Regional, non-isostatic mechanisms (e.g. groundwater withdrawal – see Rule 1995) probably account for most of this trend (cf. Kearney and Stevenson 1991). Nevertheless, the absolute rate of sea level rise does not necessarily tell us much about its impacts on the physical and biological systems of any coast. These are best understood by assessing whether the present rate represents a significant inflection in the historical sea level tendency and, thus, a significant departure from those sea level–coastal system linkages that evolved over time.

The classic deceleration in sea level rise that typifies the late Holocene sea level curve (Figure 10.2) is associated with a regional overall rate of sea level rise in the Chesapeake region of 1.2–1.8 mm a^{-1} (Gornitz and Lebedeff 1987, Douglas 1991). This rate, significantly slower than the modern trend, masks an even slower rate (and further deceleration) when considered over the time-scale of the last millennium (Figure 10.2; Kearney, in press). Peltier (1985), on the basis of theoretical calculations, gives the overall rate of sea level rise for the last thousand years in this region as about 0.4 mm a^{-1}. Paleo-sea level indicators, drawn mainly from the mid-Bay region, appear to confirm Peltier's figure, portraying a rate of sea level rise of 0.56 mm a^{-1} during this period (Kearney, in press) More importantly, the sea level changes reconstructed from these data, suggests that sea levels since AD 1850 have risen by as much as they did in the preceding 500 years (Kearney, in press).

Kearney and Stevenson (1991) have recently shown, based on changes in historical land loss and marsh accretion rates, that the present rate of sea level rise

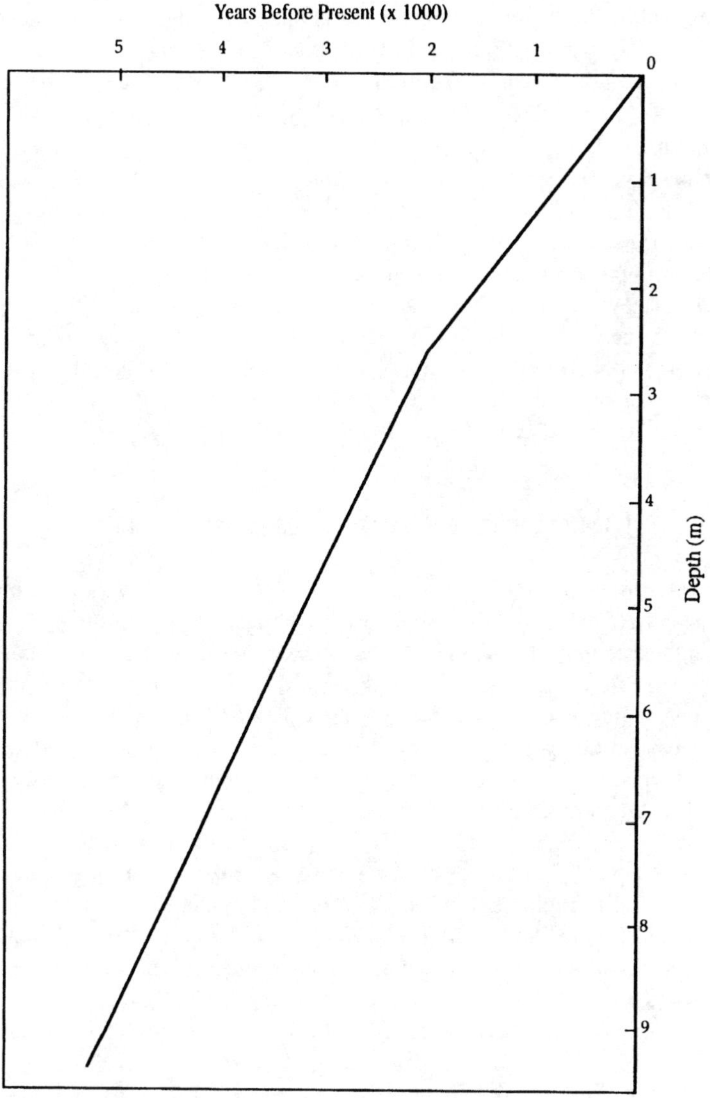

Figure 10.2 Holocene sea level curve for the Chesapeake Bay. The pre-1000 BP part is based on data summarized in Douglas (1985) for the middle and lower Bay. The post-1000 BP part of the curve is based on Kearney (in press)

in the Bay appears to have began after AD 1850. Most of the acceleration in this recent sea level trend has occurred since the 1920s. Braatz and Aubrey (1987) demonstrated, from analysis of tide-gauge records from the US Atlantic Coast, that most areas show a sharp inflection in the rate of sea level rise after ca. 1920, coinciding with widespread erosion and abandonment of islands in the Chesapeake (Kearney and Stevenson 1991). Prior to this date, decadal variations in the rate of sea level rise are not known with certainty in the Chesapeake despite the fact that first systematic tide records were made at Annapolis beginning in 1844 (Hicks 1964), continuous tide-gauge records for the Bay do not extend back far enough to assess changes before 1900.

The available data indicate that the onset of the present sea level trend, when viewed across the span of the last millennium, has been both historically sudden and perhaps unprecedented in magnitude (Kearney, in press). This accelerated transgression may be unique in the late Holocene history of the Chesapeake Bay. Any dramatic and abrupt departure in a driver of coastal change so fundamental as sea level rise is bound to reverberate through the network of interlinked processes that characterize any coast. Thus, sea level rise is critical in understanding coastal dynamics in Chesapeake Bay and many other estuaries.

10.3 IMPACT OF SEA LEVEL RISE ON MARSHES

Sea level rise drives diverse changes occurring across the wide variety of physical and biological components that comprise the Chesapeake Bay. In this section, we describe the particular effects of rising sea level on coastal marshes and shorelines, focusing on the processes and rates of changes that characterize these systems.

Chesapeake Bay contains the greatest area of tidal marshes along the middle Atlantic Coast, totaling some 149 877 ha (374 693 acres) (Tiner 1985). From a purely ecological standpoint, these marshes play an important habitat role for the traditional species associated with the Chesapeake Bay: blue crabs, terrapin turtles, waterfowl of every description, as well as a great variety of fish. Yet, coastal marshes are currently disappearing in the Chesapeake at rates (on an annual percentage basis) as rapid as those of the more widely publicized marsh losses in the Mississippi Delta (Stevenson et al. 1985b; Kearney et al. 1988). Accelerating sea level rise, as in southern Louisiana, will undermine the health of Chesapeake Bay marshes, which trigger a cascade of ecological and physical processes that, once initiated, may continue independently of sea level. Studies (Stevenson et al. 1985a, Kearney and Ward 1986, Kearney et al. 1994, see Table 10.1) of the vertical accretion rates in submerging marshes in the Chesapeake Bay underscore their very low accretionary potential, with an upward limit of perhaps 3 mm a^{-1} (unless they can capture sediment from the relatively rare overwash process). Historically, when rates of sea level rise were low (as appears to have characterized most of the last millennium), even the limited accretionary potential of these marshes was a sufficient buffer against rising water levels. In this century, the rate of sea level rise has clearly been outpacing the accretionary budgets of submerged upland marshes (especially in the mid-Bay region), and the most dramatic marsh losses in the Bay have occurred in these systems.

Table 10.1 Accretion rates for several submerged upland marshes on Maryland's Eastern Shore

Site	Accretion rate (mm a^{-1})		Relative SLR
	Mean	Range	
Blackwater Wildlife Refuge	2.65	1.7–3.6[a]	3.9
Lower Nanticoke River	2.1	1.8–2.4[a]	3.9
Monie Bay	3.0	1.5–6.3[a]	3.9
Lower Pocomoke River	1.5	——[b]	<3.9

Source: Blackwater (Stevenson et al. 1985a); Nanticoke River (Kearney and Ward 1986); Monie Bay (Kearney et al. 1994); Pocomoke River (Douglas 1985).
[a] determined by ^{210}Pb; [b] determined by pollen dating.

Not all marshes in the Chesapeake are equally vulnerable, and any prediction on the fate of marshes in this large estuary must be based on an understanding of how the evolution of certain marsh types preconditions them to rapid disappearance if sea level rises sharply, whereas other marsh types may initially remain stable.

Tidal marshes in the Chesapeake Bay essentially comprise four principal types (Stevenson et al. 1985b; 1986), including coastal (high salinity), submerged upland (brackish), estuarine meander (brackish), and tidal fresh marshes. Although vertical accretion in all marshes is the result of trapping mineral material brought in from upland streams or with the tides, as well as accumulation of on-site organic detritus, the balance of these two principal accretionary components can vary widely between marshes. The variations chiefly reflect how divergent evolutionary pathways have controlled the development of marsh tidal drainage networks and, thus by extension, tidal dynamics and influx of mineral sediments to marsh substrates. Herein lies the key to understanding their differing vulnerability to sea level rise.

10.3.1 Submerged upland marshes

These extensive marshes originate from the gradual submergence of low-lying, flat terrain which characterizes much of the Bay's Eastern Shore (Darmody and Foss 1979, Stevenson et al. 1985b). One important aspect of the development of submerged upland marshes is the absence of a well integrated tidal creek network; most networks generally consist of a few large tidal channels, with a limited number of first-order creeks feeding into them. As these marshes evolve, spreading across former upland surfaces, interior marsh areas become progressively isolated from major creek systems, mineral sediment influx drops, and vertical accretion becomes increasingly a function of peat accumulation (Stevenson et al. 1985a). Concurrently, (in an eventuality that perhaps typifies all marshes; Stevenson et al. (1988)) the lengthening reaches of the few extant tidal creeks in the expanding marsh promote stronger ebb velocities than flood velocities, and more sediment begins to be transported out of the marsh than into it during the tidal cycle. Often, these marshes end up sediment 'starved' (Stevenson et al. 1988).

The environmental conditions that give rise to marsh anoxia and eventual erosion, create an inherent evolutionary framework for marsh loss in many Chesapeake Bay marshes. Older sediments in marshes where accretion rates have been historically slow, like at the Blackwater Wildlife Refuge (Stevenson et al. 1985a), are dominated largely by waterlogged, highly organic horizons with negligible structural integrity. The mechanisms behind the formation of these oozes are not fully understood, possibly being linked to sulfate reduction and nitrate additions via groundwater. Moreover, the lack of any intact organic structure in the older sediments makes them highly erodible once they become exposed to wave activity. By comparison, in marshes where accretion rates have matched the tidal record of sea level, older sediments tend to be intact, firm peats.

Thus, marsh loss has historical antecedents stretching back to the formative stages of development. Marshes following such an evolutionary path are predisposed to loss if rates of sea level rise depart from the relatively flat trend of the last several millennia (as they have since the middle of the last century). In essence, such marshes may be phenomena of unique circumstances that may no longer prevail. Unfortunately, submerged upland marshes, the largest marsh systems in the Chesapeake, fit into this category, and an irrevocable change in sea level rise may mean extinction. The conclusions to be drawn from the trends of this century will be discussed in the last section.

10.3.2 Estuarine meander and tidal fresh marshes

These marshes are conspicuous features in the meanders of major tributaries on both the western and eastern shores of Chesapeake Bay (Ahnert 1960). In contrast to submerged upland marshes, which are largely confined to the Eastern Shore, these marshes usually contain well integrated tidal creek networks displaying the classic dendritic pattern commonly associated with coastal marshes (Kearney et al. 1988). Because of this, although both marsh types generally have lower vegetated area:creek ratios than marshes elsewhere, isolated, wide expanses of interior marsh away from tidal creeks (and, therefore, sources of sediment input) are less common than in submerged upland marshes. In addition, during high river flows, estuarine meander marshes may often become completely inundated, and receive luxuriant supplies of riverine sediment. Higher mineral sediment inputs (demonstrated by the greater ash content of these marsh peats) produce more stable sediments (Kearney et al. 1988). Not surprisingly, vertical accretion rates of these marshes often equal or exceed the present rate of submergence of the Chesapeake region (Kearney and Ward 1986). This is particularly true of tidal freshwater marshes, which generally occur in the sediment trap portion of the estuaries, and thereby benefit from higher suspended loads than those that typify lower estuarine reaches.

Nevertheless, estuarine meander and tidal freshwater marshes are not always entirely stable or immune from sea level rise. The higher accretionary potential of these marshes may enable them to accommodate moderately high rates of sea level rise compared to submerged upland marshes. As long as these marshes have access to large supplies of inorganic sediment, and are in high tidal amplitude areas (see Figure 10.3), they should be able to maintain themselves vis-à-vis sea level rise, if present rates continue. None the less, any change which diminishes their ability to trap sediment (e.g. building of roads across them, drastic reductions in upstream sediment inputs, etc.) may also diminish even their capacity to keep pace. Continued negative accretionary imbalances versus sea level rise eventually lead to the chain of events outlined above, terminating in marsh erosion. The implications of such a scenario only become clear when it is realized that once the fundamental links in this chain are made, marsh loss accelerates rapidly and, perhaps more importantly, independently of any further changes in sea level rise. Indeed, sea levels could fall and marsh losses may well continue as expanding 'rotten spots' (cf. Redfield 1972). The reasons for this grim prognosis for marsh survival involve understanding the processes that initiate marsh break-up via interior pond formation ('interior ponding'), when accretionary mechanisms are overwhelmed by sea level rise, and the physical processes that follow exploit the inherent structural instability of fragmenting marshes.

10.3.3 Processes of interior pond formation

A conspicuous feature of the marshes of Maryland's lower Eastern Shore is the multitude of 'rotten spots' and larger ponds within the marshes. Mendelssohn and co-authors (1981) have shown that progressive submergence of marshes by rising sea level eventually leads to anoxic conditions in the living substrate, with initial decline in vigor and ultimately death of the plants. Break up of the substrate often follows, perhaps accelerated when muskrats and nutria clear too much vegetation, resulting in 'eat-outs'. These interior low spots become devoid of vegetation, enlarge, and coalesce, forming shallow ponds over time.

Areas of marsh where accretionary budgets have been outpaced by sea level rise are easily recognizable in the field by thin aerial biomass, numerous 'potholes', and incipient ponds. Once ponds reach about one acre (0.4 ha) in size, initially from the slow coalescence of smaller ponds, further expansion is driven by wave erosion of the substrate, especially during storms (Stevenson et al. 1985a). At this point, further pond enlargement is chiefly dependent on the frequency of storms, not sea level rise, and elongation to diameters of several kilometers can take place in only several decades, as exemplified by the dramatic expansion of interior ponds at the Blackwater Wildlife Refuge between 1938 and 1979 (Stevenson et al. 1985a). Pond enlargement may only abate when the fringes of the pond begin to encroach on areas of young marsh where the peat is more intact, or when debris is deposited on shorelines (from ice rafting and other processes) which helps stabilize the mat , i.e. acting like natural rip-rap. Eventually, if marsh erosion progresses, the largest

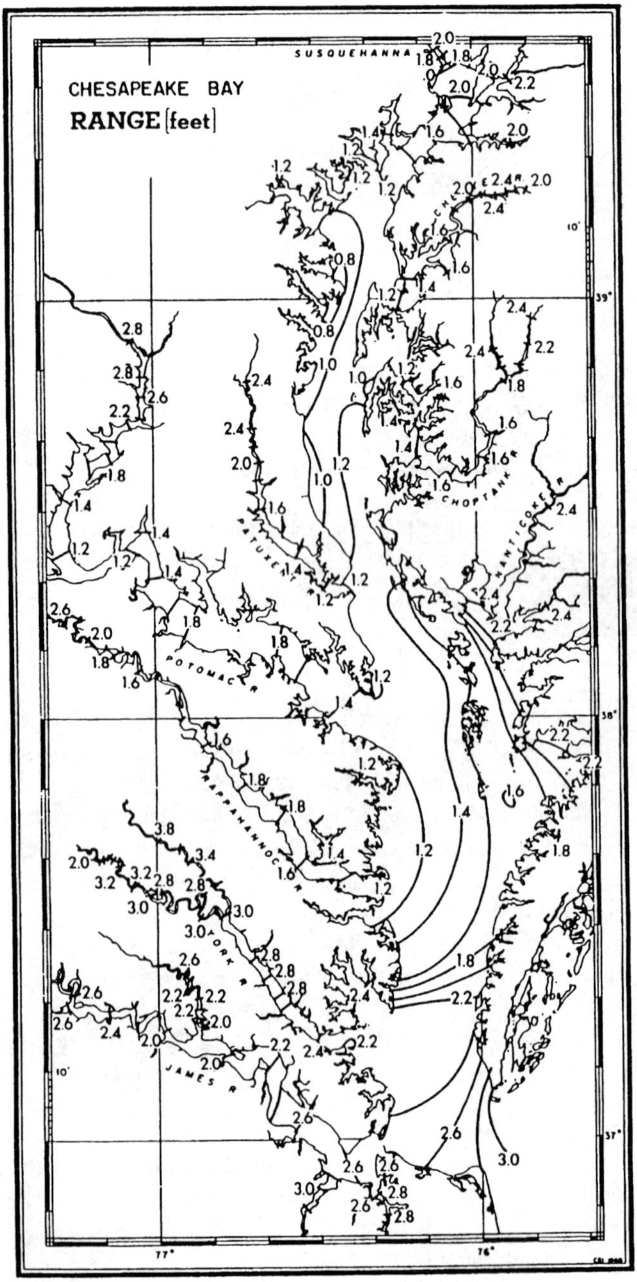

Figure 10.3 Tidal range along the shorelines of Chesapeake Bay. (Taken from Hicks 1964 used by permission of the Estuarine Research Federation, © Estuarine Research Federation

ponds can become semi-enclosed coves, which are so characteristic of the shoreline on the Eastern Shore (e.g. the most likely origin of Fishing Bay).

10.4 RATES OF SHORE EROSION

Chesapeake Bay has been likened to a narrow shallow pan and, indeed, throughout most of its length, the Bay seldom exceeds 8 km in width, with average water depths of only 8.2 m. Under such conditions wave activity is relatively low, especially in the many protected coves on the Eastern Shore. Average wave heights for most of the Bay average 0.3 m (Ward et al. 1989). During storms, depending on wind direction and speed, wave heights can approach 3 m (Ward et al. 1989). Apocryphal 19th century stories (Shomette 1982) report waves allegedly reaching heights of 15 m or more. However, the largest documented waves occurred during Hurricane Connie in 1955, when storm swells exceeded 8 m off Sharp's Island Light near the mouth of the Choptank River (Shomette 1982).

Shoreline recession, nevertheless, is pervasive along the Bay's shores in spite of generally low wave conditions. This phenomenon often typifies estuarine shorelines, where shore erosion rates can outpace those of adjacent open coasts (Nordstrom 1992). Rosen (1967) pointed out that the most vulnerable shorelines are those with low tidal amplitude because they have lower elevation beaches. One of the more vulnerable areas on the mainstem of the Bay is Calvert Cliffs, which is located in a reach north of the mouth of the Patuxent River and has less than 0.3 m tidal range (Figure 10.4). Fully 20% of the Chesapeake Bay shoreline is retreating at rates approaching 2 m a^{-1}, and most areas are experiencing annual retreat rates of at least 0.5 m (Figure 10.4). Over the past century, approximately 18 000 ha (45 000 acres) have disappeared due to shore erosion, an area about the size of the District of Columbia (USACE 1990) In selected areas, shoreline retreat is 'galloping along'. Byrne and Anderson (1978) indicate that the northwest end of Tangier Island appears to be retreating at rates of perhaps 10 m a^{-1}. In one stormy three-month period in the 1970s, this part of the island was further diminished by over 5 m of shoreline retreat (Bowes 1991). Evidence of such high rates of shoreline recession is inescapable, ranging from the presence of large numbers of trees in the surf marking the former position of a recently collapsed low sea cliff at the south end of Tilghman Island (Figure 10.5) to the precarious future of an abandoned residential trailer at Hoopers Island (Figure 10.6). Equally compelling is the loss of Sharp's Island at the mouth of the Choptank River, which washed away in the 1950s (Figure 10.7).

Like the acceleration in the sea level trend that is driving it, rapid shoreline retreat is a recent phenomenon. Analyses of historical records and maps reveal that rates of land loss among Bay islands were slow in the middle Bay prior to the mid-19th century (Kearney and Stevenson 1991). Since then, land loss rates have jumped dramatically, and formerly stable, populated islands have either

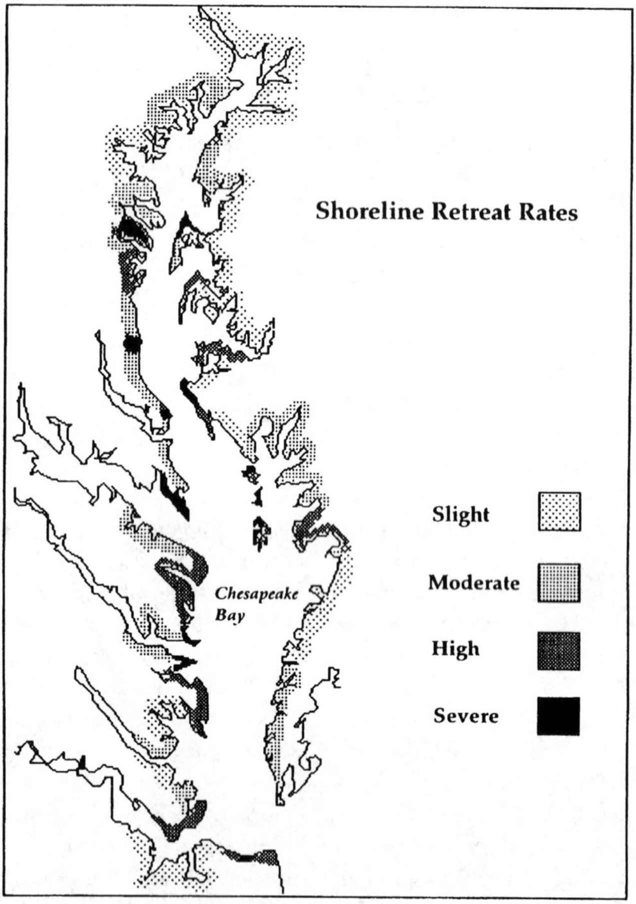

Figure 10.4 General rates of shoreline retreat for Chesapeake Bay. For Virginia, slight = 0–0.3 m a^{-1}; moderate = 0.3–1 m a^{-1}; high = 1–2 m a^{-1}; severe = 2 m a^{-1}. For Maryland, slight = 0–0.6 m a^{-1}; moderate = 0.6–>1 m a^{-1}; high = >1–>2.5 m a^{-1}. Data taken from Ward et al. (1989), Maryland Geological Survey (1975), Singewald and Slaughter (1949), and Byrne and Anderson (1978)

disappeared or shrunk beyond the point of being inhabitable. Such changes mirror the general trend in sea level rise in this period, and it is clear that the rising tidal prism underlies the increasing transgression of Bay shorelines. The rate of shoreline retreat at Tangier Island, in the middle of the Bay, is a particularly compelling example. At the beginning of the century, parts of the island were experiencing shoreline retreat rates of ca. 3 m a^{-1} (Byrne and Anderson 1978). By mid-century, these rates had risen to ca. 6 m a^{-1}. Twenty-five years later (mid-1970s), the same areas were disappearing at a rate of near 10 m a^{-1} (Byrne and Anderson 1978).

Figure 10.5 Evidence of active shore erosion along the southern tip of Tilghman Island, Maryland Eastern Shore. Rates of shoreline retreat in this area approach $3\,m\,a^{-1}$. (Photograph taken Spring 1988)

Figure 10.6 A residential trailer, placed on Lower Hoopers Island less than a decade ago, now abandoned and almost lost due to shore erosion. (Photograph taken 1991)

Figure 10.7 This photograph, taken around 1950, shows the last remnant of Sharp's Island, a formerly large island located near the mouth of the Choptank River on Maryland's Eastern Shore. The island disappeared some time after the early 1950s. In the early part of the century, the island supported a sizable hotel. (Photograph courtesy of Maryland Sea Grant and Douglas Hanks, Jr)

10.4.1 Processes of shoreline retreat

The rapid shoreline retreat of the Bay's shores is the result of two distinct processes, shore erosion and simple submergence, which often operate in tandem. Shore erosion is the principal mechanism in many areas, but until recently the processes involved had been little studied and mostly extrapolated from studies done on the open coast. One classic example is the appropriation of the Bruun Rule to explain shoreline retreat in the Virginia Bay, which was only moderately successful even though significantly altered from Bruun's original conception (Rosen 1981). The intrinsic flaw in all such analogies of Bay shorelines with open coast shore processes is the lack of true swell conditions in the estuary. Swell waves are the agents that facilitate the transfer back onshore of material lost from the shoreline during storms. In the Chesapeake, in the absence of swell conditions and with only moderate wave set-up from day-to-day winds, it is questionable whether sediments stripped from the shoreline during storms return in quiescent periods. The apparent regeneration of beaches and other shorelines in the Chesapeake noted in some areas after storm events may not reflect onshore transport of eroded sediment at all, but re-establishment of the littoral sediment transport system from an eroding headland. Littoral sediment transport rates in the Bay can be quite high, approaching several hundred thousand cubic meters annually (cf. Downs 1993).

Considerable work remains to be done on shore erosion processes in Chesapeake Bay, especially on the actual sediment budgets of the upper shoreface. However, the unequivocal tie of shoreline stability in most areas to sediments from littoral transport does have important implications for the type of shoreline protection features that will be built in coming years.

Shore erosion along the high cliff shorelines that comprise the Bay's western shore is a special case. In these shorelines, shore erosion is the linchpin of a host of processes that lead to cliff failure. Typically, during winter when the relatively small beaches fronting many cliff slopes further narrow ('winter beach' conditions), storm waves overtopping the beach and breaking against the base of slopes initiate toe erosion, ultimately destabilizing the cliff (Figure 10.8). Slope failure, when critical shear stress thresholds are finally reached, may take the form of minor slumping, or in more cohesive sediments, 'block' failure. In the event that a major slope failure occurs, the new slope equilibrium profile may entail considerable landward displacement of the slope summit. In contrast, the base of the slope where the debris accumulates may actually prograde.

Often abetting the cliff failure process is sapping, especially in the Tertiary age sands and muds composing the Calvert Cliffs of southern Maryland. Leatherman (1986) described sapping and failure of Miocene sands near the Patuxent Naval Station. Here, Miocene marls underlying the Calvert sands impede percolation and form a locally perched water table, causing subsurface water to discharge at the contact exposed in the cliff face and undermining the capping sands. This process and others underscore the fact that shore erosion along much of the western shore is simply an integral component (as well as instigator) of a much larger problem, cliff failure.

If shore erosion predominates in the western shores of the Chesapeake, simple

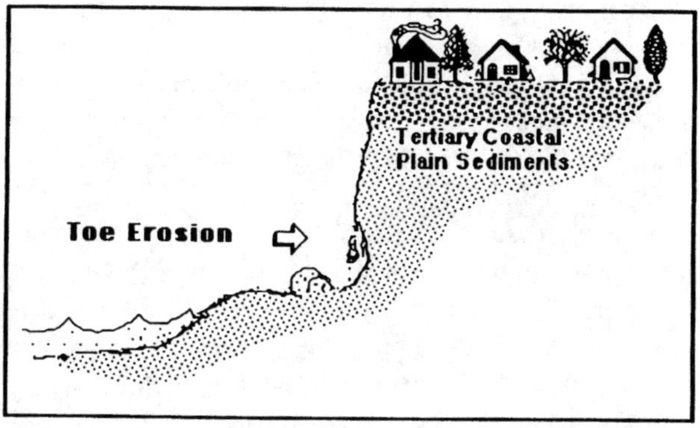

Figure 10.8 Major processes of cliff retreat along the cliff shorelines of the western shore of the Chesapeake Bay

coastal submergence is equally prevalent in many areas of the low-lying Eastern Shore. The wide expanses of submerged upland marshes fringing the shoreline of the southern Delmarva Peninsula amply testify to the progressive submergence of this coast. Yet, though the results of coastal submergence are most visible in wave-protected areas through the creation of coastal marshes, it probably operates simultaneously with (and indistinguishably from) shore erosion elsewhere. Only the rapidly disappearing islands of the Bay show clearly the separate roles that both processes play in the consumption of the mainland by sea level rise. Bayward-facing (especially those facing northwest) shorelines of these islands are generally actively eroding, being often exposed to the onslaught of waves built up over some of the largest fetches in the Bay, whereas the more protected shorelines facing the mainland are submerging and converting gradually to coastal marsh.

10.4.2 Processes of island formation

Classically, islands form in depositional environments like the heads of estuaries and deltas by accretionary processes. For example, in the Mississippi Delta, a large, rapidly growing island has developed at the mouth of the Atchafalaya (Van Heerden and Roberts 1980), as sediment inputs have increased significantly over the past half century. Such processes are not particularly evident in the Chesapeake Bay. Even at the Susquehanna Flats in the Bay headwaters, which Schubel (1981) characterizes as the expansion of the delta, sedimentation has been too slow to support island formation (particularly since the construction of Conowingo and other dams on the Susquehanna River which trap large volumes of sediment upstream).

More typically, islands form in the Chesapeake Bay by shoreline erosion when a peninsula is breached as sea level rises. Based on patterns in the present bathymetry, many of the larger peninsulas appear to have formed from the submergence of interfluves between the ancestral Susquehanna River and its tributaries. Almost always the last juncture with the mainland is a marsh. As the tides progressively inundate the marsh, it becomes more vulnerable to seawater intrusions which accelerate peat degradation, possibly through sulfate reduction. The chain of islands running south from the southern tip of Dorchester County on Maryland's Eastern Shore (lower Hoopers Island through Smith Island) exemplifies this process (Figure 10.9).

Perhaps the most dramatic example of a current shoreline change is in Dorchester County where the largest island in Chesapeake Bay is being carved out of a flat landscape. The Blackwater River is now contiguous with the Little Choptank River during high tides. Fragmentary evidence suggests that the deterioration of a marsh which used to operate as a plug, has allowed high-salinity waters to inundate the head of the Blackwater. This has led to a widening and deepening of the Blackwater channel and, if left unchecked, will effectively form the largest island in Chesapeake Bay. This appears to be the process by which the

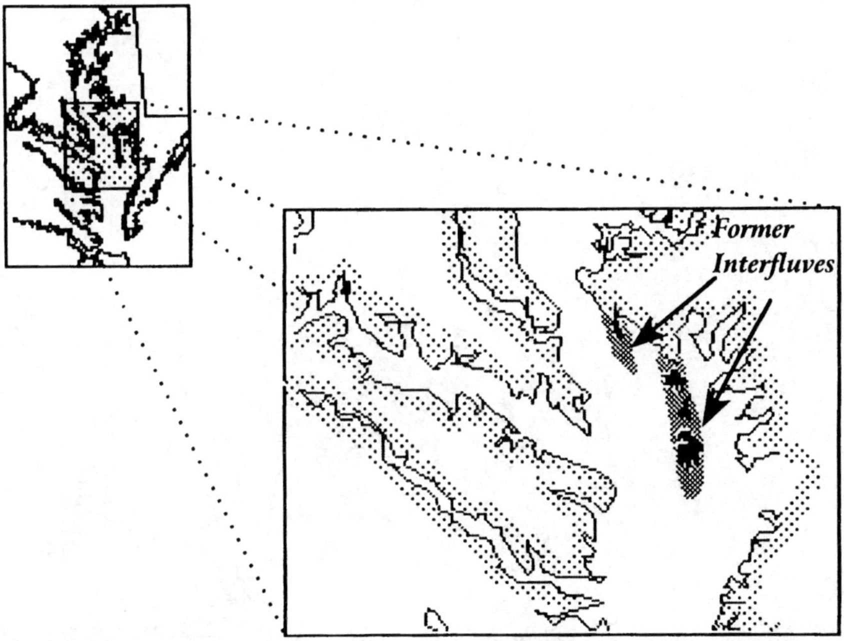

Figure 10.9 Formation of Bay islands at the tip of Dorchester County, Maryland. The present islands, indicated in black, are probably not yet submerged topographic highs of interfluves (shown in the gray dot pattern) between the ancestral Nanticoke River and the ancestral Susquehanna River

nearby James Island Archipelago was formed. On 18th century maps, James Island was connected to Taylors Island by a marsh. At present, the place that once was clearly portrayed as marshland connecting James Island to the mainland is now open water almost a mile wide and easily navigable by skiff (Figure 10.10). Kent Island was undoubtedly formed this way, as it was pinched off from the mainland. Its place of last attachment was to the east at Kent Narrows. In the 17th century this area was called the 'wading place', but now it is a deep channel which is navigable by deep draft sailboats.

10.5 CHESAPEAKE SEDIMENT BALANCE: AN ESTUARY AT THE CROSSROADS

In the next century, the escalating changes being wrought by sea level rise may find the Chesapeake Bay an appreciably different estuary than today in terms of shorelines. Subsidence contributions to the local sea level of the Bay account for two-thirds of the secular trend. It is increasingly clear that world sea levels may soon outpace present subsidence rates (Hohdahl and Morrison 1974) which are in

Shore line changes of James Island, Dorchester County, Maryland.

year	island area (acres)	map symbol
1847	1253	– – –
1942	423	——
1994	106	**——**

├──┤ =1000 feet

Figure 10.10 Formation of James Island. (Data for 1847 and 1942 from Singewald and Slaughter (1949)

themselves very variable (Figure 10.11). Over the past 100 years, there perhaps has been a more marked shift than at any time in the past 5000 years. The climatic linkage behind these changes is becoming inescapable, with surprisingly very little lag between the changes in global climate and the response of sea level (Schneider, 1989). Given the pressure for the development of the Chesapeake's shores – which

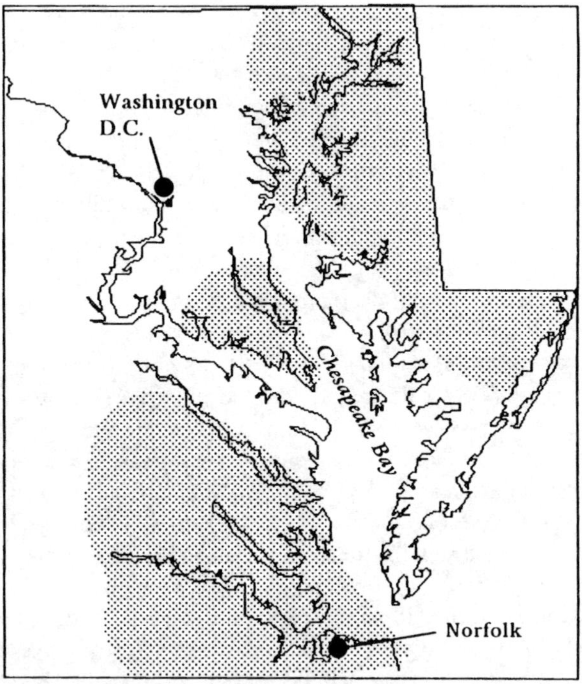

Figure 10.11 Subsidence in the Chesapeake Bay system. Shaded pattern indicates areas subsiding at rates of (up to 4.0 mm a^{-1} near Norfolk); unshaded areas are subsiding at 1.2–1.6 mm a^{-1}. (Modified after Hohdahl and Morrison 1974)

can be reasonably expected to continue well into the next century – concerns are increasing that a major disaster may be lurking in the future. However, a clear insight into basic Bay shore processes is critical if any adequate measures are to be formulated to mitigate potential hazards.

One perspective that needs amending is the concept that the Chesapeake Bay is rapidly 'geologically aging' from the high sediment inputs, and how this potentially affects rates of shore erosion. It might be argued that a rapidly shallowing Bay theoretically would lead to lessening wave activity and lower rates of shore erosion. But there is no reason to believe that this is occurring. Even using Schubel and Carter's (1976) own estimates, there is a huge shortfall in the sediment entering the Bay needed to offset sea level rise. Although it is not entirely clear how they estimated shore erosion and oceanic inputs, they concluded that the mainstem average sedimentation rate was about 1.75×10^6 tons a^{-1} or 0.8 mm a^{-1}. If this estimate for the rate of infilling is accepted, it is obvious that the Bay, with a local rate of sea level rise of 3 mm a^{-1}, is in no danger of filling in. Instead, it appears to be actually deepening, but soundings have been so sporadic that it is impossible to corroborate this from existing measurements. A similar problem affects even the most up-to-date information (USACE 1990) for all riverine and erosion sediment sources. Estimation of sediments transported into the Bay mouth by bottom currents is especially problematic. Large amounts of sand have been hypothesized as entering the lower Bay through the Virginia Capes to be deposited in shoals. Schubel and Carter (1976) placed oceanic inputs at 0.22×10^6 tons a^{-1}, or one-fifth of the average Susquehanna River input (1.07×10^6 tons a^{-1}). Nevertheless, even if oceanic inputs were confirmed to be high as those from the land (including erosion), there would still be a substantial gap in the sediment budget.

Despite the new information on the Bay's sediment budget, and the ominous signs of increasing shore erosion, the sea level future of any particular reach of the Chesapeake shoreline is hardly certain. Predictions for changes in the rate of global sea level rise are highly controversial (Schneider 1989), and are subject to regional variations such as the elevation of the Gulf Stream which modulates sea levels along the eastern US seaboard. At the local scale, indigenous factors (e.g. subsidence), are often equally poorly understood, and can temper or exacerbate any global trend. None the less, we can present some reasonable shoreline scenarios based on our understanding of present sea level effects. Indeed, some of these changes may be within our control should we choose to exercise it.

One strategy for the stabilization of marshes on the Eastern Shore that should be considered is augmenting their mineral sediment inputs, as has been tried in Louisiana . Currently, the US Army Corps of Engineers dredges large amounts (in the thousands of metric tonnes annually) of sediment from the Bay's channels, especially to keep the Port of Baltimore competitive with East Coast facilities. Historically, much of the spoil was deposited in either artificial islands or uplands. At present, there are proposals for disposal of this material in the deep trenches that comprise the former thalweg of the ancestral Susquehanna River, or for the stabilization and restoration of eroding islands. Another option may be to dispose of the dredge material on declining marshes. Techniques have been developed for spraying sediment on wetland surfaces (Cahoon and Cowan 1988), and these could no doubt be adapted for use in Chesapeake Bay.

The accelerating pace of shoreline retreat in most areas of the Bay over this century, reflecting the rate in sea level rise driving it, shows no signs of abating in the future. Indeed, the prospect of a signal leap in the rate of global sea level rise may produce a transgression of the bay's shores not seen in millennia. But though the threat is real enough, the prediction of what may happen will be difficult for any particular shoreline. Along the western shore of the Bay (and substantial areas of the Eastern Shore as well), the complicating factors of slope processes alluded to earlier make any simple linear extrapolation of historical shoreline trends risky. Compounding this is the hidden unknown of enhanced rates of longshore transport of sand liberated from rapidly eroding headlands. As transport rates increase to reaches down drift, will retreat rates slow from projected figures as increasing sediment inputs buffer shoreline response? Currently, there is simply too little known about sediment sources and longshore transport systems in most of the Bay to serve as any basis for prediction.

Prediction of shoreline behavior in the sea level future will be less complex for areas where simple submergence is the major or sole agent of shoreline retreat. In this instance, hypsometric models can provide a reliable window on possible sea level outcomes. Especially low-lying areas like those that comprise much of the shoreline of the lower Eastern Shore of Maryland will undoubtedly be largely submerged, and the hypsometry of the submergence may be developed for any sea level scenario. Giese and Aubrey (1989) undertook such an exercise for the Massachusetts coast, where rocky shores in several areas limit the straightforward application of shore erosion projections in a rapidly transgressing sea. Such analyses have yet to be done for the Chesapeake.

The need for these analyses is urgent. Over the past several decades, development along the Bay's shores has grown by 17% (Year 2020 Panel 1988). By the year 2020, population along the shoreline may increase by another 1.9 million people if present trends hold (Year 2020 Panel 1988). Unfortunately, much of this anticipated new development could not be less fortuitously placed than if it had been done by design: those shores of the Bay experiencing the greatest rates of shoreline retreat today (and in the future) will also experience the greatest population growth (Table 10.2). The future economic costs of these decisions, mounting year-to-year as shore erosion continues and even accelerates – let alone the additional havoc wreaked by storms – have yet to be calculated. Unquestionably, they will be high.

Even a rudimentary estimate, extrapolating from present coastal property values around the Bay (somewhere between $30 000 and $100 000 per acre, depending

Table 10.2 Comparison of projected population increases (1990–2020) and shore erosion trends in Maryland by County

County	Population increase (%)	Erosion rate
Eastern Shore		
Kent	0–24	Slight to moderate
Queen Annes[a]	25–99	High to severe
Talbot[a]	50–99	Slight to severe
Dorchester	0–24	Moderate to high
Wicomico	0–24	Slight to high
Somerset	0–24	High
Western Shore		
Cecil	0–24	Slight
Harford	0–24	Moderate
Baltimore	0–24	Moderate to severe
Anne Arundel	0–24	Moderate to severe
Calvert[a]	50–74	Moderate to severe
St Marys[a]	25–49	Moderate to severe

Sources: shore erosion data (Maryland Geological Survey 1975; Waed et al. 1989); population data (Year 2020 PANEL 1988).

[a] Counties with high erosion rates and future population growth.

upon locale) suggests that property losses could average \$90 to \$120 million a^{-1} if rates of shore erosion were to double. By the end of the coming century, these losses could easily total more than \$10 billion (present dollars), even without any further increase in the rate of sea level rise. Escalating property values will almost certainly inflate the absolute dollar figure, perhaps to levels beyond comprehension today. The human toll if a major storm hit the crowded shorelines of the Chesapeake in the next century is incalculable. Long decades with few major storms have lulled people in the Bay into a complacency that is not shared by people inhabiting the more open coast of the middle and southeastern Atlantic seaboard.

Better planning for the sea level future could mitigate its consequences for shoreline development, especially for rapidly developing low-lying areas like the West River of Maryland's western shore (Figure 10.12). Here, as elsewhere, shore protection structures are being built in piecemeal fashion. This typical approach, however, is hardly an adequate substitute for a more widespread and comprehensive adoption of a shoreline protection plan. A comprehensive shoreline protection strategy could also yield the added benefit of decreasing turbidity and its adverse impact on living resources, since shore erosion supplies most of the sediment to the middle and lower Bay as well as the lower reaches of tributary estuaries (USACE 1990; Marcus and Kearney 1991). However, shore protection will not be cheap, particularly in areas where high wave activity necessitates costly concrete bulkheads (Table 10.3). In today's dollars, construction costs of these structures could exceed several million dollars per linear kilometer of shoreline.

Figure 10.12 Typical development of low-lying fastland along the West River, Maryland. Note the wooden bulkhead fronting part of the shoreline. (Photograph taken April 1991)

Table 10.3 Representative costs of shore erosion measures (cost in US dollars per linear 0.3m). *** = not recommended

Measure	Low energy	Medium energy	High energy
Stone revetments	125	235	380
Timber bulkheads	220	250	275
Timber groins	140	160	175
Headland breakwaters	135	235	520
Slope regrading/ Vegetation planting	105	120	***
Vegetation	30	30	***

Modified from the USACE (1990). Low = 0.6 m waves, medium = 1.2 m waves, high = 1.8 m waves.

Any comprehensive plan for shore protection must also take into account sediment longshore transport processes, which, if disrupted, could exacerbate shore erosion in many areas. Lack of planning for such effects historically has been the root of litigation on shoreline modification along the open coast (e.g. Long Island). A fuller understanding of sediment transport pathways and budgets of various shorelines is desperately needed. Moreover, the political acceptability of drastically altering presently unencumbered vistas of the Bay's wild and pristine shores by the spread of concrete, stone and wood erosion control structures will have to be addressed. The fact that most of the Bay shoreline is privately owned does not lessen the potential for political friction, and as elsewhere on the nation's coasts, some accommodation will have to be made if the average taxpayer is to continue to bear the brunt of subsidizing coastal property through federal flood insurance. We suggest that the sea level rise/subsidence problems underlying shore erosion are driven as much (or more) by development inland, and that perhaps a larger proportion of the population ought to bear the burden of shore protection.

Shoreline protection does not stop flooding, however. Planning for the increasing risk of this threat from rising sea level will involve employing hypsometric scenarios of future potential flooding from projected rates of sea level rise, and making decisions about the course of shoreline development. In Maryland, the strict limitations on shoreline development imposed by the Critical Areas Legislation for control of the amount of pollutants entering the Bay (generally no development within 33 m (100 feet) of the shoreline), have also had the consequence of restricting future construction in the most flooded 33 m buffer zone. None the less, the accidental benefits of unrelated environmental legislation are poor substitutes for real farsighted approaches to minimizing the risks of future shoreline development.

Future plans should embrace a wider agenda than simply the flooding/erosion hazards to nearshore construction in coming decades. In particular, human impacts have considerable potential to influence the rate of sea level rise in the Bay, possibly as much as any changes in the global sea level tendency. In this regard, the most immediate threat to the sea level future of the Chesapeake may be local subsidence, as rates of groundwater withdrawal increase to accommodate the burgeoning population growth around its shores (Kearney and Stevenson 1991). Such possibilities need not prove intractable if there are foresight and regional consensus building towards solutions.

ACKNOWLEDGEMENTS

We would like to express appreciation for past collaboration and discussions with William Boicourt, George Oertel, Andrew Marcus, and Larry Ward. Sara Bellmud, Suzanne Bricker and Karen Sundberg reviewed the manuscript and made helpful suggestions. This paper is a synthesis of results from several research projects funded by National Oceanic and Atmospheric Administration: C-CAP, Estuarine Research Reserve and the Coastal Zone Programs; from the Maryland Department of Natural Resources; the University of Maryland Water Resources Research Center; and the University Maryland Graduate School. This is contribution 2768 from the Center for Environmental and Estuarine Studies of the University of Maryland System.

REFERENCES

Ahnert, F. 1960. Estuarine meanders in the Chesapeake Bay. *Geographical Review* 50: 390–401.

Bowes, D.P. 1991. Islands in time. *Mid-Atlantic Country* 12: 24–31.

Braatz, B.V. and D.G. Aubrey. 1987. Recent relative sea-level change in eastern North America, p. 29–48. *In* D. Nummedal, O.H. Pilkey and J.D. Howard (eds.), *Sea-level Fluctuation and Coastal Evolution*, Society of Economic Paleontologists and Mineralogists Special Publication No. 41. Tulsa, Oklahoma.

Byrne, R.J. and G.L. Anderson. 1978. Shoreline erosion in tidewater Virginia. Virginia Institute of Marine Science, Special Report on Applied Marine Science and Ocean Engineering No. 111, Gloucester Point, Virginia.

Cahoon, D. and J.H. Cowan, Jr. 1988. Environmental impacts and regulatory policy implications of spray disposal of dredged material in Louisiana wetlands. *Coastal Management* 16: 341–362.

Darmody, R. G. and J. E. Foss. 1979. Soil–landscape relationships of the tidal marshes of Maryland. *Soil Science of America Journal* 43: 534–541.

Douglas, B.C. 1991. Global sea level rise. *Journal of Geophysical Research* 96: 6981–6992.

Douglas, P.G. 1985. Evolution of a brackish estuarine marsh system in the Pocomoke River Estuary. M.A. thesis, University of Maryland, College Park, Maryland.

Downs, L.L. 1993. Historical shoreline analysis and determination of littoral cells, Rockhold Creek to Solomons Island, Chesapeake Bay. M.A. thesis, University of Maryland, College Park, Maryland.

Dugger, S.D. 1990. Recent sedimentation in the Port Tobacco River. M.A. thesis, University of Maryland, College Park, Maryland.

Durand of Dauphine. 1686. A Frenchman in Virginia: the memoirs of a Huguenot refugee in 1686. 1923 translation by Fairfax Harrison, Richmond, Virginia.

Ellison, R.L. and M.M Nichols. 1976. Modern and Holocene foraminifera in the Chesapeake Bay Region, pp. 131–151. *In* C T. Schafer and B. R. Pelletier (eds.), *Maritime Sediments Special Publication No. 1*, Halifax, Nova Scotia.

Froomer, N.L. 1980. Sea level changes in the Chesapeake Bay during historic times. *Marine Geology* 36: 289–305.

Giese, G. and D.G. Aubrey. 1989. The relationship between relative sea-level rise and coastal retreat in New England, p. 5–6. *In* J.C. Topping, Jr. (ed.), *Coping with Climate, Proceedings of the Second North American Conference on Preparing for Climate Change: A Cooperative Approach.* Climate Institute, Washington, D.C.

Gornitz, V. and S. Lebedeff. 1987. Global sea-level changes during the past century, p. 3–16. *In* D. Nummedal, O.H. Pilkey, Jr. and J.D. Howard (eds.), *Sea-Level Fluctuation and Coastal Evolution.* Society of Economic Paleontologists and Mineralogists Special Publication No. 41, Tulsa, Oklahoma.

Gottschalk, L.C. 1945. Effects of soil erosion on navigation in the upper Chesapeake Bay. *Geographical Review* 35: 219–238.

Harrison, S.A. 1915. *History of Talbot County*, Vol. 2. Williams and Wilkins, Baltimore, Maryland.

Hicks, S.D. 1964. Tidal wave characteristics of Chesapeake Bay. *Chesapeake Science* 5: 103–113.

Hobbs, C.H. III, D.B. Blanton, R.A. Gammisch and J. Broadwater. 1994. A marine archaeological reconnaissance using sidescan sonar, Jamestown Island, Virginia, U.S.A. *Journal of Coastal Research* 10: 351–359.

Hohdahl, S.R. and N.L. Morrison. 1974. Regional investigations of vertical crustal in the U.S., using precise relevelings and mareograph data. *Tectonophysics* 23: 373–390.

Hunter, J.F. 1914. Erosion and sedimentation in the Chesapeake Bay around the mouth of the Choptank River. US Geological Survey Professional Paper 90-B.

Kearney, M.S. in press. Sea-level change during the last thousand years in Chesapeake Bay. *Journal of Coastal Research* (accepted for publication in 1997).

Kearney, M.S. and J.C. Stevenson. 1991. Island land loss and marsh vertical accretion rate evidence for historical sea-level changes in Chesapeake Bay. *Journal of Coastal Research* 7: 403–416.

Kearney, M.S. and L.G. Ward. 1986. Accretion rates in brackish marshes of a Chesapeake Bay estuarine tributary. *Geo-Marine Letters* 6: 41–49.

Kearney, M.S., R.E. Grace and J.C. Stevenson. 1988. Marsh loss in Nanticoke Estuary, Chesapeake Bay. *Geographical Review* 78: 205–220.

Kearney, M.S., J.C. Stevenson and L.G. Ward. 1994. Spatial and temporal changes in vertical accretion rates at Monie Bay: implications for sea-level rise. *Journal of Coastal Research* 10: 1010–1020.

Leatherman, S.P. 1986. Cliff stability along western Chesapeake Bay, Maryland. *Marine Technology Society Journal* 20: 28–36.

Lubbers, L., W.R. Boynton and W.M. Kemp. 1990. Variations in structure of estuarine fish communities in relation to abundance of submersed vascular plants. *Marine Ecology Progress Series* 65: 1–14.

Malone, T., W. Boynton, T. Horton and J.C. Stevenson. 1993. Nutrient loadings to surface waters: Chesapeake Bay case study, p. 8–38. *In* Uman, M. (ed.), *Keeping Pace With Science And Engineering: Case Studies In Environmental Regulation*. National Academy of Sciences Press, Washington, DC.

Marcus, W.A. and M.S. Kearney. 1991. Upland and coastal sediment sources in a Chesapeake Bay tributary. *Annals of the American Association of Geographers* 81: 408–424.

Maryland Geological Survey. 1975. *Historical Shorelines and Erosion Rates*. Maryland

Department of Natural Resources, Baltimore.
Mendelssohn, I.A., K.L. McKee and W.H. Patrick, Jr. 1981. Oxygen deficiency in *Spartina alterniflora* roots. *Science* 214: 439–441.
Nixon, S.W. 1980. Between coastal marshes and coastal waters – A review of twenty years of speculation and research on the role of salt marshes in estuarine productivity and water chemistry, p. 437–525. *In* P. Hamilton and K. MacDonald (eds.), *Estuarine and Wetland Processes*. Plenum, New York.
Nordstrom, K.F. 1992. *Estuarine Beaches*. Elsevier Science Publishers, London.
Orth, R.J., K.L. Heck and J. van Montfrans. 1984. Faunal communities in seagrass beds: A review of the influence of plant structure and prey characteristics on predator–prey relationships. *Estuaries* 7: 339–350.
Peltier, W.R. 1985. Climatic implications of isostatic adjustment constraints on current variations of eustatic sea level, p. 92–103. *In Glaciers, Ice Sheets and Sea Level: Effects of a CO_2 – Induced Climatic Change*, US Department of Energy DOE/ER60235-1 Attachment 2.
Redfield, A.C. 1972. Development of a New England salt marsh. *Ecological Monographs* 42: 201–237.
Rosen, P. 1967. Increasing shoreline erosion rates with decreasing tidal range in Virginia Chesapeake Bay. *Chesapeake Science* 18: 383–386.
Rosen, P.S. 1981. A regional test of the Bruun Rule on shoreline erosion. *Marine Geology* 26: M7–M16.
Rule, T.C. 1995. Groundwater withdrawal, land subsidence, and marsh loss in the Chesapeake Bay region of Maryland. M.S. thesis, University of Maryland, College Park Maryland.
Scatena, F.N. 1987. Sediment budgets and delivery in a suburban watershed: Anacostia watershed. Ph.D. dissertation, Johns Hopkins University, Baltimore, Maryland.
Schneider, S. 1989. *Global Warming: Are We Entering the Greenhouse Century?* Sierra Club Books, San Francisco.
Schubel, J.R. 1981. *The Living Chesapeake*. Johns Hopkins University Press, Baltimore Maryland.
Schubel, J.R. and H.H. Carter. 1976. Suspended sediment budget for Chesapeake Bay, p. 48-62. *In* Wiley, M. (ed.), *Estuarine Processes Vol. 2*. Academic Press, New York.
Shomette, D.G. 1982. *Shipwrecks on the Chesapeake Bay and Its Tributaries, 1608–1978*. Tidewater Publishers, Centreville, Maryland.
Singewald, J. and T.H. Slaughter. 1949. Shore erosion and measurement in tidewater Maryland. Maryland Department of Geology, Mines, and Mineral Resources Bulletin 6 Baltimore, Maryland.
Stevenson, J.C. 1988. Comparative ecology of submersed grass beds in fresh, brackish and marine environments. *Limnology and Oceanography* 33: 867–893.
Stevenson, J.C. and N.M. Confer. 1978. Summary of available information on Chesapeake Bay submerged vegetation. US Fish and Wildlife Service, Office of Biological Programs Slidell, Louisiana.
Stevenson, J.C., M.S. Kearney and E.C. Pendleton. 1985a. Sedimentation and erosion in a Chesapeake Bay brackish marsh system. *Marine Geology* 67: 213–235.
Stevenson, J.C., L.G. Ward, M.S. Kearney and T.E. Jordan. 1985b. Sedimentary processes and sea level rise in tidal marsh systems of Chesapeake Bay, p. 37–62. *In* H.A. Groman et al. (eds.), *Wetlands of the Chesapeake*. Environmental Law Institute, Washington, DC.
Stevenson, J.C., L.G. Ward and M.S. Kearney. 1986. Vertical accretion rates in marshes with varying rates of sea-level rise, p. 241–259. *In* Wolfe, D.A. (ed.), *Estuarine Variability*. Academic Press, New York.
Stevenson, J.C., L.G. Ward and M.S. Kearney. 1988. Sediment transport and trapping in

marsh systems: implications of tidal flux studies. *Marine Geology* 80: 37–59.

Tiner, R.W. 1985. Wetlands of the Chesapeake Bay: an overview, p. 16–24. *In* H.A. Groman et al. (eds.), *Wetlands of the Chesapeake*. The Environmental Law Institute, Washington, DC.

USACE (US Army Corps of Engineers). 1990. Chesapeake Bay Shoreline Erosion Study, Vol. 2. Baltimore District, Baltimore, Maryland.

Van Heerden, I.L. and H.H. Roberts. 1980. The Atchafalaya delta: rapid progradation along a traditionally retreating coast (South–Central Louisiana). *Zietschrift fur Geomorphology* 34: 188–201.

Ward, L.G., P.S. Rosen, W.J. Neal, O.H. Pilkey, Jr, O.H. Pilkey, Sr, G. Anderson and S.G. Howie. 1989. *Living With the Chesapeake Bay and Virginia's Ocean Shores*. Duke University Press, Durham.

Wolman, G.M. 1968. The Chesapeake Bay: geology, and geography, p. 117–149. *In Proceedings of the Governors Conference on Chesapeake Bay*. Annapolis, Maryland.

Year 2020 Panel. 1988. Population Growth and Development in the Chesapeake Bay Watershed to the Year 2020. Report of the Year 2020 Panel to the Chesapeake Bay Executive Panel, Annapolis, Maryland.

Chapter Eleven

The role of seagrasses in nearshore sedimentary processes: a review

MARK S. FONSECA
National Marine Fisheries Service

11.1 INTRODUCTION

There are few data that relate the role of seagrasses, which occur mostly in subtidal but sometimes in intertidal settings, to sedimentary processes of the emergent shoreline. Most studies of seagrass bed hydrodynamic or sedimentary processes are limited to the confines of the seagrass bed itself, and usually to only a small portion of that bed. Therefore for this chapter, the concept of shoreline is broadened to include the estuarine and seafloor adjacent to emergent shorelines, i.e. the nearshore environment. Although studies of seagrass beds and shorelines *per se* have historically been separate, this separation is likely more the result of historically isolated scientific disciplines than a recognition of functional separation of seagrass beds from any other part of the littoral zone.

Seagrasses are unique marine flowering plants. With the exception of some species that occur in the rocky intertidal, they bind millions of acres of shallow underwater sediments in the coastal waters of the US with their roots and rhizomes while simultaneously baffling waves and currents with their leafy canopy (Ginsberg and Lowenstam 1958; Taylor and Lewis 1970; den Hartog 1971; Fonseca et al. 1983). The leafy canopy inhibits resuspension of fine particles and traps water column-borne material (Ward et al. 1984), cleansing the water column both of sediment and water column nutrients (Bulthuis et al. 1984; Short and Short 1984). Nutrient uptake by seagrass blades and their associated epiphytes and macroalgae

Estuarine Shores: Evolution, Environments and Human Alterations. Edited by K. F. Nordstrom and C. T. Roman.

readily incorporates dissolved nutrients into plant biomass, which can improve water quality. The positive influence of the canopy on sediment stabilization is enhanced by the presence of a root and rhizome mat. Overall, the ecological function of seagrasses in many coastal ecosystems has been extensively documented (see reviews by Phillips 1984; Thayer et al. 1984; Zieman and Zieman 1989). I will not review those functions in detail here because they have received extensive documentation worldwide. This review is limited to plants restricted to near-marine salinities (another tacit definition of seagrasses), yet many of the interactions of these plants with their physical environment are similar in many ways to non-marine submerged aquatic vegetation.

11.2 THE PRESENCE OF SEAGRASS COVER AND ITS GEOMORPHOLOGICAL CONSEQUENCES

Although seagrass beds are well recognized for their ability to modify sediment quality and quantity, only limited information exists as to any effects of either the presence or absence of a seagrass bed on shorelines. The influence of seagrasses in modifying emergent shoreline processes, while indirect, has the potential to be highly significant. The existence in many coastal areas of hundreds or thousands of hectares of seagrass beds seaward of the shoreline that can dampen waves, capture sediment, promote fining of sediments and export tons of organic material which modify sediment composition, provides all the attributes to modify shoreline processes both qualitatively and quantitatively. Despite this apparent capacity for shoreline modification, most observations of seagrass–shoreline interactions came as a result of the decline of eelgrass (*Zostera marina*) in the North Atlantic during the early 1930s. Wilson's (1949) narrative on eelgrass loss in England revealed that areas of the seafloor previously colonized by eelgrass dropped by 0.6–0.9 m after the decline. Interestingly, Wilson reported that the dead plants left their root and rhizome system intact in the sediment and that that material maintained the mounded bedforms for 2 years until it decayed, leading to erosion of the bedform. With the loss of the entire plant system, both Wilson (1949) and Rasmussen (1973) reported a substantial shift from mud–sand sediments to coarse sand and even cobbles, as the finer-grained material was washed away (Figure 11.1). Shoreward of the lost seagrass, substantial changes occurred. Wilson (1949) reported up to 1.2 m of sand accumulating on the beach, a phenomenon he attributed to destabilization of the seagrass beds. Rasmussen (1973) reported that once fine sediments were removed, macroalgae colonized the exposed cobbles. The added drag of the algae caused the cobbles to be moved ashore during storms, making the high-water line of the beaches considerably more rocky. Rasmussen (1973) also reported dramatic changes in circulation around Vorso island, Denmark where the loss of eelgrass resulted in less diffuse flow across broad flats, leading to increased channelization and higher tidal current speeds. In the most detailed study of

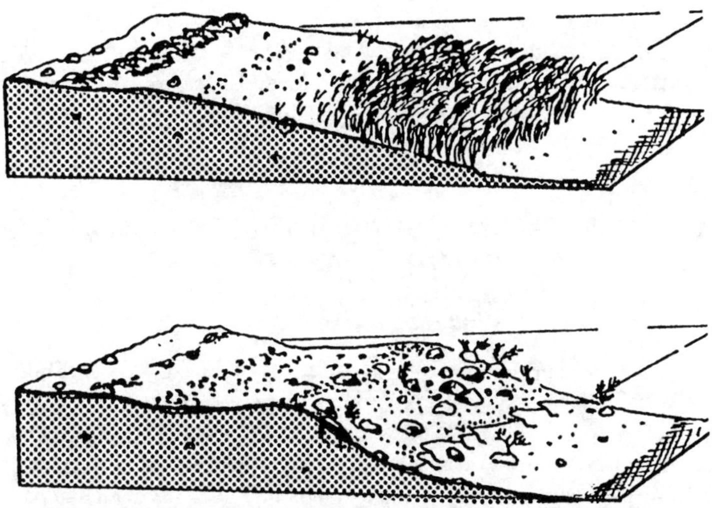

Figure 11.1 Changes in the Danish fjords before (above) and after (below) the die-off of *Zostera marina* in the 1930s. Note the change in wrack abundance and the increase in both exposed rocks and associated macroalgae. Redrawn from Rasmussen (1973)

shoreline effects in recent history, Hine and co-authors (1987; and references therein) followed the geomorphologic changes in a Florida barrier island over 30 years in relation to changes in seagrass faciation. When offshore seagrass beds declined, landward-migrating sand bars formed, resulting in a 30% lengthening of the island, a distance of ~1200 m in a time period representing 2% of the total age of the island. They reported that nearshore sands were being carried to the surf zone and transported mostly to the north by the long-shore current system. The nearshore sands only became available for transport when the seagrass coverage declined *en masse*. This comparatively rapid change in sedimentary processes (in a geologic sense) was also shown to greatly complicate interpretation of sedimentary facies.

Seagrass beds also have been shown to contribute indirectly to coastal dune formation. Hemminga and Nieuwenhuize (1990) examined soil profiles and found that seagrass that had been washed ashore (wrack) was often buried by offshore aeolian transport of desert sand. This alternating wrack deposition and burial formed the basis for a coastal dune system. Wrack formation by seagrass is a regular part of many shorelines; although wrack has had other uses such as insulation or garden mulch, its role in sediment formation and erosion reduction appears to have received little study.

These limited descriptions constitute most of what is known about the direct linkage of seagrass beds with the emergent shoreline, yet their message is consistent and compelling. In many coastal areas the submerged seagrass beds are a critical component of shoreline composition and structural integrity. For most of the remainder of this chapter I will review the reciprocal interaction between seagrass and water motion, while emphasizing associated sedimentary processes. Finally I will discuss the utility of viewing seagrass beds at various spatial scales as a means of identifying and understanding relevant scales of associated sedimentary processes.

11.3 FACTORS AFFECTING SEAGRASS COVERAGE: CONSEQUENCES FOR SEDIMENTARY PROCESSES

11.3.1 A short history: studies of seagrass coverage and hydrology

Waves and tidal currents have long been recognized as contributing to the configuration of seagrass beds. Variation in seagrass bedform can easily be visually detected from low-level aerial reconnaissance (Figure 11.2) and appears to be correlated with exposure to waves and currents. For example, Molinier and Picard (1952) described extraordinary vertical stands of *Posidonia* which apparently form when the edge of the seagrass bed is slowly eroded and the rhizome mat folds down. The roots apparently continue to grow horizontally into the underlying sand, continuing the adhesion of the seagrass to the unconsolidated sediment. I have observed similar formations occurring in Rhode Island (Figure 11.3) which reveal the extent to which seagrass can reduce sediment erosion.

For approximately 30 years after Molinier and Picard's work, interest in the physical processes occurring in seagrass beds was confined largely to qualitative descriptions of their geological role (e.g. Schubel 1973) and, to a lesser degree, the implications of seagrass bed stability on animal utilization. For example, Ginsberg and Lowenstam (1958) reported on the stabilization and accumulation of the Florida Bay mudbanks by seagrass. O'Gower and Wacasey (1967) stratified their faunal sampling to account for current-induced differences in sediment deposition and, thus, particle size in seagrass beds. Scoffin (1970) examined the erosion threshold afforded by seagrass in a Bahamian study, but also described in detail how *Thalassia testudinum* enhanced sediment accumulation and bedform development. Taylor and Lewis (1970) recognized that sediment ripples (several decimetres in height and 1–2 m in length) occurred with seagrasses on the ripple crest. They also depicted the arrangement of these ripples in detail over a wider spatial scale (~20 m) than previous work. Simultaneously, Wood and colleagues (1969) concluded that the sediment buildup under seagrass cover was the result of resistance to erosion and subsequent accumulation. den Hartog (1971) synthesized the extant literature and recognized that sediment accumulation by seagrasses ultimately led to an equilibrium between deposition and erosion, a phenomenon further described by Fonseca and co-authors (1983). Basan (1973) also noted dune-like formations 'stabilized by the marine grasses'.

Curiously, up to 1973, all of the above papers explicitly recognized the effect of seagrasses on tidal current reduction and sediment stabilization (and notably to a

Figure 11.2 Aerial photograph of mixed *Halodule wrightii, Ruppia maritima, Syringodium filiforme*, and *Thalassia testudinum* beds on the western margin of Tampa Bay, Florida. In the foreground at the bayward edge of the shoal are what appear to be wave-sculpted beds while further landward, in shallow water are more continuous cover beds. Reduction in wave energy from both the shelving shoal and the grass itself is thought to be responsible for the resultant seagrass bed landscape pattern

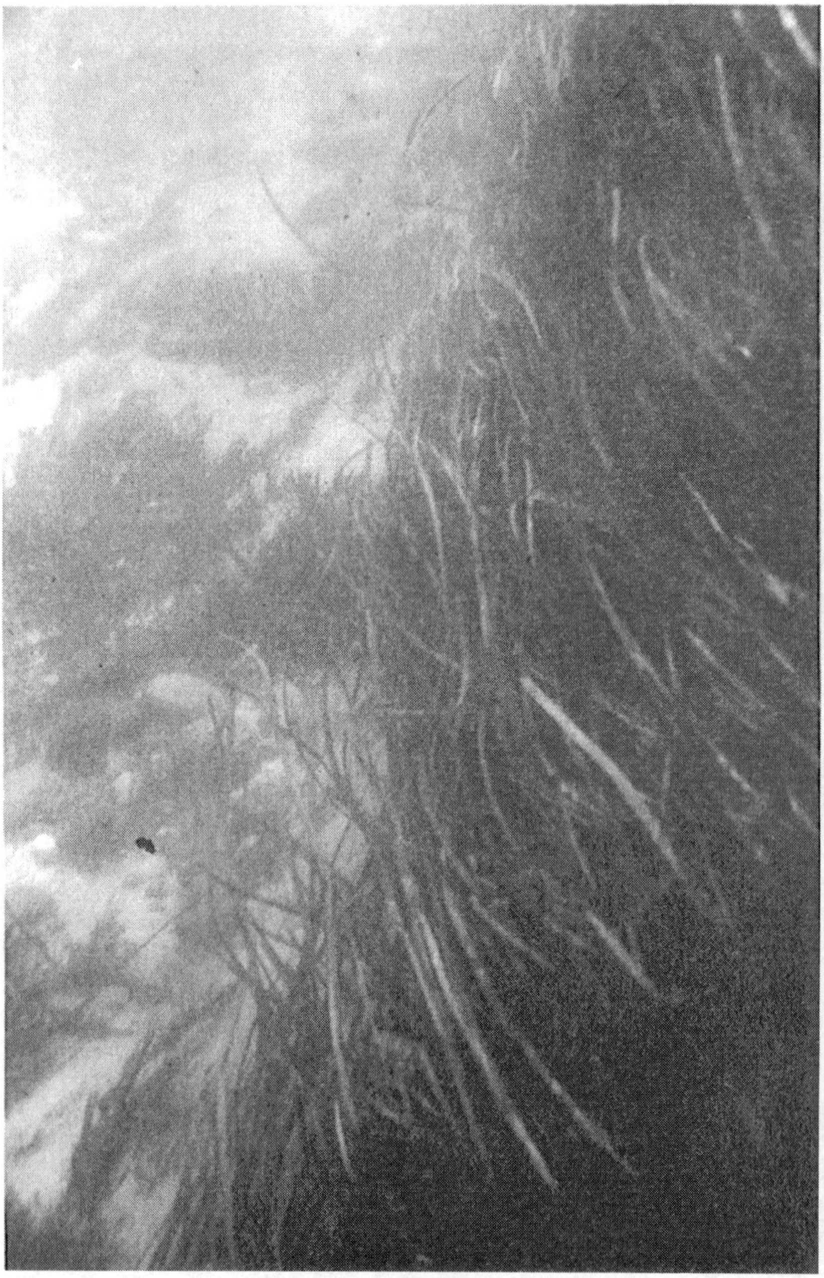

Figure 11.3 Approximately 2 m high wall of *Zostera marina* in the Charlestown Pond breachway, along the southern coast of Rhode Island. Tidal current speed here exceed $1.5\,m\,s^{-1}$

much lesser extent on waves), but did not uniformly connect those effects with development of the bedforms themselves. Scoffin (1970), after describing *T. testudinum* mounds in a channel, stated: 'It is not known what initiated the uneven surface in the channel.' Taylor and Lewis (1970) and Basan (1973) concluded that formation of sand ripples ~5 m long by 0.5 m high sand ripples preceded colonization by seagrass while Basan (1973) suggested that periods of low current were required for seagrasses to colonize these ripples. It was unclear whether Basan meant that colonization could progress incrementally during slack tide or that longer-term periods of low current would be required, as might be found when tidal channels shift position, leaving some areas with slower currents. Although both scenarios are certainly possible, it is equally possible that the seagrass itself may have spread even in the high-current areas and thus been the agent of bedform development through differential reduction of sediment erosion inside as opposed to outside the plant canopy and by enhancing deposition. At about the same time, den Hartog (1971) used the term 'leopard-skin pattern' to describe the apparent increase in seagrass bed patchiness with increased water motion. However, while acknowledging the bottom-raising ability of *Posidonia*, den Hartog questioned the capacity of other widely distributed genera, such as *Zostera*, to accumulate sediments and elevate the bottom, apparently because he believed *Zostera* rhizomes to be incapable of vertical growth. den Hartog (1971) concluded that the precise mechanisms responsible for these elevated seagrass formations were not yet known and may have resulted from sand ripple colonization, as was suggested by Taylor and Lewis (1970) and later by Basan (1973). After this time, questions regarding the sediment accumulation capacity of seagrasses faded from the literature and became tacitly accepted as part of their existence, when in fact some of the scenarios of ripple colonization (e.g. Taylor and Lewis 1970) likely describe opportunistic colonization events by faster-spreading seagrass species. The conclusion here is that these strategies are not mutually exclusive: seagrasses may cause dune-like formations or may colonize them after they are formed.

In 1975, Patriquin's work shifted attention back to large (>10 m) spatial scale assessment of the pattern of seagrass beds resulting from physical disturbance. What differentiated Patriquin's work from past work was that, aside from documenting patchy bed patterns, he proposed a specific mechanism of formation. He demonstrated that sand patches (type I: open uncolonized areas within a larger, colonized substrate, *sensu* Sousa 1985) were consistent with wave- and current-exposed areas. The open areas were termed 'blowouts' and were shown to migrate up to $2\,m\,a^{-1}$. Marba and colleagues (1994a) also documented seagrass dune migration of ~$13\,m\,a^{-1}$. The ecological relevance of bed or blowout migration was strengthened when Orth (1977) provided quantitative evidence of the importance of seagrass-derived sediment stability for associated fauna (increased instability was equated with decreased faunal abundance and diversity). Wanless (1981) expanded on Patriquin's work, and published aerial photographs much like those seen in O'Gower and Wacasey (1967), Taylor and Lewis (1970), and Basan (1973), and specifically ascribed the seagrass landscape pattern to effects of tidal currents and waves. Fonseca and colleagues (1983) conducted more quantitative field surveys and provided evidence that isolated mounds of seagrass (*Zostera marina* and *Halodule wrightii*) were positively correlated with tidal current speed, and produced a conceptual model regarding the influence of tidal currents in seagrass bed development at multiple temporal scales. More recently, Kirkman (1985), Hine and co-workers (1987) and Kirkman and Kuo (1990) extended the spatial scale of consideration to whole coastal regions (many kilometres) by qualitatively linking seagrass landscape patterns across this range of spatial scales to frequency and intensity of storm events and changes in seagrass coverage.

Aside from reviews (Phillips 1984; Thayer et al. 1984; Fonseca and Kenworthy 1987; Zieman and Zieman 1989), and laboratory studies (Fonseca et al. 1982; Fonseca and Fisher 1986; Gambi et al. 1990; Fonseca and Cahalan 1992), I am aware of no additional published quantitative evaluations linking seagrass landscapes to water motion. However, ongoing work (author's unpublished data) is attempting to equate the pattern of seagrass landscapes with water motion. Large areas of the estuarine seafloor have been surveyed at the 1 m scale to yield detailed, three-dimensional maps of seagrass beds (Figure 11.4) which were then correlated with exposure to waves (Keddy 1982; 1983) and tidal current speed. As predicted by Fonseca and colleagues; (1983), mounding of seagrass beds increased with increasing water motion while seagrass cover decreased (Figure 11.5).

There is additional evidence that seagrass landscapes may be influenced by water motion. I have observed seagrass shoots forming distinct rows, spaced ~200 mm apart, normal to both waves (site in Figure 11.2) and tidal current flow (site in Figure 11.3). These formations seem to form not only under relatively high water motion conditions but at sites where water motion impinges on the beds generally from one direction. While the effects of these shoot arrangements may influence bed pattern at larger spatial scales, I cannot offer any mechanical or architectural model (*sensu* Callaghan et al. 1990) to explain these phenomena.

11.3.2 Alternative sources of bed pattern

There are ways that seagrasses can form patterns other than those imposed by hydrodynamic forces. The basic centrifugal, vegetative growth and mortality pattern of seagrasses and most clonal organisms (Harper 1985) confers a characteristic form of patch development. The frequency and spacing of seed colonization or vegetative fragment interment, as well as subsequent vegetative growth strategies (Williams 1990), have been described as the basis for the shape of seagrass beds across scales of tens of meters. Rate of asexual shoot reproduction, mortality, rhizome elongation and branching architecture (Cook 1985; Waller and Steingraeber 1985; Callaghan et al. 1990) have successfully described clonal plant growth. Thus, historical contingency (e.g. time since colonization or recolonization after disturbance) could result in isolated patches that are simply intermediary stages of bed expansion leading towards coalescence of individual patches (Fonseca et al. 1983) yet mimic hydrodynamically derived formations. Marba and Duarte (1995) clearly demonstrated that subaqueous dune migration maintained *Cymodocea* 'in a continuous state of colonization ... which is ultimately responsible for the characteristic patchy landscape ...' of a seagrass-dominated bay in Spain.

Bioturbation is another mechanism, presumably independent of water motion, which confers pattern in seagrass beds. den Hartog (1971) not only described in some detail the dynamics of seagrass beds as a whole, but focused attention on the shifting margins of beds and did not suggest bioturbation as a cause of bed pattern. Since then Callianasid shrimp (Taylor and Lewis 1970; Suchanek 1983), echinoderms (Valentine and Heck 1991; Valentine et al. 1994; T. Backman, personal communication) and rays (Townsend and Fonseca, unpublished) have been described as sources of disruption and, potentially, the cause of alternating zones of sand and seagrass several meters in size in many seagrass beds. Smaller animals such as portunid crabs and sediment deposit feeders also may prevent the expansion of bed margins (Philippart 1994). Other sources of seagrass bed pattern are sediment depth and physiological stress. In many subtropical and tropical systems, unconsolidated sediment depth is a principal source of seagrass pattern especially where *T. testudinum* dominates (Wood et al. 1969; Scoffin 1970; Zieman 1972; Basan 1973). Depressions in bedrock fill with unconsolidated sediment and provide suitable rooting. The spatial distribution of these sediment pockets (*sensu* Zieman 1972) creates a mosaic of seagrass patches that is independent of water motion, growth pattern, or bioturbation. Moreover, physiological stress through emersion can also be influential in producing patterns of seagrass distribution. Except for Cooper and McRoy (1988), Perez-Llorenss and Niell (1993), and Adams and Bate (1994), there has been little documentation of emersion effects on seagrass distribution despite its long acknowledgement as a limiting factor (den Hartog 1971). Rather, attention has been focused on limitations of seagrass at the lower end of its distribution through light limitation (see review in Kenworthy and Haunert 1991). I know of no attempts to associate light limitation with patterns of seagrass growth across a range of depths (i.e. light quality or quantity), and this work is needed to model seagrass demographic processes effectively.

11.4 HYDRODYNAMICS OF SEAGRASS BEDS AND ASSOCIATED PROCESSES

11.4.1 Effect of fluid motion on seagrasses and sedimentary processes

Water motion, usually in the form of unidirectional (and reversing) tidal flow, has been described in relation to many seagrass bed attributes. For example,

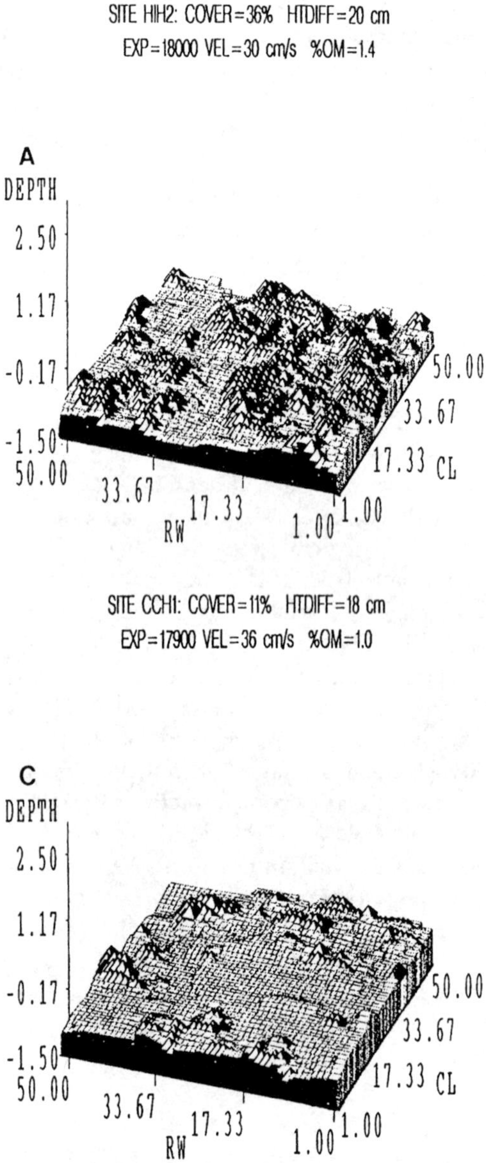

Figure 11.4 Map of four different seagrass beds in the Beaufort, NC region. Maps are 50 m on a side with 1 m pixels. Dark, pointed areas are sections of the seafloor with seagrass; flat, non-pointed areas are unvegetated sand. All surfaces are of the seafloor and do not reflect the canopy height. Vertical exaggeration is 6×. (A) High wave and current site; large, dune-like formations which migrate at up to 1 m a^{-1}; (B) low wave and current site with low,

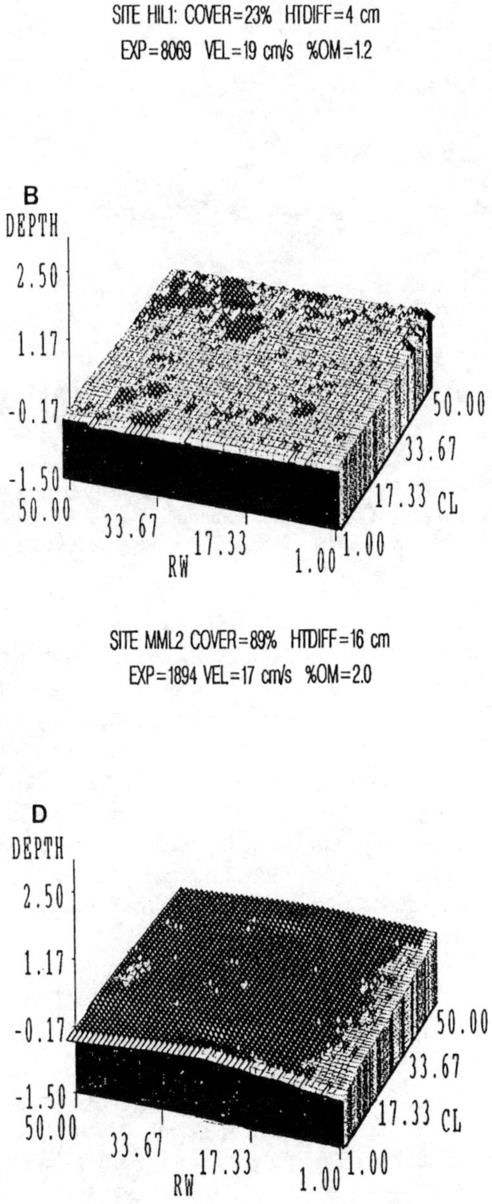

patchy coverage; patchiness is seasonal and fluctuates between the level shown and 100% cover due to seedling recruitment and dieback of a conspecific seagrass; (C) high wave and current site, shallower than site A by ~0.3 m and as a result, has less cover; (D) low wave and current site on edge of channel (to right in figure); note almost unbroken cover away from channel

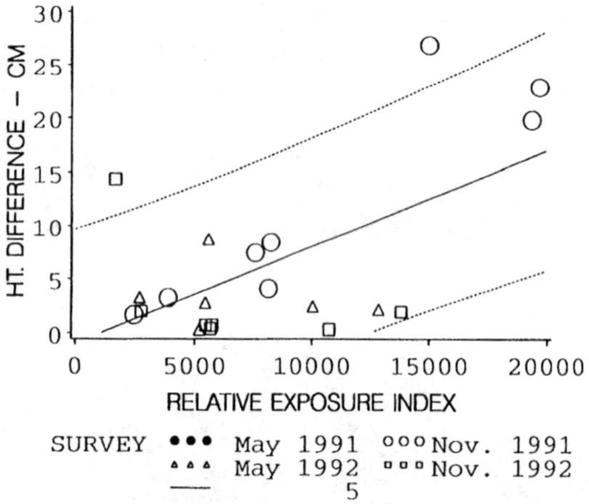

Figure 11.5 Regression of height difference between vegetated and unvegetated areas of seagrass beds near Beaufort, NC (r^2=0.18, F=18.098, pF <0.0001). Exposure index is a product of fetch, maximum wind speed and frequency of wind by compass direction (see Keddy 1982, 1983; Murphey and Fonseca 1995). Open circles = Nov. 1991, triangles = May 1992, squares = Nov. 1992

productivity (Fonseca and Kenworthy 1987), animal feeding and distribution (Peterson et al. 1984; Pihl 1986; Cahalan et al. 1989; Irlandi 1994), depositional processes (Scoffin 1970; Patriquin 1975; Orth 1977; Fonseca et al. 1983; Peterson et al. 1984; Fonseca and Fisher 1986 and references therein), unidirectional flow modification (Fonseca et al. 1982; 1983; Fonseca and Fisher 1986; Fonseca 1989; Gambi et al. 1990), wave attenuation (Fonseca and Cahalan 1992), and ecosystem structure and development (den Hartog 1971; Patriquin 1975; Fonseca et al. 1983) have all been described as being influenced by hydrodynamic conditions. Despite this attention, work regarding the mechanics of seagrass and water motion interactions comprises a small fraction of the over 6000 papers regarding seagrass (~600 since 1988 alone) in the open literature.

Given that air movement is recognized as being a significant structuring element of terrestrial landscapes (Sprugel 1976), and seawater is ~1000 times more dense and 60 times more viscous than air, water motion should have an enormous impact on the physical structure of seagrass beds. Propagule dispersal and colonization, drag on seagrass shoots and anchoring systems, including seagrass bed sediment erosion, accumulation and bed migration, material exchange across the canopy–water column interface, and seagrass nutrient uptake are but a few of the relevant processes that may be affected by water motion (*sensu* Koehl 1986).

However, distinctions need to be drawn regarding water motion as it affects seagrass ecosystems. One is the nature of the water motion, whether it is unidirectional flow, oscillatory flow, or some complex combination of the two. Another is identifying the relevant spatial and temporal scales at which a particular form of water motion and sedimentary process is being measured. I will discuss water motion at spatial and temporal scales consummate with forces acting on individual shoots which are sometimes in isolation but, more frequently, as a group of individual shoots which together form a canopy, because they may provide the basis for expression of bed patterns at all larger scales of consideration.

11.4.1.1 Unidirectional flow

The influence of seagrass on fluid motion has been the target of quantitative research in only a limited number of instances in the field (Conover 1964; Scoffin 1970; Fonseca et al. 1983; Peterson et al. 1984; Fonseca 1989; Ackerman and Okubo 1993; Worcester 1995), and in the laboratory (Fonseca et al. 1982; Fonseca and Fisher 1986; Fonseca and Kenworthy 1987; Gambi et al. 1990; Powell and Schaffner 1991). Only four of the above papers included species other than *Z. marina*. Conover (1964) expressly implied a positive relationship of current flow with seagrass growth, distribution and biomass, but quantitative data were limited. Scoffin (1970) employed a flume in situ and evaluated erosion in turtlegrass (*Thalassia testudinum*) at velocities up to $1.50\,\mathrm{m\,s^{-1}}$ to determine initiation of sediment motion and erosion. Both Conover (1964) and Scoffin (1970) made qualitative observations regarding the effect of shoot density on current velocity; both reported that shoot density negatively affected velocity and that the highest densities of seagrass shoots were found in high-current areas.

The shoot density-dependent findings of Conover (1964) and Scoffin (1970) conflict somewhat with later, more detailed laboratory studies of unidirectional flow (Fonseca et al. 1982; Fonseca and Fisher 1986; Fonseca and Kenworthy 1987; Gambi et al. 1990). Moreover, data have not always been reported in easily comparable formats. Fonseca and Fisher (1986) and Ackerman and Okubo (1993) reported shear velocity measurements as drag coefficients (Darcy coefficient, *f*, and *Cd*, coefficient of drag, respectively) to provide a quantitative basis for comparison with plant communities in other flow settings. Only Fonseca and Fisher (1986), and Gambi and colleagues (1990) computed Reynolds numbers ($Re = LU/v$: L = roughness length, U = mean velocity, v = kinematic viscosity), allowing comparative study through dynamic scaling among experiments. Although Fonseca and Fisher's (1986) findings were dismissed by Gambi and colleagues (1990) because of scaling problems, Gambi and co-authors (1990) failed to note the similarities among the studies. Part of the reason for this similarity may be that the *Re* values among the two studies were of the same order of magnitude; both studies yielded identical conclusions regarding density effects, flow stratification and downstream flow field restructuring. Surprisingly, these few papers have yielded relatively consistent results in terms of velocity reduction:

1. In low-relief beds (no down-current vertical change in bed elevation) flow fields rapidly reconstitute along the down-current axis of flow.
2. Flow stratification is intense at the canopy–water column interface.
3. Substantial deflection and compression of the leaf canopy occurs with

increasing current velocity, a strategy that substantially reduces drag ove upright forms (Vogel 1981).

4. Increasing seagrass shoot density (within most natural ranges) has little effect on current velocity reduction.

Simple measures of flow reduction by a seagrass canopy do not go far enough in describing relevant hydrodynamic conditions. The work on flow turbulence by Gambi and colleagues (1990) was the first to compare flow within the seagrass canopy with the entire water column. Evaluation of turbulence is critical to understanding sediment resuspension events as well as biologically relevant events such as larval dispersal and settling (Jumars and Nowell 1984; Cahalan et al. 1989 Eckman et al. 1989; Judge et al. 1992). Only work by Ackerman (1993) had previously contemplated turbulent structure in seagrass canopies. Ackerman (1986) dealt primarily with the consequences of turbulent modification on seagrass fertilization, whereas Ackerman and Okubo (1993) focused on the relation of turbulent rescaling to plant movement (waving: Honami, *sensu* Inoue (1955). These examinations revealed that turbulent intensity was significantly reduced within the canopy of seagrass beds, irrespective of shoot density (but see Worcester 1995).

In a study of flow conditions at a much larger scale, Powell and Schaffner (1991) examined the influence of *T. testudinum* and *H. wrightii* on flow reduction off mudbanks at low tide and extremely slow current speed. When water levels had dropped well below the height of the canopy and plants were lying over at an extreme deflection (*sensu* Dennison 1979), water seepage off the banks (horizontal distance of 300–500 m) was retarded for at least eight hours. At the resulting low flow speed, removal of sediment from the water column and deposition both among and upon the plants were likely very high, as reported by Bulthuis and co-workers (1984), who demonstrated suspended solid levels in water moving off seagrass banks in Australia to be up to five times lower than in water moving off unvegetated banks.

So far I have primarily discussed the effect of seagrass on flow conditions. The reciprocal effect of flow on the plants is relevant to understanding hydrodynamic and sedimentary processes as well. The response of seagrass to fluid motion is complex and likely comprises several advantageous mechanical responses to fluid motion. The flexibility or stiffness of individual plants, their size (particularly surface area), cross-sectional area (implying ability to withstand flow-induced stress), and their existence as a clonal population, are all factors that interact in a complex fashion, presumably as trade-offs to optimize light-capturing ability while minimizing stress on the plant (*sensu* Charters et al. 1969; Koehl 1986; Koehl and Alberte 1988; Holbrook et al. 1991). Fonseca and colleagues (1982) and Fonseca and Kenworthy (1987) found that seagrass bending angle increased linearly as a function of flow speed. Fonseca and Fisher (1986) found that canopy friction decreased as the canopy was deflected and Reynolds number (*Re*) increased although deflection decreased with downstream distance. These studies all revealed the generally high flexibility of seagrasses which, as pointed out by Koehl (1986) for algae, is a means of reducing drag on the plants and their anchoring system especially when bending over near the substratum, which further reduces drag. Superimposed on these factors is the synchronous movement of the canopy under unidirectional flow. This movement acts to reduce skin and form drag in a manner consistent with that observed in terrestrial crops (Inoue (1955), reviewed for seagrasses by Fonseca and Kenworthy (1987)) and demonstrated for macroalgae by Koehl and Alberte (1988). Waving may also act as a peristaltic device to pump material through the seagrass vascular system. But the flexibility of seagrasses is also clearly a factor in their survival. Dennison (1979) measured how *Z. marina* bending tracks water depth, and the ability to bend over and mesh with neighboring shoots has value both during periods of water motion and during near-emersion at low tides, avoiding desiccation during emersion.

On a smaller spatial scale, some information exists regarding the physical or structural limits of seagrasses in response to the hydrodynamic setting. Scoffin (1970) conducted limited experiments where an underwater flume was used to determine qualitative measures of leaf damage to *T. testudinum*, and reported that at ~$0.70\,m\,s^{-1}$ blades began to flap violently and break. However, Fonseca and colleagues (1983) reported *Z. marina* existing at current speeds of 1.20–$1.50\,m\,s^{-1}$ (Figure 11.3), and those plants were largely intact. But there are some areas where data are sorely needed. I am not aware of any quantitative information regarding the effects of added drag from epiphytization of seagrass shoots, which can be significant in eutrophicated areas. Moreover, the importance of shoot drag may well change with tidal stage when waves may change from deep to shallow water waves and, perhaps, even with time of day with changes in lacunar gas storage and buoyancy (Dennison 1979). Beyond demonstrating the flexible nature of seagrasses and some rudimentary data on structural limitations, no comparative work has been performed among natural beds in different velocity regimes to identify the existence of any feedback effects of water motion on morphology or structure (e.g. flexural stiffness, root and rhizome abundance or architecture) that could constitute an advantage that improves local seagrass bed survival and persistence.

Beyond mechanical limits of individual plants, their arrangement as a larger structure and the consequent aggregate roughness as it affects water motion and sedimentary processes have been studied only a little. Work conducted on the effect of shoot density on alteration of unidirectional flow (Fonseca et al. 1982; Fonseca and Fisher 1986; Gambi et al. 1990) and waves (Fonseca and Cahalan 1992) showed little effect. Given the lack of demonstrable shoot density effects on flow, other responses by seagrass to current flow, such as shoot arrangement and morphological differences, should be examined, but I am aware of no studies which emphasize the possible role of seagrass shoot organization either as a response to, or as an effect of, unidirectional flow and/or waves. Although seagrass morphology can change dramatically in response to other physical factors, such as light, the consequences of the highly variable morphology among seagrasses in producing different effects on the flow field have received only limited attention in seagrass systems (Fonseca and Fisher 1986; Fonseca 1989; Fonseca and Cahalan 1992, but see wave effects on life history of freshwater plants: Kautsky 1987; 1988). Moreover, seagrass shoot morphology exhibits broad interspecific variation, with leaves varying in shape from rod-like, nearly cylindrical leaves to small, quasi-oval forms. Substantial intraspecific variation also occurs either as the result of phenotypic plasticity (Backman 1991) or genetic polymorphisms (Laushman 1993). For example, in the space of 20 m, eelgrass (*Zostera marina*) can exhibit blade lengths ranging from 0.15 to 1.50 m and concomitant blade widths of 4 to 15 mm (Puget Sound, personal observation). These morphological differences complicate intra- and interspecific comparisons of hydrodynamic changes affected by seagrass.

11.4.1.2 Oscillatory flow

The influence of seagrasses on oscillatory flow is even less well known than their influence on unidirectional flow. As seen with unidirectional flow, only a few seagrass species have been considered: *Halodule wrightii, Syringodium filiforme, Thalassia testudinum* and *Zostera marina.* Wayne (1975) alluded in an unpublished thesis to wave reduction capabilities by turtlegrass (*T. testudinum*) but with few specifics. Ward and colleagues (1984) conducted measures of wave reduction in and out of seagrass beds, focusing on sediment resuspension. They found that when seagrass occupied most of the water column, sediment resuspension was reduced as compared with unvegetated areas. Heller (1987), in the most complete, as yet unpublished, work to my knowledge, also targeted sediment movement; she found significant differences in wave reduction and orbital paths among different seagrass species. Fonseca and Cahalan (1992), using a small, non-orthogonal data set, determined wave reduction capabilities to be similar among the seagrass species tested and equivalent to that of saltmarshes, or ~40% reduction in wave height per meter of seagrass bed travelled. They also found significant wave height reduction when seagrasses occupied most of the water column, as seen by Ward and colleagues (1984) for sediment resuspension.

The influence of seagrass on oscillatory flow is of course defined by the structural attributes of the plants themselves. Mechanical constraints in marine plant form have been documented in a number of studies on algal growth (Koehl and Wainwright 1977; 1986; Jackson 1984; Armstrong 1987; Biedka et al. 1987). For example, Armstrong (1987) detected greater streamlining of kelp among wave-exposed as compared with wave-sheltered areas. The only interspecific comparison of mechanical adaptation in seagrass of which I am aware was that by Cooper and McRoy (1988) of sediment-rooted *Zostera* with *Phyllospadix* attached in rocky littoral zones. They found greater root hair development, thickened rhizomes and smaller lacunae in *Phyllospadix*, the latter apparently unnecessary for maintaining buoyancy in such a highly turbulent setting while constituting a means of reducing volume and thus susceptibility to acceleration-reaction force. Whereas Cooper and

McRoy (1988) could not detect structural differences among *Phyllospadix* spp., they found that this genus had substantial anatomical adaptations useful for wave-swept environments that were not found in a non-rocky intertidal seagrass, *Z. marina*. There may be no other published data regarding the mechanical effects of oscillatory water motion on seagrass growth.

11.4.2 Role of root-rhizome system in seagrass bed roughness/stability

Most work on plant anchoring mechanics has been conducted in agricultural settings (Ennos 1989, but see Brewer and Parker 1990). Unfortunately, relatively little attention has been given to rhizomatous species, an area with more direct application to seagrass rhizosphere mechanics than isolated plants such as maize. Although models presented by Fritter and Ennos (1989) indicate that resistance to horizontal forces such as those experienced by seagrass during maximum extension under waves and unidirectional flow require additional anchorage above that provided by tensile roots, nowhere is there mention of interconnected shoots or the resistance of embedded rhizomes themselves.

Although quantitative information on anchoring mechanics in seagrass is limited, their capabilities in this regard are easily observed and have been recognized for many years (Wilson 1949; Molinier and Picard 1952; Ginsberg and Lowenstam 1958; Wood et al. 1969; Scoffin 1970; Patriquin 1975; Orth 1977; Wanless 1981; Fonseca et al. 1983; Fonseca and Fisher 1986; Fonseca 1989). Seagrass beds which form patchy, dune-like formations (Fonseca et al. 1983) often have such fibrous root-rhizome mats that it is extremely difficult even to work one's fingers into the structure. When eroded, the mat often behaves as a carpet and rolls up and fragments in large (up to ~1 m pieces) together with the attached leafy shoots. This mat cohesiveness is often not found in quiescent areas, but for *Z. marina*, I have found that only belowground biomass can be positively correlated with wave exposure (unpublished data). This correlation is similar to those observed by Coops and colleagues (1991) for freshwater littoral vegetation.

11.4.3 Effects of sedimentation on seagrasses

Most seagrasses can withstand moderate deposition and grow vertically, maintaining light capture ability. The deposition rate that a given plant can withstand must vary as a function of the original size of the plant as well as species. However, the ability to respond to sediment deposition with vertical growth or simply survive deposition has been observed in *Thalassia* (Patriquin 1975) and *Cymodocea* (Marba et al. 1994a,b; Marba and Duarte 1995). More recently, I have found that *Zostera* beds only suffer significant shoot mortality when 50% of the plant's height is buried under sand or silt–sand sediments for approximately three weeks (unpublished data).

Previous limited work (Fonseca et al. 1985) examined the effect of seagrass rhizomes on sediment stabilization, although Peterson (1982) compared difficulty of sediment penetration in and out of *H. wrightii*. Fonseca and colleagues (1985) removed *Z. marina* leaves and stems from mounds 0.08 m high by 1.00 m diameter and examined subsequent mound destabilization. Approximately 5 mm of sediment was lost almost immediately but no further net loss of sediment was detected over the following ten days. Over a much longer term, Wilson (1949) observed that *Z. marina* mounds persisted for 2 years after shoot loss. In a concomitant experiment, Fonseca and colleagues (1985) found that sediment accumulation from transplanted *Z. marina* ranged from 11 to 92 $m^3 ha^{-1}$ over a year. Almasi and colleages (1987) found that *T. testudinum* beds enhanced deposition ~60% more than unvegetated areas.

Except for the previously described findings by Scoffin (1970) and Fonseca and colleagues (1985), all reports of seagrass anchoring capabilities have been anecdotal (Thayer et al. 1984; Zieman and Zieman 1989, and references therein). Although severe damage can be wrought in seagrass beds during severe storms, such as hurricanes, there is ample evidence that established beds of some species

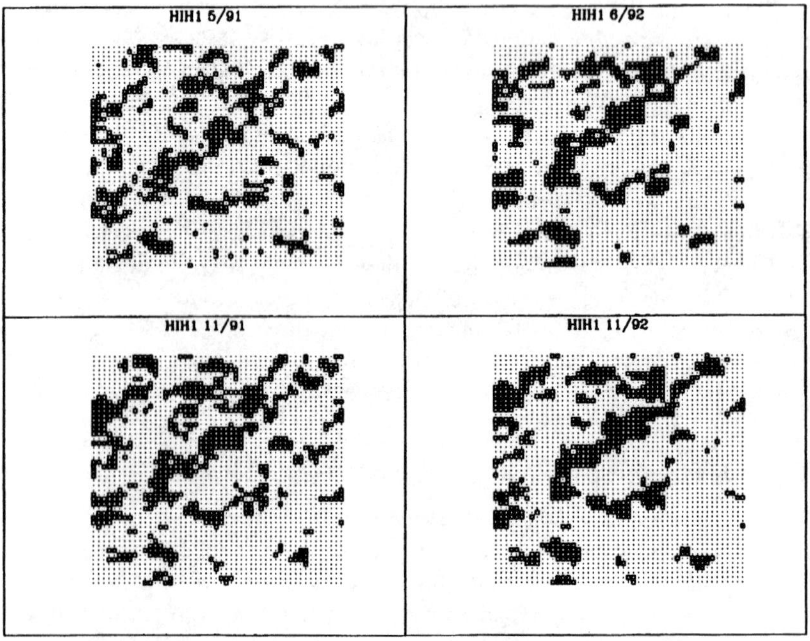

Figure 11.6 A four-year chronology of seagrass coverage in a 50 × 50 m section of the estuary near Beaufort, NC. Each square represents a 1 m pixel with seagrass. Each dot is a 1 m pixel without seagrass. A hurricane-force storm (average maximum sustained hourly winds of 91 kph) struck in March 1993, resulting in a lasting reduction of seagrass coverage. The site represents a comparatively exposed location in the estuary; less exposed sites and deeper sites were largely unaffected

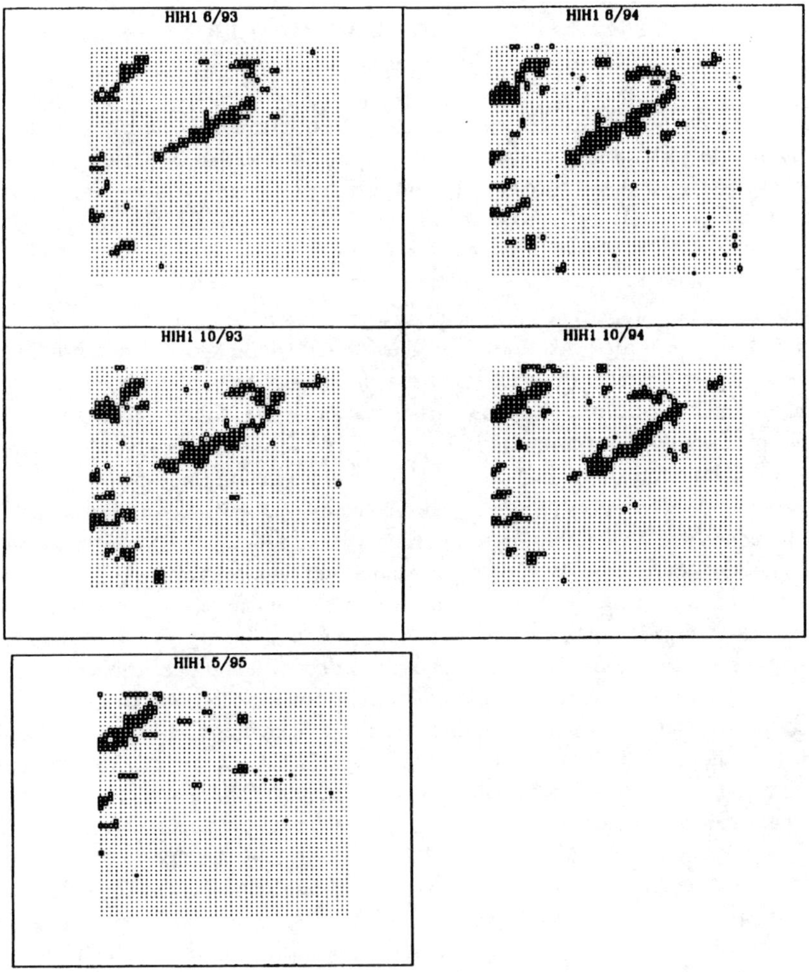

Figure 11.6 *(continued)*

can persist under hurricane conditions (Eleuterius and Miller 1976) although others cannot (Kenyon and Poiner 1987). I have found that naturally patchy beds may be more susceptible to severe storm erosion (Figure 11.6). Fonseca and co-authors (1982) and Fonseca and Fisher (1986) demonstrated that the ability of seagrass canopies to reduce currents can be overcome at moderate unidirectional flow velocities, and Fonseca and Cahalan (1992) show that seagrasses are highly effective in reducing waves only when occupying most of the water column. Those results, taken with the observations on root-rhizome stability made by Wilson (1949), Fonseca and colleagues (1985), and Hine and co-workers (1987) lead me to conclude that much of the erosion prevention by seagrasses is due to the effect of the root-rhizome system.

11.5 NEW DIRECTIONS: CONTRIBUTIONS FROM LANDSCAPE ECOLOGY

The burgeoning field of landscape ecology offers some guidance in how to integrate the wide range of spatial and temporal scales that influence processes in seagrass ecosystems. However, the ability to link habitat pattern and associated sedimentary processes is a consequence of the scale at which we conduct our examinations (Dayton and Tegner 1984). For example, the diffusion of nutrients across a seagrass leaf surface is not governed by the same physical processes acting to distribute those nutrients across an estuarine basin. Similarly, the relation of an animal's body size to its ability to take shelter in the spaces among individual shoots is not a relevant predictor of its vulnerability to predators when it crosses unvegetated gaps among seagrass patches. Thus, the ability to detect detail in habitat structure, to measure its change (and to imply function) over time depends on a combination of the range over which we are making our measurements and resolution of our sensors (e.g. from eyes to satellites) (*sensu* Duarte and Kalff 1990; Schneider 1994). Therefore in turn, our concepts of habitat structure, function and change are framed by our choices of spatial and temporal resolution. Despite the clear relationship of water motion to seagrass bedform, we have only begun to investigate organization of their spatial or temporal hierarchies related to water motion regimes. Otherwise seagrass beds have consistently been treated as a black box at the landscape scale. To build on information accumulated from other ecosystems and apply this to seagrass systems, research emphasis must include not only the normative 1 m scale study, but scales that are relevant to the formation, maintenance, and function of whole systems, such as sediment transport pathways or an organism's range. Such studies are underway and indicate that demographics of seagrass bed fauna (Murphey and Fonseca 1995) and trophic pathways may be altered by seagrass landscape heterogeneity (Irlandi 1994). Through attention to landscape pattern, the time seems ripe for a rapid advance in our understanding of seagrass systems because the patterns of seagrass beds that form in high wave and/or current settings are probably highly deterministic (*sensu* Fonseca et al. 1983) and therefore predictable across a much wider temporal and spatial range than previously enjoyed. Finally, studies of seagrass physiological ecology must shift from an emphasis on photosynthetic efficiency and depth limits, to incorporating the influence of both physical and chemical microclimatological conditions on the plants as well. These changes must occur in order to comprehend fully the role that seagrasses play in the nearshore environment.

If the pattern of distribution observed in seagrass beds is the result of physical processes whose effects vary with spatial scale, then it follows that the relationships between pattern and process are in turn dependent upon spatial scale as well (Bian and Walsh 1993). Unlike true fractals (patterns which display detail at all scales of observation), most patterns in nature display spatial dependence over a limited range of scales (Bian and Walsh 1993). Therefore, knowing the range of these scales is potentially valuable if, after gathering empirical evidence, one can infer structural attributes at other scales of interest, especially scales that may be less expensive to obtain than ground-based surveys (e.g. aerial photography). It is often taken for granted that a relationship between ecological phenomena and spatial dependence is realistic and often intuitive or is at least statistically inescapable (Rossi et al. 1992). The notion that interactions at one scale (spatial or temporal) affect that which is expressed on another scale provides the basis for hypothesizing scale-dependent effects. Therefore, many spatial and temporal patterns seen in seagrass ecosystems are the result of physical processes acting both on individual plants and the local population level (individual patch). Responses of individual plants as well as ramets to water motion accumulate and may be the basis of seagrass landscape patterns perceived at coarser scales of resolution. To summarize, examples of the importance of deriving scale dependence in seagrass beds include identification of:

1. The scale at which samples taken in the landscape are independent of one another and improve sampling stratification.
2. Their effect on animal utilization and distribution.
3. The relevant scales over which sedimentary processes are controlled, providing a better prediction of alterations in current patterns, interception (or lack thereof) of wave energy and sedimentary processes as the result of altering the seagrass landscape.

REFERENCES

Ackerman, J.D. 1986. Mechanistic implications for pollination in the marine angiosperm, *Zostera marina. Aquatic Botany* 24: 343–354.

Ackerman, J.D. and A. Okubo. 1993. Reduced mixing in a marine macrophyte canopy. *Functional Ecology* 7: 305–309.

Adams, J.B. and G.C. Bate. 1994. The tolerance to desiccation of the submerged macrophytes *Ruppia cirrhosa* (Petagna) Grande and *Zostera capensis* Setchell. *Journal of Experimental Marine Biology and Ecology* 183: 53–62.

Almasi, M.N., C.M. Hoskin, J.K. Reed and J. Milo. 1987. Effects of natural and artificial *Thalassia* on rates of sedimentation. *Journal of Sedimentary Petrology* 57: 901–906.

Armstrong, S.L. 1987. The behavior in flow of the morphologically variable seaweed *Hedophyllum sessil* (C.Ag.) Setchell. *Hydrobiologia*. 183: 115–122.

Backman, T.W.H. 1991. Genotypic and phenotypic variability of *Zostera marina* on the west coast of North America. *Canadian Journal of Botany* 69: 1361–1371.

Basan, P.B. 1973. Aspects of sediment deposition and development of a carbonate bank in the Barracuda Keys, south Florida. *Journal of Sedimentary Petrology* 43: 42–53.

Bian, L. and S.J. Walsh. 1993. Scale dependencies of vegetation and topography in a mountainous environment of Montana. *Professional Geographer* 45: 1–11.

Biedka, R.F., J.M. Gosline and R.E. De Wreede. 1987. Biomechanical analysis of wave-induced mortality in the marine alga *Pterygophora californica. Marine Ecology Progress Series* 36: 163–170.

Brewer, C.A. and M. Parker. 1990. Adaptations of macrophytes to life in moving water: upslope limits and mechanical properties of stems. *Hydrobiologia* 194: 133–142.
Bulthuis, D.A., G.A. Brand and M.C. Mobley. 1984. Suspended sediments and nutrients in water ebbing from seagrass-covered and denuded tidal mudflats in a southern Australian embayment. *Aquatic Botany* 20: 257–266
Cahalan, J.A., S.E. Siddall and M.W. Luckenbach. 1989. Effects of flow velocity, food concentration and particle flux on growth rates of juvenile bay scallops *Argopecten irradians. Journal of Experimental Marine Biology and Ecology* 129: 45–60.
Callaghan, T.V., B.M. Svensson, H. Bowman, D.K. Lindley and B.A. Carlsson. 1990. Models of clonal plant growth based on population dynamics and architecture. *Oikos* 57:257–269.
Charters, A.C., M. Neushul and C. Barilotti. 1969. The functional morphology of *Eisenia arborea. Proceedings International Seaweed Symposium*. 6: 89–105.
Conover, J.T. 1964. Environmental relationships of benthos in salt ponds (plant relationships). University of Rhode Island, Graduate School of Oceanography, Tech. Rep. No. 3.
Cook, R.E. 1985. Growth and development in clonal plant populations, p. 259–296. *In* J.B.C. Jackson, L.W. Buss and R.E. Cook (eds.), *Population Biology and Evolution of Clonal Organisms*. Yale University Press.
Cooper, L.W. and C.P. McRoy. 1988. Anatomical adaptations to rocky substrates and surf exposure by the seagrass genus *Phyllospadix. Aquatic Botany* 32: 365–381.
Coops, H., R. Boeters and H. Smit. 1991. Direct and indirect effects of wave attack on helophytes. *Aquatic Botany* 41: 333–352.
Dayton, P.K. and M.J. Tegner. 1984. The importance of scale in community ecology: A kelp forest example with terrestrial analogs, p. 457–482. *In* P.W. Price, C.N. Slobodchikoff, and W.S. Gaud (eds.), *A New Ecology, Novel Approaches to Interactive Systems*. John Wiley and Sons, New York.
den Hartog, C. 1971. The dynamic aspect in the ecology of seagrass communities. *Thalassia Jugoslavica*. 7: 101–112.
Dennison, W.C. 1979. Light adaptations of plants: a model based on the seagrass *Zostera marina* L. MS. Thesis, University of Alaska, Fairbanks.
Duarte, C.M. and J. Kalff. 1990. Patterns in the submerged macrophyte biomass of lakes and the importance of the scale of analysis in the interpretation. *Canadian Journal of Fisheries and Aquatic Science* 47: 357–363.
Eckman, J.E., C.H. Peterson and J.A. Cahalan. 1989. Effects of flow speed, turbulence, and orientation on growth of juvenile bay scallops *Argopecten irradians concentricus* (Say). *Journal of Experimental Marine Biology and Ecology* 132: 123–140.
Eleuterius, L.N. and G.J. Miller. 1976. Observations on seagrasses and seaweeds in Mississippi Sound since Hurricane Camille. *Journal of Mississippi Academy of Science* 21: 58–63.
Ennos, A.R. 1989. The mechanics of anchorage in seedlings of sunflower, *Helianthus annuus* L. *New Phytologist* 113: 185–192.
Fonseca, M.S. 1989. Sediment stabilization by *Halophila decipiens* in comparison to other seagrasses. *Estuarine Coastal and Shelf Science* 29: 501–507.
Fonseca, M.S. and J.A. Cahalan. 1992. A preliminary evaluation of wave attenuation by four species of seagrass. *Estuarine, Coastal and Shelf Science* 35: 565–576.
Fonseca, M.S. and J.S. Fisher. 1986. A comparison of canopy friction and sediment movement between four species of seagrass with reference to their ecology and restoration. *Marine Econoly Progress Series* 29: 15–22.
Fonseca, M.A. and W.K. Kenworthy. 1987. Effects of current on photosynthesis and distribution of seagrasses. *Aquatic Botany* 27: 59–78.

Fonseca, M.S., J.S. Fisher, J.C. Zieman and G.W. Thayer. 1982. Influence of the seagrass *Zostera marina* L., on current flow. *Estuarine, Coastal and Shelf Science* 15: 351–364.

Fonseca, M.S., J.C. Zieman, G.W. Thayer and J.S. Fisher. 1983. The role of current velocity in structuring seagrass meadows. *Estuarine, Coastal and Shelf Science*. 17: 367–380.

Fonseca, M.S., W.J. Kenworthy, G.W. Thayer, D.Y. Heller and K.M. Cheap. 1985. Transplanting of the seagrasses *Zostera marina* and *Halodule wrightii* for sediment stabilization and habitat development on the east coast of the United States. US Army Corps of Engineers, Waterways Experiment Station, Vicksburg, MS, Tech. Rep. EL-85-9.

Fritter, A.H. and A.R. Ennos. 1989. Architectural constraints to root system function, p. 15–22 *In Roots and the Soil Environment. Aspects of Applied Biology*. v. 22. 25–28 Sept, Univ. St. Andrews, Wellsborne, UK.

Gambi, M.C., A.R.M. Nowell and P.A. Jumars. 1990. Flume observations on flow dynamics in *Zostera marina* (eelgrass) beds. *Marine Ecology Progress Series* 61: 159–169.

Ginsberg, R.N. and H.A. Lowenstam. 1958. The influence of marine bottom communities on the deposition environment of sediments. *Journal of Geology* 66: 310–318.

Harper, J.L. 1985. Modules, branches and the capture of resources, p. 1–34. *In* J.B.C. Jackson, L.W. Buss and R.E. Cook (eds.), *Population Biology and Evolution of Clonal Organisms*. Yale University Press.

Heller, D.Y. 1987. Sediment transport through seagrass beds. M.S. Thesis, University of Virginia, Charlottesville.

Hemminga, M.A. and J. Nieuwenhuize. 1990. Seagrass wrack-induced dune formation on a tropical coast (Banc d'Arguin, Mauritania). *Estuarine Coastal and Shelf Science* 31: 499–502.

Hine, A.C., M.W. Evans, R.A. Davis and D.F. Belknap. 1987. Depositional response to seagrass mortality along a low-energy, barrier-island coast: west-central Florida. *Journal of Sedimentary Petrology* 57: 431–439.

Holbrook, N.M., M.W. Denny and M.A.R. Koehl. 1991. Intertidal 'trees': consequences of aggregation on the mechanical and photosynthetic properties of sea-palms *Postelsia palmaeformis* Ruprecht. *Journal Experimental Marine Biology and Ecology*. 146: 39–67.

Inoue, E. 1955. Studies of the phenomena of waving plants (honami) caused by wind. Part 1: Mechanism and characteristics of waving plants phenomena. *Journal Meterological Society (Japan)* 41: 317–326.

Irlandi, E.A. 1994. Large and small-scale effects of habitat structure on rates of predation: how percent coverage of seagrass affects rates of predation and siphon nipping on an infaunal bivalve. *Oecologia* 98: 176–183.

Jackson, G.A. 1984. Internal wave attenuation by coastal kelp stands. *Journal of Physical Oceanography* 14: 1300–1306.

Judge, M.L., L.D. Coen and K.L. Heck. 1992. The effect of long-term alteration of *in-situ* water currents on the growth of *Mercenaria mercenaria* in the northern Gulf of Mexico. *Limnology and Oceanography* 37: 1550–1559.

Jumars, P.A. and A.R.M. Nowell. 1984. Fluid and sediment dynamic effects on marine benthic community structure. *American Zoologist* 24: 45–55.

Kautsky, L. 1987. Life-cycles of three populations of *Potamogeton pectinatus* L., at different degrees of wave exposure in the Asko area, northern Baltic proper. *Aquatic Botany* 27: 177–186.

Kautsky, L. 1988. Life strategies of aquatic soft bottom macrophytes. *Oikos* 53: 126–135.

Keddy, P.A. 1982. Quantifying within-lake gradients of wave energy: interrelationships of wave energy, substrate particle size and shoreline plants in Axe Lake, Ontario. *Aquatic Botany* 14: 41–58.

Keddy, P.A. 1983. Shoreline vegetation in Axe Lake, Ontario: effects of exposure on zonation patterns. *Ecology* 64: 331–344.

Kenworthy, W.J. and D. Haunert. 1991. The light requirements of seagrasses: proceedings of a workshop to examine the capability of water quality criteria, standards and monitoring programs to protect seagrasses. NOAA Tech. Memo. NMFS-SEFC-287.

Kenyon, R. and I. Poiner. 1987. Seagrass and cyclones in the western Gulf of Carpentaria. CSIRO Fact Sheet No. 1., Cleveland, Queensland, Australia.

Kirkman, H. 1985. Community structure in seagrasses in southern Western Australia. *Aquatic Botany* 21: 363–375.

Kirkman, H. and J. Kuo. 1990. Pattern and process in southern Western Australian seagrasses. *Aquatic Botany* 37: 367–382.

Koehl, M.A.R. 1986. Seaweeds in moving water: form and mechanical function. pp. 603–634, *In* T.J. Givnish (ed.), *On the Economy of Plant Form and Function*. Cambridge University Press, Cambridge.

Koehl, M.A.R. and R.S. Alberte. 1988. Flow, flapping, and photosynthesis of *Nereocystis luetkeana*: a functional comparison of undulate and flat blade morphologies. *Marine Biology* 99: 435–444.

Koehl, M.A.R. and S.A. Wainwright. 1977. Mechanical design of a giant kelp. *Limnology and Oceanography* 22: 1067–1071.

Koehl, M.A.R. and S.A. Wainwright. 1986. Biomechanics, p. 292–313, *In* M.L. Littler and D.S. Littler (eds.), *Handbook of Phycological Methods. Ecological Field Methods: Macroalgae*. Cambridge University Press, Cambridge.

Laushman, R.H. 1993. Population genetics of hydrophilous angiosperms. *Aquatic Botany* 44: 147–158.

Marba, N. and C.M. Duarte. 1995. Coupling of seagrass (*Cymodocea nodosa*) patch dynamics to subaqueous dune migration. *Journal of Ecology* 83: 381–389.

Marba, N., J. Cebrian, S. Enriquez and C. Duarte. 1994a. Migration of large-scale subaqueous bedforms measured with seagrasses (*Cymodocea nodosa*) as tracers. *Limnology and Oceanography* 39: 126–133.

Marba, N., M.E. Gallegos, M. Merino and C. Duarte. 1994b. Vertical growth of *Thalassia testudinum*: Seasonal and interannual variability. *Aquatic Botany* 47: V1–11.

Molinier, R. and J. Picard. 1952. Recherches sur les herviers, de phanerogames marines du littoral Mediterranean Fran[181]ais. *Annales de l'Institut Oceanographique* 27: 157–234.

Murphey, P.L. and M.S. Fonseca. 1995. Role of high and low energy seagrass beds as nursery areas for *Penaeus duorarum* in North Carolina. *Marine Ecology Progress Series* 121: 91–98.

O'Gower, A.K. and J.W. Wacasey. 1967. Animal communities associated with *Thalassia, Diplanthera*, and sand beds in Biscayne Bay. I. Analysis of communities in relation to water movement. *Bulletin of Marine Science* 17: 175–210.

Orth, R.J. 1977. The importance of sediment stability in seagrass communities, p. 281–300. *In Ecology of Marine Benthos*. University of South Carolina Press, Columbia, B.C. Coull (ed.).

Patriquin, D.G. 1975. 'Migration' of blowouts in seagrass beds at Barbados and Carriacou, West Indies and its ecological and geological applications. *Aquatic Botany* 1: 163–189.

Perez-Llorens, J.L. and F.X. Niell. 1993. Temperature and emergence effects on the net photosynthesis of two *Zostera noltii* Hornem. morphotypes. *Hydrobiologia* 254: 53–64.

Peterson, C.H. 1982. Clam predation by whelks (*Busycon* spp.): experimental tests of the importance of prey size, prey density and seagrass cover. *Marine Biology* 66: 159–170.

Peterson, C.H., H.C. Summerson and P.B. Duncan. 1984. The influence of seagrass cover on population structure and individual growth rate of a suspension-feeding bivalve *Mercenaria mercenaria*. *Journal of Marine Research* 42: 123–138.

Philippart, C.J.M. 1994. Interactions between *Arenicola marina* and *Zostera noltii* on a tidal flat in the Wadden Sea. *Marine Ecology Progress Series* 111: 251–257.

Phillips, R.C. 1984. The ecology of eelgrass meadows in the pacific northwest: a community profile. US Fish Wildl. Serv. FWS/OBS-84/24.

Pihl, L. 1986. Exposure, vegetation, and sediment as primary factors for mobile epibenthic faunal community structure and production in shallow marine soft bottom areas. *Netherlands Journal of Sea Research* 20: 75–83.

Powell, G.V.N. and F.C. Schaffner. 1991. Water trapping by seagrasses occupying bank habitats in Florida Bay. *Estuarine, Coastal and Shelf Science* 32: 43–60.

Rasmussen, E. 1973. Systematics and ecology of the Isef Fjord marine fauna (Denmark) with a survey of the eelgrass (*Zostera*) vegetation and its communities. *Ophelia* 11: 1–507.

Rossi, R.E., D.J. Mulla, A.G. Journel and E.H. Franz. 1992. Geostatistical tools for modeling and interpreting ecological spatial dependence. *Ecological Monographs* 62: 277–314.

Schubel, J.R. 1973. Some comments on seagrasses and sedimentary processes. Chesapeake Bay Institute, Johns Hopkins University Special Report, 33.

Schneider, D.C. 1994. *Quantitative Ecology: Spatial and Temporal Scaling*. Academic Press, New York.

Scoffin, T.P. 1970. Trapping and binding of subtidal carbonate sediments by marine vegetation in Bimini Lagoon. *Journal of Sedimentary Petrology* 40: 249–273.

Short, F.T. and C.A. Short. 1984. The seagrass filter: purification of estuarine and coastal waters, p. 395–414. *In* V.S. Kennedy (ed.), *The Estuary as a Filter*. Academic Press, New York.

Sousa, W.P. 1985. Disturbance and patchy dynamics on rocky intertidal shores, p. 101–124. *In* P.S. White and S.T.A. Pickett (eds.), *The Ecology of Natural Disturbance and Patch Dynamics*. Academic Press, New York.

Sprugel, D.G. 1976. Dynamic structure of wave-regenerated *Abies balsamea* forests in the north-eastern United States. *Journal of Ecology* 64: 889–911.

Suchanek, T.H. 1983. Control of seagrass communities and sediment distribution by *Callianassa* (Crustacea, Thalassinidea) bioturbation. *Journal of Marine Research* 441: 281–298.

Taylor, J.D. and M.S. Lewis. 1970. The flora, fauna and sediments of the marine grass beds of Mahe, Seychelles. *Journal of Natural History* 4: 199–220.

Thayer, G.W., W.J. Kenworthy and M.S. Fonseca. 1984. The ecology of eelgrass meadows of the Atlantic coast: A community profile. US Fish Wildl. Serv. FWS/OBS-84/02.

Valentine, R. and K.A. Heck. 1991. The role of sea urchin grazing in regulating subtropical seagrass meadows: evidence from field manipulations in the northern Gulf of Mexico. *Journal of Experimental Marine Biology and Ecology* 154: 215–230.

Valentine, J.F., K.L. Heck, P. Harper and M. Beck. 1994. Effects of bioturbation in controlling turtlegrass (*Thalassia testudinum* Banks ex Konig) abundance: evidence from field enclosures and observations in the northern Gulf of Mexico. *Journal of Experimental Marine Biology and Ecology* 178: 181–192.

Vogel, S. 1981. *Life in Moving Fluids*. Willard Grant, Boston.

Waller, D.M. and D.A. Steingraeber. 1985. Branching and modular growth: theoretical models and empirical patterns, p. 225–258. *In*, J.B.C. Jackson, L.W. Buss and R.E. Cook (eds.), *Population Biology and Evolution of Clonal Organisms*. Yale University Press.

Wanless, H.R. 1981. Fining-upwards sedimentary sequences generated in seagrass beds. *Journal of Sedimentary Petrology* 51: 445–454.

Ward, L.G., W.M. Kemp and W.R. Boynton. 1984. The influence of waves and seagrass communities on suspended particles in an estuarine embayment. *Marine Geology* 59: 85–103.

Wayne, C.J. 1975. Sea and marsh grasses: their effect on wave energy and nearshore sand transport. M.S. Thesis, Florida State University, Tallahassee.

Williams, S.L. 1990. Experimental studies of Caribbean seagrass bed development. *Ecological Monographs* 60: 449–469.

Wilson, D.P. 1949. The decline of *Zostera marina* L. at Salcombe and its effects on the shore. *Journal of the Marine Biological Association of the United Kingdom* 28: 395–412.

Wood, E.J.F., W.E. Odum and J.C. Zieman. 1969. Influence of seagrasses on the productivity of coastal lagoons, p. 495–502. *In* A.A. Castanares and F.B. Phleger (eds.), *Coastal Lagoons, a Symposium*. Ciudad Universitaria, Universidad Nacional Autonoma de Mexico.

Worcester, S.E. 1995. Effects of celgrass beds on advection and turbulent mixing in low current and low shoot density environments. *Marine Ecology Progress Series* 126: 223–232.

Zieman, J.C. 1972. Origin of circular beds of *Thalassia testudinum* in south Biscayne Bay, Florida and their relationships to mangrove hammocks. *Bulletin of Marine Science* 22: 559–574.

Zieman, J.C. and R.T. Zieman. 1989. The ecology of the seagrass meadows of the west coast of Florida: a community profile. US Fish. Wildl. Serv. Biol. Rep. 85(7.25).

Chapter Twelve

Interception of freshwater and nitrogen transport from uplands to coastal waters: the role of saltmarshes

BRIAN L. HOWES
Woods Hole Oceanographic Institution
PETER K. WEISKEL
US Geological Survey

and

DALE D. GOEHRINGER and JOHN M. TEAL
Woods Hole Oceanographic Institution

12.1 INTRODUCTION

Saltmarshes perform a variety of functions within the coastal landscape. In addition to their roles as storm buffers and aesthetic resources, saltmarshes have been frequently cited as important to coastal food chains. Since only a few per cent of the plant biomass is grazed (Teal 1962; Vince et al. 1981), the high organic matter production within marshes is thought to form the basis of detrital food chains both within the marsh (supporting the marsh nursery concept) and within adjacent coastal waters via the export of marsh-derived organic matter (Odum 1980; Werme 1981). While the role of saltmarshes as potential organic matter exporters was presented as early as the First Salt Marsh Conference at Sapelo Island, Georgia in 1958, the annual import/export balance of nutrients and organic matter within tidal wetlands has been difficult to quantify. Over the intervening decades, saltmarshes

Estuarine Shores: Evolution, Environments and Human Alterations. Edited by K. F. Nordstrom and C. T. Roman.

from almost every coastal state in the US have been studied to address the issue of saltmarsh export and associated roles in coastal fisheries (Nixon 1980; Valiela 1983). While it has become clear that saltmarshes generally export organic matter (cf. Jordan and Correll 1991), export in the form of fish and shrimp is now frequently considered to be of greatest importance (Adam 1990).

The finding of nitrogen limitation of productivity of phytoplankton in coastal waters (Ryther and Dunstan 1971), saltmarshes (Valiela and Teal 1974; Sullivan and Daiber 1974) and more recently nearshore macroalgal production (Fujita et al. 1989), resulted in an additional focus on the nitrogen balance of saltmarshes. The suggestion that nitrogen was a key nutrient of component ecosystems within the coastal margin helped to form the hypothesis that marshes may 'buffer' adjacent coastal waters against excessive nitrogen loading. In this role, saltmarshes take up nitrogen from flood waters with removal through sediment accretion or through denitrification within the marsh or, alternatively, transform and later release nitrogen in fall and winter when the assimilative capacity of the adjacent waters is likely to be greatest (cf. Valiela et al. 1978; Valiela and Teal 1979). In a major review, Nixon (1980) supported the concept of saltmarshes as nutrient transformers and concluded that the net nutrient exchanges 'are too small to influence the annual nitrogen budget of most coastal systems', although there may be some 'local importance attached to the marsh-estuarine nitrogen flux in some areas'. Almost all studies focusing on whole-wetland tidal exchanges (cf. Valiela 1983; Jordan and Correll 1991) as well as the vegetated marsh (Whiting et al. 1989) are consistent with this view. However, recent recognition that eutrophication of nearshore waters resulting from nutrient inputs from upland sources is a major threat to estuarine and coastal ecosystems (Jaworski 1984; Rosenberg 1985; Nixon et al. 1986) has renewed interest in the potential role of wetlands in the nitrogen economy of the coastal zone.

Nitrogen from terrestrial sources enters tidal saltmarshes indirectly, after entering the coastal and estuarine waters which flood them, or directly, through non-point surface water and groundwater inflows. Precipitation is an additional route of direct entry but is generally a small part of the allochthonous nitrogen input. The indirect tidal pathway has received the majority of study to date; the primary issue is now to what extent tidal saltmarshes increase their removal (and to a lesser extent transformations) of nitrogen as concentrations in flood waters increase (Valiela and Teal 1979). Until recently, the direct pathways have received considerably less study. In coastal ecosystems with highly permeable soils in the adjacent upland, groundwater can be the principal mechanism for the delivery of terrestrially derived nutrients (Johannes 1980; Valiela et al. 1990; Weiskel and Howes 1992). In trying to manage coastal systems we need to understand the ability of tidal marshes, which lie between upland and coastal waters, to intercept this seaward transport of freshwater and the nutrients which it carries.

Knowledge of the fate of terrestrially derived nutrients directly entering coastal waters (Capone and Bautista 1985; Turner and Rabalais 1994) versus that directly entering saltmarshes (Harvey and Odum 1990) is essential to understanding the effect of increasing terrestrial nitrogen loads on subtidal systems. In this paper we focus on the interception by saltmarshes of nitrogen transported via groundwater flows from upland sources to coastal waters, the controlling hydrologic and biogeochemical processes, and potential alterations under increasing watershed nitrogen loading. Our discussion centers upon two New England saltmarshes: Great Sippewissett Marsh where we have previously conducted tidal exchange, nitrogen retention and hydrologic studies (cf. Valiela and Teal 1979; White and Howes 1994a,b; Howes and Goehringer 1994); and Namskaket Marsh where detailed aquifer/marsh studies are underway (DeSimone et al. 1995; DeSimone and Howes 1995; Weiskel et al. 1996).

12.2 SITE DESCRIPTIONS

Both Great Sippewissett Marsh (Falmouth, Massachusetts) and Namskaket Marsh (Orleans, Massachusetts) are situated in highly permeable watersheds comprising primarily glacial outwash. These marshes are typical of "pocket" marshes along the Atlantic coast of the US, forming in coastal indentations behind barrier dunes. Freshwater enters these marshes through groundwater discharge and through small dispersed surface water flows, which tend to be groundwater fed (Valiela and Teal 1979; Millham and Howes 1994). Both marshes are 'mature' with tall *Spartina alterniflora* along creekbanks, short *S. alterniflora* inland and with significant high marsh areas. Within Namskaket Marsh the study area is the inlandmost portion of the marsh with brackish water wetland plants dominating the upland/marsh border on an areal basis. Flood tidal waters from Cape Cod Bay entering inner Namskaket Marsh are ca. 17‰ (median), fresher than those entering Great Sippewissett 32‰ directly from Buzzards Bay. The fresher waters of inner Namskaket result from dilution of flooding Bay waters by outflowing freshwaters within the outer, more extensive portion of the marsh system. Both marshes are underlain by thick peats (15–25% carbon, porosity 85–90%) typical of marshes which receive little inorganic sediment loading and must accrete largely through organic matter accumulation. Marsh deposits within both systems are typically 2–3 m thick with maximum accumulations of 4–5 m. The interior saltmarsh deposits overlie freshwater wetland sediments, a pattern typical of pocket marshes in the region (Orson and Howes 1992). More detailed descriptions are available for both Great Sippewissett Marsh (Valiela and Teal 1979; Howes et al. 1986) and inner Namskaket Marsh (Weiskel et al. 1996).

12.3 WHOLE MARSH NITROGEN BALANCE

To evaluate interception of terrestrially derived nitrogen by entire wetlands, we can compare upland inputs to the whole system against net tidal imports or exports

Table 12.1 Nutrient concentrations in interstitial waters and freshwater inflows to Gre Sippewissett Marsh. Values are means (S.E.), N=number of samples.

Source	Depth (m)	Salinity (‰)	Nitrate (µM)	Ammonium (µM)	N
Surface water	0.00	0.3 (0.1)	140.1 (13.7)	1.5 (0.25)	1C
Groundwater					
a) Upland edge springs*	0.00	0.2 (0.1)	63.8 (6.2)	4.3 (0.7)	45
b) Creekbottom porewater:					
High seepage zone	0.05	1.6 (0.2)	41.2 (15.1)	9.6 (0.5)	5
Low seepage zone	0.05	9.6 (4.0)	7.5 (1.7)	9.0 (3.9)	8
c) 'Deep' groundwater	12.20	0.0 (0.0)	5.1 (0.5)	2.1 (0.4)	4
Porewater					
Short *S.alterniflora* zone	0.20	29.4 (0.6)	<0.1 (0.0)	82.2 (4.1)	28

* Calculated from Valiela et al. 1978 using only samples with <1‰ salinity

throughout the year. While this does not necessarily yield direct information o processes, it does yield insight into the potential magnitude of nitrogen fluxes. W have re-examined the published nitrogen balance study for Great Sippewisse Marsh (Valiela and Teal 1979) to evaluate specifically interception of terrestriall derived nitrogen before entry into adjacent Buzzards Bay waters. Nitrate + nitrit (hereafter refered to as nitrate) comprise ca. 90% of the inorganic nitrogen i freshwater inflow to Great Sippewissett Marsh (Table 12.1). Nitrate is th predominant nitrogen species in aquifers receiving nitrogen loading fron agricultural practices or on-site domestic sewage disposal (NRC 1978; Capon and Bautista 1985; Weiskel and Howes 1992) and is thus a primary 'eutrophying nutrient in coastal waters. In the present analysis we do not include groundwater transported dissolved organic nitrogen, since it appears generally to be refractor (DeSimone and Howes, in press) and appears to be transported through the mars (Valiela and Teal 1979).

Net tidal exchange of each nitrogen species was determined at monthly interval throughout the year in Great Sippewissett Marsh. Net exchange was determine throughout complete tidal cycles from volumetric measures of tidal flow and hourl samples of nutrient concentrations. Measurements were made at the entrance of th single tidal channel exchanging with Buzzards Bay and include the net integrate import or export of nitrogen from both the vegetated marsh and creekbottom areas Annual inputs of nitrogen in freshwater inflows were determined fron measurements of dissolved nitrogen in freshwater discharges at the marsh/uplan border (springs) and the volume of freshwater required to produce the decrease i salinity in consecutive flood to ebb tides (Valiela et al. 1978).

Great Sippewissett Marsh, like shallow coastal embayments (Cherkauer an McKereghan 1989), is likely to be a focus of groundwater discharge from th surrounding upland. The rate of freshwater flow to the marsh with its associate nitrogen is nearly constant throughout the year. Almost all of the freshwater entering the marsh is ultimately derived from the groundwater flow system, with surface water inflows being primarily groundwater fed (cf. Millham and Howes 1994). The higher concentration of inorganic nitrogen entering the marsh in surface water (streams) versus direct groundwater discharge (Table 12.1) is likely the result of higher nitrogen concentrations in the shallow groundwater which is discharged to the streams. Surface freshwater inflows ($140\mu\, NO_3^-$) were at the high end of the range measured in groundwater discharging along the marsh/upland boundary (2–161 $\mu M\, NO_3^-$). Dissolved inorganic nitrogen levels in deep groundwater (12 m below watertable) are indicative of relatively uncontaminated groundwater and are consistent with the rates and patterns of groundwater flow and the areal distribution of nitrogen loading within the adjacent watershed (cf. Weiskel and Howes 1991). Since saltmarshes are most likely to intercept groundwater within the shallower flow system of an aquifer which generally contains the the highest nitrate concentrations, their potential ability to intercept N is disproportionately high compared to the volume of intercepted freshwater.

Although Great Sippewissett Marsh has the 'reputation' of high groundwater inputs, there is no reason to conclude that it receives greater ground and surface water inflows (per kilometer of upland edge) than other marshes surrounded by permeable watersheds. The salinity of ebb tide waters is typically less than 2‰ lower than flooding waters. Its reputation likely results from the difference in areal scale between pocket and more extensive riverine or barrier island saltmarshes. While groundwater discharge may only rarely be significant in wetland areas kilometers from the marsh/upland boundary, the landward regions of larger saltmarsh systems are frequently receiving upland flows qualitatively similar to Great Sippewissett Marsh. It is therefore likely that the hydrologic and biogeochemical processes controlling interception of nitrogen from upland sources are similar among the various saltmarsh types.

As reported previously (Valiela and Teal 1979), Sippewissett Marsh appears to import oxidized nitrogen and export reduced forms (Figures 12.1 and 12.2). Just as nitrate delivery is nearly continuous throughout the year, so generally is the marsh's uptake of nitrate (Figure 12.1). Although it was suggested that the reduced forms of nitrogen exported are products of nitrate uptake, transformation and loss by the marsh plants, this is unlikely. Nitrate entering the marsh must be transported to the rooting zone for uptake. Because most of the nitrate enters in freshwater inflows through the tidal creeks, it must first be added to the flood tidal water volume to reach plant roots under the vegetated marsh surface for uptake (presumably after infiltration). The low infiltration rates of vegetated sediments (ca. $4\,l\,m^{-2}\,d^{-1}$; Dacey and Howes 1984) and dilution with tidal water minimize the flux through this indirect pathway. Direct transport of nitrate via upward groundwater discharge through the sediments to the rooting zone (0–200 mm) is not supported by the negligible nitrate levels in porewater of vegetated marsh (Table 12.1) or the hydrologic data (discussed below). Although of uncertain magnitude, we suggest

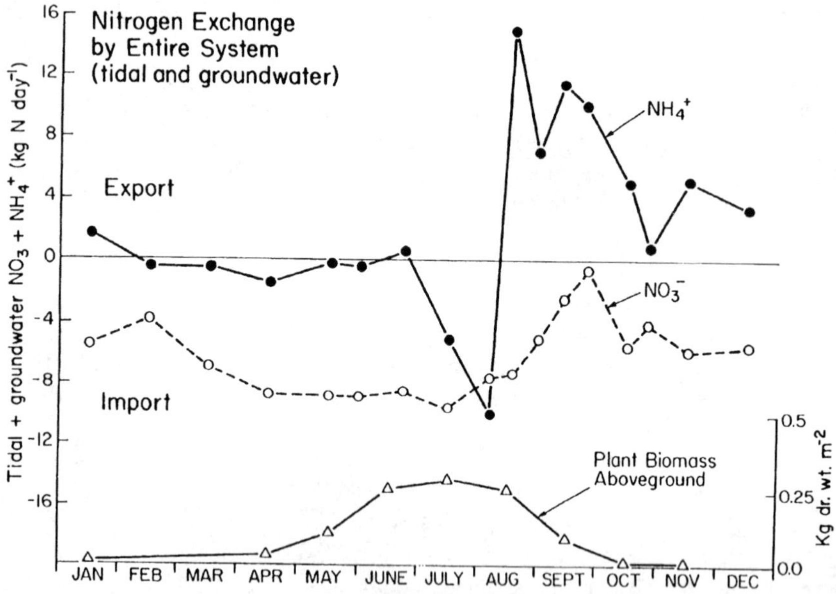

Figure 12.1 Annual cycle of net import or export of NH_4^+ and NO_3^- from tidal water and groundwater flows to Great Sippewissett Marsh (calculated from Valiela and Teal 1979). Ammonium fluxes are primarily via tidal exchanges, while import of NO_3^- is almost wholly via groundwater inflows. The growth cycle of the major plant type on an areal basis in this marsh is indicated by live aboveground dry biomass (m^{-2}) of short *S. alterniflora*

that the role of the vegetated marsh in the observed interception of nitrate by the entire marsh system is small.

Ammonium shows no net exchange throughout half of the year, but a brief import in mid-summer followed in fall and early winter by a large export. Given the coincidence of NH_4^+ import with maximum plant biomass (Figure 12.1), it has been speculated that plant uptake is involved. As with NO_3^-, this pathway would require plant access to tidal NH_4^+, which should be small given the small amount of infiltration (ca. 4 mm; Dacey and Howes 1984) to the rooting zone compared to the large tidal volume (150–400 mm flood height over vegetated marsh). While algal uptake and other transformations may be responsible for the import, temporary incorporation into the litter layer of the marsh is more likely. Rapid uptake of ammonium by litter has been observed in freshwater tidal marshes (Bowden 1986). Similarly, ^{15}N studies have shown that about half of the nitrogen in *S. alterniflora* litter in mid-summer has been incorporated from external sources (White and Howes 1994b). The litter layer has a high potential uptake capacity and is regularly flooded by tidal waters.

The large NH_4^+ export, on the other hand, most likely results from leaching losses from aboveground plant biomass during senescence (Valiela et al. 1978).

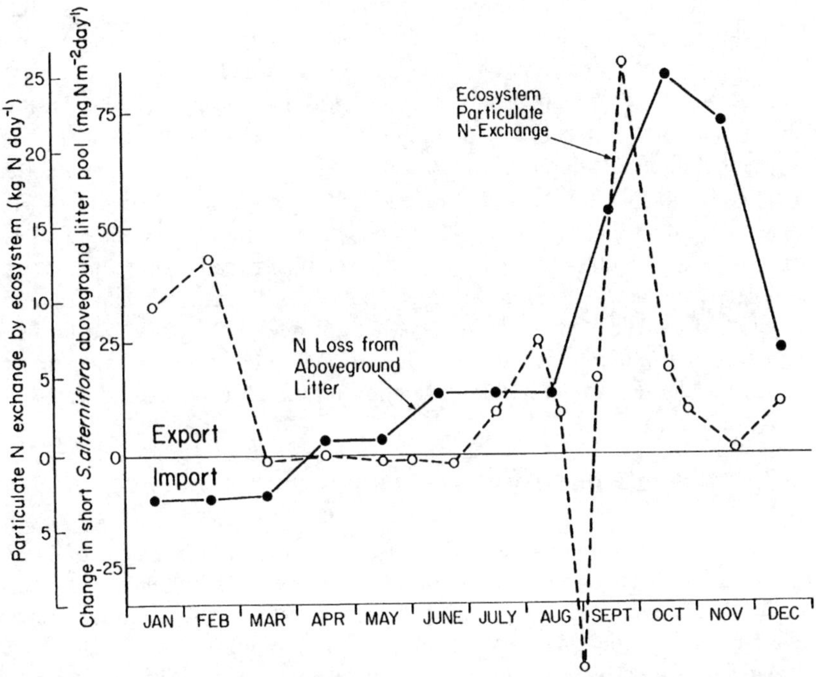

Figure 12.2 Annual cycle of net import or export of particulate nitrogen in tidal exchanges between Great Sippewissett Marsh and adjacent Buzzards Bay (calculated from Valiela and Teal 1979). Monthly changes in the pool of total nitrogen in aboveground *S. alterniflora* litter (m^{-2}) is shown for comparison to tidal fluxes

The ca. 900 kg NH4+–N exported during fall and winter represents about 3 g N m^{-2} from the vegetated marsh. This compares well with estimates of total N loss in a Louisiana marsh (Turner 1993) but is higher than estimates based on microcosms (White and Howes 1994c). What additional sources of NH_4^+ contribute to this export are at present unclear. However, loss of NH_4^+ regenerated in creekbottom sediments, possibly as a result of increased organic matter delivery during plant senescence and decay, is certainly one source (see below). Measurements of drainage of high NH_4^+ porewater concentrations through creekbanks indicate that rates are insufficient to impact the measured NH_4^+ export. Inorganic nitrogen loss through creekbank drainage accounts for only about 3% of the total annual inorganic nitrogen export from this marsh (Howes and Goehringer 1994).

Particulate nitrogen export from Great Sippewissett Marsh is most likely associated with the loss of above-ground litter from the vegetated areas of the marsh (Figure 12.2). Although the timing of whole ecosystem export does not exactly coincide with losses from plant litter in the marsh (October to February), much of the particulate matter lost from litter may be deposited in creekbottoms and only later be resuspended (e.g. winter storms) and exported from the marsh. We have measured increased organic matter deposition in creekbottoms in the fall anc early winter, both as *Spartina* macro-detritus and as an increase in surficial organic carbon content of creekbottom sediments (unpublished data).

A recent nitrogen budget of short *S. alterniflora* grass stands measured using ^{15}N indicates a system in balance, with outputs through denitrification, burial and tida losses equaling inputs through rainfall and N_2 fiixation. The budget is supported by previous independent measurements of each pathway (cf. White and Howes 1994a). It appears that the losses of reduced nitrogen from the marsh are related to losses associated with the plants, both for particulate N and ammonium. However the tidally exported, reduced nitrogen enters the marsh via N_2 fixation rather than from upland sources (e.g. nitrate import). For this model of reduced nitrogen expor to be correct it must be consistent with both the patterns of freshwater transpor through the marsh and the processes controlling the observed nitrate import.

12.4 PATHWAYS OF FRESHWATER ENTRY

Freshwater flows represent the principal mechanism for transport of nutrients across the land/sea margin. Therefore, the ability of a saltmarsh to intercept terrestrially derived nutrients is directly related to its interception of freshwater flows. There are four major potential pathways for upland freshwater to enter saltmarshes: via surface streams, and groundwater discharge through the interior vegetated marsh, the marsh/upland boundary, and marsh creekbottoms. Surface-water inflows have been the most intensively studied as they are visible and readily measured. Groundwater flow paths through intertidal wetlands have only recently become the subject of intensive investigation. Since the fate of groundwater-transported nitrate depends, in part, upon its flow path through the marsh system, we must first address each of the potential pathways.

Vegetated areas of Great Sippewissett Marsh represent almost two-thirds of the total marsh area, with creekbottoms making up almost all of the remainder. We have measured sediment porewater turnover throughout Great Sippewissett Marsh using in situ lysimeters. The approach is to isolate large volumes of intact, vegetated surficial sediments, measure changes in water volume through time, and calculate the associated fluxes. Within the lysimeters all water exchange is through the surface, due to their sealed bottoms which prevent both downward drainage of porewater and upward movement of groundwater. Throughout the central vegetated marsh areas, water loss (water table excursion) inside lysimeters is equivalent to that in the surrounding marsh (Figure 12.3 top, Dacey and Howes 1984; Howes et al. 1986; Howes and Goehringer 1994). In creekbank grass stands, the greater water loss outside versus inside the lysimeters results from drainage of porewater through the creekbanks into the creeks at low tide (Figure 12.3 bottom). Parallel studies of creekbank water loss using lysimeters and direct measurements of creekbank drainage losses using chambers, indicate that the water draining from creekbanks is

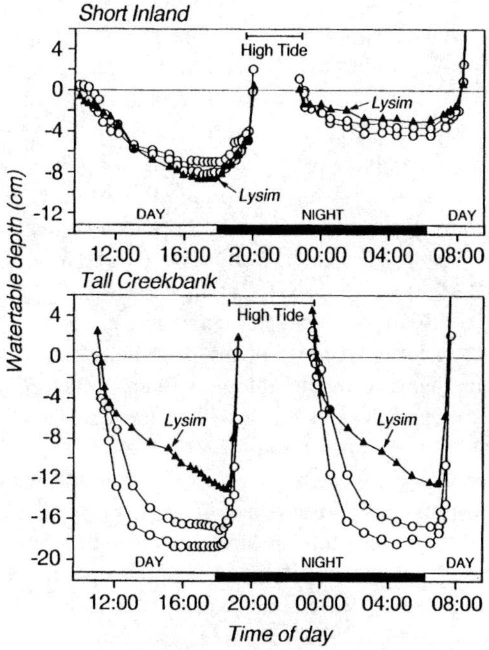

Figure 12.3 Time course of water table excursion within a lysimeter and two adjacent wells in both inland short (upper panel) and creekbank tall (lower panel) *S. alterniflora* grass stands at peak above-ground biomass. The greater water table drop outside versus inside lysimeters in the tall grass zone results from porewater drainage through the adjacent creekbank. Plant water uptake occurs both day and night. (Reproduced by permission from Howes and Goehringer 1994)

indeed porewater (not groundwater). The average annual drainage loss throughout Great Sippewissett Marsh is ca. $8.9\,l\,m^{-1}$ of creekbank per tide (Howes and Goehringer 1994). Groundwater influence was not apparent in creekbank drainage waters from central marsh areas. It appears that in this marsh upward flow of groundwater through the sediments underlying the central vegetated regions is negligible. Using an approach based upon Darcy's Law, Nuttle (1988) obtained similar results in another New England marsh underlain by clayey peat sediments.

Measurements of interstitial salinities from the interior marsh area also support the concept that groundwater inputs comprise only a very small portion of the water balance in Great Sippewissett Marsh. While upward flow could occur without causing low-salinity porewaters in surficial sediments (Harvey and Nuttle 1995; Nuttle and Harvey 1995), it does affect the water balance of the surficial sediments. Preventing groundwater discharge to surficial sediments with lysimeters should result in an increase in porewater salinities until a new steady-state is reached. We have measured the salinity of the porewaters within lysimeters and adjacent sediments for several weeks (August to September) 3 months after placement and then 15 months after placement (White and Howes, unpublished). No significant increases (> 1‰) were found between lysimeters and controls over the initial several months, but we did observe an increase in the surface 50 mm (but not from 50–300 mm) in measurements 15 months after placement. Whether prevention of groundwater flow or artifact is responsible for this salinity increase after long-term placement is unclear. However, the volume of 'missing' freshwater was only ca. 14 l m^{-2}, equal to 3–4 days of evapotranspiration in August.

The above data are consistent with the 2–3 m thick peat deposit of relatively low hydraulic conductivity in the surficial sediments (0–300 mm) and at a depth near the base of the peat (Knott et al. 1987). Lateral movement of groundwater through the higher conductivity mid-depth sediments or even small vertical flows are possible (Hemond and Fifield 1982) but the potential rates are slow, and it is the total volume of flow that controls the nitrate delivery. Integrating all of the data, it is clear that upward groundwater flow within the central marsh area is small relative to the total water balance.

In contrast, upward movement or discharge of groundwater through vegetated marsh sediments is potentially significant in areas close to an upland boundary where the organic deposits are thin or have higher hydraulic conductivity. These criteria appear to be met at a study site in a Virginia marsh (ca. 100 m from upland, 2 m of silty loam sediment, 4% organic matter, with a bulk density and porosity approaching sandy aquifers). In the Virginia study, groundwater discharge through vegetated sediments comprised a significant part of the sediment water balance (Harvey and Nuttle 1995), although the areal extent of the discharge is unknown, since other studies within the same marsh using Darcian modelling indicated a negligible role for groundwater (Chambers et al. 1992).

Groundwater discharge near the marsh/upland boundary is significant in Great Sippewissett Marsh (Valiela et al. 1978) and Namskaket Marsh (Weiskel et al. 1996). While interstitial salinities do not yield rates of groundwater discharge, the occurrence of nearly fresh water in the surficial sediments periodically inundated by salt water can be used as an indicator of the distribution of discharge (Vanek 1993; Millham and Howes 1994). We have mapped the salinity distribution of interstitial waters within the marsh and underlying aquifer of the inner Namskaket Marsh system. Inner Namskaket Marsh is flooded by tidal waters of ca. 17‰. There is a consistent horizontal gradient of interstitial salinity (at 250–350 mm depth) from the marsh/upland boundary to the central marsh area. Groundwater discharge through the marginal marsh sediments is apparent both by the continuous, near-fresh porewaters (<4‰) fringing the marsh and in the visible zone of seepage in the boundary areas (Figure 12.4). The role of groundwater discharge in controlling the horizontal salinity gradient is somewhat confounded by the flow of seepage waters across the marsh surface after breakout at the marsh margin. The effect of groundwater inflow through the marsh sediments at the upland boundary is nevertheless clear. These data, although more dramatic than at Great Sippewissett Marsh (due to the larger ratio of upland boundary to

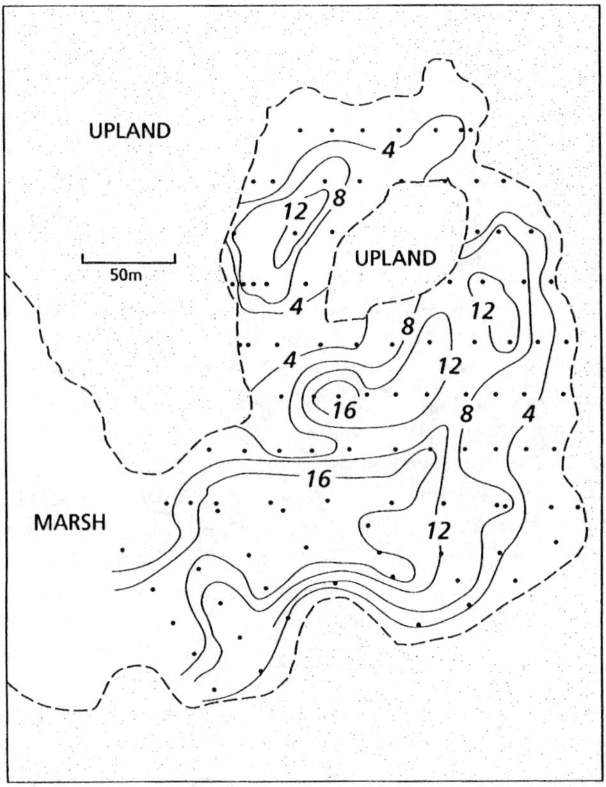

Figure 12.4 Summer salinity distribution in the surficial sediments within the most inland portion of an irregularly flooded New England salt marsh, Namskaket Marsh, Orleans, MA. Median floodwater salinities are 17 (parts per thousand). Contours represent equal salinities (ppt) in porewater from just below the rooting zone (300 mm). Dots represent porewater samplers (Howes et al. 1985), creeks are not shown. The salinity gradient from the surrounding upland in the marsh results from a nearly continuous groundwater seepage zone at the upland/marsh boundary

marsh area at inner Namskaket), are consistent with observed seepage areas and their associated salinities and nitrate levels (cf. Table 12.1, 'Upland Edge Springs').

The horizontal extent of the marsh/upland seepage zone in any marsh is directly related to the thickness and hydraulic conductivity of the marsh sediments and the upward hydraulic gradient between the underlying aquifer and the marsh. However, since marsh deposits are typically thinner at the upland boundary than in the central vegetated marsh areas in transgressing marshes (i.e. landward migration as relative sea levels rise, Redfield 1972), and since upward gradients are typically greatest near the marsh/upland boundary (Harvey and Odum 1990), a marsh/upland seepage zone should be found in most coastal marshes.

A cross-section of the stratigraphy and porewater salinity through Namskake Marsh based upon cores collected at each of the well sites (Figure 12.4) an probing with a rod at 2–10 m intervals along the well transects also supports th concept of marginal flows. More importantly, it suggests a mechanism for the thir pathway of groundwater inflow, creekbottom seepage. Namskaket Mmarsh forme upon fine glacio-lacustrine sands that now comprise the underlying aquifer. A laye of clay was deposited over these sands in some areas, generally in the deepe basins. However, as the creeks formed and maintained their position as mars elevation rose, the sediments accumulating in the creekbottoms were composed o coarser material (sands, silts and reworked plant material) than the organi sediments of the vegetated marsh (primarily peat). These more permeabl sediments beneath the major channels of the inner marsh appear to extend to th underlying aquifer. This phenomenon was previously described by Redfield (1972 based upon the lower water content of sediments underlying high-mars creekbottoms versus vegetated marsh sediments. Such relatively coarse deposit can occur in organically -dominated systems, due to winnowing of the creekbotton sediments by the flow of tidal waters. At Namskaket, the permeable zone appears t have a relatively vertical profile in the marsh creek (Figure 12.5). However, creek that have meandered can still have a continuous or near-continuous coarse sedimen 'channel' beneath them but with lateral migrations in the profile (cf. Redfiel 1972). These relatively high-conductivity channels allow enhanced groundwate seepage through creekbottoms. Springs in creekbottoms are likely an extreme cas of this mechanism. In Great Sippewissett Marsh high seepage zones (Table 12.1 have been observed in creekbottoms with nitrate concentrations similar to spring at the upland edge (Table 12.1). We have measured rates of groundwater discharg through the creekbottoms of inner Namskaket Marsh by determining the increase i the volume of freshwater flow (total flow corrected for salinity) at various distance from the creek origins to the main channel (Weiskel et al. 1996). Discharge throug the main creekbottom shown in Figure 12.5 was ca. $12 \, l \, m^{-2} \, h^{-1}$, with smalle tributary creeks extending closer to the upland edge having seepage rates up to tw times higher. These creekbottom seepage rates represent the upper range for th whole of Namskaket Marsh, since all of the creeks within the inner marsh stud area are located adjacent to the upland or terminate near the upland/mars boundary.

We have summarized our current understanding of groundwater entry int mature saltmarshes with organic sediment in a simplified schematic (Figure 12.6) Groundwater discharge is focused at the boundary seepage zone (shallow flow path, Figure 12.6) and the creekbottom zone (intermediate flow path). It is in thes two zones that the marsh deposits are thinnest, the hydraulic conductivity i greatest, and the upward hydraulic gradients are largest. Analogous to the situatior in lakes (McBride and Pfannkuch 1975), upward gradients between the underlying aquifer and the marsh sediment reach their maximum at the topographic break between marsh and upland, and tend to decline exponentially with distance away

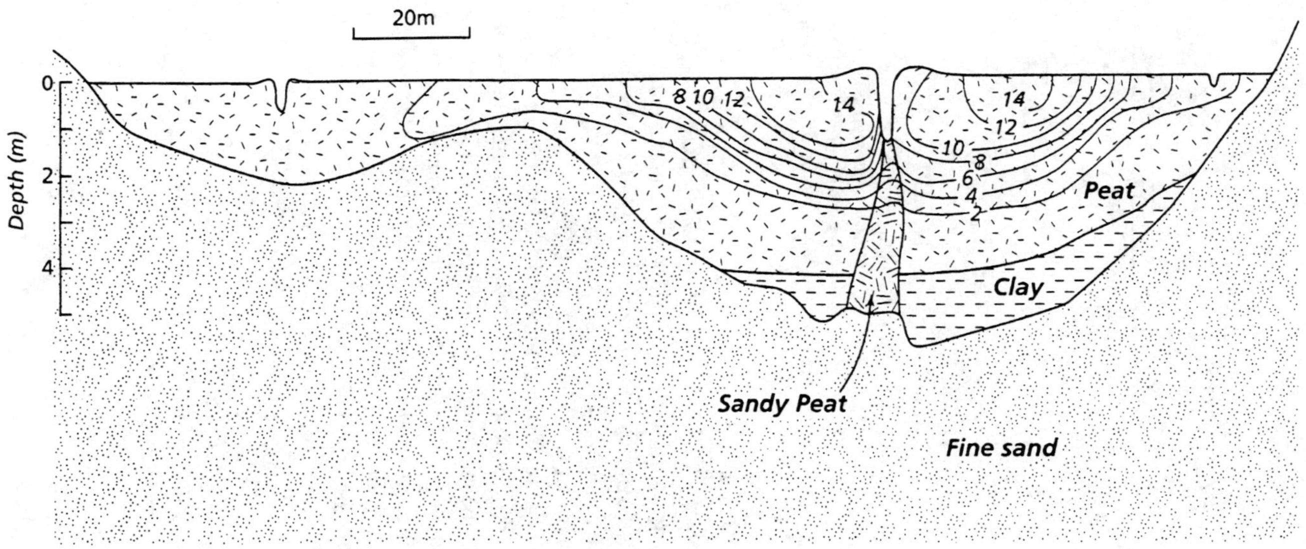

Figure 12.5 Vertical cross-section through inner Namskaket Marsh (vertical exaggeration 8×) taken at the east-west transect just below the island in the marsh. Contours represent equal salinities as determined by multilevel porewater samplers, stratigraphy was determined from cores. Depressions in marsh surface are tidal creeks. The upward distortion of the salinity contours beneath the main creek is consistent with observed groundwater seepage through the inland reach of the creekbottom

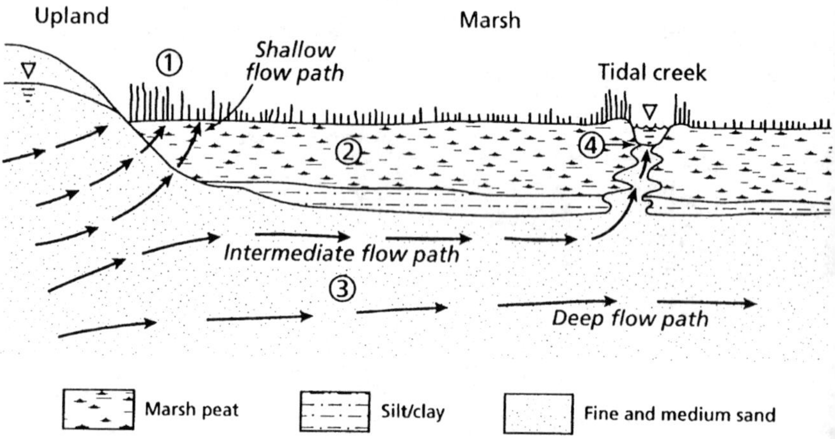

Figure 12.6 Schematic of groundwater flow through the upland–saltmarsh boundary Principal zones are: (1) boundary seepage zone; (2) interior vegetated marsh zone; (3) sub marsh aquifer zone; and (4) creek bottom zone. Arrows depict groundwater flow paths. Sub marsh aquifer conducts groundwater flow to potential high conductivity zones of inlandmos creek bottoms (intermediate flow path) and deeper flow paths which discharge beyond th marsh/bay boundary.

from the marsh boundary (cf. Harvey and Odum 1990). In the creekbottom zone however, the general trend of declining gradients is interrupted, and substantia upward flows can be driven by the differences in hydraulic head between th aquifer and the creek. The upward head gradients are greatest during the ebb tide and are particularly large in the deeply incised creeks of macrotidal marshes. A Namskaket Marsh, for example, upward gradients below the creek bottoms durin the ebb are typically ca. 0.30 m m^{-1}, or 10–20 times larger than in the adjacen interior marsh zone. In the interior marsh zone, groundwater discharge is limited b the relatively small aquifer-to-marsh hydraulic gradients, and the low conductivit of the marsh peat, which often includes basal clay deposits. Depending upon th aquifer thickness, aquifer hydraulic properties (e.g. the anisotropy o horizontal:vertical ratio in hydraulic conductivity), and upland recharge area, significant fraction of the groundwater in the sub-marsh aquifer zone may flow under the marsh directly to adjacent coastal waters (deep flow path, Figure 12.6)

12.5 NITROGEN DYNAMICS OF MARSH SUBSYSTEMS AND MARSH NITROGEN BALANCE

Since groundwater enters the marsh predominantly through creekbottoms and the marsh/upland boundary (where it quickly enters creeks), most interception o groundwater-transported nitrate occurs in these areas. Given the distribution o groundwater entry (hence access to nitrate) and recent N balance studies o vegetated marsh (White and Howes 1994a), denitrification within the central vegetated marsh appears to be independent of terrestrially derived nitrate. At present we have only limited quantitative data on the fate of nitrate at the marsh/upland boundary (Howes and DeSimone, unpublished data), although denitrification assays using the acetylene block method suggest significant activity within the marsh deposits. In contrast, denitrification within the creekbottom sediments of Great Sippewissett Marsh has been quantified using both laboratory and in situ techniques. Hence we can directly address the role of creekbottoms in the interception of nitrate through freshwater inflow as well as interpret the system-wide tidal exchange data (Figure 12.1).

Annual rates of denitrification in Great Sippewissett Marsh were measured using N_2 production of small volumes of creekbottom sediment held under helium atmosphere by Kaplan and colleagues (1979), who argued that the measured rates were the sum of denitrification and nitrogen fixation occuring within the samples. However, since application of the technique required the removal of N_2, fixation would be effectively prevented within the incubation vessel. The recalculated rates are only 9% lower than previously reported for creekbottoms.

In situ rates of denitrification have also been measured in Great Sippewissett Marsh creekbottoms as dark nitrate uptake in chambers (method detailed in Howes and Teal 1995). Chambers were deployed at low tide and were fitted with a side port with an attached plastic bag to allow equalization of the head inside and outside of the chamber as a result of tidal ebb or seepage. Chambers (ca. 500 cm^2) were aerated continuously to maintain oxygen levels and to mix the water overlying the sediments. Water volume, nitrate (Wood et al. 1967), ammonium (Scheiner 1976) and salinity were monitored in the chamber waters, expansion bag and surrounding creek. Additional measures of dark nitrate uptake (hereafter refered to as denitrification) were made by incubating cores (65 mm i.d., 300 mm long) at in situ temperature with overlying water at ca. 50μM nitrate and constantly aerated. These latter assays were conducted for comparison to the earlier study (Kaplan et al. 1979, similar locations in creeks but not the same sites) and to allow parallel examination of the effect of the nitrification inhibitor, N-Serve (2-chloro-6-(trichloro-methyl) pyridine), on nitrate uptake and ammonium release rates from creekbottom sediments. N-Serve selectively inhibits the oxidation of ammonium to nitrite (Goring 1962). N-Serve was added to the water overlying the cores (40 $\mu g\,ml^{-1}$; Caffrey et al. 1993) and changes in dissolved inorganic nitrogen concentrations measured in time course to 24 hours.

Measurement of denitrification rates in the cores incubated in the laboratory, 236 (26) to 220 (29) $\mu mol\,N\,m^{-2}\,h^{-1}$ (Table 12.2), were similar to the net rates of Kaplan and colleagues (1979), 280 $\mu mol\,N\,m{-}^{2}\,h{-}^{1}$ (August, mean (SE)). No significant differences were found in nitrate uptake between cores with and without N-Serve. Similarly, in incubations both with high and low nitrate in the overlying water there was no significant difference ($p<0.05$) in rates of either nitrate uptake or ammonium release (Table 12.2). Since nitrate reduction to ammonium is thought

Table 12.2 Rates of nitrate uptake and ammonium release from control and N-serve-amended cores. Values are means (S.E.), N = number of samples

Location	Date	Temp (°)	N	Initial $[NO_3^-]$ (μM)	N-Serve NO_3^- Uptake (μmol/m^{2}h)	Control NO_3^- Uptake (μmol/m^{2}h)	N-Serve NH_4^+ Efflux (μmol/m^{2}h)	Control NH_4^+ Efflux (μmol/m^{2}h)	Ratio: N-serve/Control Uptake	Ratio: N-serve/Control Efflux
Site 1	7/11/84	13	4	38.8	158.3 (27.9)	168.8 (29.9)	48.1 (2.3)	47.0 (6.9)	0.94	1.02
Site 2	4/26/88	7	5	49.0	103.3 (6.3)	106.5 (14.0)	5.2 (0.7)	5.3 (0.6)	0.97	0.99
	8/11/88	13	5	46.5	236.9 (26.0)	220.3 (28.6)	NA	NA	1.08	NA
	11/19/88	10	5	49.4	116.6 (6.3)	134.6 (3.7)	5.3 (1.5)	9.1 (3.1)	0.87	0.58
Site 2	7/29/88	13	5	0.6	3.8 (0.7)	5.4 (1.0)	34.7 (7.4)	34.9 (5.3)	0.70	0.99

to occur at low nitrate concentration, it is not likely to be a major part of these results. It appears that either nitrification is not a major fraction of denitrification or that the N-Serve is inhibiting ammonification within the sediments (Malhi and Nyborg 1983). If nitrification were a major source of the nitrate for denitrification then an increase in the ammonium efflux in the N-Serve treatments (versus controls) would be expected over 24 h, but this did not occur. Most of the nitrate being reduced in these sediments is probably coming from the high concentration of nitrate (50 μM) in the overlying water.

Comparing the annual rate of creekbottom denitrification (recalculated from Kaplan et al. 1979) with the whole system nitrate import, it appears that for most of the year creekbottom denitrification is a sufficient sink for the observed import (Figure 12.7). Only during the winter does creekbottom denitrification not match or exceed nitrate import to the whole system, probably due to low-temperature depression of denitrifier activity (Kaplan et al. 1977; Seitzinger 1988). The nitrate sink causing the import during the winter may be related to the increase in creekbottom algae in winter. We have found that nitrate uptake by creekbottom epibenthic algae (in light) doubles the rate of dark uptake (Howes et al., unpublished data). In winter, shading of the creekbottom is decreased due to the death of the creekbank grasses. As grazing also decreases in winter, increases in the epibenthic algal population may be important in the uptake of nitrate from overlying waters. It is also possible that winter denitrification rates are higher than observed previously. During our field chamber study we observed that creekbottom temperatures were several degrees higher in seepage zones than further downstream. This appears to be due to heat transported by the discharging groundwater. Groundwater on Cape Cod is generally ca. 10°C year-round, significantly higher than the winter creekbottom temperatures (2°C away from seepage zones) reported by Kaplan and colleagues (1977). The magnitude of increased denitrification within in these zones cannot at

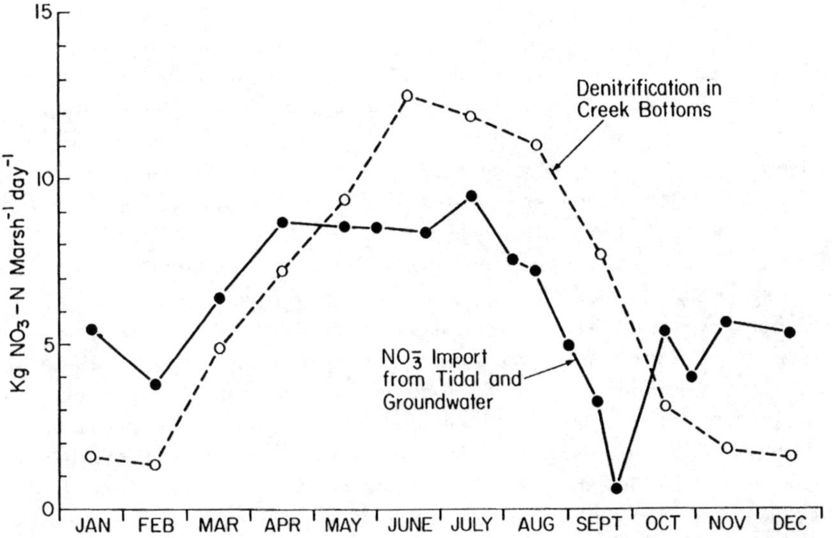

Figure 12.7 Annual cycle of denitrification within the creekbottom sediments and net import of NO_3^- from tidal and groundwaters by Great Sippewissett Marsh. Denitrification was measured by N_2 production of creekbottom sediments (Kaplan et al. 1979). NO_3^- import is based upon groundwater inflow and whole system tidal exchange measurements (Valiela and Teal 1979). Winter NO_3^- import in excess of creekbottom denitrification may result from greater algal uptake in winter

present be assessed but it appears that the winter marshwide estimate (Figure 12.7) of denitrification is too low.

The significant dip in nitrate import observed in September (Figure 12.7) is coincident with the pulse of dissolved inorganic nitrogen export (Figure 12.1). It is likely that nitrate import through the creekbottoms has not changed, but there is a new source of nitrate within the marsh obscuring the creekbottom effects and causing the observed reduction in net import. This source is probably the new litter decomposing, releasing nitrogen as ammonium, some of which is nitrified and exported.

Denitrification rates (mean May–November, five dates) within the field chambers vary considerably. Rates were consistently more than five-fold higher nearest the head of the creek adjacent to groundwater seepage zones, where nitrate concentrations are highest, than at downstream sites (Figure 12.8A). Higher denitrification rates were associated with higher nitrate concentrations (Figure 12.8B) in outflowing waters (low tide). This is consistent with other systems where the highest rates of denitrification are generally found in ecosystems receiving external nitogen inputs (Seitzinger 1988). Similarly, microcosm experiments with tidal flat sediments showed nearly linear increases in nitrate uptake with increasing nitrate levels in overlying water (Taylor 1992). Chamber measurements in

Namskaket Marsh gave results similar to those from Great Sippewissett Marsh. In Namskaket, nitrate additions to chambers resulted in near proportional increases in dark nitrate uptake rates. It appears that the rate of denitrification in saltmarsh creekbottom sediments is controlled by the nitrate levels in the overlying water. The relationship of nitrate concentration to creekbottom denitrification rate suggests that, as nitrate concentrations in groundwaters entering saltmarsh creeks increase due to continuing nitrogen loading to the adjacent upland (through development activities), the annual mass of nitrate import to the marsh will also increase. Within the creekbottoms a continuous increase in nitrate concentration in seepage waters is likely to expand the high denitrification region of the creeks farther downstream. These results also suggest that marshes are more efficient at removing nitrate entering via direct groundwater and stream inflows versus indirectly after mixing with flooding waters, since direct freshwater entry is undiluted resulting in a ca.10-fold higher nitrate concentration in waters overlying creekbottom sediments (Table 12.1, Figure 12.8) compared with tidal water concentrations (Valiela et al. 1978). Nitrate interception will be ultimately limited by the rate at which nitrate can diffuse into the sediments (versus advective removal by ebb-tidal waters) and the organic carbon supply to the creekbottoms.

The organic matter fueling heterotrophic respiration in creekbottoms appears to be primarily 'allochthonous', derived from particulate export from the vegetated marsh surface (Teal 1962 and unpublished data), although epibenthic algae are also important. This transfer of fixed carbon from the marsh surface to the creekbottoms represents the link between the two major subsystems of the marsh.

Based upon the above studies we can put forward a new conceptual model of saltmarsh carbon and nitrogen cycling relative to interception of terrestrially derived nitrogen. While it still appears that marshes import oxidized fixed nitrogen and export reduced forms, plant uptake and denitrification within vegetated marsh are generally not major factors in the observed import. Instead, it appears that saltmarshes consist of two separate but linked subsystems: vegetated marsh and creekbottom. The vegetated marsh has primarily a closed nitrogen cycle, in which the loss of fixed nitrogen to tidal waters, accretion and denitrification are balanced by nitrogen fixation and rain inputs. Inorganic nitrogen leaching from senescing

---

Figure 12.8 (*opposite*) (A) Mean (May–Nov, $N = 5$ dates) in situ rates of dark uptake by creekbottom sediments (1330–140 μmol $m^{-2} h^{-1}$) and NO_3^- concentration (36–9 μmol $m^{-2} h^{-1}$) with increasing distance (meters) from the groundwater high seepage area (indicated by bar) within the inlandmost reach of a tidal creek. Decrease in NO_3^- concentrations with increasing distance is the result of both denitrification and dilution of inflowing groundwater by low NO_3^- (and high-salinity) runoff from the vegetated marsh surface and from creekbank drainage. Uptake rates outside the groundwater high seepage zone are comparable to denitrification rates measured by Kaplan et al. (1979). (B) Dark NO_3^- uptake versus NO_3^- concentration from in situ chamber incubations, April to November. Highest uptake was consistently associated with the areas of groundwater seepage where NO_3^- concentrations were highest

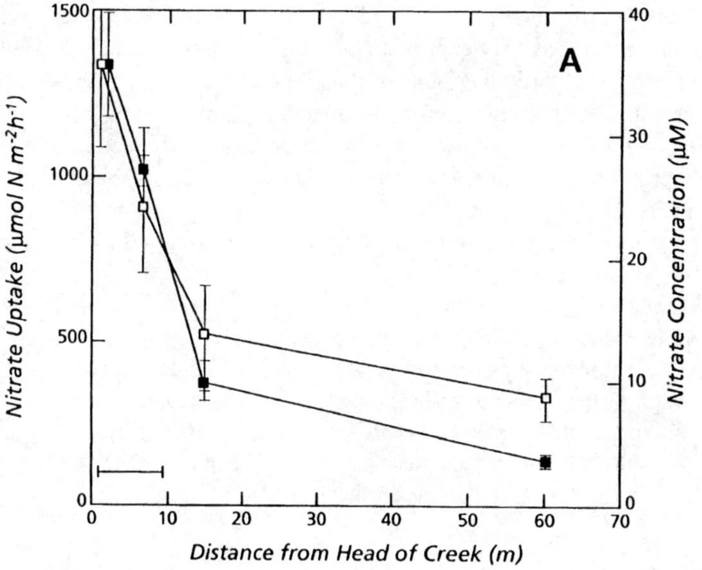
A
Nitrate Uptake (μmol N m-2h-1)
Nitrate Concentration (μM)
Distance from Head of Creek (m)
1500
1000
500
0
40
30
20
10
0
0
10
20
30
40
50
60
70

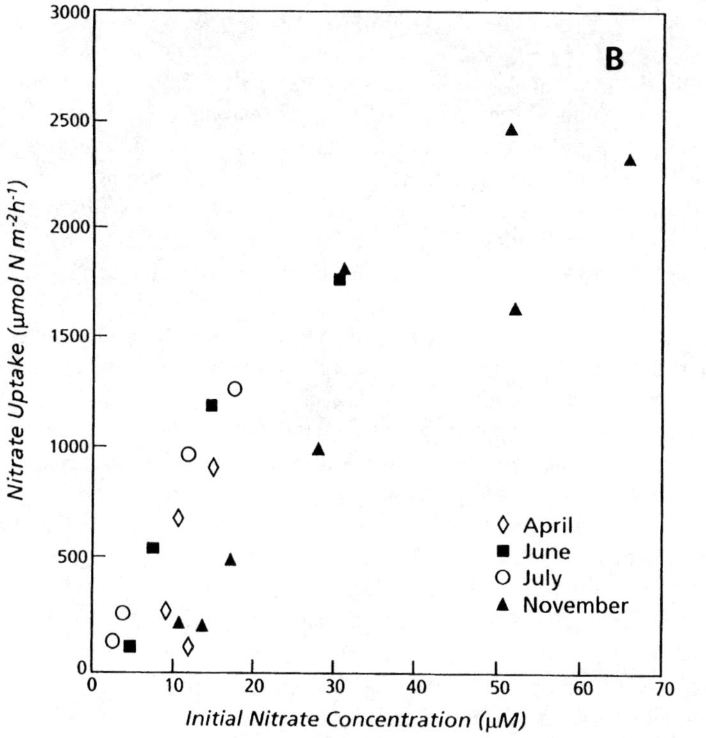
B
Nitrate Uptake (μmol N m-2h-1)
Initial Nitrate Concentration (μM)
April
June
July
November
3000
2500
2000
1500
1000
500
0
0
10
20
30
40
50
60
70

and dead plants results in a large export of ammonium from the marsh in fall. Particulate organic matter exported during the fall and winter is deposited in creekbottoms and is also exported from the marsh and observed as whole system export. Both the particulate nitrogen and ammonium exports have as their source primarily nitrogen fixation rather than externally derived fixed nitrogen. Particulate organic carbon from the vegetated marsh deposited in creekbottoms (and in situ algal production) supports heterotrophic microbial respiration within those sediments. Denitrification in creekbottoms is fueled primarily by the 'imported' carbon from the vegetated marsh and by nitrate entering at low tide through upland surface water and groundwater inflows. Denitrification within the creekbottoms is sufficient to account for most of the annual nitrate import by the whole marsh system. Creekbottom denitrification appears to be primarily limited by nitrate inputs. We predict that as nitrate concentrations increase in non-point source freshwater inflows to saltmarshes, the marshes will increase the mass of nitrate that they intercept and denitrify. The result is that saltmarsh systems are capable of reducing nitrogen loading to coastal waters only to the extent that they can intercept the freshwaters transporting the nitrate from land to sea.

12.6 CONCLUSIONS

Overall, it appears that marshes adjacent to highly permeable watersheds intercept significant amounts of groundwater flow, resulting in a significant reduction in the amount of upland nitrate reaching coastal waters. In addition, it is likely that denitrification of upland nitrate within marsh creekbottoms will increase with increasing concentration of nitrate in inflowing waters. The ability of saltmarshes to intercept terrestrially derived nutrients is ultimately controlled by their ability to intercept freshwater flows and the denitrification capacity of marsh creekbottom sediments.

In highly permeable watersheds, the ability of saltmarshes to intercept nutrients in transit from upland to coastal waters is directly related to the interception of groundwater flow as controlled by the subsurface hydrology of the marsh/aquifer setting. Groundwater appears to enter mature saltmarshes with their several meter thick organic rich deposits (i.e. peat) through two areas: a seepage zone a few meters wide at the marsh/upland boundary where peat thins; and the thinner higher hydraulic conductivity deposits beneath inner-marsh creekbottoms, especially at low tide when upward flow is driven by the large hydraulic gradient that develops. Groundwater entry through the several meter thick, poorly conductive sediments of the vegetated marsh areas appears to be negligable. It is likely that saltmarshes with thinner or more conductive sediment may have a marsh/upland seepage zone extending further into the interior of the vegetated marsh.

Groundwater and non-point surface water inflows transport significant amounts of nitrate into Great Sippewissett Marsh. Denitrification within creekbottom sediments is sufficient to reduce most (ca. 80%) of the NO_3 entering in inflowing freshwater. NH_4^+ and PN export is derived from the vegetated marsh, but most likely is predominantly derived from N_2 fixation (on an annual basis) rather than groundwater or tidal water sources.

Nitrogen interception from freshwater likely requires organic carbon transfers from vegetated marsh to creekbottoms to supply the reducing power to denitrifiers. The effect is two distinct nitrogen cycles: a vegetated marsh cycle dominated by recycling, nitrogen fixation and export, and a creekbottom cycle dominated by import of organic matter from vegetated marsh and nitrate from upland which fuels high rates of N removal through denitrification.

It appears that the denitrification capacity of the creekbottom zone is not fully utilized and rates will increase as nitrate concentrations in freshwater inflows (especially groundwater) increase. While this indicates that interception of nitrate by saltmarshes should increase as loading increases, the extent to which the efficiency of interception will change requires additional study.

ACKNOWLEDGEMENTS

We thank I. Valiela for insights and discussions, R. Van Etten, L. DeSimone, C. Hee, B. Zinn and J. Waples for field assistance, S. Brown-Leger and R. Van Etten for analytical assistance, C. Taylor and L. DeSimone for critical comments on the manuscript and Mr A. Diddel for use of his saltmarsh property. This study was supported by NSF grant BSR-87-17701, the US Geological Survey and the Massachusetts Department of Environmental Protection, Office of Watershed Management. Contribution no. 9066 of the Woods Hole Oceanographic Institution.

REFERENCES

Adam, P. 1990. *Saltmarsh Ecology*. Cambridge Studies in Ecology, University Press, Cambridge.

Bowden, W.B. 1986. Nitrification, nitrate reduction and nitrogen immobilization in a tidal freshwater marsh sediment. *Ecology* 67: 88–99.

Caffrey, J.M., N.P. Sloth, H.F. Kaspar and T.H. Blackburn. 1993. Effect of organic loading on nitrification and denitrification in a marine sediment microcosm. *FEMS Microbiology Ecology* 12: 159–167.

Capone, D.G. and M.F. Bautista. 1985. A groundwater source of nitrate in nearshore marine sediments. *Nature* 313: 214–216.

Chambers, R.M., J.W. Harvey and W.E. Odum. 1992. Ammonium and phosphate dynamics in a Virginia salt marsh. *Estuaries* 15: 349–359.

Cherkauer, D.S. and P.F. McKereghan. 1989. Groundwater discharge to lakes: Focusing in embayments. *Ground Water* 29: 72–80.

Dacey, J.W.H. and B.L. Howes. 1984. Water uptake by roots controls water table movement and sediment oxidation in short *Spartina* marsh. *Science* 224: 487–489.

DeSimone, L.A. and B.L. Howes. 1995. Hydrogeologic, water-quality, and biogeochemical data collected at a septage-treatment facility, Orleans, Cape Cod, Massachusetts, October 1988 through December 1992. US Geological Survey Open-File Report 95-439, Reston, Virginia.

DeSimone, L.A. and B.L. Howes. In Press. Groundwater denitrification and nitrogen transport in a coastal aquifer receiving wastewater discharge. *Environmental Science and Technology*.

DeSimone, L.A., P.M. Barlow and B.L. Howes. 1995. A nitrogen-rich septage-effluent plume in a glacial aquifer, Cape Cod, Massachusetts. US Geological Survey Open-File Report 95-290, Reston, Virginia.

Fujita, R.M., P.A. Wheeler and R.L. Edwards. 1989. Assessment of macroalgal nitrogen limitation in a seasonal upwelling region. *Marine Ecology Progress Series* 53: 293–303.

Goring, C.A. 1962. Control of nitrification by 2-chloro-6(trichloro-methyl) pyridine. *Soil Science* 93: 211–218.

Harvey, J.W. and W.E. Odum. 1990. The influence of tidal marshes on upland groundwater discharges to estuaries. *Biogeochemistry* 10: 217–236.

Harvey, J.W. and W.K. Nuttle. 1995. Fluxes of water and solute in a coastal wetland sediment. 2. Effect of macropores on solute exchange with surface water. *Journal of Hydrology* 164: 109–125.

Hemond, H.F. and J.L. Fifield. 1982. Subsurface flow in salt marsh peat: a model and field study. *Limnology and Oceanography* 27: 127–136.

Howes, B.L. and D.D. Goehringer. 1994. Porewater drainage and dissolved organic carbon and nutrient losses through the intertidal creekbanks of a New England saltmarsh. *Marine Ecology Progress Series* 114: 289–301.

Howes, B.L. and J.M. Teal. 1995. Nutrient balance of a Massachusetts cranberry bog and relationships to coastal eutrophication. *Environmental Science and Technology* 29: 960–974.

Howes, B.L., J.W.H. Dacey and S.G. Wakeham. 1985. Effects of sampling technique on measurements of porewater constitiuents in saltmarsh sediments. *Limnology and Oceanography* 30: 221–227.

Howes, B.L., J.W.H. Dacey and D.D. Goehringer. 1986. Factors controlling the growth form of *Spartina alterniflora*: feedbacks between above-ground production, sediment oxidation, nitrogen and salinity. *Journal of Ecology* 74: 881–898.

Jaworski, N.A. 1984. Sources of nutrients and the scale of eutrophication problems in estuaries, p. 83–110. *In* B.J. Neilson and L.E. Cronin (eds.), *Estuaries and Nutrients*. Humana Press. Clifton, New Jersey.

Johannes, R.E. 1980. The ecological significance of the submarine discharge of groundwater. *Marine Ecology Progress Series* 3: 365–373.

Jordan, T.E. and D.L. Correll. 1991. Continuous automated sampling of tidal exchanges of nutrients by brackish marshes. *Estuarine Coastal Shelf Science* 32: 527–545.

Kaplan, W., J.M. Teal and I. Valiela. 1977. Denitrification in salt marsh sediments. Evidence for seasonal temperature selection among populations of denitrifying bacteria. *Microbial Ecology* 3: 193–204.

Kaplan, W., I. Valiela and J.M. Teal. 1979. Denitrification in a salt marsh ecosystem. *Limnology and Oceanography* 24: 726–734.

Knott, J.F., W.K. Nuttle and H.F. Hemond. 1987. Hydrologic parameters of salt marsh peat. *Hydrological Processes* 1: 211–220.

Malhi, S.S. and M. Nyborg. 1983. Release of mineral N from soils: influence of inhibitors of nitrification. *Soil Biology and Biochemistry* 15: 581–585.

McBride, M.S. and H.O. Pfannkuch. 1975. The distribution of seepage within lakebeds. *Journal of Research of the US Geological Survey* 3: 505–512.

Millham, N.P. and B.L. Howes. 1994. Freshwater flow into a coastal embayment: Groundwater and surface water inputs. *Limnology and Oceanography* 39: 1928–1944.
National Research Council (NRC). 1978. *Nitrates: An Environmental Assessment.* National Academy of Sciences, Washington, DC.
Nixon, S.W. 1980. Between coastal marshes and coastal waters – a review of twenty years of speculation and research on the roles of salt marshes in estuarine productivity and chemistry, p. 437–527. *In* P. Hamilton and K. MacDonald (eds.), *Estuarine and Wetland Processes.* Plenum Press, New York.
Nixon, S.W., C.A. Oviatt, J. Frithsen and B. Sullivan. 1986. Nutrients and the productivity of estuarine and coastal marine ecosystems. *Journal of Limnological Society of South Africa* 12: 43–71.
Nuttle, W.K. 1988. The extent of lateral water movement in the sediments of a New England salt marsh. *Water Resources Research* 24: 2077–2085.
Nuttle, W.K. and J.W. Harvey. 1995. Fluxes of water and solute in a coastal wetland sediment. 1. The contribution of regional groundwater discharge. *Journal of Hydrology* 164: 89–107.
Odum, E.P. 1980. The status of three ecosystem-level hypotheses regarding saltmarsh estuaries: tidal subsidy, outwelling and detritus-based food chains, p. 485–495. *In* V.S. Kennedy (ed.), *Estuarine Perspectives.* Academic Press, New York.
Orson, R.A. and B.L. Howes. 1992. Salt marsh development studies at Waquoit Bay, MA: Influence of geomorphology on long-term plant community structure. *Estuarine Coastal Shelf Science* 35: 453–471.
Redfield, A.C. 1972. Development of a New England salt marsh. *Ecological Monographs* 42: 201–237.
Rosenberg, R. 1985. Eutrophication: the future marine coastal nuisance. *Marine Pollution Bulletin* 16: 227–231.
Ryther, J.H. and W.M. Dunstan. 1971. Nitrogen, phosphorus, and eutrophication in the coastal marine environment. *Science* 171: 1008–1013.
Scheiner, D. 1976. Determination of ammonia and Kjeldahl nitrogen by the indophenol method. *Water Research* 10: 31–36.
Seitzinger, S.P. 1988. Denitrification in freshwater and coastal marine ecosystems: ecological and geochemical significance. *Limnology and Oceanography* 33: 702–724.
Sullivan, M.J. and F.C. Daiber. 1974. Response in production of Cord Grass, *Spartina alterniflora*, to inorganic nitrogen and phosphorus fertilizer. *Chesapeake Science* 15: 121–123.
Taylor, D.I. 1992. The influence of upwelling and short-term changes in concentrations of nutrients in the water column of fluxes across the surface of a salt marsh. *Estuaries* 15: 68–74.
Teal, J.M. 1962. Energy flow in the salt marsh ecosystem of Georgia. *Ecology* 43: 614–624.
Turner, R.E. 1993. Carbon, nitrogen and phosphorus leaching rates from *Spartina alterniflora* salt marshes. *Marine Ecology Progress Series* 92: 135–140.
Turner, R.E. and N.N. Rabalais. 1994. Coastal eutrophication near the Mississippi delta. *Nature* 368: 619–621.
Valiela, I. 1983. Nitrogen in salt marsh ecosystems, p. 469–478. *In* E.J. Carpenter, O.C. Capone (eds.), *Nitrogen in the Marine Environment.* Academic Press, New York.
Valiela, I. and J.M. Teal. 1974. Nutrient limitation in salt marsh vegetation, p. 547–561. *In* R.J. Reimold and W.H. Queen (eds.), *Ecology of Halophytes.* Academic Press, New York.
Valiela, I. and J.M. Teal. 1979. Nitrogen budget of a salt marsh ecosystem. *Nature* 280: 652–656.

Valiela, I., J.M. Teal, S. Volkmann, D. Shafer and E.J. Carpenter. 1978. Nutrient and particulate fluxes in a salt marsh ecosystem: Tidal exchanges and inputs by precipitation and groundwater. *Limnology and Oceanography* 23: 798–812.

Valiela, I., J. Costa, K. Foreman, J. Teal, B. Howes and D. Aubrey. 1990. Groundwater-borne inputs from watersheds to coastal waters. *Biogeochemistry* 10: 177–198.

Vanek, V. 1993. Groundwater regime of a tidally influenced coastal pond. *Journal of Hydrology* 151: 317–342.

Vince, S.W., I. Valiela and J.M. Teal. 1981. An experimental study of the structure of herbivorous communities in a salt marsh. *Ecology* 62: 1662–1678.

Weiskel, P.K. and B.L. Howes. 1991. Quantifying dissolved nitrogen flux through a coastal watershed. *Water Resources Research* 27: 2929–2939.

Weiskel, P.K. and B.L. Howes. 1992. Differential transport of sewage-derived nitrogen and phosphorus through a coastal watershed. *Environmental Science and Technology* 26: 352–360.

Weiskel, P.K., L.A. DeSimone and B.L. Howes. 1996. Transport of wastewater nitrogen through a coastal aquifer and marsh, Orleans, MA. US Geological Survey, Open-File Report 96–111, Reston, Virginia.

Werme, C.E. 1981. Resource partitioning in a salt marsh fish community. Ph.D. Thesis, Boston University, Boston, Massachusetts.

White, D.S. and B.L. Howes. 1994a. Long-term 15N-nitrogen retention in the vegetated sediments of a New England salt marsh. *Limnology and Oceanography* 39: 1878–1892.

White, D.S. and B.L. Howes. 1994b. Nitrogen incorporation into decomposing litter of *Spartina alterniflora. Limnology and Oceanography* 39: 1878–1892.

White, D.S. and B.L. Howes. 1994c. Translocation, remineralization and turnover of nitrogen in the roots and rhizomes of *Spartina alterniflora* (Gramineae). *American Journal of Botany* 82: 1225–1234.

Whiting, G.J., H.N. McKellar, J.D. Spurrier and T.G. Wolaver. 1989. Nitrogen exchange between a portion of vegetated saltmarsh and the adjoining creek. *Limnology and Oceanography* 34: 463–473.

Wood, E.D., F.A.J. Armstrong and F.A. Richards. 1967. Determination of nitrate in seawater by cadmium-copper reduction to nitrite. *Journal of the Marine Biological Association of the United Kingdom* 47: 23–31.

Chapter Thirteen

Direct nekton use of intertidal saltmarsh habitat and linkage with adjacent habitats: a review from the southeastern United States

CAROLE C. McIVOR
National Biological Service

and

LAWRENCE P. ROZAS
National Marine Fisheries Service

13.1 DEFINITIONS, REGIONAL MARSH CHARACTERIZATION AND LARGE-SCALE PATTERNS

13.1.1 Introduction

We use the term nekton (organisms capable of active swimming) to include both fishes and macrocrustaceans (crabs, shrimp). Direct use of saltmarsh habitats by nekton is linked to geomorphology via the physical processes that determine estuarine landscapes or environments. The physical processes include sediment deposition and redistribution, delta-building, sea level change, and the action of currents and tides. Landforms in turn are influenced by climate. Climate affects salinity regimes at the scale of the estuary and the region: climate affects vegetation type and stature directly through the balance between rainfall and evaporation. Tidal range combined with slope of the intertidal zone determines both the area inundated and the depth of inundation. Tidal regime coupled with large-scale

Estuarine Shores: Evolution, Environments and Human Alterations. Edited by K. F. Nordstrom and C. T. Roman.

meteorological forcing functions determines the pattern and duration of marsh flooding. Relative sea level change combined with sediment supply determines whether shorelines are subsiding, accreting, or remaining relatively constant in terms of area.

In this review of patterns of nekton use of intertidal saltmarshes, our purpose is threefold: (1) to summarize the knowledge on taxa using the marsh surface; (2) to analyze patterns of distribution and abundance as they relate to geomorphological features of the marsh; and (3) to examine the ways in which nekton link intertidal and subtidal habitats. Geomorphological features of saltmarshes considered herein include position along a gradient of stream order (*sensu* Horton 1945), subtidal geomorphology, marsh elevation and flooding duration, drainage density, and relative amount of edge.

The review has an acknowledged North American bias because most of the literature on saltmarsh nekton is from this geographic area. Further, the majority of such studies have been conducted within the region of most extensive saltmarsh development, from the mid-Atlantic coast to the northwest coast of the Gulf of Mexico. This review therefore focuses on this region, though we refer to studies in other locations when appropriate. We have chosen to focus on the vegetated intertidal portion of saltmarshes, i.e. the marsh surface, both because of recognition that this portion of the marsh has direct habitat 'value' for nektonic organisms, and because recent sampling innovations have permitted the quantitative testing of a number of hypotheses about the relationship between nekton distribution and geomorphological and hydrological features of saltmarshes. Although the review focuses on saltmarshes, we include research from the tidal freshwater portion of the estuary where that research elucidates the geomorphological theme. Finally, we have largely restricted our discussion to natural, relatively unimpacted marshes. Consideration of impounded, constructed, and degraded marshes is beyond the scope of this review.

13.1.2 Characterization of saltmarsh development in different ecoregions

Most southeastern Atlantic estuaries (Virginia–North Carolina border to the Georgia–Florida border) are classified as coastal plain saltmarsh estuaries (Figure 13.1). Most lack a major river source, except North and South Santee rivers, South Carolina, and Charleston Harbor, South Carolina (Dardeau et al. 1992). There are extensive, well defined drainage networks, dendritically intersecting extensive coastal saltmarshes. Typically, open water area is less than 20% of the estuary (Kjerfve 1989). Saltmarshes are extensive and dominated by *Spartina alterniflora* in regularly flooded marshes. *Juncus roemerianus* marshes occur at higher elevations, where salinity of interstitial water is lowered, or where flooding frequency is irregular (Dardeau et al. 1992). Within this region, saltmarshes are most extensive in Georgia (167 400 ha) and South Carolina (165 167 ha) (Dardeau et al. 1992). Tidal regime is semidiurnal; tidal range is 2–3 m in the middle of the

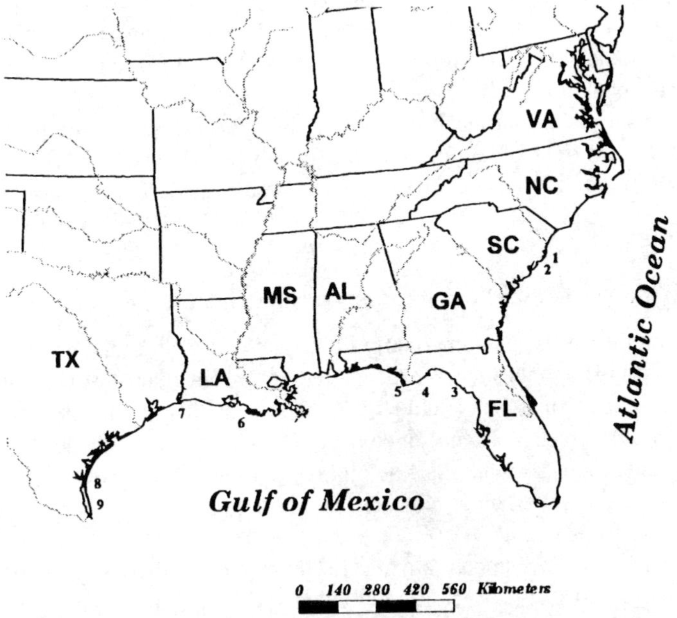

Figure 13.1 Southeastern US coastal zone. Numbers refer to geographic locations mentioned in the text: 1, North and South Santee Rivers; 2, Charleston Harbor; 3, Suwanee River; 4, Apalachicola Bay; 5, St Andrews Bay; 6, Atchafalaya Bay; 7, Sabine Lake; 8, Baffin Bay; 9, Laguna Madre

Georgia coast at Sapelo Island, and decreases to <1 m both north and south (Wiegert and Freeman 1990; Dardeau et al. 1992). Greater tidal amplitude in Georgia results in more extensive development of natural levees along creek banks (Dardeau et al. 1992), a geomorphological factor that has received little attention as it relates to marsh use by nekton.

Saltmarshes of the northern Gulf of Mexico coastline (Suwannee River, Florida to the Texas–Louisiana border) are extensive, covering about 2.1 million ha (Figure 13.1) (Dardeau et al. 1992). Regularly flooded marshes are dominated by *Spartina alterniflora*. Such marshes are best developed in the Mississippi deltaic plain in Louisiana. Much of the Mississippi, Alabama, and Florida coast (within the northern Gulf) are dominated by *Juncus roemerianus* marshes. Such marshes are characterized by less frequent flooding from predominantly wind-driven tides (as opposed to predictable lunar tides). In these *Juncus* marshes, *S. alterniflora* occurs only in a narrow band adjacent to tidal creeks or other open water (Dardeau et al. 1992). Mixed tides (two unequal high and/or two unequal low waters each tidal day) occur from Suwannee River to Apalachicola Bay, Florida and from Atchafalaya Bay to Sabine Lake. Diurnal tides (one high and one low/day) occur from St Andrews Bay, Florida west through the Mississippi Delta region (Dardeau et al. 1992). Throughout the northern Gulf region, the tidal range is small, averaging approximately 0.30 m at the coast within the Mississippi deltaic plain, for example. Seasonal changes in water level are pronounced with the result that in coastal Louisiana, *S. alterniflora* marshes are flooded only about 20% of the time in January, but up to 80% of the time in September and October (Gosselink 1984). Regularly flooded saltmarsh is also locally extensive along the upper Texas coast, but decreases in areal extent in south Texas where *S. alterniflora* occurs mostly as narrow fringe marsh along major water courses. *S. alterniflora* is essentially absent along the margins of Laguna Madre and Baffin Bay (Webb 1982).

13.1.3 Major families using marsh-surface habitats

We compiled a list of numerically dominant fish and decapod crustacean species collected in marsh-surface habitats as reported from studies of Gulf of Mexico and Atlantic coastal marshes (Table 13.1). We stress that studies that collected fauna from marsh creeks are not included. We classified a species as numerically dominant in a study when it was reported as one of the five most abundant of a taxonomic group (fishes or crustaceans) and represented >1% of the total catch in that group.

Limitations to these data are several. Published studies of marsh habitat use by nekton are restricted in geographic extent. Most research has been conducted in estuaries of the northern Gulf of Mexico. A few studies of Atlantic coastal marshes, including those in tidal freshwater reaches of the estuary, have been completed. However, other than Chamberlain and Barnhart (1994), we are unaware of any published studies of nekton use of the vegetated surface of natural marshes of the Pacific coast. In none of the geographic regions have the nekton associated with the marsh surface been described in all of the different marsh types that exist. For example, on the Gulf coast, direct use of tidal freshwater marshes by nekton has not been studied, and few studies have been conducted in oligohaline and polyhaline environments (Table 13.1). Studies of the marsh surface on the Atlantic coast are confined to tidal freshwater and polyhaline regions of the estuary (Table 13.1). In addition, published studies are almost exclusively of low intertidal marshes that are frequently inundated; we know of only one published study of high, infrequently flooded marshes (Table 13.1: Murphy 1991). Other factors that must be considered when interpreting the data presented in Table 13.1 are that these studies employed a variety of sampling methodologies, and that samples were collected in different marsh-surface microhabitats (areas differing sometimes subtly in physical, chemical or biological features). Any of these factors may bias the data. For example, in the studies employing a drop sampler, most samples were taken in a 'marsh edge' habitat, 1–2 m from the water–marsh interface (Zimmerman and Minello 1984). In contrast, Kneib (1991) and Kneib and Wagner (1994) sampled 'interior marsh' 25–90 m from the nearest subtidal habitat using the flume weir. The proximity of a marsh to a subtidal area has an effect on habitat use in that densities of most species decline with distance from the marsh–water interface

(Kneib and Wagner 1994; Peterson and Turner 1994). Therefore, caution is advised in comparing studies in which different microhabitats were sampled.

In a recent review of studies conducted in saltmarshes of the southeast region of the US, Rozas (1993) identified 51 species in 24 families of fishes and seven species in three families of decapod crustaceans that used marsh-surface habitats. The families Cyprinodontidae, Gobiidae, Sciaenidae, Palaemonidae, and Penaeidae contributed the most species to marsh-surface assemblages of saltmarshes in the southeast region (Rozas 1993). Species belonging to these families (except Sciaenidae) also numerically dominated marsh-surface assemblages in the tidal freshwater and brackish marshes that we reviewed (Table 13.1). In addition, members of the family Centrarchidae were important components of assemblages in tidal freshwater environments (Table 13.1). Daggerblade grass shrimp (*Palaemonetes pugio*) and two species of cyprinodonts (gulf killifish, *Fundulus grandis*, on the Gulf coast and mummichog, *Fundulus heteroclitus*, on the Atlantic coast) are among the most abundant macrofaunal species found on the marsh surface, and they were classified as numerically abundant in nearly all of the studies we reviewed (Table 13.2). Other cyprinodonts that were classified as numerically dominant in at least half the studies we reviewed are diamond killifish (*Adinia xenica*) and sheepshead minnow (*Cyprinodon variegatus*) on the Gulf coast, and banded killifish *Fundulus diaphanus* (in tidal freshwater only) on the Atlantic coast. Gobies (naked goby, *Gobiosoma bosc*, and darter goby, *Gobionellus boleosoma*) numerically dominated marsh-surface assemblages only on the Gulf coast (Table 13.2). Blue crab (*Callinectes sapidus*) and penaeid shrimp (mostly brown shrimp, *Penaeus aztecus*, and white shrimp, *Penaeus setiferus*) were abundant in most of the habitats sampled (Table 13.2). In the only published work on direct use of Pacific coast marsh, Chamberlain and Barnhart (1994) found threespine stickleback (*Gasterosteus aculeatus*) and gobies (arrow goby, *Clevelandia ios*, and tidewater goby, *Eucyclogobius neuberryi*) to be most abundant. However, the elevation of their natural study marsh was relatively high, a factor that may have contributed to the low densities and few species they reported.

13.1.4 Patterns of resident versus transient use among marsh microhabitats

Estuarine nekton can be classified into various groups based on ecological affinity (McHugh 1967; Deegan and Thompson 1985). Accordingly, the species listed in Table 13.1 were classified into the three affinity groups (freshwater, estuarine, and estuarine-marine) after Deegan and Thompson (1985) as shown in Table 13.2. Freshwater species generally spawn in freshwater or low salinity regions of the estuary, and because they cannot tolerate high salinities, are usually restricted to the upper estuary. Estuarine species spawn and spend most of their lives within the estuary; they are commonly found throughout the estuary because they can tolerate a wide range of salinity. Estuarine-marine species occur in the estuary primarily as

Table 13.1 List of numerically dominant fishes and crustaceans (in descending order of abundance) collected on the marsh surface compiled from studies of northern Gulf Mexico and Atlantic coast estuaries. The sampling frequency and duration, gear type, number of sampling sites and the total number of samples taken, salinity regime, dominant marsh vegetation, and the source of information are given for each data set

	Lavaca Bay, TX	Lavaca Bay, TX	Trinity Bay, TX
Fishes:	Naked goby Darter goby Silver perch Pinfish Gulf killifish	Naked goby Bay anchovy Gulf killifish Diamond killifish Pinfish	Sheepshead minnow Gulf killifish Striped mullet Rainwater killifish Bayou killifish
Crustaceans:	Daggerblade grass shrimp Brown shrimp Marsh grass shrimp Blue crab White shrimp	Daggerblade grass shrimp Brown shrimp Blue crab White shrimp Pink shrimp	Daggerblade grass shrimp Blue crab Pink shrimp Brown shrimp
Sampling Frequency & Duration:	2/year (fall and spring) for 1 year	2/year (fall and spring) for 1 year	3/year (spring, summer, and fall) for 1 year
Gear Type:	2.6 m squared drop sampler	2.6 m squared drop sampler	2.6 m squared drop sampler
Sampling Sites (Total Number):	24 (24)	24 (24)	24 (24)
Salinity Regime:	Mesohaline	Mesohaline	Oligohaline
Dominant Vegetation:	*Spartina alterniflora*	*Juncus roemerianus*	*Scirpus* spp
Source:	Zimmerman et al. (1990a)	Zimmerman et al. (1990a)	Zimmerman et al. (1990b)

	Galveston Bay, TX	Christmas Bay/West Bay, TX	Terrebonne Bay, LA
Fishes:	Naked goby Speckled worm eel Pinfish Gulf killifish Blackcheek tonguefish	Darter goby Sheepshead minnow Pinfish Inland silverside Naked goby	Striped mullet Sheepshead minnow Gulf killifish Diamond killifish Inland silverside
Crustaceans:	Daggerblade grass shrimp Blue crab Marsh grass shrimp Brown shrimp White shrimp	Daggerblade grass shrimp Brown shrimp Blue crab White shrimp Pink shrimp	Daggerblade grass shrimp Blue crab White shrimp Brown shrimp
Sampling Frequency & Duration	3/year (spring, summer, and fall) for 1 year	3/year (spring, summer, and fall) for 1 year	1–3/month for 8 months (April–November)
Gear Type	2.6 m squared drop sampler	2.6 m squared drop sampler	6 m squared bottomless lift net
Sampling Sites (Total Number):	24 (24)	24 (24)	6 (90)
Salinity Regime:	Mesohaline	Polyhaline	Mesohaline
Dominant Vegetation:	*Spartina anterniflora*	*Spartina anterniflora*	*Spartina anterniflora*
Source:	Zimmerman et al. (1990b)	Zimmerman (1990b)	Rozas and Reed (1993)

Table 13.1 *(continued)*

	Terrebone, Bay, LA	Terrebone Bay, LA	Terrebone Bay, LA
Fishes:	Gulf killifish Sheepshead minnow Diamond killifish Striped mullet Bayou killifish	Naked goby Gulf killifish Diamond killifish Bayou killifish Darter goby	Gulf killifish Diamond killifish Sheepshead minnow Naked goby Inland silverside
Crustaceans:	Daggerblade grass shrimp Blue crab Brown shrimp	*Grass shrimp Blue crab	Daggerblade grass shrimp Blue crab Brown shrimp
Sampling Frequency & Duration:	1-3/month for 8 months (April–November)	1/month for 11 months (January–November)	2/month for 1 year
Gear Type:	6 m squared bottomless lift net	Flume	Flume
Sampling Sites (Total Number):	3 (45)	15 (165)	6 (120)
Salinity Regime:	Mesohaline	Mesohaline	Mesohaline
Dominant Vegetation:	*Distichlis spicata*	*Spartina alterniflora*	*Spartina alterniflora*
Source:	Rozas and Reed (1993)	Peterson and Turner (1994)	Rozas (1992)
	Barataria Bay/Caminada Bay, LA	**Duplin River, GA**	**Newport River, NC**
Fishes:	Naked goby Darter goby Inland silverside Bay anchovy Gulf menhaden	Mummichog Spotfin killifish **Mullets Spot Sailfin molly	Mummichog Spot Striped killifish Sheepshead minnow Spotfin mojarra
Crustaceans:	Daggerblade grass shrimp Blue crab Brown shrimp White shrimp	Daggerblade grass shrimp White shrimp	***Grass shrimp ***Penaeid shrimp (3 spp) Blue crab
Sampling Frequency & Duration:	1–2/month for 3.5 year (fewer samples in winter)	1–10 (mean = 6)/mo for 2 year (fewer in winter)	1/month for 1 year
Gear Type:	0.5 and 1.2 m squared drop samplers	100 m squared flume weir	Block net
Sampling Sites (Total Number):	Fishes = 309 (309)/crustaceans = 97 (97)	2 (271)	8 (96)
Sanity Regime:	Mesohaline	Polyhaline	Polyhaline
Dominant Vegetation:	*Spartina alterniflora*	*Spartina alterniflora*	*Spartina alterniflora*
Source:	D. M. Baltz, personal communication	Kneib (1991)	Hettler (1989)

Table 13.1 *(continued)*

	Chickahominy River, VA	Morris Creek, VA	Webhenet River/Little River, ME
Fishes:	Mummichog Banded killifish Inland silverside Spottail shiner Pumpkinseed	Banded killifish Mummichog Bluegill Pumpkinseed Mosquitofish	****Mummichog
Crustaceans:	Daggerblade grass shrimp Blue crab	Daggerblade grass shrimp Blue crab	
Sampling Frequency & Duration:	4/month for 4 month (July–October)	2–4/month for 6–7 months (spring–fall) over 2 year	36–44/5 month (May–September) for 1 year
Gear Type:	Flume net	Flume net	Flume net
Sampling Sites (Total Number):	9 (144)	6 (204)	6 (232)
Salinity Regime:	Tidal freshwater	Tidal freshwater	Polyhaline
Dominant Vegetation:	*Peltandra virginica*	*Peltandra virginica*	*Spartina patens*
Source:	Rozas and Odum (1987)	McIvor and Odum (1988)	Murphy (1991)

* Peterson and Turner (1994) did not identify grass shrimp (Palaemonetes spp) to species
** Knieb (1991) did not identify mullets (Mugil spp) to species
*** Hettler (1989) did not enumerate grass shrimp and identified penaeids to genus only
**** Murphy (1991) collected few fish other than mummichogs (six ninespine stickleback, two Atlantic silverside, and one fourspine stickleback), and grass shrimp were only infrequently taken (S. C. Murphy, personal communication)

Table 13.2 List of numerically dominant fishes and crustaceans collected on the marsh surface of Gulf of Mexico and Atlantic coastal marshes. The frequency with which each species was reported as abundant (FOA) from studies listed in Table 13.1 and its ecological affinity (after Deegan and Thompson 1985) are given

Common name	Scientific name	FOA (%)	Ecological Affinity
Gulf of Mexico			
Fishes			
Gulf killifish	*Fundulus grandis* Baird & Girard	80	Estuarine
Naked goby	*Gobiosoma bosc* (Lacepede)	60	Estuarine
Sheepshead minnow	*Cyprinodon variegatus* Lacepede	50	Estuarine
Diamond killifish	*Adinia xenica* (Jordan & Gilbert)	50	Estuarine
Inland silverside	*Medinia beryllina* (Cope)	40	Estuarine
Pinfish	*Lagadon rhomboides* (Linnaeus)	40	Estuarine
Darter goby	*Gobionellus boleosoma* (Jordan & Gilbert)	40	Estuarine
Striped mullet	*Mugil cephalus* Linnaeus	30	*Estuarine–Marine
Bayou killifish	*Fundulus pulvereus* (Evermann)	30	Estuarine
Bay anchovy	*Anchoa mitchilli* (Valenciennes)	20	Estuarine–Marine
Blackcheek tonguefish	*Symphurus plagiusa* (Linnaeus)	10	*Estuarine–Marine
Speckled worm eel	*Myrophis punctatus* Lutken	10	Estuarine–Marine
Rainwater killifish	*Lucania parva* (Baird & Girard)	10	Estuarine
Silver perch	*Bairdiella chrysoura* (Lacepede)	10	Estuarine–Marine
Gulf menhaden	*Brevoortia patronus* Goode	10	Estuarine–Marine
Crustaceans			
Daggerblade grass shrimp	*Palaemonetes pugio* Holthius	100	Estuarine
Blue crab	*Callinectes spidus* Rathbun	100	Estuarine–Marine
Brown shrimp	*Penaeus aztecus* Ives	90	Estuarine–Marine
White shrimp	*Penaeus setiferus* (Linnaeus)	60	Estuarine–Marine
Pink shrimp	*Penaeus duorarum* Burkenroad	30	Estuarine–Marine
Marsh grass shrimp	*Palemontes vulgaris* (Say)	20	Estuarine
Atlantic coast			
Fishes			
Mummichog	*Fundulus heteroclitus* (Linnaeus)	100	Estuarine
Spot	*Leiostomus xanthurus* Lacepede	40	Estuarine–Marine
Pumpkinseed	*Lepomis gibbosus* (Linnaeus)	40	Freshwater
Banded killifish	*Fundulus diaphanus* (Lesueur)	40	Freshwater
Spotfin killifish	*Fundulus luciae* (Baird)	20	Estuarine
Inland silverside	*Menidia beryllina* (Cope)	20	Estuarine
Sheepshead minnow	*Cyprinodon variegatus* Lacepede	20	Estuarine
Striped killifish	*Fundulus majalis* (Walbaum)	20	Estuarine
Sailfin molly	*Poecilia latipinna* (Lesueur)	20	Freshwater
Bluegill	*Lepomis macrochirus* Rafinesque	20	Freshwater
Spotfin mojorra	*Eucinostomus argentus* Baird & Girard	20	Estuarine–Marine
Spottail shiner	*Notropis hudsonius* (Clinton)	20	Freshwater
Mosquitofish	*Gambusia holbrooki* (Girard)	20	Freshwater
Crustaceans			
Daggerblade grass shrimp	*Palaemonetes pugio* Holthius	80	Estuarine
Blue crab	*Callinectes sapidus* Rathbun	60	Estuarine–Marine
White shrimp	*Penaeus setiferus* (Linnaeus)	20	Estuarine–Marine

* In marsh–surface habitats, these fishes fit the definition of Estuarine–Marine species, but were previously classified as Marine species by Deegan and Thompson (1985)

young of the year; they usually spawn in nearshore or marine areas. Many of the species that belong to the estuarine-marine group support important coastal fisheries.

Marsh-surface assemblages of fishes on both coasts were dominated by estuarine species, except in tidal freshwater environments of the Atlantic coast, where freshwater species predominated (Table 13.2). Most species in these two affinity groups (estuarine and freshwater) are year-round residents of tidal marshes. Another estuarine species, the daggerblade grass shrimp, was the most abundant crustacean using the marsh surface. However, most other numerically dominant crustaceans associated with marsh-surface habitats (e.g. blue crab and penaeid shrimp) belong to the estuarine-marine group.

13.2 SMALLER SCALE CORRELATES OF PATTERNS OF COMPOSITION AND NUMERICAL ABUNDANCE

13.2.1 Subtidal geomorphology

Because nekton using the marsh surface must leave this habitat with the ebb tide, features of the subtidal environment are also relevant to understanding fish assemblages using marsh-surface habitat. Stream meanders and associated depositional and erosional banks are a universal feature of streams (Morisawa 1968; Garofalo 1980). Depositional and erosional banks resulting from stream sinuosity differ most obviously in depth (Figure 13.2), but also in slope, current velocity (Morisawa 1968), substrate particle size (Hynes 1970), concentration of detrital size fractions (Pickral and Odum 1977), and presence or absence of submerged aquatic vegetation (personal observation of the authors). Working in the tidal freshwater portion of the Chickahominy River, Virginia (tidal range 0.7 m, McIvor and Odum (1988) used flume nets to sample nekton at marsh-surface sites adjacent to shallow depositional and steeper erosional banks. They found that marsh-surface sites adjacent to depositional banks supported significantly higher numbers of nekton. Both lower encounter rates with piscivorous predators (assessed from tethering experiments) and higher food availability (assessed from feeding experiments) in the depositional subtidal zone are likely mechanisms producing this pattern. Analogous studies have not been conducted in more saline portions of estuaries, or in other regions where tidal ranges are smaller. McIvor and Odum (1988) postulated that because meanders and associated depositional and erosional banks are a universal geomorphological feature of streams (Morisawa 1968; Garofalo 1980), the results of their study were likely to apply to tidal creeks and associated estuarine marshes in general. However, the results may not hold in Gulf coast marshes where tidal ranges seldom exceed 0.30 m, where the differences between depositional and erosional banks are less pronounced, and where the marsh–water interfaces are more likely due to pond formation from subsidence

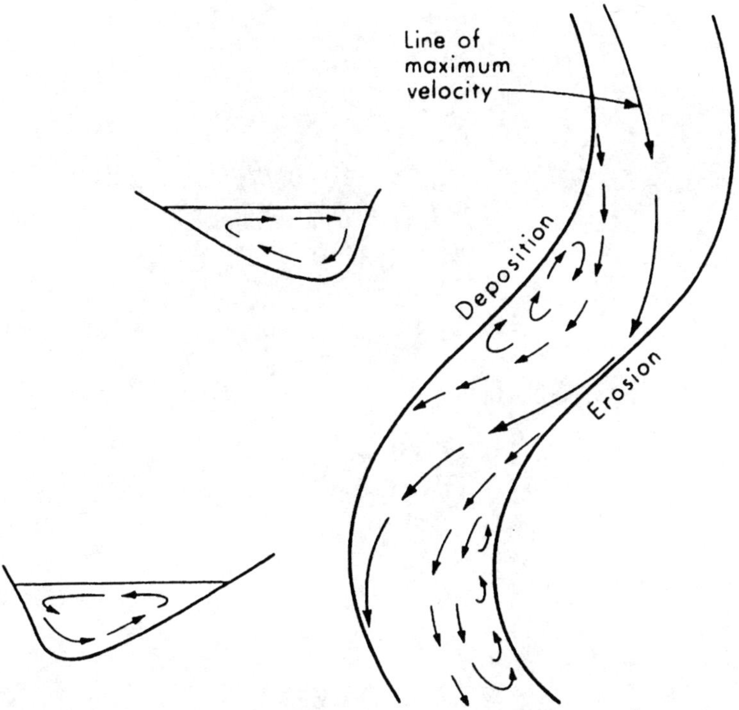

Figure 13.2 Meandering stream reach, showing depositional and erosional banks. Transverse sketches show lateral movement of water in the bends. (Reproduced by permission from Morisawa 1968)

(Sasser et al. 1986; Reed 1991) than from classic channel development. More research is needed to ascertain the generality of the relative importance of bank profile in influencing intertidal marsh use by nekton.

13.2.2 Stream order

Weinstein (1979), in an analysis of nursery areas for nekton, first proposed that 'marshes fill up backwards during recruitment'. This statement might be interpreted to imply that smaller order streams (*sensu* Horton 1945, Figure 3) have higher abundances of nekton. Rozas and Odum (1987a) tested this hypothesis in tidal freshwater marshes of the Chickahominy River, Virginia. Using flume nets, they sampled nekton on intertidal marshes adjacent to stream orders 2 (headwater), 3 (main creek), and 4 (river). Though crustacean numbers did not differ significantly between marshes, fish numbers were significantly greater on headwater and main creek marshes than on river marshes. These investigators hypothesized that the relationship between marsh stream order and fish abundance may be due in part to the distribution of submerged aquatic vegetation (SAV) in tidal marsh creeks. SAV

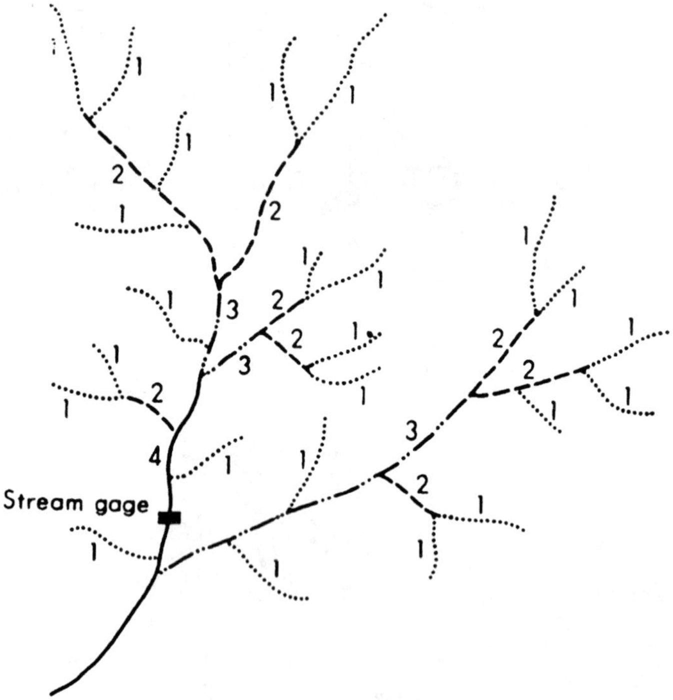

Figure 13.3 Hierarchy of stream ordering after Horton 1945. Numbers and pattern indicate order of respective segments. The watershed is fourth-order. (Reproduced by permission from Morisawa 1968)

decreased in abundance (as did fish abundance) with increasing stream order in these particular streams. The presence of SAV in tidal creeks may enhance the refuge value and food resources of those creeks, and thus, the habitat value of adjacent tidal marshes (Rozas and Odum 1987a).

Even smaller (and of lower stream order) than headwater creeks are intertidal rivulets, small creeks or channels that drain the marsh surface. Though tens of centimeters lower than the general marsh surface, rivulets begin to flood earlier on the rising tide and remain inundated longer than the remainder of the marsh surface on the ebb tide. Flumes placed along rivulets in a tidal freshwater marsh contained three times the number and 14 times the biomass of fishes as flumes placed on depositional creekbanks (Rozas et al. 1988). Rivulets may serve as preferred corridors between marsh surface and tidal channel habitats for natant organisms because of the benefits associated with a longer period of inundation, i.e. more extended access to food in periodically flooded habitats, and protection from predators requiring deeper waters for effective foraging.

One other investigator has documented patterns of nekton use of marsh surfaces that support the stream order hypothesis, and extend it from tidal freshwater to saline marshes. Hettler (1989) used block nets (essentially flume nets without side or back partitions) to compare nekton use of two polyhaline marsh sites in the Newport River estuary, North Carolina. Channel sites of order 3 were bordered by steeply sloping subtidal areas, whereas rivulet sites (order 1) were bordered by shallow depositional subtidal profiles. Significantly more individuals were found in rivulet sites, although more species were found in channel sites. During summer, the number of individuals from the rivulet habitat was an order of magnitude greater than the number collected from the channel marsh, due to the abundance of resident killifishes (Hettler 1989).

13.2.3 Submergence time (elevation)

Flooding duration or marsh submergence time obviously influences its use by nekton, because aquatic organisms can occupy the habitat only when it is flooded. (Exceptions include larval or postlarval forms that can survive in small water-filled depressions or crab burrows.) Submergence time is affected by several factors including marsh-surface elevation, tidal regime, and meteorological events. However, within the same marsh system, surface elevation has the most influence on flooding duration and habitat availability. Therefore, elevation may be an important factor controlling the distribution of animals among marsh-surface microhabitats (Minello et al. 1991; Rozas and Reed 1993). Rozas and Reed (1993) examined the influence of submergence time on habitat selection by nekton by comparing densities of natant organisms in three habitats having different elevations (high *Distichlis* marsh, medium *Spartina alterniflora* marsh, and low *S. alterniflora* marsh). Penaeid shrimp (white shrimp and brown shrimp) were most abundant in the low *Spartina* marsh, which flooded longer and deeper than the other habitats that were sampled. Minello and colleagues (1994) reported densities of daggerblade grass shrimp and brown shrimp 1.2–4.3 times higher on low than high *S. alterniflora* marsh, but found no effect of elevation on the abundance of white shrimp. Similarly, Kneib and Wagner (1994), working in a *Spartina alterniflora* marsh in Georgia, found apparent habitat preference for most species for sites of lower elevation that were closer to permanent tidal channels. An exception was the spotfin killifish *Fundulus luciae*, a high marsh specialist.

Extended periods of marsh submergence should benefit organisms that use the marsh surface, especially when food and cover are less abundant in subtidal habitats (Weisberg and Lotrich 1982; Minello and Zimmerman 1991). Kneib (1993) found that for larval mummichogs, growth was positively related and mortality rates negatively related to flooding duration in a Georgia saltmarsh. He attributed the higher growth rates and greater survival during longer flooding periods to increased prey availability, primarily benthic harpacticoid copepods. Mummichogs had largely unrestricted access to prey when the marsh surface was flooded, which was 16–32% of a given 24 h period at this location, depending on marsh elevation and tidal conditions.

Surface elevation may also indirectly affect habitat use by influencing prey densities and structural characteristics of the marsh surface. The distribution of benthic infauna in marshes is closely tied to surface elevation (Teal 1958; Cammen 1976; Fell et al. 1982; Kneib 1984; Bishop and Hackney 1987; Kneib 1992). Even subtle differences in marsh-surface topography may affect the abundance of infaunal prey and their availability to fish and crustacean predators that exploit the habitat. In addition, plant stem density is generally inversely related to flooding duration. Therefore, stem density in low marsh is relatively sparse (Mendelssohn and McKee 1988), a factor that may enhance the foraging opportunity for small natant predators (West and Williams 1986; Rozas and Reed 1993). Sparse vegetation may provide more foraging surface than unvegetated areas, yet may interfere less with the movement and foraging activity of predators than thick vegetation (Vince et al. 1976; Van Dolah 1978; West and Williams 1986).

Although high marsh floods infrequently and for shorter periods of time, it is exploited by some natant species. During a rising tide, killifishes follow the advancing edge of flooding water across the marsh surface and are concentrated in high marsh at high tide (Kneib 1976, 1984; Rozas and Reed 1993). If the risk of stranding can be overcome, exploiting the high marsh may have several advantages. High marsh environments may contain more food than low marsh because prey are exposed to fewer predators and for shorter periods of time (Kneib 1984; 1993). High marsh also has denser vegetation and shallower water than low marsh, factors that may offer greater protection from piscivorous predators. In addition, resident estuarine species that spawn on the marsh surface may select high marsh for suitable spawning sites (Greeley and MacGregor 1983).

High marsh pools differ from surrounding vegetated high marsh in providing permanently flooded microhabitat for resident fishes and crustaceans. Such pools are common features in New England marshes (Redfield 1972), where they act as habitat, primarily for resident fishes (several species of killifish, silversides, a goby, freshwater eels (*Anguilla rostrata*)) (Able et al. 1995, Chapter 14). Additionally, such pools serve as overwintering habitat for mummichogs (Smith and Able 1994). Mummichogs and blue crabs occupied an analogous pool in a North Carolina salt marsh (Kneib 1982).

13.2.4 Edge (drainage density)

The proximity of a marsh to subtidal habitat is important because most aquatic organisms using the marsh surface are confined to subtidal areas at low tide, and there seems to be a limit to the distance some organisms will travel into the marsh from open-water (Peterson and Turner 1994). Edge habitat, i.e. the interface between marsh and open-water, often originates and is configured in fundamentally different ways in Atlantic coast marshes (and other environments with moderate tidal ranges), and in Gulf coast marshes. In the former, edge is related primarily to drainage density or the complexity of the channel network of the system, itself a function of marsh age. Findings from saltmarshes around Sapelo Island, Georgia suggest that drainage density decreases as marshes become older (Frey and Basan 1978). Additionally, high drainage density habitats tend to be associated with headwaters, whereas areas of low drainage density usually occur near the mouths of tidal channels (Figure 13.4) (Kneib 1994). On the Gulf coast, especially in areas undergoing coastal submergence, much of the edge habitat, which occurs in the marsh interior away from channels, is created by pond formation (Sasser et al. 1986; Reed 1991).

Nekton are not evenly distributed over the marsh surface at high tide. Most species are concentrated in the emergent vegetation near the marsh–water interface (edge habitat) (Baltz et al. 1993; Peterson and Turner 1994). Species using interior marsh remote from subtidal habitats are mostly estuarine species belonging to two families, Cyprinodontidae and Palaemonidae (Peterson and Turner 1994). These two families include some of the most abundant species exploiting the marsh surface (e.g. daggerblade grass shrimp, gulf killifish, mummichog, and sheepshead minnow). Few species in the estuarine-marine affinity group seem to venture very far from the marsh edge. Striped mullet and blue crab may be the only species in this group that use interior marsh on the Gulf coast (Rozas and Reed 1993; Peterson and Turner 1994), whereas mullet, spot, white shrimp, and blue crab exploit such habitat on the Atlantic coast (Kneib 1991). Even so, many of the species that have been collected from interior marsh may be found in greater densities in edge habitats. Distance traveled and the timing of movement across the marsh surface by nekton may be dependent on organism size and swimming ability (Kneib and Wagner 1994). Relatively large animals and strong swimmers incur less risk of stranding by exploiting the marsh surface than small organisms and weak swimmers, and therefore may venture farther away from the edge into interior marsh. Relatively large, strong swimmers would also be expected to reach interior marsh habitats sooner during the tidal cycle (Kneib and Wagner 1994).

Peterson and Turner (1994) identified four patterns of marsh surface use based on the frequency of occurrence of nekton in portions of a Louisiana marsh differing in elevation, distance from open water, vegetation density, and proximity to marsh pools or potholes (Figure 13.5). Their data strongly suggest that nekton species differ substantially in their use of different portions of the marsh surface. This question of use of interior, often higher elevation habitat versus edge, generally low elevation habitat has not been fully addressed.

Because most species (and almost all fishery species) exploit marsh edge habitat, secondary productivity may be related to the amount of edge habitat contained in a marsh system (Zimmerman and Minello 1984; Zimmerman et al. 1991; Peterson and Turner 1994). Kneib (1994) used flume weirs to compare fish densities at a high and a low drainage density site along the Duplin River, Georgia. Over 19 months, he consistently found higher fish densities at the site with high drainage density. Minello and colleagues (1994) experimentally addressed the question of the importance of edge in determining standing stocks of nekton in a

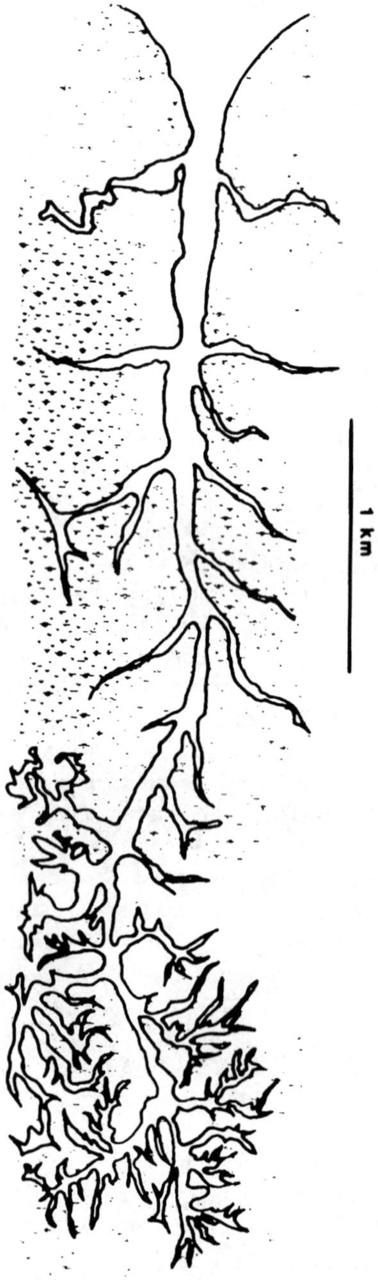

Figure 13.4 Geomorphology of drainage channels in a saltmarsh on Sapelo Island, GA. Note the higher density of drainages in the headwater portion. (Reproduced by permission from Kneib 1994)

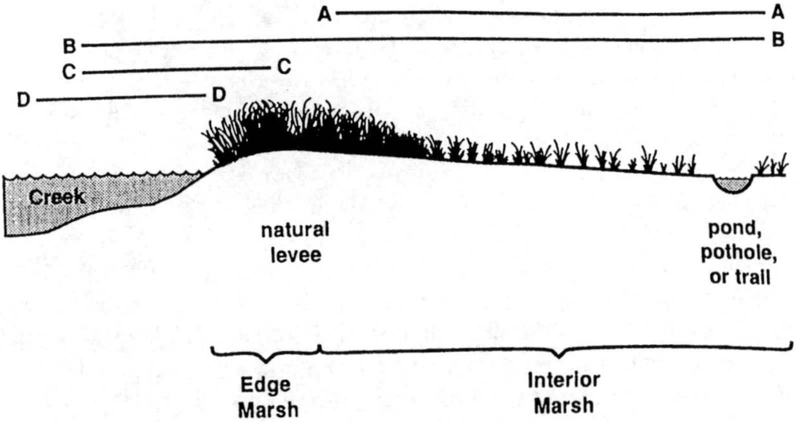

Figure 13.5 Patterns of nekton use of marsh habitats in a Louisiana marsh. A = interior marsh residents, species that use the marsh interior and retreat to potholes, muskrat trails, or small ponds at low tide; B = interior marsh users, species that use interior marshes but retreat to the creek edge at low tide; C = marsh edge users, species that use only marsh edge habitat; D = marsh subtidal group, open-water species not utilizing the marsh surface. (Reproduced by permission from Peterson and Turner 1994)

Texas marsh. They compared nekton densities at sites within a transplanted *Spartina alterniflora* marsh where channels were added, versus control sites without added channels. Adding channels (and increasing edge habitat) in the *Spartina* marsh dramatically increased densities of brown shrimp and white shrimp near the channels by a factor of 4.6 to 13 (Minello et al. 1994). Adding channels also significantly raised the densities of polychaete worms and daggerblade grass shrimp in the marsh edge. These animals are preyed on by small fishes, blue crabs, and brown shrimp (Harrington and Harrington 1961; Gleason and Wellington 1988; Minello et al. 1989; Thomas 1989; McTigue and Zimmerman 1991).

At a regional scale, densities of penaeid shrimp were much higher in marsh surface vegetation in Galveston Bay in the Gulf of Mexico (Zimmerman and Minello 1984) than in saltmarsh habitats in South Carolina (Wenner and Beatty 1993). Such differences are consistent with an explanation of greater edge in the reticulated marshes in Galveston Bay (Zimmerman and Minello 1984). However, other differences, including greater duration of marsh flooding (seasonal in Galveston Bay), are likely also involved (Wenner and Beatty 1993; Rozas 1995).

13.3 LANDSCAPE-LEVEL PROCESSES

13.3.1 Linkage of adjacent habitats

Ecologists and managers are increasingly interested in linkages between adjacent ecosystems or habitats. Linkages between intertidal marshes and adjacent subtidal habitats can be viewed in two fundamentally different ways: (1) the systems are coupled by the exchange of energy and nutrients, and (2) the systems are linked because many organisms make direct use of both as habitat.

The focus on habitat linkages derives from the predictable use by some nekton of two or more adjacent habitats, and the inference that both habitats contribute to the growth and survival of these organisms. This view is based on studies of nekton abundance in adjacent habitats under different tidal regimes. Some organisms common in permanent subtidal habitats at ebb tide are found primarily within intertidal habitats on flood tides. Such distributions are indicative of active movement between adjacent habitats as the intertidal habitats are alternately flooded and drained by tidal fluctuation. A wide variety of fishes and crustaceans exhibit this pattern of habitat use (see 13.1.2). Of these species, the mummichog and brown shrimp have been experimentally shown to benefit from use of intertidal habitat. For instance, mummichogs caged in permanent subtidal waters grew more slowly than others allowed access to the intermittently flooded marsh surface (Weisberg and Lotrich 1982). In a similar experiment, brown shrimp held in cages containing *S. alterniflora* had significantly higher growth rates than those in cages devoid of vegetation (Minello and Zimmerman 1991).

It can be concluded that despite only being available to nekton for a few hours once or twice a day (or sometimes seasonally), the vegetated intertidal marsh surface has high habitat 'value' for a wide variety of species of nekton. Whereas some of the killifishes spawn here (Taylor et al. 1979; Able 1984), most organisms probably gain food or shelter, or both, from this highly structured, relatively shallow environment. Gulf killifish (Rozas and LaSalle 1990), mummichogs (Butner and Brattstrom 1960; Rozas et al. 1988), freshwater killifish (Rozas et al. 1988), and spot (Miller and Dunn 1980) contain more food when leaving than when entering the marsh surface. Gulf killifish feed on a wide range of benthic invertebrates and detritus while on the marsh surface (Harrington and Harrington 1961; Rozas and LaSalle 1990). Blue crabs in a Georgia marsh feed intertidally on fish and non-portunid crabs (Fitz and Wiegert 1991). Additional indirect evidence of the probable use of flooded intertidal habitats is provided by a comparison of the state of the tidal cycle in which marsh nekton contain the most food in their gut. Both juvenile silver perch (*Bairdiella chrysura*) and blue crabs contain fuller stomachs during or immediately following peak tidal inundation than individuals of those species taken at other times in the tidal cycle (Kleypas and Dean 1983; Ryer 1987).

Quantitative data on the refuge value of intertidal saltmarshes compared with that of adjacent habitats are largely lacking. However, Minello (1993) determined experimentally that tethered brown shrimp exhibited higher survival rates in saltmarsh (and seagrass habitats) than on unvegetated sand bottom in a Texas bay. Most of our hypotheses about the refuge value of saltmarsh habitat are presently extrapolated from assessments of refuge value derived from other habitat comparisons, e.g. submersed aquatic vegetation versus bare substrate (Zimmerman and Minello 1984; Rozas and Odum 1987b), or from laboratory experiments (e.g. Minello and Zimmerman 1983).

13.3.2 Nekton as agents of export from the intertidal marsh

Coupling of adjacent habitats by the exchange of materials may be controlled strictly by physical processes, may be mediated by the movement of biological organisms, or a combination of both. Physical linkages are far better documented. The best understood examples include the one-way downslope movement of dissolved and particulate matter mediated by physical factors including runoff and groundwater flow (e.g. Fisher and Likens 1973; Valiela et al. 1992). Bidirectional movement of inert materials, nutrients and carbon, on and off the marsh surface by the tides and exchange with estuarine waters are also reasonably well known, though spatially and temporally complex (e.g. Odum et al. 1979; Nixon 1980).

Mobile organisms can also be nutrient and carbon vectors. However, the best estimates of movement of nutrients and organic matter by fishes come from other ecosystems. Bray and colleagues (1981) showed that planktivorous damselfish (*Chromis punctipinnis*) imported $8\,g\,C\,m^{-2}\,a^{-1}$ from the pelagic zone to a California rocky reef community. Meyer and Schultz (1985) demonstrated that French and white grunts (*Haemulon flavolineatum, H. plumieri*) imported an average of 76 g POC (particulate organic carbon) $m^{-2}\,a^{-1}$ from seagrass beds to a coral reef community. Analogous quantitative data on nekton as vectors of nutrient movement between subtidal and intertidal marsh habitats are lacking, though often hypothesized or inferred (McIvor and Odum 1988; Rozas and LaSalle 1990; Fitz and Wiegert 1991; Kneib and Wagner 1994). At the estuarine scale, Deegan (1993) estimated that gulf menhaden (*Brevoortia patronus*) export $22.5\,g\,C\,m^{-2}\,a^{-1}$ from Fourleague Bay, LA to the Gulf of Mexico. This is roughly 5–10% of the primary production of this estuarine area.

13.4 SUMMARY

Geomorphological and hydrological features of saltmarshes (subtidal profile, stream order, flooding duration, elevation, drainage density and amount of edge) directly affect both the degree and patterns of use of marsh surface habitat by nekton. Whereas we understand the broad general patterns of each of these factors alone in specific locations, we have a poor understanding of how the factors interact to affect marsh surface use and microhabitat selection. Most importantly, we know little of how the patterns and interactions (and rankings) of factors differ regionally. Further, our understanding of the linkages between intertidal and subtidal habitats is in its infancy. Future research should address how these linkages affect the population dynamics (recruitment success, growth, mortality) of the most abundant resident and transient species. More studies of direct marsh use are needed, especially on the Pacific and Atlantic coasts, and future investigatior should include regional comparisons of similar microhabitats using identic quantitative sampling methods. Research is also needed on the relative importanc of nekton as vectors of carbon and nutrient exchange between adjacent intertid and subtidal habitats.

ACKNOWLEDGEMENTS

We wish to acknowledge the support of the National Biological Service and th National Marine Fisheries Service during preparation of this document. We than D. Baltz for providing unpublished data for inclusion in Table 13.1. R. Kneib, T Minello, W. Richards, C. Roman and P. Sheridan provided valuable comments o earlier versions of the manuscript. We dedicate this chapter to the memory of W.E Odum who supported and encouraged our research in tidal marshes.

REFERENCES

Able, K.W. 1984. Variation in spawning site selection of the mummichog, *Fun heteroclitus*. *Copeia* 1984: 522–525.

Able, K.W., D.A. Witting, R.S. McBride, R.A. Roundtree and K.J. Smith. 1996. Fishes o polyhaline estuarine shores in Great Bay–Little Egg Harbor, New Jersey: A case study o seasonal and habitat influences. *In*: Nordstrom, K.F. and C.T. ... *Estuarin Shores*. John Wiley, Chichester.

Baltz, D.M., C. Rakocinski and J.W. Fleeger. 1993. Microhabitat use by marsh- a Louisiana estuary. *Environmental Biology of Fishes* 36: 109–126.

Bishop, T.D. and C.T. Hackney. 1987. A comparative study of the mollu two oligohaline intertidal marshes: spatial and temporal distributions of th and biomass. *Estuaries* 10: 141–152.

Bray, R.N., A.C. Miller and G.G. Geesey. 1981. The fish connection: A trophic link bet planktonic and rocky reef communities? *Science* 214: 204–205.

Butner, A. and B.H. Brattstrom. 1960. Local movement in *Menidia* and *Fundulus*. *Copei* 1960: 139–141.

Cammen, L.M. 1976. Macroinvertebrate colonization of *Spartina* marshes artifi established on dredge spoil. *Estuarine, Coastal and Marine Science* 4: 357–372.

Chamberlain, R.H. and R.A. Barnhart. 1994. Early use by fish of a mitigation sal Humbolt Bay, California. *Estuaries* 16: 769–783.

Dardeau, M.R., R.F. Modlin, W.W. Schroeder and J.P. Stout. 1992. Estuaries, *In*: Hackney, C.T., S.M. Adams and W.H. Martin (eds.), *Biodiversity of the Sou United States: Aquatic Communities*. John Wiley, New York.

Deegan, L.A. 1993. Nutrient and energy transport between estuaries and coastal marine eco- systems by fish migration. *Canadian Journal of Fisheries and Aquatic Sciences* 50: 74–79

Deegan, L.A., and B.A. Thompson. 1985. The ecology of fish communities in the Mississippi River deltaic plain, p. 35–56. *In* Yanez-Arancibia, A. (ed.), *Fish Communit Ecology in Estuaries and Coastal Lagoons: Towards an Ecosystem Integration*. DR (R UNAM Press, Mexico.

Fell, P.E., N.C. Olmstead, E. Carlson, W. Jacob, D. Hitchcock and G. Silber. 1982. Distribution and abundance of macroinvertebrates on certain Connecticut tidal marshes, with emphasis on dominant molluscs. *Estuaries* 5: 234–239.

Fisher, S.G. and G.E. Likens. 1973. Energy flow in Bear Brook, New Hampshire: An integrative approach to stream ecosystem metabolism. *Ecological Monographs* 43: 421–439.

Fitz, H.C. and R.G. Wiegert. 1991. Utilization of the intertidal zone of a salt marsh by the blue crab *Callinectes sapidus*: density, return frequency, and feeding habits. *Marine Ecology Progress Series* 76: 249–260.

Frey, R.W. and P.B. Basan. 1978. Coastal salt marshes, p. 101–169. *In* Davis, R.A. Jr. (ed.), *Coastal Sedimentary Environments*. Springer-Verlag, New York.

Garofalo, D. 1980. The influence of wetland vegetation on tidal stream channel migration and morphology. *Estuaries* 3: 258–270.

Gleason, D.F. and G.M. Wellington. 1988. Food resources of postlarval brown shrimp (*Penaeus aztecus*) in a Texas salt marsh. *Marine Biology* 97: 329–337.

Gosselink, J.G. 1984. The ecology of delta marshes of coastal Louisiana: a community profile. US Fish & Wildlife Service. FWS/OBS-84/09.

Greeley, M.S. Jr. and R. MacGregor III. 1983. Annual and semilunar reproductive cycles of the Gulf killifish, *Fundulus grandis*, on the Alabama Gulf coast. *Copeia* 1983: 711–718.

Harrington, R.W. Jr. and E.S. Harrington. 1961. Food selection among fishes invading a high subtropical salt marsh: from onset of flooding through the progress of a mosquito brood. *Ecology* 42: 646–666.

Hettler, W.F. 1989. Nekton use of regularly-flooded saltmarsh cordgrass habitat in North Carolina, USA. *Marine Ecology Progress Series* 56: 111–118.

Horton, R.E. 1945. Erosional development of streams and their drainage basins: hydrophysical approach to quantitative morphology. *Bulletin of the Geological Society of America* 56: 275–370.

Hynes, H.B.[illegible] *The Ecology of Running Waters*. Liverpool University Press, Liverpool.

[illegible] [illegible]uarine geomorphology and physical oceanography, Chapter 2, p. 47–78. [illegible] J.W. Jr., C.A.S. Hall, W.M. Kemp and A. Yanez-Arancibia (eds). *Estuarine* [illegible]. J. Wiley and Sons, New York.

[illegible] J.M. Dean. 1983. Migration and feeding of the predatory fish, *Bairdiella* [illegible] Lacepede, in an intertidal creek. *Journal of Experimental Biology and Ecology* [illegible]2: 199–209.

Kneib, R.T. 1976. Feeding, reproduction, growth and movements of killifishes on a North Carolina salt marsh. M.A. Thesis, University of North Carolina, Chapel Hill.

Kneib, R.T. 1982. The effects of predation by wading birds (Ardeidae) and blue crabs (*Callinectes sapidus*) on the population size structure of the common mummichog, (*Fundulus heteroclitus*). *Estuarine, Coastal and Shelf Science* 14: 159–165.

Kneib, R.T. 1984. Patterns of invertebrate distribution and abundance in the intertidal salt marsh: causes and questions. *Estuaries* 7: 392–412.

[illegible] R.T. 1991. Flume weir for quantitative collection of nekton from vegetated intertidal [illegible]. *Marine Ecology Progress Series* 75: 29–38.

[illegible], R.T. 1992. Population dynamics of the tanaid *Hargeria rapax* (Crustacea:Peracarida) in a tidal marsh. *Marine Biology* 113: 437–445.

Kneib, R.T. 1993. Growth and mortality in successive cohorts of fish larvae within an estuarine nursery. *Marine Ecology Progress Series* 94: 115–127.

Kneib, R.T. 1994. Spatial pattern, spatial scale, and feeding in fishes, p. 171–185. *In* Fresh, K., D. Stouder and R. Feller (eds), *Theory and Application in Fish Feeding Ecology*. University of South Carolina Press, Columbia.

Kneib, R.T. and S.L. Wagner. 1994. Nekton use of vegetated marsh habitats at differen stages of tidal inundation. *Marine Ecology Progress Series* 106: 227–238.

McHugh, J.L. 1967. Estuarine nekton, p. 581–620. *In* Lauff, G.H. (ed.), *Special Publicatio 83*, American Association for the Advancement of Science, Washington, DC.

McIvor, C.C. and W.E. Odum. 1988. Food, predation risk, and microhabitat selection in marsh fish assemblage. *Ecology* 69: 1341–1351.

McTigue, T.A. and R.J. Zimmerman. 1991. Carnivory versus herbivory in juvenile *Penaeu setiferus* (Linnaeus) and *Penaeus aztecus* (Ives). *Journal of Experimental Marine Biolog and Ecology* 151: 1–16.

Mendelssohn, I.A. and K.L. McKee. 1988. *Spartina alterniflora* die-back in Louisiana: tim course investigations of soil waterlogging effects. *Journal of Ecology* 76: 509–521.

Meyer, J.L. and E.T. Schultz. 1985. Migrating haemulid fishes as a source of nutrients an organic matter on coral reefs. *Limnology and Oceanography* 30: 146–156.

Miller, J.M. and M.L. Dunn. 1980. Feeding strategies and patterns of movement in juvenil estuarine fishes, p. 437–448. *In* Kennedy, V.S. (ed.), *Estuarine Perspectives*, Academi Press, New York.

Minello, T.J. 1993. Chronographic tethering: a technique for measuring prey survival tim and testing predation pressure in aquatic habitats. *Marine Ecology Progress Series* 101 99–104.

Minello, T.J. and R.J. Zimmerman. 1983. Fish predation on juvenile brown shrimp, *Penaeu aztecus* Ives: the effect of simulated *Spartina* structure on predation rates. *Journal o Experimental Marine Biology and Ecology* 72: 211–231.

Minello, T.J. and R.J. Zimmerman. 1991. The role of estuarine habitats in regulating growt and survival of juvenile penaeid shrimp, p. 1–16. *In* DeLoach, P., W.J. Dougherty an M.A. Davidson (eds.), *Frontiers of Shrimp Research*. Elsevier Science Publishers Amsterdam.

Minello, T.J., R.J. Zimmerman and E.X. Martinez. 1989. Mortality of young brown shrim *Penaeus aztecus* in estuarine nurseries. *Transactions of the American Fisheries Societ* 118: 693–708.

Minello, T.J., R.J. Zimmerman, R.B. Wooten, J.L. Martinez, T.J. Baumer and M.C. Pattillo 1991. Habitat availability and utilization by benthos and nekton in Hall's Lake and Wes Galveston Bay. NOAA Technical Memorandum, NMFS-SEFC-275.

Minello, T.J., R.J. Zimmerman and R. Medina. 1994. The importance of edge for natan macrofauna in a created salt marsh. *Wetlands* 14: 184–198.

Morisawa, M. 1968. *Streams: Their Dynamics and Morphology*. McGraw-Hill, New York

Murphy, S.C. 1991. The ecology of estuarine fishes in southern Maine high salt marshes access corridors and movement patterns. M.S. Thesis, University of Massachusetts Amherst.

Nixon, S.W. 1980. Between coastal marshes and coastal waters – a review of twenty years o speculation and research on the role of salt marshes in estuarine productivity and wate chemistry, p. 438–525. *In* Hamilton, P. and K.B. McDonald (eds), *Estuarine and Wetlan Processes*, Plenum, New York.

Odum, W.E., J.S. Fisher, and J.C. Pickral. 1979. Factors controlling the flux of particulat organic carbon from estuarine wetlands, p. 69–80. *In* Livingston, R.J. (ed), *Ecologica Processes in Coastal and Marine Systems*. Plenum Press, New York.

Peterson, G.W. and R.E. Turner. 1994. The value of salt marsh edge vs interior as a habita for fish and decapod crustaceans in a Louisiana tidal marsh. *Estuaries* 17: 235–262.

Pickral, J.C. and W.E. Odum. 1977. Benthic detritus in a salt marsh tidal creek, p. 280–292 *In* Wiley, M.L. (ed.), *Estuarine Processes*, Vol. II. Academic Press, New York.

Redfield, A.C. 1972. Development of a New England salt marsh. *Ecological Monograph* 42: 201–237.

Reed, D.J. 1991. Ponds and bays: natural processes of coastal marsh erosion in the Mississippi deltaic plain, Louisiana, USA. *Z. Geomorph. N. F. Suppl.-Bd.* 81: 41–51.

Rozas, L.P. 1992. Comparison of nekton habitats associated with pipeline canals and natural channels in Louisiana salt marshes. *Wetlands* 12: 136–146.

Rozas, L.P. 1993. Nekton use of salt marshes of the Southeast region of the United States, p. 528–537. *In* Magoon, O., W.S. Wilson, H. Converse and L.T. Tobin (eds.), *Coastal Zone '93, Volume 2.* Proceedings of the 8th Symposium on Coastal and Ocean Management. American Society of Civil Engineers, New York.

Rozas, L.P. 1995. Hydroperiod and its influence on nekton use of the salt marsh: a pulsing ecosystem. *Estuaries* 18: 579–590.

Rozas, L.P. and W.E. Odum. 1987a. Use of tidal freshwater marshes by fishes and macrofaunal crustaceans along a marsh stream-order gradient. *Estuaries* 10: 36–43.

Rozas, L.P. and W.E. Odum. 1987b. Fish and crustacean use of submerged plant beds in tidal freshwater marsh creeks. *Marine Ecology Progress Series* 38: 101–108.

Rozas, L.P. and M.W. LaSalle. 1990. A comparison of the diets of Gulf killifish, *Fundulus grandis* Baird and Girard, entering and leaving a Mississippi brackish marsh. *Estuaries* 13: 332–336.

Rozas, L.P. and D.J. Reed. 1993. Nekton use of marsh-surface habitats in Louisiana (USA) deltaic salt marshes undergoing submergence. *Marine Ecology Progress Series* 96: 147–157.

Rozas. L.P., C.C. McIvor and W.E. Odum. 1988. Intertidal rivulets and creekbanks: corridors between tidal creeks and marshes. *Marine Ecology Progress Series* 47: 303–307.

Ryer, C.H. 1987. Temporal patterns of feeding by blue crabs (*Callinectes sapidus*) in a tidal-marsh creek and adjacent seagrass meadow in the lower Chesapeake Bay. *Estuaries* 10: 136–140.

Sasser, C.E., M.D. Dozier, J.G. Gosselink and J.M. Hill. 1986. Spatial and temporal changes in Louisiana's Barataria Basin marshes, 1945–1980. *Environmental Management* 10: 671–680.

Smith, K.J. and K.W. Able. 1994. Salt-marsh pools as winter refuges for the mummichog, *Fundulus heteroclitus*, in New Jersey. *Estuaries* 17: 226–234.

Taylor, M.H., G.J. Leach, L. McMichele, W.M. Levitan and W.F. Jacob. 1979. Lunar spawning cycle in the mummichog, *Fundulus heteroclitus* (Pisces: Cyprinodontidae). *Copeia* 1979: 291–297.

Teal, J.M. 1958. Distribution of fiddler crabs in Georgia salt marshes. *Ecology 39: 185–193.*

Thomas, J.L. 1989. A comparative evaluation of *Halodule wrightii* Aschers, *Spartina alterniflora* Loisel and bare sand as nursery habitats for juvenile *Callinectes sapidus* (Rathbun). M.S. Thesis, Texas A & M University, College Station.

Valiela, I., K. Foreman, M. LaMontagne, D. Hersh, J. Costa, P. Peckol, B. DeMeo-Andreson, C. D'Avanzo, M. Babione, C. Sham, J. Brawley and K. Lajtha. 1992. Couplings of watersheds and coastal waters: sources and consequences of nutrient enrichment in Waquoit Bay, Massachusetts. *Estuaries* 15: 443–457.

Van Dolah, R.F. 1978. Factors regulating the distribution and population dynamics of the amphipod (*Gammarus palustris*) in an intertidal salt marsh community. *Ecological Monographs* 48: 191–217.

Vince, S., I. Valiela, N. Backus and J.M. Teal. 1976. Predation by the salt marsh killifish *Fundulus heteroclitus* (L.) in relation to prey size and habitat structure: consequences for prey distribution and abundance. *Journal of Experimental Marine Biology and Ecology* 23: 255–266.

Webb, J.W. Jr. 1982. Salt marshes of the western Gulf of Mexico, p. 89–109. *In* Lewis, R.R. (ed.), *Creation and Restoration of Coastal Plant Communities*, CRC Press, Boca Raton, FL.

Weinstein, M.P. 1979. Shallow marsh habitats as primary nurseries for fishes and shellfish Cape Fear River, North Carolina. *Fishery Bulletin* 77: 339–357.

Weisberg, S.B. and V.A. Lotrich. 1982. The importance of an infrequently flooded intertida marsh surface as an energy source for the mummichog *Fundulus heteroclitus*: a experimental approach. *Marine Biology* 66: 307–310.

Wenner, E.L. and H.R. Beatty. 1993. Utilization of shallow estuarine habitats in Sout Carolina, U.S.A., by postlarval and juvenile stages of *Penaeus* spp. (Decapoda Penaeidae). *Journal of Crustacean Biology* 13: 280–295.

West, D.L. and A.H. Williams. 1986. Predation by *Callinectes sapidus* (Rathbun) withi *Spartina alterniflora* (Loisel) marshes. *Journal of Experimental Marine Biology an Ecology* 100: 75–95.

Wiegert, R.G. and B.F. Freeman. 1990. Tidal salt marshes of the Southeast Atlantic coast: A community profile. Biological Report 85(7.29), US Fish & Wildlife Service, Washington DC.

Zimmerman, R.J. and T.J. Minello. 1984. Densities of *Penaeus aztecus, Penaeus setiferu* and other natant macrofauna in a Texas salt marsh. *Estuaries* 7: 421–433.

Zimmerman, R.J., T.J. Minello, D.L. Smith and J. Kostera. 1990a. The use of *Juncus* an *Spartina* marshes by fisheries species in Lavaca Bay, Texas, with reference to effects o floods. NOAA Tech. Memo. NMFS-SEFC-251.

Zimmerman, R.J., T.J. Minello, M.C. Castiglione and D.L. Smith. 1990b. Utilization o marsh and associated habitats along a salinity gradient in Galveston Bay. NOAA Tech Memo. NMFS-SEFC-250.

Zimmerman, R.J., T.J. Minello, E.F. Klima and J.M. Nance. 1991. Effects of accelerated sea level rise on coastal secondary production, p. 110–124. *In* Bolten, H.S. and O.T. Magoo (eds.), *Coastal Zone '91, Coastal Wetlands*. American Society of Civil Engineers, Ne York.

Chapter Fourteen

Fishes of polyhaline estuarine shores in Great Bay–Little Egg Harbor, New Jersey: a case study of seasonal and habitat influences

KENNETH W. ABLE and DAVID A. WITTING
Rutgers University
RICHARD S. McBRIDE
Florida Marine Research Institute
RODNEY A. ROUNTREE
National Marine Fisheries Service

and

KELLY J. SMITH
Rutgers University

14.1 INTRODUCTION

In recent years ecological research has recognized the role of boundaries or ecotones in ecological processes both within and between adjacent ecosystems (di Castri et al. 1988). As part of this re-examination, it has been suggested that the spatial organization of the land–sea margin has an important effect on the linkages between the terrestrial and aquatic components of the margin (Vernberg 1993). This may be especially true of shallow estuarine shores, and even more so for saltmarshes (McIvor and Rozas, this volume Chapter 13), because they are extremely productive systems that provide a variety of aesthetic and economic functions (Nixon 1980; Daiber 1982; Seneca and Broome 1992; Vernberg 1993).

Estuarine Shores: Evolution, Environments and Human Alterations. Edited by K. F. Nordstrom and C. T. Roman.

At the same time estuarine shores are heavily impacted by current and planne development (Spinner 1969; Kennish 1992) and may undergo large-scale chang if the potential impacts of accelerated sea level rise (Varekamp et al. 1992) a realized (Kennedy 1990; Zimmerman et al. 1991). Among the ecologically an economically important components of the land–sea margin fauna are the fish (Haedrich 1983; Francis 1990; Frank et al. 1990). Despite their importance we hav an inadequate understanding of the specific habitat types used by fishes (Hoss an Thayer 1993), including the spatial and temporal scales of the faunal assembla structure. As a result, it is often difficult to determine the functional significance these estuarine shore habitats and how natural and man-made changes wi influence them.

Estuarine shores, at least in relatively unaltered systems on the east coast of th US, are composed of an array of habitat types that vary along vertical an horizontal gradients with tidal flooding and other associated physical and flor factors. Most previous studies of estuarine shore fishes in the Mid-Atlantic Bigh have focused primarily on one level or portion of the tidal range with a single gea type. As a result, fish assemblages in this region have not been examined along th entire gradient of estuarine shore habitats from intertidal to subtidal.

In Great Bay–Little Egg Harbor in southern New Jersey, an intensive samplin effort over the past eight years has provided a general understanding of season and annual variation of the fish fauna in the deeper waters of the estuar (Szedlmayer and Able 1996) including some shore habitats (Sogard and Able 199 Rountree and Able 1992a, 1993; Smith 1995), and several autecological studie (see Discussion). Here we synthesize the information on the seasonal availability o fishes and their habitat use patterns in an attempt to gain insights into how fish lif histories, movements, and trophic interactions link estuarine shores to the rest o the estuary.

14.2 METHODS

The Great Bay–Little Egg Harbor estuarine systems are polyhaline, shallo (average depth at mean low water 1.7 m (Durand 1984)) embayments that shar many characteristics with other estuaries in the Mid-Atlantic Bight, including broad seasonal temperature range (−2°C to 28°C), moderate tidal rang (approximately 1 m) (Able et al. 1992) and a relatively diverse fauna of reside and migratory fish species that are enriched by northern and southern componen (Able 1992). Almost the entire margin (approximately 283 km of estuarin shoreline; R. Lathrop, personal communication) of Great Bay–Little Egg Harbo consists of extensive marshes. The marsh vegetation is typical of the northeaster US (Chapman 1960; Nixon 1982) and is dominated by the infrequently floode 'short form' of *Spartina alterniflora*.

This estuarine system differs from many others in the Mid-Atlantic Bight in tha

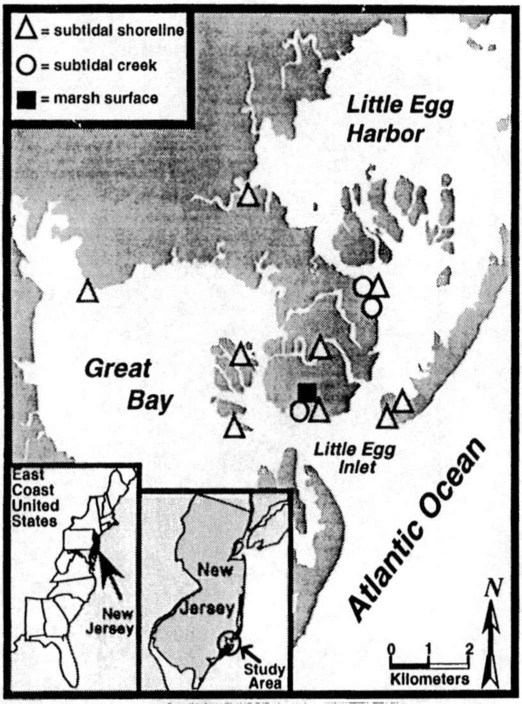

Figure 14.1 Great Bay–Little Egg Harbor study area in southern New Jersey with location of sampling sites for different shore habitats

it is relatively unaltered. Most of the upper portion of the drainage basin (Figure 14.1) is in a relatively unimpacted state and part of the Pinelands National Reserve (Good and Good 1984). The extensive marshes in the estuary have few alterations with the exception of mosquito ditches, although unditched marshes are extensive as well. For all of these reasons the water quality is exceptional. This system is the site of long-term environmental measurements (Durand and Nadeau 1972; Durand 1988; Able et al. 1992), and is in the process of being approved for inclusion in the National Estuarine Research Reserve program in recognition of its nearly pristine status (Psuty et al. 1993).

14.2.2 Habitat evaluation methods

The synthesis of information on the fish fauna of the Great Bay–Little Egg Harbor estuarine shore habitats (Figures 14.1 and 14.2) incorporates published and unpublished data from a variety of sources (Table 14.1). The locations of the intertidal and shallow subtidal sampling sites for this study are depicted in Figure 14.1 and the range of habitat types are characterized in Table 14.2. The faunal data

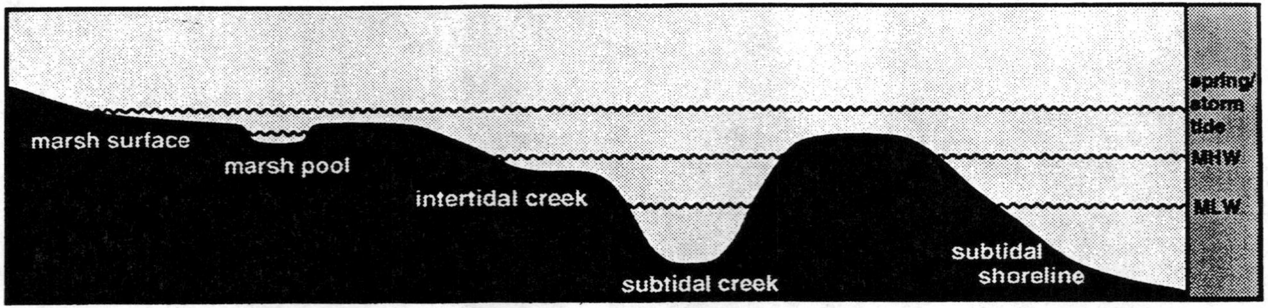

Figure 14.2 Diagrammatic illustration of representative estuarine shore habitat types relative to tidal range

are the basis for the characterization of each species with regard to mode of estuarine utilization and life history stage represented in these habitats (Table 14.3). The sampling gear, of necessity, varied with habitat type (Table 14.1). We consider each habitat as the composite of the structural (vegetation, sediment type, etc.) and environmental (temperature, salinity, etc.) components, in a manner similar to that of Peters and Cross (1992).

14.2.2.1 Intertidal habitats

We examined three major intertidal habitat types: surface of the marsh, permanent pools on the surface, and intertidal creeks (Figure 14.2). The marsh pools always contain water, while the marsh surface is only flooded on spring or storm tides. The marsh surface (Table 14.1) was sampled with nine 1 m^2 throw traps set at three distances from an intertidal creek (2, 16 and 32 m) with three replicates at each site. Collections occurred on spring high tides at dusk as soon as the marsh surface was covered with water. The marsh pools (Table 14.1) were sampled with a 0.25 m^2 aluminum throw trap. After deployment from the edge of a marsh pool the contents were removed with a bar seine (1.5 mm mesh), identified, measured and released into the same pool. The three replicates for each pool were staggered over the course of a day in order for the fish to redistribute within a pool before it was resampled. Additional details of this sampling approach are available in Smith (1995). Intertidal creeks (Figure 14.2) were sampled with a weir (1.8 m × 1.2 m high, with 4.5 m × 1.2 m wings) of either 3.2 or 6.4 mm mesh. Each weir was set across the mouth of two intertidal creeks at high tide and retrieved when the creek had drained completely at low tide (Rountree et al. 1992). These intertidal creeks were connected to a larger subtidal creek that was also sampled as part of this effort.

Table 14.1 Summary of habitat–specific data sources for daytime collections of estuarine shore fishes in the Great Bay – Little Egg Harbor estuary. Sampling sites are indicated on Figure 14.1

Habitats	Sampling gear	Sampling period	Sample collections	Sample locations	Number of fish collected
Marsh surface	Throw trap (1 m^2)	May–Sept 1991	54	1	245
Marsh surface pools	Throwtrap (0.25 m^2)	June–Oct 1990	147	30	6610
Intertidal creek	Weir	May–Oct 1989	14	2	9033
Subtidal creek	Weir and seine (6.4 mm mesh)	July–Oct 1988	10	2	15 587
Subtidal shoreline	Seine (6.1 m, 4 mm mesh)	May–Nov 1990	310	9	43 387

Table 14.2 Characterization of the estuarine shore habitats sampled during July to September

Habitat	Depth (MLW, m)	Flooding frequency (number/year)	Flood duration (hours/year)	Substrate	Vegetation	Temperature (range °C)	Salinity (range ‰)
Marsh surface	0	33	85	Peat	*Spartina alterniflora*	21–23	28–31
Marsh surface pools	0.5–1.0	33	Continuous	Detritus and peat	*Ruppia maritima* and filamentous algae	19–31	18–36
Intertidal creek	0	520	1177	Mud with peat	*Ulva lactuca*, *Spartina alterniflora*	24–32	21–30
Subtidal creek	0.3–1.2	Continuous	Continuous	Mud with shell	*Ulva lactuca*	19–28	25–33
Subtidal shore	0.5–1.0	Continuous	Continuous	Poorly to well sorted sand	Predominantly *Ulva lactuca*, *Enteromorpha* spp, some *Fucus* and *Gracilaria* (0–1.7 l/m^2)	22–30	25–31

Table 14.3 Characteristics of estuarine shore fishes by habitat and life history mode based on concurrent sampling during the summer (July–September). Each life history mode includes the number of species collected followed by the percentage of the total within that habitat

Habitat	Species richness	Resident species	Migratory species: Nursery	Migratory species: Seasonal residents	Strays
Marsh surface	5	5 (100)	0 (0)	0 (0)	0 (0)
Marsh surface pools	6	5 (83)	0 (0)	1 (17)	0 (0)
Intertidal creek	15	8 (53)	4 (27)	2 (13)	1 (7)
Subtidal creek	31	8 (26)	13 (42)	6 (19)	4 (13)
Subtidal shoreline	44	8 (18)	17 (39)	10 (23)	9 (20)
All habitats	58	12	22	11	13

14.2.2.2 Subtidal habitats

Two shallow subtidal habitats, marsh creeks and subtidal shores, the latter of which included open bay shores as well as large marsh channels (Figure 14.2), were sampled with large weirs and seines. The weir was set to block the mouth of a creek at high tide and fish were captured by retrieving the net at low tide. The weir and wings were 6.4 mm mesh. The seine (18.3 m long by 1.2 m high with 6.4 mm mesh) was set above the weir and retrieved while using the weir as a block net. Rountree and Able (1992a) give further details of both sampling techniques. Subtidal shorelines (Figure 14.2) were sampled with a bag seine (6.1 m with 4 mm mesh equipped with a tickler chain) in three standardized hauls that each covered 150 m^2. Sampling was synchronized to spring low tides. For all of the above sampling, all fish, or a representative subsample, were identified, counted and measured live.

Environmental variables (Table 14.2) measured at the time of sampling included depth, temperature (hand-held thermometer) and salinity (refractometer). Values for flooding frequency and flood duration are based on Halupa and Howes (1995) for the same marsh system. The amount of submerged aquatic vegetation at subtidal shore stations was measured volumetrically.

14.3 RESULTS AND DISCUSSION

14.3.1 Seasonal dynamics

All of the sampling sites (Figure 14.1) were in the high-salinity portion of the estuary, and salinity varied relatively little across them (Table 14.2). The temperatures were strongly seasonal (Figure 14.3) with a broad range typical of Mid-Atlantic Bight estuaries. Both water depth and frequency of flooding varied considerably with elevation of each habitat type (Table 14.2). The marsh surface was infrequently flooded, i.e. only during regular spring tides or irregularly during storms. The intertidal creeks were flooded during the twice-daily high tides and thus the frequency and duration of flooding was considerably greater than on the marsh surface. The marsh surface pools differed from the other intertidal habitats in that they consistently maintained water depths (Smith 1995) for fish populations during all stages of the tides and seasons but only exchanged water and, thus, fishes during spring tides and storms (i.e. approximately the same frequency and duration as the marsh surface). Substrate type varied from organic peat and muds at the higher elevations to larger quantities of sand at the lower subtidal habitats (Table 14.2).

The fish fauna of Great Bay–Little Egg Harbor is highly seasonal in shallow (Able et al. 1989) and deep (Szedlmayer and Able 1996) waters. Most species are abundant during the summer, migrating into the ocean through late summer and fall; they are absent during the winter, and they return again in the spring and early summer. For this estuarine system, the seasonal migration from estuarine shores to the ocean in winter has been demonstrated for *Syngnathus fuscus* (Lazzari and Able 1990), *Paralichthys dentatus* (Szedlmayer and Able 1993; Able and Kaiser 1994), *Centropristis striata* (Able et al. 1995), *Prionotus carolinus* and *P. evolans* (McBride and Able 1994) and *Scophthalmus aquosus* (Morse and Able 1995).

Overall abundance, and thus habitat use, is increased by reproduction by seasonal and year-round residents during the spring and summer. The seasonal pattern of abundance is evident for subtidal shores with greatest abundance from June through November, which corresponds closely to the seasonal high in temperatures both in

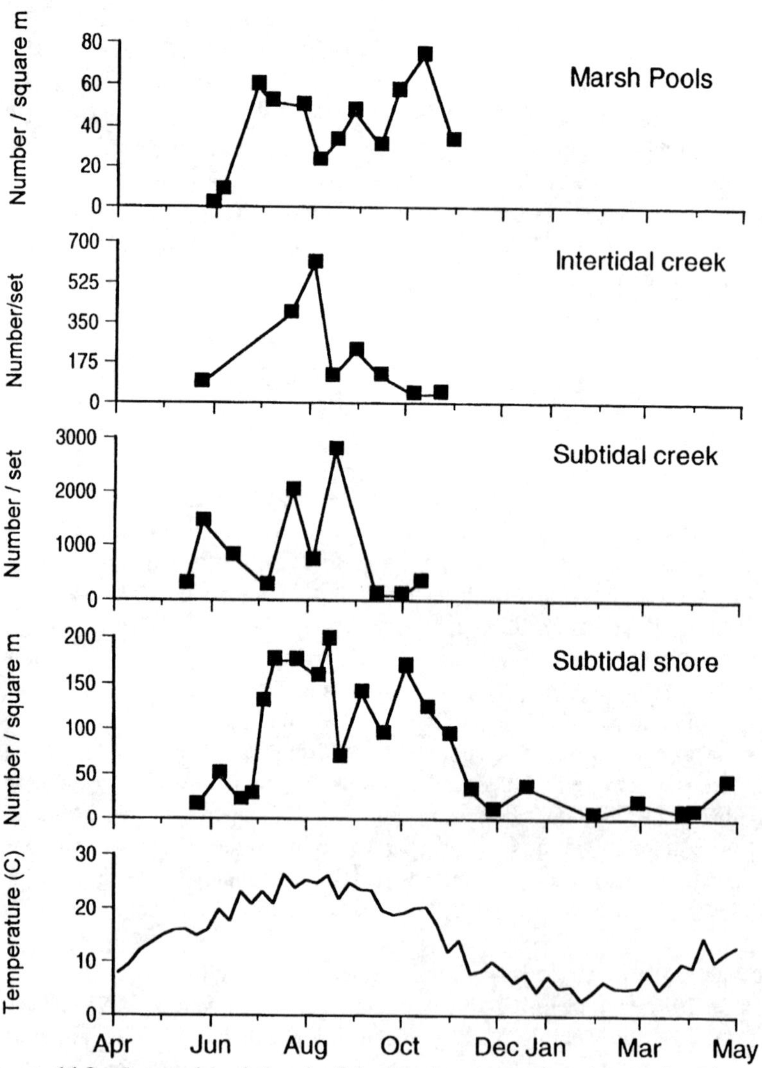

Figure 14.3 Seasonal variation in fish abundance at selected estuarine shore habitats (Table 14.1) including marsh pools, subtidal creeks, and subtidal shorelines and seasonal temperature changes in the Great Bay–Little Egg Harbor study area. Temperature data are from the estuarine shore of Great Bay during 1990–1991 (see Able et al. (1992) for further details)

the short (Figure 14.3) and long term (Able et al. 1992). This same trend in seasonal abundance and high temperature is evident for some of the other estuarine shore habitats based on more restricted seasonal sampling (Figure 14.3). The seasonal pattern of movement out of the estuary in the winter to warmer average temperatures on the continental shelf may be a response to prevent mortality due to low temperatures. Clearly, portions of the estuarine shore fauna, including some dominant members such as *Paralichthys dentatus* (Szedlmayer and Able 1993; Keefe and Able 1993), *Tautoga onitis, Tautogolabrus adspersus* and *Centropristis striata* (Hales and Able in press), are susceptible to winter mortality due to ambient estuarine winter temperatures. This response does not occur for all components of the fauna because one of the dominant species, *Fundulus heteroclitus* (Table 14.3), overwinters in marsh surface pools where the temperatures may average somewhat higher than adjacent subtidal habitats (Smith and Able 1994).

14.3.2 Faunal composition

The majority of species found along estuarine shores, represented by the over 70 000 individuals in this study (Table 14.1), migrate in and out of the estuary (57%). These are composed of either species that use the estuary extensively (i.e. all life history stages can be found there during some season but migrate out of the estuary for the winter – Seasonal Residents, 19%) while others occur primarily as young-of-the-year or older juveniles (Nursery, 38%) (Table 14.3). Both groups are dominated by species that overwinter on the continental shelf. The remaining species are true residents (i.e. spend their entire life cycle in estuarine shore habitats (21%)) or strays (22%).

The fauna is phylogenetically diverse with 58 species from 35 families and 11 orders represented across all estuarine shore habitats (Table 14.4). The diversity of the fauna is increased by frequent visitors from southern regions, usually in late summer (i.e. *Hyporhamphus meeki, Lutjanus griseus, Chaetodon ocellatus*) or those from more northern areas in late winter or early spring (i.e. *Myoxocephalus aenaeus*). A portion of the fauna is of economic importance in commercial and recreational fisheries. These include such species as *Anguilla rostrata, Brevoortia tyrannus, Morone americana, Centropristis striata, Pomatomus saltatrix, Leiostomus xanthurus, Menticirrhus saxatilis, Micropogonias undulatus, Tautoga onitis, Paralichthys dentatus, Pleuronectes americanus* and *Sphoeroides maculatus*. The biomass that occurs in estuarine shores is high although not well quantified across habitats. For subtidal creeks, the biomass of several resident (*Fundulus heteroclitus*) and migratory (*Menidia menidia, Pomatomus saltatrix, Paralichthys dentatus, Cynoscion regalis*) fishes is especially high and for most of these species (with the exception of *C. regalis*) this is due, almost exclusively, to young-of-the-year individuals (Rountree and Able 1992a). In general, the fauna identified from Great Bay–Little Egg Harbor estuarine shores is typical of those found in Mid-Atlantic Bight estuaries to the north, including Connecticut (Pearcy and Richards 1962) and Massachusetts (Werme 1981; Teal 1986) but not Maine (Ayvazian et al. 1992), and to the south, including Delaware (Derickson and Price 1973), Virginia (Richards and Castagna 1970; Weinstein and Brooks 1983) and North Carolina (Weinstein 1979; Weinstein et al. 1980; Hettler 1989) and to a lesser extent in Georgia (Dahlberg 1972).

Table 14.4 Abundance (percentage of total number) of each species within each habitat type along the marsh surface to subtidal creek gradient based on the sampling periods indicated in Table 14.1. See Figure 14.2 for diagram of habitat types. Life history mode defined as: resident (R) = species that spend the entire life in the estuary; seasonal residents (SR) = species that may be represented in the estuary by all life history stages but migrate offshore for the winter; nursery (N) = species that use the estuary primarily as a habitat for young-of-the-year (YOY), strays (S) = species that are infrequently found. These interpretations are augmented by gill net sampling that provided data for larger adults of some species (Rountree and Able, unpublished data)). YOY = individuals less than age 1 year; Juv = individuals age 1+ but not sexually mature; Adult = individuals that have reached sexual maturity; All = life history stages from YOY to adult

Taxonomic group	Marsh surface	Marsh surface pools	Intertidal creek	Subtidal creek	Subtidal shoreline	Life stage	Life history mode
Order Anguilliformes							
Family Anguillidae (freshwater eels)							
Anguilla rostrata (American eel)		0.03	0.022	0.16	0.018	YOY, Juv	N
Family Congridae (conger eels)							
Conger oceanicus (conger eel)					0.028	YOY, Juv	N
Order Clupeiformes							
Family Clupeidae (herrings)							
Alosa pseudoharengus (alewife)				0.01		YOY	N
Brevoortia tyrannus (Atlantic menhaden)				0.06	0.507	YOY	N
Clupea harengus (Atlantic herring)					0.002	YOY	N
Sardinella aurita (Spanish sardine)				0.06		YOY	N
Family Engraulidae (anchovies)							
Anchoa hepsetus (striped anchovy)				4.71		Adult	S
Anchoa mitchilli (bay anchovy)				4.50	5.474	All	SR
Engraulis eurystole (silver anchovy)					0.025	YOY	S
Order Aulopiformes							
Family Synodontidae (lizardfishes)							
Synodus foetens (inshore lizardfish)					0.046	YOY	N
Order Gadiformes							
Family Gadidae (codfishes)							
Urophycis regia (spotted hake)					0.016	YOY	N
Family Ophidiidae (cusk–eels)							
Ophidion marginatum (striped cusk–eel)					0.002	YOY, Adult	SR?
Order Batrachoidiformes							
Family Batrachoididae (toadfishes)							
Opsanus tau (oyster toadfish)				0.03	0.055	All	R
Order Atheriniformes							
Family Exocoetidae (flyingfishes)							
Hyporhamphus meeki					0.002		S
Family Belonidae (needlefishes)							
Strongylura marina (Atlantic needlefish)				0.24	0.005	All	SR
Family Cyprinodontidae (killifishes)							
Cyprinodon variegatus (sheepshead minnow)		13.5	2.026	0.14	0.106	All	R
Fundulus heteroclitus (mummichog)	54.69	75.2	53.77	6.86	6.756	All	R
Fundulus luciae (spotfin killifish)	0.41	4.19	0.443			All	R
Fundulus majalis (striped killifish)			1.317	0.22	12.62	All	R
Lucania parva (rainwater killifish)	5.31	3.19	0.066	0.19		All	R
Family Atherinidae (silversides)							
Menidia beryllina (inland silverside)	12.24	3.66	2.502	0.01		All	R
Menidia menidia (Atlantic silverside)	21.63	0.121	36.29	43.27	66.79	All	SR

Table 14.4 *(continued)*

Taxonomic group	Marsh surface	Marsh surface pools	Intertidal creek	Subtidal creek	Subtidal shoreline	Life stage	Life history mode
Order Gasterosteiformes							
Family Gasterosteidae (sticklebacks)							
Apeltes quadracus (fourspine stickleback)					0.044	All	R
Gasterosteus aculeatus (threespine stickleback)		0.061			0.039	All	SR
Family Syngnathidae (pipefishes and seahorses)							
Hippocampus erectus (lined seahorse)				0.01	0.207	All	SR
Syngnathus fuscus (northern pipefish)				0.81	1.588	All	SR
Order Scorpaeniformes							
Family Triglidae (searobins)							
Prionotus carolinus (northern searobin)					0.014	All	SR
Prionotus evolans (striped searobin)					0.023	All	SR
Family Cottidae (sculpins)							
Myoxocephalus aenaeus (grubby)					0.032	All	R?
Order Perciformes							
Family Percichthyidae (temperate basses)							
Morone americana (white perch)					0.002	All	R
Family Serranidae (sea basses)							
Centropristis striata (black sea bass)				0.01	0.002	YOY, Juv	N
Family Pomatomidae (bluefishes)							
Pomatomus saltatrix (bluefish)			0.166	2.33	0.044	All	SR
Family Rachycentridae (cobias)							
Rachycentron canadum (cobia)				0.01		YOY	S
Family Carangidae (jacks)							
Caranx hippos (crevalle jack)				0.14	0.002	YOY	N
Selene vomer (lookdown)				0.01		YOY	S
Trachinotus falcatus (permit)					0.014	YOY	S
Family Lutjanidae (snappers)							
Lutjanus griseus (grey snapper)						YOY	S
Family Gerreidae (mojarras)			0.410		0.014	YOY	S
Family Sciaenidae (drums)							
Bairdiella chrysoura (silver perch)					0.007	YOY	N
Leiostomus xanthurus (spot)				30.11	0.141	YOY	N
Menticirrhus saxatilis (kingfish)				0.01	0.145	YOY	N
Micropogonias undulatus (Atlantic croaker)					0.002	YOY	N
Family Chaetodontidae (butterflyfishes)							
Chaetodon ocellatus (spotfin butterflyfish)			0.001		0.007	YOY	N
Family Mugilidae (mullets)							
Mugil spp. (primarily *M. curema*)			2.712	0.153	0.745	YOY	N
Family Sphyraenidae (barracudas)							
Sphyraena borealis (northern sennet)			0.111	2.75		YOY	N
Family Labridae (wrasses)							
Tautoga onitis (tautog)				0.12	0.177	YOY	N
Tautogolabrus adspersus (cunner)					0.007	YOY	N
Family Uranoscopidae (stargazers)							
Astroscopus guttatus (northern stargazer)					0.007	YOY	S
Family Gobiidae (gobies)							
Gobionellus boleosoma (darter goby)					0.014	All	S?
Gobiosoma sp. (primarily *G. bosc*)		0.03	0.089	2.63	0.982	All	R
Family Stromateidae (butterfishes)							
Peprilus triacanthus (butterfish)				0.01		YOY	S

Table 14.4 (continued)

Taxonomic group	Marsh surface	Marsh surface pools	Intertidal creek	Subtidal creek	Subtidal shoreline	Life stage	Life history mode
Order Pleuronectiformes							
Family Bothidae (lefteye flounders)							
Etropus crossotus (fringed flounder)					0.002	YOY	S
Etropus microstomus (smallmouth flounder)					0.475	All	N
Paralichthys dentatus (summer flounder)				0.02	0.021	All	N
Scophthalmus aquosus (windowpane)					0.076	All	SR
Family Pleuronectidae (righteye flounders)							
Pleuronectes americanus (winter flounder)			0.078	0.38	2.823	All	R
Family Soleidae (soles)							
Trinectes maculatus (hogchoker)					0.018	All	S
Order Tetraodontiformes							
Family Tetraodontidae (puffers)							
Sphoeroides maculatus (northern puffer)				0.03	0.053	YOY	N

14.3.3 Faunal composition by habitat

The marsh surface and marsh pools were dominated by a few species of cyprinodontids and atherinids (Tables 14.3 and 14.4). The depauperate nature of the marsh surface fauna, in general, reflects the extreme conditions encountered, with high temperatures in the summer, low dissolved oxygen, and variable salinity (Smith 1995). The dominant group, the cyprinodontids, are among the most tolerant fishes to these conditions (Weisburg 1986).

The intertidal creek habitats were slightly enriched over the marsh surface habitats by a variety of small species, or young-of-the-year, from a variety of families (Table 14.4). These species were more abundant in the adjacent subtidal creek, with the possible exception of *Mugil* spp. (primarily *M. curema*) and gerreids, which were much more abundant in the intertidal creek. The subtidal creek had a much more diverse fauna that was enhanced by clupeids, engraulids, syngnathids, carangids, sciaenids, and a variety of other families (Table 14.4). The subtidal shores had the most diverse fauna and substrate types (Table 14.2), which included almost all of the species found in the other habitats as well as a number of species that were typical of deeper estuarine waters.

Many of the estuarine shore habitats are important as nurseries for juvenile fishes in this estuary. Specific examples for this system include those for *P. dentatus* (Rountree and Able 1992b; Szedlmayer et al. 1992; Keefe and Able 1993), *Pleuronectes americanus* (Sogard and Able 1992; Sogard 1992; Witting 1995), *Tautoga onitis* (Sogard et al. 1992), *Centropristis striata* (Able et al. 1995), and *Mustelus canis* (Rountree and Able 1996). A comparison with deeper habitats (1–8 m) in the Great Bay–Little Egg Harbor system, sampled with an otter trawl (4.9 m, 6 mm mesh), suggests that many species are shared (Szedlmayer and Able

1996). However, 13 species (22%) of the estuarine shore fauna were not collected in these deeper habitats. Many of these were strays but of notable importance are those associated with shallow saltmarsh habitats such as *Cyprinodon variegatus, Fundulus luciae, Lucania parva* and *Mugil* spp.

Some species were such dominant components of the estuarine shore collections that they were present across all habitats (Table 14.4). These included the cyprinodontid, *Fundulus heteroclitus*, and the atherinid, *Menidia menidia*. Others were found in the highest shore habitats (i.e. from the marsh surface to the subtidal creek) including *M. beryllina* and *Lucania parva*, while others (*Anguilla rostrata, Cyprinodon variegatus* and *Gobiosoma* spp. – primarily *G. bosc*) were found across the four habitats lowest in elevation. Many species were only found in subtidal habitats, including *Opsanus tau, Hippocampus erectus* and *Syngnathus fuscus*. Only a few species were limited to two or fewer habitats (Table 14.4) including *F. luciae*, which occurred only in the marsh surface and intertidal creeks, and *Sphyraena borealis*, which was only collected in intertidal and subtidal marsh creeks. The general lack of habitat fidelity for most of the fauna largely reflects the mobility of many of these species. As an example, movement of *Paralichthys dentatus* in the study area occurs throughout the subtidal and occasionally intertidal portions of marsh creeks, and occasionally the adjacent bay shore, on diel and tidal cycles during the summer (Szedlmayer and Able 1992; Rountree and Able 1992b).

Some marsh surface fishes such as adult *F. heteroclitus* are possible exceptions. They are highly localized within seasons and have been found in the same marsh surface pools for long periods of time, at least during the winter (Smith and Able 1994) and have small home ranges in tidal creeks during the summer (Lotrich 1975). Other species in subtidal habitats, in these instances young-of-the-year, are also highly localized, at least during the summer, including *Centropristis striata* (Able and Hales in press) and *Tautoga onitis* and *Tautogolabrus adspersus* (Hales and Able unpublished results). These seasonal and tidal movements provide a link between the estuarine shore habitats that are arrayed across the horizontal and vertical gradients. The timing and degree of the tidal linkage is largely a function of the frequency and duration of flooding that varies markedly from the marsh surface to subtidal waters (Table 14.2).

The mode of habitat use is reflected to some degree in the life histories. The resident species along estuarine shores are annual species (Able 1990) that dominate on the marsh surface and in marsh surface pools (Tables 14.3, 14.4). Some of these may make seasonal migrations from the creeks to the pools. At least this is true for the most abundant species, *F. heteroclitus* (Smith and Able 1994). The seasonal residents, which are most abundant in the subtidal habitats, range from small, annual species such as *Syngnathus fuscus* (Lazzari and Able 1990) and *Anchoa mitchilli* (Vouglitois et al. 1987) to larger species which are dominated by 0+ or 1+ age individuals such as *Paralichthys dentatus* (Rountree and Able 1992b; Szedlmayer and Able 1993) and *Pomatomus saltatrix* (McBride and Conover 1991). Those species that migrate into estuarine shores and use them as nurseries are most abundant in subtidal creeks and shorelines where they comprise 39–42% of the species collected there (Table 14.3). The same is true, but to lesser extent (27%) for intertidal creeks. Both of the subtidal habitats have the highest percentage of seasonal residents (19–23%) (Table 14.3).

14.3.4 Linkages across the estuarine shore

The linkages across estuarine shore habitats are, in part, also a function of the life histories. For example, *Fundulus heteroclitus*, one of the dominant members of the estuarine shore assemblage, spawns in intertidal habitats along the edge of intertidal creeks (Taylor et al. 1977; Able 1984). After hatching, the larvae are carried or swim to the marsh surface where they develop in depressions or pools (Talbot and Able 1984; Kneib 1993). As they grow during the first summer the juveniles disperse from the marsh surface, probably by way of the intertidal creeks to the subtidal creeks. At the end of the summer or in early fall when they are nearly adult size they move back onto the marsh surface, at least in the study estuary, to spend the winter (Smith and Able 1994).

The movements and life histories of other species link estuarine shores to the deeper waters of the adjacent estuary and the ocean (Weinstein 1979; Weinstein et al. 1980; see also McIvor and Rozas, this volume Chapter 13). Thus while trophic linkages across estuarine shore habitats are seldom demonstrated for fishes these linkages can be assumed when the life histories place them in different habitats at different life history stages. As an example from the study area, *Paralichthys dentatus*, one of the major predators in the study site, spawns in the ocean (Able et al. 1990), the metamorphosing larvae enter estuaries (Keefe and Able 1993) where they settle (Keefe and Able 1994) and spend much of the first summer in subtidal creeks or along subtidal shorelines where they attain one of the fastest growth rates for estuarine fishes (Szedlmayer et al. 1992; Rountree and Able 1992b; Szedlmayer and Able 1993) and then migrate to the edge of the continental shelf for the winter (Able and Kaiser 1994). Another example is that for *Mustelus canis* in which the pups are born in the estuary, spend much of the first summer in subtidal creeks, move into deeper waters as they grow and eventually into the ocean for the winter (Rountree and Able 1996, unpublished data).

14.4 CONCLUSIONS

Despite the general taxonomic similarities between estuarine shore assemblages along the east coast of the US (see Rountree and Able (1992a) for more discussion of comparisons for subtidal creeks and McIvor and Rozas, this volume, for intertidal creeks) there are a number of factors that make comparisons difficult. First, the observations reported here are for a high-salinity system; those for a system with lower salinity elsewhere in the Mid-Atlantic Bight such as the Rhodes

River estuarine system in upper Chesapeake Bay, have a much greater contribution of mesohaline and freshwater forms (Hines et al. 1987) and differ in structure and possibly function. Second, most studies, including this one, are based on daytime surveys, although there are pronounced differences in species composition and relative abundances when day and night collections are compared (Rountree and Able 1993; Sogard and Able 1994). Third, variation within the major estuarine shore habitat types examined here may be as important as the variation between these habitats. In this system we have previously identified differences in species composition and relative abundance between marsh surface pools with and without vegetation (Smith 1995), between subtidal creeks with different topography (Rountree 1992), and in growth rates between adjacent and distant subtidal vegetated and unvegetated habitats (Sogard 1992). These distinctions need further attention before we can understand what structural components of estuarine shore fish assemblages contribute most to ecological function. However, despite these caveats, it is clear that many species provide linkages across the vertical and horizontal gradients of these estuarine shore habitats as a function of the life histories. In other instances, fish life histories link the estuarine shores with the deeper portions of the estuary as well as the adjacent continental shelf and, as a result, are an important pathway for the transfer of energy from the land margin to the sea, as has been previously recognized (Weinstein 1984; Deegan 1993).

ACKNOWLEDGMENTS

This work is the result, in part, of research sponsored by NOAA, Office of Sea Grant, Department of Commerce under Grant Nos. NA89AA-D-SG057 (Project No. R/F-42) and NA-36-RG0505 (Project No. R/F-65). The US Government is authorized to produce and distribute reprints for governmental purposes, notwithstanding any copyright notation that may appear hereon. Support for personnel was provided from the Rutgers University Institute of Marine and Coastal Sciences (IMCS). Numerous individuals from the Rutgers University Marine Field Station participated in the sampling under a variety of conditions. R. Lathrop provided estimates of shoreline distance based on satellite images. We are grateful to all of the above. This paper is IMCS Contribution No. 95-30.

REFERENCES

Able, K.W. 1984. Variation in spawning site selection of the mummichog, *Fundulus heteroclitus*. *Copeia* 1984: 522–525.

Able, K.W. 1990. Life history patterns of New Jersey salt marsh killifishes. *Bulletin of the New Jersey Academy of Science* 35: 23–30.

Able, K.W. 1992. Checklist of New Jersey saltwater fishes. *Bulletin of the New Jersey Academy of Science* 37: 1–11.

Able, K.W. and L.S. Hales, Jr. (in press). Movements and growth of juvenile black sea bass, *Centropristis striata*, in a southern New Jersey estuary. *Journal of Experimental Marine Biology and Ecology*

Able, K.W. and S.C. Kaiser. 1994. Synthesis of summer flounder habitat parameters. NOAA Coastal Ocean Program Decision Analysis Series No. 1. NOAA Coastal Ocean Office, Silver Spring, Maryland.

Able, K.W., K.A. Wilson and K.L. Heck, Jr. 1989. Fishes of vegetated habitats in New Jersey estuaries: Composition, distribution and abundance based on quantitative sampling. Rutgers University, Center for Coastal and Environmental Studies Tech. Rept. No. 1041, New Brunswick, New Jersey.

Able, K.W., R.E. Matheson, W.W. Morse, M.P. Fahay and G.R. Shepherd. 1990. Patterns of summer flounder *Paralichthys dentatus* early life history in the Mid-Atlantic Bight and New Jersey estuaries. *Fishery Bulletin US* 88: 1–12.

Able, K.W., R. Hoden, D.A. Witting and J.B. Durand. 1992. Physical parameters of the Great Bay-Mullica River Estuary with a list of research publications. Rutgers University, Institute of Marine and Coastal Sciences, Tech. Rept. 92-06, New Brunswick, New Jersey.

Able, K.W., M.P. Fahay and G.R. Shepherd. 1995. Early life history of black sea bass *Centropristis striata* in the Mid-Atlantic Bight and a New Jersey estuary. *Fishery Bulletin US* 93: 429–445.

Ayvazian, S.G., L.A. Deegan and J.T. Finn. 1992. Comparison of habitat use by estuarine fish assemblages in the Acadian and Virginian zoogeographic provinces. *Estuaries* 15: 368–383.

Chapman, V.J. 1960. *Salt Marshes and Salt Deserts of the World.* Interscience, New York.

Dahlberg, M.D. 1972. An ecological study of Georgia coastal fishes. *Fishery Bulletin* 70: 323–353.

Daiber, F.C. 1982. *Animals of the Tidal Marsh.* Van Nostrand Reimbold, New York.

Deegan, L.A. 1993. Nutrient and energy transport between estuaries and coastal marine ecosystems by fish migration. *Canadian Journal of Fisheries and Aquatic Sciences* 50: 74–79.

Derickson, W.K. and K.S. Price Jr. 1973. The fishes of the shore zone of Rehoboth and Indian River Bays, Delaware. *Transactions of the American Fisheries Society* 102: 552–562.

di Castri, F., A.J. Hansen and M.M. Holland (eds.). 1988. A new look at ecotones: Emerging international projects on landscape boundaries. *Biology International 17.* International Union of Biological Sciences, Paris.

Durand, J.B. 1984. Nitrogen distribution in New Jersey coastal bays. *In* M.J. Kennish and R.A. Lutz (eds.), *Ecology of Barnegat Bay, New Jersey.* Springer-Verlag, New York.

Durand, J.B. 1988. Field studies in the Mullica River–Great Bay estuarine system. Vol. I. Technical Report, Center for Coastal and Environmental Studies, Rutgers University, New Brunswick, NJ.

Durand, J.B. and R.J. Nadeau. 1972. Water resources development in the Mullica River basin: Part I. Biological evaluation of the Mullica River–Great Bay Estuary. Rutgers University Water Resources Research Institute.

Francis, R.C. 1990. Climate change and marine fisheries. *Fisheries* 15: 7–9.

Frank, K.T., R.I. Perry and K.F. Drinkwater. 1990. Predicted response of northwest Atlantic invertebrates and fish stocks to CO_2–induced climate change. *Transactions of the American Fisheries Society* 119: 353–365.

Good, R.E. and N.F. Good. 1984. The Pinelands National Reserve: An ecosystem approach to management. *Bioscience* 34: 169–173.

Haedrich, R.L. 1983. Estuarine fishes, p. 183–207. *In*: B.H. Ketchum (ed.), *Estuaries and Enclosed Seas. Ecosystems of the World, Vol. 26*. Elsevier Scientific Publishing, New York.

Hales, L.S. Jr. and K.W. Able. (in press). Overwinter mortality, growth and behavior of young-of-the-year of four coastal marine fishes in New Jersey (USA) waters. *Marine Biology*.

Halupa, P.J. and B.L. Howes. 1995. Effects of tidally mediated litter moisture contents on decomposition of *Spartina alterniflora* and *Spartina patens*. *Marine Biology* 123: 379–391.

Hettler, W.F. 1989. Nekton use of regularly flooded saltmarsh cordgrass habitat in North Carolina. *Marine Ecology Progress Series* 56: 111–118.

Hines, A.H., P.J. Haddon, J.J. Miklas, L.A. Wiechert and A.M. Haddon. 1987. Estuarine invertebrates and fish: sampling design and constraints for long-term measurements of population dynamics, p. 140–164. *In*: T. Boyle (ed.), *Longterm Monitoring in Aquatic Ecosystems*. American Society for Testing of Materials, Philadelphia, Pennsylvania.

Hoss, D.E. and G.W. Thayer. 1993. The importance of habitat to the early life history of estuarine dependent fishes. *American Fisheries Society Symposium* 14: 147–158.

Keefe, M. and K.W. Able. 1993. Patterns of metamorphosis in summer flounder, *Paralichthys dentatus*. *Journal of Fish Biology* 42: 713–728.

Keefe, M. and K.W. Able. 1994. Contributions of abiotic and biotic factors to settlement in summer flounder, (*Paralichthys dentatus*). *Copeia* 1994: 458–465.

Kennedy, V.S. 1990. Anticipated effects of climate change on estuarine and coastal fisheries. *Fisheries* 16: 16–24.

Kennish, M.J. 1992. *Ecology of Estuaries: Anthropogenic Effects*. CRC Press, Ann Arbor, Michigan.

Kneib, R.T. 1993. Growth and mortality in successive cohorts of fish larvae within an estuarine nursery. *Marine Ecology Progress Series* 94: 115–127.

Lazzari, M.A. and K.W. Able. 1990. Northern pipefish, *Syngnathus fuscus*, occurrences over the Mid-Atlantic Bight continental shelf: Evidence of seasonal migration. *Environmental Biology of Fishes* 27: 177–185.

Lotrich, V.A. 1975. Summer home range and movements of *Fundulus heteroclitus* in a tidal creek. *Ecology* 56: 191–198.

McBride, R.S. and D.O. Conover. 1991. Recruitment of young-of-the-year bluefish *Pomatomus saltatrix* to the New York Bight: variation in abundance and growth of spring- and summer-spawned cohorts. *Marine Ecology Progress Series* 78: 205–216.

McBride, R.S. and K.W. Able. 1994. Reproductive seasonality, distribution, and abundance of *Prionotus carolinus* and *P. evolans* (Pisces: Triglidae) in the New York Bight. *Estuarine, Coastal and Shelf Science* 38: 173–188.

McIvor, C.C. and L.P. Rozas. 1996. Nekton use of intertidal salt marshes: a review. *In*: Nordstrom, K.F. and C.T. Roman, eds., *Estuarine Shores: Evolution, Environments and Human Alterations*, John Wiley, Chichester, England.

Morse, W.W. and K.W. Able. 1995. Distribution and life history of windowpane, *Scophthalmus aquosus*, off the north eastern United States. *Fishery Bulletin* 93: 675–693.

Nixon, S.W. 1980. Between coastal marshes and coastal waters – a review of twenty years of speculation and research on the role of salt marshes in estuarine productivity and water chemistry, p. 437–525. *In*: P. Hamilton and K.B. Macdonald (eds.), *Estuarine and Wetland Processes with Emphasis on Modeling*. Plenum Press, New York.

Nixon, S.W. 1982. The ecology of New England salt marshes: a community profile. US Fish and Wildlife Services Office of Biological Services, FWS/OBS-81/55.

Pearcy, W.G. and S.W. Richards. 1962. Distribution and ecology of fishes of the Mystic River Estuary, Connecticut. *Ecology* 43: 248–259.

Peters, D.S. and F.A. Cross. 1992. What is coastal fish habitat? p. 17–22. *In*: R.H. Strou (ed.), *Stemming the Tide of Coastal Fish Habitat Loss. Proceedings of the Symposium o Conservation of Coastal Fish Habitat*, Baltimore, Maryland. Marine Recreationa Fisheries No. 14. National Coalition for Marine Conservation, Inc. Savannah, Georgia.
Psuty, N.P., M.P. De Luca, R. Lathrop, K.W. Able, S. Whitney and J.F. Grassle. 1993. Th Mullica River–Great Bay Estuarine Research Reserve: A unique opportunity for researcl preservation and management. *Proceedings of Coastal Zone '93*, July 1993, Nev Orleans, Louisiana.
Richards, C.E. and M. Castagna. 1970. Marine fishes of Virginia's eastern shore (inlet an marsh, seaside waters). *Chesapeake Science* 11: 235–248.
Rountree, R.A. 1992. Fish and macroinvertebrate community structure and habitat us patterns in salt marsh creeks of southern New Jersey, with a discussion of marsh carbo export. Ph.D dissertation. Rutgers University, New Brunswick.
Rountree, R.A. and K.W. Able. 1992a. Fauna of polyhaline subtidal marsh creeks in souther New Jersey: Composition, abundance and biomass. *Estuaries* 15: 171–185.
Rountree, R.A. and K.W. Able. 1992b. Foraging habits, growth, and temporal patterns o salt-marsh creek habitat use by young-of-year summer flounder in New Jersey *Transactions of the American Fisheries Society* 121: 765–776.
Rountree, R.A. and K.W. Able. 1993. Diel variation in decapod crustaceans and fis assemblages in New Jersey polyhaline marsh creeks. *Estuarine, Coastal and Shel Science* 37: 181–201.
Rountree, R.A., K.J. Smith and K.W. Able. 1992. Length frequency data for fishes an turtles from polyhaline subtidal and intertidal marsh creeks in southern New Jersey Rutgers University, Institute of Marine and Coastal Sciences Tech. Rept. 92-34, Nev Brunswick, New Jersey.
Rountree, R.A. and K.W. Able. 1996. Seasonal abundance, growth and foraging habits o juvenile smooth dogfish, *Mustelus canis*, in a New Jersey estuary. *Fishery Bulletin US* 94 522–534.
Seneca, E.D. and S.W. Broome. 1992. Restoring tidal marshes in North Carolina and France p. 53–78. *In*: G.W. Thayer (ed.), *Restoring the Nation's Marine Environment*. Marylan Sea Grant, College Park, Maryland.
Smith, K.J. 1995. Processes regulating habitat use by salt marsh nekton in a southern Nev Jersey estuary. Ph.D. Thesis. Rutgers University, New Brunswick, New Jersey.
Smith, K. J. and K.W. Able. 1994. Salt-marsh tide pools as winter refuges for th mummichog, *Fundulus heteroclitus*, in New Jersey. *Estuaries* 17: 226–234.
Sogard, S.M. 1992. Variability in growth rates of juvenile fishes in different estuarin habitats. *Marine Ecology Progress Series* 85: 35–53.
Sogard, S.M. and K.W. Able. 1991. A comparison of eelgrass, sea lettuce macroalgae, an marsh creeks as habitats for epibenthic fishes and decapods. *Estuarine, Coastal and Shel Science* 33: 501–519.
Sogard, S.M. and K.W. Able. 1992. Growth variation of newly settled winter flounde (*Pseudopleuronectes americanus*) in New Jersey estuaries as determined by otolit microstructure. *Netherlands Journal of Sea Research* 29: 163–172.
Sogard, S.M. and K.W. Able. 1994. Diel variation in immigration of fishes and decapo crustaceans to artificial seagrass habitat. *Estuaries* 17: 622–630.
Sogard, S.M., K.W. Able and M.P. Fahay. 1992. Early life history of the tautog *Tautog onitis* in the Mid-Atlantic Bight. *Fishery Bulletin US* 90: 529–539.
Spinner, G.P. 1969. The wildlife, wetlands and shellfish areas of the Atlantic coastal zone Serial Atlas of the Marine Environment Folio 18, American Geographical Society NewYork.
Szedlmayer, S.T. and K.W. Able. 1992. Validation studies of daily increment formation fo

larval and juvenile summer flounder, *Paralichthys dentatus. Canadian Journal of Fisheries and Aquatic Sciences* 49: 1856–1862.

Szedlmayer, S.T. and K.W. Able. 1993. Ultrasonic telemetry of age-0 summer flounder, *Paralichthys dentatus*, movements in a southern New Jersey estuary. *Copeia* 1993: 728–736.

Szedlmayer, S.T. and K.W. Able. 1996. Patterns of seasonal availability and habitat use by fishes and decapod crustaceans in a southern New Jersey estuary. *Estuaries* 19: 697–709.

Szedlmayer, S.T., K.W. Able and R.A. Rountree. 1992. Growth and temperature-induced mortality of young of the year summer flounder (*Paralichthys dentatus*) in southern New Jersey. *Copeia* 1992: 120–128.

Talbot, C.W. and K.W. Able. 1984. Composition and distribution of larval fishes in New Jersey high marshes. *Estuaries* 7: 434–443.

Taylor, M.H., L. DiMichele and G.J. Leach. 1977. Egg stranding in the life cycle of the mummichog *Fundulus heteroclitus. Copeia* 1977: 397–399.

Teal, J.M. 1986. The ecology of regularly flooded salt marshes of New England: A community profile. US Fish and Wildlife Service, Biological Report 85(7.4).

Varekamp, J.C., E. Thomas and O. Van de Plassche. 1992. Relative sea-level rise and climate change over the last 1500 years. *Terra Nova* 4: 293–304.

Vernberg, F.J. 1993. Salt-marsh processes: a review. *Environmental Toxicology and Chemistry* 12: 2167–2195.

Vouglitois, J.J, K.W. Able, R.J. Kurtz and K.A. Tighe. 1987. Life history and population dynamics of the bay anchovy in New Jersey. *Transactions of the American Fisheries Society* 116: 141–153.

Weinstein, M.P. 1979. Shallow marsh habitats as primary nurseries for fishes and shellfish, Cape Fear River, North Carolina. *Fishery Bulletin* 77: 339–357.

Weinstein, M. P. 1984. Epilogue – comparative ecology of macrofauna of saltmarshes: toward an ecosystem synthesis. *Estuaries* 7: 469–470.

Weinstein, M.P. and H.A. Brooks. 1983. Comparative ecology of nekton residing in a tidal creek and adjacent seagrass meadow: community composition and structure. *Marine Ecology Progress Series* 12: 15–27.

Weinstein, M.P., S.L. Weiss and M.F. Walters. 1980. Multiple determinants of community structure in shallow marsh habitats, Cape Fear River estuary, North Carolina, USA. *Marine Biology* 58: 227–243.

Weisburg, S.B. 1986. Competition and coexistence among four estuarine species of *Fundulus. American Zoology* 26: 249–257.

Werme, C.E. 1981. Resource partitioning in a salt marsh community. Ph.D. dissertation. Boston University, Boston, Massachusetts.

Witting, D.A. 1995. Settlement of winter flounder, *Pleuronectes americanus*, in a southern New Jersey estuary: spatial and temporal dynamics and the effect of decapod predation. Ph.D. Thesis. Rutgers University, New Brunswick, New Jersey.

Zimmerman, R.J., T.J. Minello, E.F. Klima, and J.M. Nance. 1991. Effects of accelerated sea-level rise on coastal secondary production. *Proceedings of Coastal Zone '91*, July 1991, Long Beach, California.

Chapter Fifteen

Ecological dynamics of a tropical intertidal mudflat community

JOSÉ A. VARGAS
Universidad de Costa Rica

15.1 INTRODUCTION

The Gulf of Nicoya is a tectonic estuary on the Pacific coast of Costa Rica (10°N, 85°W), extending about 100 km from the Tempisque River to the 500 m isobath (Figure 15.1). The estuary is the most important fishing ground in Costa Rica, and the main Pacific ports are located here. Owing to its importance for the country, and the wide variety of ecosystems present, the Gulf of Nicoya has been the subject of a long-term research program. To date more than 100 papers have been published, making the Gulf one of the better known tropical estuaries worldwide (Vargas 1995).

A three-year survey of the macrofauna (organisms retained on a 500 μm mesh sieve) and meiofauna (those organisms passing through a 500 μm mesh sieve but retained on a 62 μm mesh) was conducted on the Gulf of Nicoya (De la Cruz and Vargas 1987; Emig and Vargas 1990; Vargas 1987, 1988a,b, 1989a,b). This paper will discuss these data, with a particular focus on seasonal variability of the benthic community and a comparison with temperate estuaries. Parallel to the intertidal flat studies, other scientists working in the Punta Morales region focused on topics such as the biology of blue crabs and mantis shrimp (Dittel et al. 1985; Dittel 1991), hermit crabs (Borjesson and Szelistowski 1989), fish (Szelistowski 1990), birds (Barrantes and Pereira 1992), coastal vegetation and associated fauna (Perry 1988), and pollution (De la Cruz 1989). These and other studies will be discussed in relation to identifying linkages between the intertidal flats, adjacent habitats, and trophic levels. Some of the relevant features of the flat are included in Table 15.1.

Estuarine Shores: Evolution, Environments and Human Alterations. Edited by K. F. Nordstrom and C. T. Roman.

Table 15.1 General characteristics of the Punta Morales intertidal mudflat in the Gulf of Nicoya estuary, Pacific coast of Costa Rica, Central America

Location	10°04 min N 84°58 min W.
Tidal range	Mean 2.3 m; spring tide 3.3 m
Seasons	Dry (December–April), rainy (May–November)
Temperature fluctuation	25°C to 41°C at sediment surface at low tide. Daily fluctuation may exceed seasonal range
Seasonal salinity range	24‰ (rainy season) to 36‰ (dry season)
Sediment type (mean)	65% sand 32% silt and clay 2% organic matter
Dominant trophic group	Deposit feeders
Macrofaunal density range	3787 ind. m^{-2} to 41 086 ind. m^{-2}
Meiofaunal density range	1.17 to 4.25 million ind. m^{-2}
More abundant polychaete species	*Mediomastus californiensis* (a sub–surface deposit feeder) *Paraprionospio pinnata* (a surface deposit feeder)
Most conspicuous biogenic structure	Tubes of the polychaete worm *Diopatra ornata,* with attached leaves of mangrove trees
Conspicuous motile fauna	Shrimp of the genus *Penaeus* The crab *Callinectes arcuatus* Stomatopod mantis shrimp Hermit crabs of the genus *Clibanarius*
Fish at high tide	About 120 species. Most abundant are catfish
Birds at low tide	27 species of Charadriiform.
Neighborhood habitats	Red and black mangrove forests (*Rhizophora mangle* and *Avicennia germinans,* respectively) Mid upper Gulf of Nicoya estuary (less than 20 m deep)
Human impacts	Pesticide residues and trace metals in low concentration *Anadara* ark clam extraction

The Gulf of Nicoya differs from its temperate counterparts in that much of the nitrogen entering the estuary is from deep water offshore that is transported across the shelf into the Gulf. It is similar to other tropical estuaries in that it is subjected to extreme seasonal variations in riverine flow resulting from dry (December–April) and rainy (May–November) seasons (Epifanio et al. 1983; Voorhis et al. 1983). At low tide, horizontal distribution of salinity (‰) in the upper Gulf may range from 10 to 30 during the rainy season (Lizano and Vargas 1993). Most of the upper Gulf of Nicoya is surrounded by mangrove forests, sandy beaches, and intertidal flats. The intertidal flats are defined as the area between the shoreline and the water interface at mean low tide. For the inner Gulf of Nicoya (the area located west of the 85°W meridian, Figure 15.1) mangroves cover 11 837 ha, and intertidal flats 3390 ha (Kapetsky et al. 1987). The latter area represents about one-third of the total intertidal flat area in the Gulf.

Punta Morales, where the intertidal mud flat community reported herein is located, is a small peninsula on the mid upper Gulf. It is bordered by mangrove swamps, rocky shores, and a white sand beach. Immediately offshore, most of the peninsula is fringed by an intertidal flat (Figure 15.1).

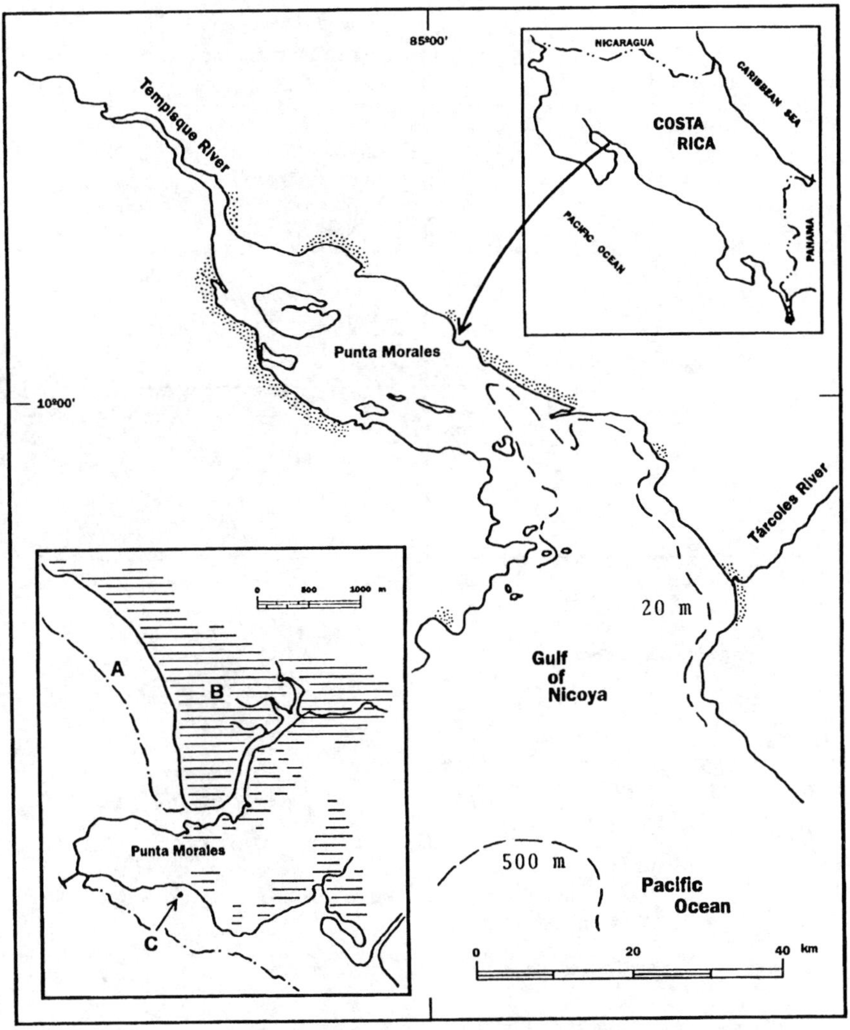

Figure 15.1 Gulf of Nicoya estuary, Pacific coast of Costa Rica, Central America. Location of the Punta Morales peninsula on the upper Gulf. Stippled areas around the Gulf indicate mangrove forests and intertidal flats. A: mudflats at low tide, B: mangrove forests, C: sampling site. Modified from Vargas (1988a)

15.2 SEASONAL TRENDS IN POPULATION AND COMMUNITY DYNAMICS

Surveys of the intertidal flat yielded 98 species the first year, 78 species the second year, and 95 in the third year for an overall total of 107 species (Vargas 1989a). The 30 numerically most important species collected in the first year were included in

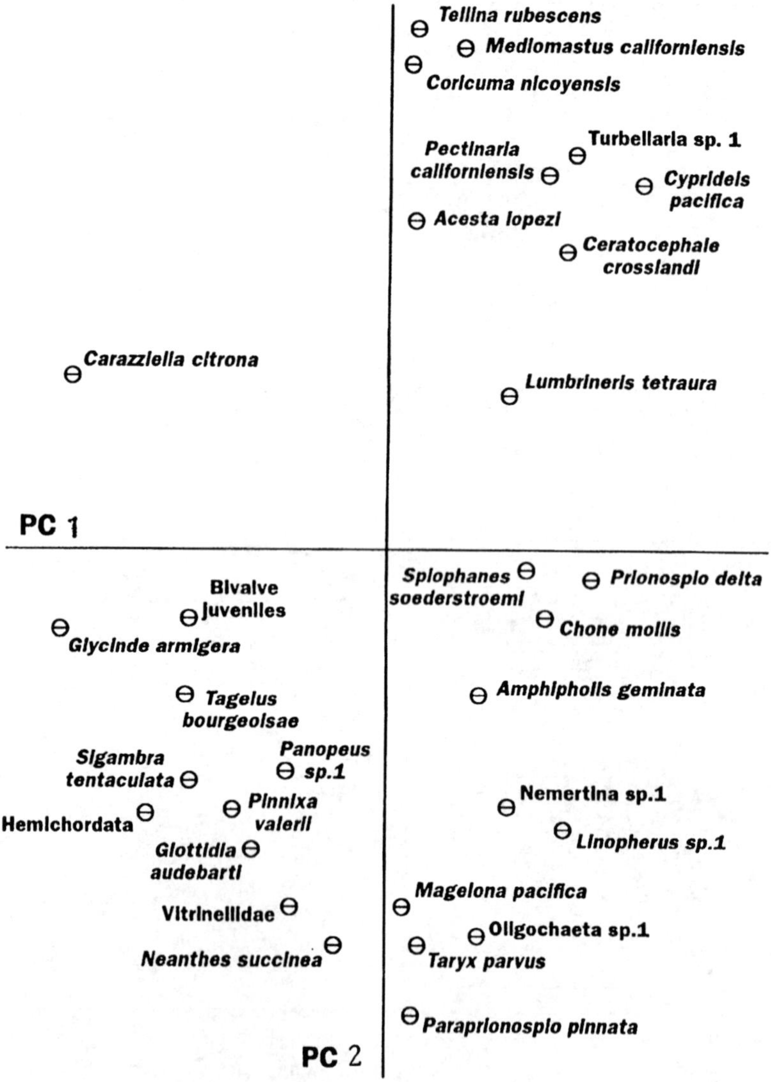

Figure 15.2 First and second Principal Components (PC) calculated from the number of individuals per species over the sampling period. Punta Morales intertidal mudflat, Gulf of Nicoya, Costa Rica, February 1984 to February 1985

Principal Component Analysis (PCA). This multivariate technique places those species having similar peaks in abundance during the sampling period in proximity. The results of this analysis indicated a roughly clockwise change in species abundances with time (Figure 15.2). Those species with peak abundance early in the year (January to May, Table 15.2) were placed in the top right quadrant, while

Table 15.2 Number of individuals of selected species collected monthly at the Punta Morales intertidal flat in the Gulf of Nicoya estuary, Pacific coast of Costa Rica, March 1984 – February 1985. A total of 98 species was collected

Species	M	A	M	J	J	A	S	O	N	D	J	F
Mediomastus californiensis	271	195	149	143	88	83	101	86	79	135	163	208
Cyprideis pacifica	608	1703	597	81	29	90	21	37	23	16	37	174
Platyhelminthes	196	648	68	10	0	1	0	0	0	0	1	145
Coricuma nicoyensis	643	846	51	38	95	47	28	58	41	58	397	964
Tellina rubescens	76	50	6	8	11	10	3	7	24	14	48	48
Spiophanes soederstroemi	122	21	4	77	84	40	25	31	1	0	3	2
Paraprionospio pinnata	10	23	21	124	163	76	82	96	41	72	30	26
Lumbrineris tetraura	50	116	63	58	36	8	57	34	36	25	42	32
Neanthes succinea	5	5	5	0	21	40	37	22	12	17	15	7
Magelona pacifica	0	0	0	0	13	19	4	6	3	3	3	4
Glottidia audebarti	0	0	2	2	0	0	10	36	32	14	4	2
Hemichordata	1	0	0	2	8	13	19	29	13	28	14	9
Pinnixa valerii	1	9	2	8	10	4	22	20	10	15	11	7
Tagelus bourgeoisae	1	0	0	2	8	15	7	7	5	1	7	18
Encope stokesi	0	0	0	3	1	1	0	2	3	6	8	9
Natica unifasciata	1	3	1	2	0	1	1	2	4	1	0	0
Nassarius luteostoma	4	1	3	3	0	4	6	3	1	0	5	1
Amphipholis geminata	4	3	7	8	1	5	4	2	9	4	3	4
Diopatra ornata	1	1	0	0	2	0	1	3	1	0	0	0
Carazziella citrona	0	0	0	0	0	0	0	0	0	10	14	8

[a] Samples collected by coring to a depth of 0.15 m into the sediment. Cores originally collected at semimonthly intervals (28 cores per date). Data in this table include total per month (56 cores pooled for a total area of 0.1 m^2)

those whose peak abundance occurred in mid-year or later were found in the lower right and left quadrants (Figure 15.2). *Carazziella citrona*, the spionid polychaete, is found alone in the upper left quadrant due to its ocurrence only towards the end of the sampling period.

The seasonal changes in populations indicated by PCA above are further illustrated by the patterns included in Table 15.2. The capitellid polychaete, *Mediomastus californiensis*, and the cumacean, *Coricuma nicoyensis*, show a marked seasonal pattern, with low abundances during the rainy season months and increased abundance during the dry season. A similar pattern is presented by the bivalve, *Tellina rubescens*. A contrasting seasonal pattern is illustrated by three polychaetes, the spionid *Paraprionospio pinnata*, the nereid *Neanthes succinea*, and the magelonid *Magelona pacifica*. Strong population oscillations are present in the ostracod *Cyprideis pacifica*, with a maximum of 1703 individuals in April, in an unidentified flat worm (Platyhelminthes, Turbellaria) with 648 individuals also in April, and in the spionid polychaete, *Spiophanes soederstroemi* with peaks in March and June (Table 15.2). Species characterized by moderate changes in abundance include the brachiopod *Glottidia audebarti*, an unidentified acorn worm (Hemichordata), the crab *Pinnixa valerii*, and the bivalve *Tagelus bourgeoisae*. Small changes in abundance throughout the sampling period are illustrated by the sand dollar *Encope stokesi*, the snails *Natica unifasciata* and *Nassarius luteostoma*, the brittle star, *Amphipholis geminata*, and the onuphid polychaete, *Diopatra ornata* (Table 15.2). Similar populational oscillations were found the second and third years (Vargas 1988a, 1989a,b), except for the spionid, *Carazziella citrona*, which presented a marked peak of abundance during the second year, when it represented close to 22% of the total number of individuals collected (Vargas 1988a, 1989a).

The results of PCA performed on abundance data collected at semi-monthly intervals (25 sampling dates) the first year, and at monthly intervals (12 sampling dates) the second year are included in Figure 15.3A and 15.3B, respectively, and further support the idea that the community shows a rainy season phase and a dry season phase. The 25 sampling dates are arranged into two main groups by PCA. The upper group includes samples taken from February through May of 1984, and from January and February of 1985 (dry seasons). The lower group covers the period corresponding to the rainy season of 1984 (Figure 15.3A). For the second year the results of PCA, performed on half the number of sampling dates used in the first year, clearly separate the dry season phase of 1985, which comprises the period from March through June (Figure 3B, dots 1–4). A sharp separation between the rainy season samples of 1985 (Figure 3B, dots 5–8) from those of the following dry season (dots 9–12) is not as evident as for the first year, at least in the two dimensions represented by PC1 and PC2. However, dots 9–12 form a more coherent group than those of the rainy season. Other factors that may have influenced the difference in position between these dry season periods is related to the drastic increase in abundance of *C. citrona*, and the presence of seven new species in the samples. Strong seasonal fluctuations in populations were also observed during the remainder of 1986 and early 1987 (Vargas 1989a,b).

Three facts evident from the data presented above are noteworthy as they are contrary to popular conceptions regarding tropical benthic communities. The total number of species collected is less than expected, as well as the presence of marked oscillations in abundance of certain species and less evidence of seasonality at the population and community levels. These expectations have their origin in early generalizations considering the tropics as a species-rich, environmentally stable region of the biosphere (e.g. MacArthur 1972). Recent evidence, however, indicates that this paradigm is in need of revision for tropical soft-bottom communities (Maurer and Vargas 1984; Warwick and Ruswahyuni 1987; Maurer et al. 1988; Vargas 1987, 1988a, 1989a; Alongi 1989, 1990).

15.3 ECOLOGICAL ROLES AT DIFFERENT LATITUDES

The existence of seasonality at the population and community levels, the presence of strong population fluctuations and a modest number of species, make this tropical intertidal flat suitable for additional comparisons with similar

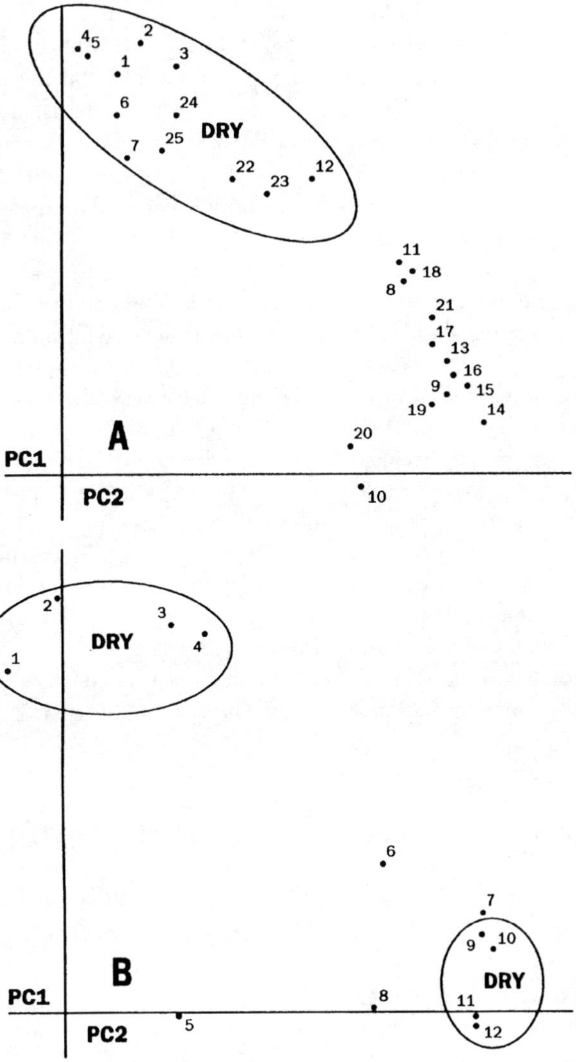

Figure 15.3 First and second Principal Components (PC) calculated from the number of individuals collected per date. A: 25 sampling dates (February 1984 to February 1985), B: 12 sampling dates (March 1985 to March 1986). Groups of dots corresponding to dry season dates are within ellipses, and the remaining correspond to rainy season dates

environments in temperate latitudes. Intertidal flats worldwide are sedimentary environments whose local characteristics are defined by a particular combination of the percentages of sand, silt, clay, and organic matter content. Moving on or through the sediments requires morphological adaptations that are necessary regardless of latitude. Thus, the fauna of tidal flats from different geographical regions may show similarities in the types of main body-designs present. Organisms such as tube-building polychaete worms, burrowing crabs, crawling snails, and sediment-reworking hemichordates, are some of the life-forms to be expected in tidal flats. Recent evidence, however, reveals that these similarities can be traced at lower taxonomic levels. Reise (1991) has noted that comparisons between benthic faunas of tidal flats from different latitudes can be made by looking at closely related species occupying similar niches and unrelated species performing analogous ecological functions.

A comparison of the intertidal flat in the Gulf of Nicoya with a similar environment in a temperate latitude estuary (Georgia, USA) shows more than ten congeneric species playing similar ecological roles (Table 15.3). On the other hand unrelated species such as microcrustaceans, amphipods in the temperate site and cumaceans at the tropical site, may be playing analogous roles as modifiers of sediment characteristics. These roles are important since intertidal benthic community structure has been related to the intensity and frequency of biological and physical disturbance (Thistle 1981; Grant 1983; Reise 1985). Organisms create small disturbances during the processes of feeding and moving on or through the sediments. The relative importance of these roles in structuring tropical communities remains to be investigated. For instance, how different are the rates and tactics of predation by the tropical naticid snail *Natica unifasciata* from those of its temperate counterpart, the naticid *Polinices duplicatus*? Though this search for analogous ecological roles is related to functional aspects (energy flow) of the community, very little information exists on energy transfer in tropical marine benthic systems (Alongi 1989, 1990).

15.4 EXPERIMENTAL MANIPULATION OF THE COMMUNITY

Caging experiments in different latitudes have produced, with a few exceptions, a significant increase in the number of individuals and species inside caged areas (Vargas 1979, 1988a, 1995). The usual explanation for these results is that the cages exclude predators, thus allowing some of the prey species to increase in abundance inside the 'predator-free' area. However, many species of benthic invertebrates, which remain as part of the caged community, are themselves active predators, as discussed by Ambrose (1984). Thus, the impact of disturbance during feeding and other activities might be more important than predation on a community where most of the species are sediment reworkers, such as in the tropical flat (Table 15.3). In spite of these interpretative problems, caging experiments are valuable tools for the study of biological interactions (Virnstein 1978; Vargas 1995). However, only one experimental caging study has been published for a tropical intertidal mudflat (Vargas 1988a). A comparison of the results of this caging experiment with one the author conducted in a temperate location (Vargas 1979) yields several interesting observations (Table 15.4): A

Table 15.3 List of selected species and their inferred ecological roles (other roles are possible) at a tropical (Gulf of Nicoya, Costa Rica)and a temperate (Sapelo Island, USA) intertidal flat. Data from Vargas (1988a), and Howard and Dörges (1972), respectively

Punta Morales (10°N)	Nannygoat (31°N)	Roles*
POLYCHAETA		
Mediomastus californiensis	*Heteromastus filiformis*	2, 6, 9
Paraprionospio pinnata	*Spiophanes bombyx*	1, 7, 10
Carazziella citrona	*Polydora websteri*	1, 7
Diopatra ornata	*D. cuprea*	4, 7, 10
Lumbrineris tetraura	*L. impatiens*	9
Nephtys monroi	*N. picta*	3, 9
Glycera americana	*G. dibranchiata*	3, 9
CRUSTACEA		
Pinnixa valerii	*P. chaetopterana*	4, 9
Albunea lucasia	*A. pareti*	9
Corophium sp.	*C. simile*	7, 10
Coricuma nicoyensis	*Oxyurostylis*	1, 9
MOLLUSCA		
Tagelus bourgeoisae	*Solen virdis*	5, 9
Nassarius luteostoma	*N. obsoletus*	4, 9, 10
Natica unifasciata	*Polynices duplicatus*	3, 9, 11
Dosinia dunkeri	*D. discus*	5
Tellina rubescens	*T. texana*	1, 5, 9
ECHINODEMATA		
Encope stokessi	*Mellita quinquiesperforata*	9, 10, 11
HEMICHORDATA (Entero–pneusta) sp.	*Balanoglossus*	2, 8, 9

	Punta Morales	Sapelo Island**
Total number of specimens	4412	3462
Total number of species	78	50
Polychaete species (%)	38	38
Crustacean species (%	27	36)
Molluscan species (%)	19	18
Other groups (%)	16	8

* 1: surface deposit feeders, 2: subsurface deposit feeders, 3: carnivores, 4: scavengers, 5: filter feeders, 6: fecal pellet production, 7: tube builders, 8: fecal mounds, 9: sediment reworking, 10: sediment stabilizer, 11: feeding tracks

** A 500 μM mesh sieve was used in Punta Morales and an 800 μM sieve in Sapelo Island

significant increase in the number of individuals occurred at the cage site in the temperate environment, although the mesh size of the cages was 7 mm coarser than that used in the tropical site. The number of species making up more than 70% of the total abundance is less than ten in both studies. The relative positions of the dominant species changed in response to caging in both studies. The Shannon–Weiner diversity index (H) was found to be of little use when comparing data

Table 15.4 Results from caging experiments conducted from 19 July to 19 November 1977, at a temperate shallow (<1 m deep) muddy sand site (Rehoboth Bay, USA), and from 10 August to 16 December, 1985, at a tropical intertidal mudflat (Gulf of Nicoya, Costa Rica). Data from Vargas (1979, 1988a). Numbers are percentages of total number of individuals. N=total number of individuals; S=total number of species; H=Shannon-Weiner Diversity

Rehoboth Bay (38°N)			Punta Morales (10°N)		
UNCAGED			UNCAGED		
Heteromastus filiformis	P*	34.20	*Caraziella citrona*	P	0.48
Streblospio benedicti	P	31.34	*Bivalvia* juveniles	P	15.48
Limnodriloides monothecus	O	8.42	*Mediomastus californiensis*	P	14.90
Gemma gemma	B	6.20	*Pinnixa valerii*	D	7.32
Polydora ligni	P	3.35	*Paraprionospio pinnata*	P	3.66
Corophium insidiosum	A	2.96	*Coricuma nicoyensis*	C	2.58
Ampelisca vadorum	A	1.40	*Tagelus bourgeoisae*	B	2.58
Erichtonus brasiliensis	A	1.35	*Ceratocephale crosslandi*	P	2.33
Glycinde solitaria	P	1.20	*Panopeus* sp.	D	2.25
Nassarius vibex	M	0.98	*Prionospio delta*	P	2.25
		91.40			73.83
N = 1924			N = 1201		
S = 39			S = 66		
H'= 1.37–3.40			H' = 2.53–3.12		
CAGED (12 mm mesh)			CAGED (5 mm mesh)		
S. benedicti		50.09	*Bivalvia,* juveniles		0.52
H. filiformis		36.70	*M. californiensis*		12.51
P. ligni		2.45	*P. citrona*		10.39
L. monothecus		1.88	*P. valerii*		4.66
G. gemma		1.13	*T. bourgeoisae*		3.25
Capitella capitata	P	1.04	*P. pinnata*		2.40
A. vadorum		0.77	*P. delta*		2.26
G. solitaria		0.57	*Glycinde armigera*	P	2.05
Diopatra cuprea	P	0.52	*Panopeus* sp.		1.62
Eteone heteropoda	P	0.49	*Spiophanes soederstroemi*	P	1.55
		95.64			81.21
N = 8791			N = 1414		
S = 46			S = 58		
H' = 1.15–2.27			H' = 1.54–3.08		

* P = Polychaeta, O = Oligochaeta, B = Bivalvia, A = Amphipoda, G = Gastropoda, D = Decapoda, C = Cumacea
At both sites macropredators considered to be excluded were: benthic fish, crabs of the genus *Callinectes*, other crustaceans, and shore birds

among and between environments. The sub-surface deposit-feeding capitellid polychaetes (*Heteromastus filiformis* and *Mediomastus californiensis*, respectively) failed to show significant increases in abundance in response to caging. In contrast, the surface deposit-feeding spionid polychaetes (*Streblospio benedicti* and *Carazziella citrona*, respectively) presented important changes in abundance between caged and uncaged sites. Several of these observations are similar to those obtained by Reise (1977), who deployed cages of 5 mm mesh size and smaller, in an intertidal flat in the North Sea. He also found that the relative positions of the numerically important species changed from the uncaged to the caged site, that the diversity index (*H*) failed to reflect changes in abundance, and that the polychaete *Heteromastus filiformis*, which has a mode of life similar to the tropical *Mediomastus californiensis*, showed no increase in abundance within the cages. Moreover, the spat of young bivalves settled in his cages, a result also obtained in the tropical flat experiment (Table 15.4).

From the experiment in the tropical site, it is concluded that the relative importance of birds, fish, and crabs as predators regulating community structure was probably negligible (Vargas 1988). From the point of view of trophic links between the intertidal flat and nearby environments (Table 15.1), it may also be argued that the energy contribution via macropredators to these environments is probably of a small magnitude. The author arrived at this preliminary conclusion after observing only 13 species of birds in low numbers foraging at low tide on the intertidal flat. Among these species, the herons, *Casmerodius albus* and *Egretta caerulea*, the plover, *Pluvialis squatarola*, and the sandpipers, *Numenius phaeopus, Catoptrophorus semipalmatus* and, *Limnodromus griseus*, were observed frequently (Vargas 1988a). The results obtained several years later by Barrantes and Pereira (1992) indicate, however, that shorebirds use tidal flats intensively in the Gulf of Nicoya (Table 15.1). They found 27 species of Charadriiform birds, of which *Pluvialis squatarola, Numenius phaeopus, Charadrius wilsonia*, and *Catoptrophorus semipalmatus* were present throughout the year at the Chomes flat, next to the author's study site. Individuals of *Charadrius semipalmatus, Limnodromus griseus, Calidris mauri*, and *Arenaria interpres* migrate to North America during the reproductive season. The remainder of the birds were present in low abundance.

Nearly 214 species of fish have been reported in the Gulf of Nicoya estuary (Bartels et al. 1984). Campos and Corrales (1986) examined the stomach contents of 21 species of fish from the Gulf of Nicoya and found that crustaceans represented 58% of the prey organisms identified, followed by fish (18.2%), ophiuroids (7.8%), polychaete worms (7.4%), and molluscs (3.6%). Phillips (1983) collected 61 species of fish using a seine on the intertidal flat discussed in this chapter. Of those species the catfish, *Arius furthii* was among the more common. Szelistowski (1990) found 120 species of fish associated with the mangroves next to this intertidal flat (Figure 15.1). Among these fish were microalgal grazers, detritivores, and several types of carnivores. Crustacean predators such as the blue crab, *Callinectes arcuatus*, and mantis shrimp of the genus *Squilla* are common in this environment (Dittel et al. 1985; Dittel 1991). Diets of *C. arcuatus* were investigated in the upper Gulf of Nicoya by Dittel (1993). She identified 12 main food categories, and crustaceans constituted 29% of the diet, while mussels such as *Mytella guyanensis* and other bivalves represented 18% of the 556 foregut contents examined. Snails, brittle stars and plant material were also present. The relative percentages of food items varied between sites and seasons, suggesting that spatial and temporal changes in diet are related to prey availability. Considering the abundance of potential predators at the Punta Morales intertidal flat, the weak effects observed in the caging experiment (Table 15.4) appear contradictory. In a review of the problems associated with the use of cages, Hall and colleagues (1990) have stated that one cannot talk about weak or zero effects of the predators on non-responsive species unless the latter are known to be prey. Hall and colleagues (1990) also emphasize that published results of experiments in soft sediments generally give the impression that predators have large effects on benthic community structure. However, the extent to which these results reflect the prevalence of strong trophic links in benthic food webs is open to debate.

15.5 TIDAL FLATS AND LINKAGES TO NEARBY ECOSYSTEMS

The link between intertidal flats and nearby environments may be established via the feeding activities of crustaceans, fish, and birds as discussed earlier. Reise (1985) has stated that for temperate tidal flats, these environments serve as an ecological turntable between two adjacent ecosystems. The flats import organic matter from the open waters and saltmarshes, produce their own matter via benthic photosynthetic bacteria and microalgae, recycle part of the matter through the benthic food web, and export some of this energy via non-resident predators. In temperate tidal flats the primary productivity shows seasonal peaks due to strong oscillations of light intensity and temperature. For tropical environments it may be theorized that because photosynthesis is possible the year around, and because of the high turnover rates of benthic diatoms, the primary productivity of the microalgae on tropical mudflats may exceed that of mangrove trees. No comparative studies of this nature are available. The importance of epibenthic microalgae (mostly motile pennate diatoms) in the fixation of energy in intertidal flats has been emphasized in early studies, as the one conducted in the Ythan estuary, Scotland, by Leach (1970). Moreover, the primary production of well mixed shallow estuarine waters is often greatly augmented by entrainment of benthic diatoms into the water column (Shaffer and Sullivan 1988). Pennate diatoms are also important in the sediments of the Punta Morales intertidal flat in the Gulf of Nicoya, where sometimes the substrate is covered with a thin yellowish layer of microalgae (Vargas 1988b).

The flow of energy in mangroves was studied by Odum and Heald (1972), who viewed these forests as the places where large amounts of carbon are fixed by trees to support detritus-based food webs, in the mangrove swamps, the intertidal flats, and offshore waters. The best approach to trace the fate of organic matter through the system is by the use of carbon isotopes. Such a study has been conducted in a Malaysian mangrove swamp by Rodelli and colleagues (1984). They found, in agreement with Odum and Heald's (1972) model, that bivalves, shrimps, crabs, and fish obtained carbon from the mangrove trees and that a portion of this carbon was exported offshore as detritus. A similar study has been conducted in the Gulf of Nicoya by Szelistowski (1990). His mangrove creek site was near the tidal flat discussed in this chapter (Figure 15.1). Unlike the results of Rodelli and colleagues (1984) for the Malaysian mangroves, he found that stable isotope ratios (C, N, S) of fish tissues indicated that phytoplankton and benthic microalgae, rather than the mangroves, are the major producers supporting fish food webs at the site. Moreover, in studies of the ichthyoplankton conducted at the same mangrove creek, Ramírez and colleagues (1989, 1990) found no evidence indicating that the mangroves at Punta Morales serve as spawning or larval nursery grounds relative to adjacent, offshore waters. These results may indicate that one of the main sources of energy for the mangrove creek fish fauna is the nearby intertidal flats. Additional support for this inference is based on the observation that the deposit-feeding dominated macrofauna attains relatively high densities at the flat (Table 15.1). The same seems to be true for the meiofauna on the Punta Morales mudflat (Table 15.1), which is dominated by nematodes (88%) and foraminiferans (5.9%) but also includes ciliated protozoans and flatworms (De la Cruz and Vargas 1987; Vargas 1988b). Many species of nematodes, ciliates, and flatworms are known to feed on diatoms at the sediment surface (Reise 1985).

The intertidal flats in Punta Morales also include a rich fauna of decapod crustaceans, of which fiddler crabs (*Uca spp.*), blue crabs (*Callinectes arcuatus*), pinnotherids (*Pinnixa* spp.), and porcellanids (*Petrolisthes* spp.) are frequent. The crabs release larvae into the overlying waters, and these larvae may be carried by currents into the mangroves or to offshore waters. The study by Dittel and colleagues (1991), in Punta Morales, provides the first data on the direction of larval movement. They found that net larval transport of intermediate zoeal stages of most taxa was into the mangrove creek, while consistent net export of zoea 1 occurred in *Uca* spp. These results indicate that development of intermediate stages might occur in the creek. In this sense the mangroves might provide a habitat where crabs from the intertidal flat may continue larval development. Transport by floating mangrove leaves appears also to be important in this mangrove creek for recruitment of crabs, and shrimps of the genus *Penaeus*, into the mangroves (Wehrtmann and Dittel 1990). On the other hand, the transport of larvae into offshore waters may expose them to unsuitable habitats for larval development or adult existence. There is evidence that the larvae of many species of estuarine invertebrates have evolved mechanisms which allow them to be retained in their parent estuary, or to export their larvae into coastal waters where further development occurs (Epifanio 1988). Epifanio and Dittel (1982) made a comparison of larval retention mechanisms for the crabs, *Callinectes sapidus*, from the Delaware estuary (USA) and *C. arcuatus* from the Gulf of Nicoya. This comparison prompted research in the Gulf of Nicoya, where females of *C. arcuatus* were found to migrate to the lower estuary to spawn during the dry season and migrated back into the upper estuary during the rainy season (DeVries et al. 1983;

Dittel et al. 1985). The zoeal larvae of *C. arcuatus* are apparently retained in the lower Gulf of Nicoya, and larval metamorphosis seems to occur in this region Recruitment to adult populations is effected by juvenile migration into the upper Gulf of Nicoya (Dittel and Epifanio 1990). Thus, the transport of decapod larvae into the mangroves discussed above, and of *C. arcuatus* into offshore waters implies that the intertidal flats export energy to these environments in the form of larval biomass. The flats also import energy from these environments in the form of juvenile crabs. The direction of the net transport of energy remains to be investigated.

The links between the flats and nearby ecosystems are also evident when the impacts of coastal human populations are considered. Boating and fishing activities increase with the rising tide over the flats. At low tide the harvest of benthic resources, such as bivalves, takes place. In recent years overfishing of the ark clam *Anadara grandis* has contributed to drastically reduced natural populations in the Gulf of Nicoya (Fournier and De la Cruz 1987). This overfishing problem prompted recommendations by Rojas and colleagues (1988) for the management of the jackknife clam, *Tagelus peruvianus*, in the Gulf. Another resource in this environment is the giant onuphid polychaete worm *Americonuphis reesei*, which has been used as fish bait, and as a dietary supplement to accelerate gonad maturation in the mariculture of penaeid shrimp (D'Croz et al. 1988). The impact of this fishery on community structure is unknown. In addition, pesticide residues and trace metals are brought into contact with the sediments and biota (De la Cruz 1989; Fuller et al. 1990).

In spite of the commercial importance of some of the resources present in intertidal flats, these environments are generally considered worldwide as wastelands. In this sense Reise (1985) pointed out that tidal flats 'remained for long the muddy backwaters of marine ecological research. Compared to the vastness of the open sea, tidal flats seemed to be a negligible fringe, full of aberrant phenomena unworthy of attention to a science concerned with the world's largest ecosystem.' This perception is changing towards seeing the mudflats as an endangered ecotone, crucial for the survival of many species.

ACKNOWLEDGEMENTS

This research was made possible by a grant from the Universidad de Costa Rica (Projects 808-88-423 and 888-90-410). I thank C. Roman and two anonymous reviewers for comments on an earlier draft. Carmen Duran and Ruth Mora typed the final draft, and A. Leon prepared the illustrations. This is CIMAR contribution 225.

REFERENCES

Alongi, D. 1989. Ecology of soft-bottom benthos: a review with emphasis on emerging concepts. *Revista de Biología Tropical* 37: 85–100.

Alongi, D. 1990. The ecology of tropical soft-botton benthic ecosystems. *Oceanography and Marine Biology Annual Revue* 28: 381–496.

Ambrose, W.G. 1984. Role of predatory infauna in structuring marine soft-bottom communities. *Marine Ecology Progress Series* 17: 109–115.

Barrantes, G. and A. Pereira. 1992. Abundancia y fluctuaciones de aves limícolas (Charadriiformes) en una playa fangosa de Chomes, Costa Rica. *Revista de Biología Tropical* 40: 303–307.

Bartels, C., K.S. Price, M. López and W. Bussing. 1984. Ecological assessment of finfish as indicators of habitats in the Gulf of Nicoya, Costa Rica. *Hydrobiologia* 112: 197–207.

Borjesson, D.L. and W.A. Szelistowski. 1989. Shell selection, utilization, and predation in the hermit crab *Clibanarius panamensis* Stimpson in a tropical mangrove estuary. *Journal of Experimental Marine Biology and Ecology* 133: 213–228.

Campos, J. and A. Corrales. 1986. Preliminary results on the trophic dynamics of the Gulf of Nicoya, Costa Rica. *Anales del Instituto Ciencias del Mar y Limnología de la UNAM* 13: 329–334.

D'Croz, L., L. Wong, G. Justine and M. Gupta. 1988. Prostaglandins and related compounds from the polychaete worm *Americonuphis reesei* Fauchald (Onuphidae) as possible inducers of gonad maturation in Penaeid shrimps. *Revista de Biología Tropical* 36: 331–332.

De la Cruz, E. 1989. Organochlorine and total mercury in the *Anadara tuberculosa* bivalve and other species from the Nicoya Gulf, Costa Rica, M.Sc. Thesis. Vrije Universiteit, Brussels.

De la Cruz, E. and J.A. Vargas. 1987. Abundancia y distribución vertical de la meiofauna en la playa fangosa de Punta Morales, Golfo de Nicoya, Costa Rica. *Revista de Biología Tropical* 35: 363–367.

DeVries, M., C.E. Epifanio and A.I. Dittel. 1983. Reproductive periodicity of the tropical crab *Callinectes arcuatus* Ordway in the Gulf of Nicoya, Costa Rica. *Estuarine, Coastal and Shelf Science* 17: 709–716.

Dittel, A. 1991. Distribution, abundance and sexual composition of stomatopod crustacea in the Gulf of Nicoya, Costa Rica. *Journal of Crustacean Biology* 11: 269–276.

Dittel, A. 1993. Cambios en los hábitos alimentarios de *Callinectes arcuatus* (Crustacea: Decapoda) en el Golfo de Nicoya, Costa Rica. *Revista de Biología Tropical* 41: 639–646.

Dittel, A.I. and C.E. Epifanio. 1990. Seasonal and tidal abundance of crab larvae in a tropical mangrove system, Gulf of Nicoya, Costa Rica. *Marine Ecology Progress Series* 65: 25–34.

Dittel, A., Epifanio, C.E. and J.B. Chavarría. 1985. Population biology of the portunid crab *Callinectes arcuatus* Ordway in the Gulf of Nicoya, Costa Rica; Central America. *Estuarine, Coastal and Shelf Science* 20: 593–602.

Emig, C. and J.A. Vargas. 1990. *Glottidia audebarti* (Broderip) (Brachiopoda, Lingulidae) from the Gulf of Nicoya, Costa Rica. *Revista de Biología Tropical* 38: 251–258.

Epifanio, C.E. 1988. Dispersal strategies of two species of swimming crab on the continental shelf adjacent to Delaware Bay. *Marine Ecology Progress Series* 49: 243–248.

Epifanio, C.E. and A. Dittel. 1982. Comparison of dispersal of crab larvae in Delaware Bay, (USA) and the Gulf of Nicoya, Central America, pp. 477–487. *In*: V. Kennedy (ed.), *Estuarine Comparisons*. Academic Press, New York.

Epifanio, C.E., D. Maurer and A.I. Dittel. 1983. Seasonal changes in nutrients and dissolved oxygen in the Gulf of Nicoya, a tropical estuary on the Pacific coast of Central America. *Hydrobiologia* 101: 231–238.

Fournier, M.L. and E. De la Cruz 1987. Reproduction of the cockle *Anadara grandis*, i Costa Rica. *Naga* 10: 6.

Fuller, C.C., J.A. Davis, D.J. Cain, P.J. Lamothe, T.L. Fries, G. Fernández, J.A. Varga and M.M. Murillo. 1990. Distribution and transport of sediment-bound meta contaminants in the Río Grande de Tárcoles, Costa Rica (Central America). *Wate Research* 24: 805–812.

Grant, J. 1983. The relative magnitude of biological and physical sediment reworking in a intertidal community. *Journal of Marine Research* 41: 673–689.

Hall, S.J., D. Raffaelli and W.R. Turrel. 1990. Predator-caging experiments in marin systems: a reexamination of their value. *American Naturalist* 136: 657–672.

Howard, J.D. and J. Dörges. 1972. Animal–sediment relationships in two beach-related tida flats; Sapelo Island, Georgia. *Journal of Sedimentary Petrology* 42: 608–623.

Kapetsky, J.M., L. McGregor and H. Nanne. 1987. A geographical information system an satellite remote sensing to plan for aquaculture development: a FAO-UNEP/GRII cooperative study in Costa Rica. *FAO Fisheries Technical Papers* 287: 1–51.

Leach, J.H. 1970. Epibenthic algal production in an intertidal mudflat. *Limnology an Oceanography* 15: 514–521.

Lizano, O. and J.A. Vargas. 1993. Distribución espacio-temporal de la salinidad y l temperatura en la parte interna del Golfo de Nicoya. *Tecnología en Marcha* 12: 3–16.

MacArthur, R.H. 1972. *Geographical Ecology*. Princeton University Press, Princeton.

Maurer, D. and J.A. Vargas. 1984. Diversity of soft-bottom benthos in a tropical estuary Gulf of Nicoya, Costa Rica. *Marine Biology* 81: 97–106.

Maurer, D., J.A. Vargas and H.K. Dean. 1988. Polychaetous annelids from the Gulf o Nicoya, Costa Rica. *Internationale Revue ges Hydrobiologie* 73: 43–59.

Odum, W.E. and E.J. Heald. 1972. Trophic analyses of an estuarine mangrove community *Bulletin of Marine Science* 22: 671–738.

Perry, D. 1988. Effects of associated fauna on growth and productivity in the red mangrove *Ecology* 69: 1064–1075.

Phillips, P.C. 1983. Diel and monthly variation in abundance, diversity and composition o littoral fish populations in the Gulf of Nicoya, Costa Rica. *Revista de Biología Tropica* 31: 297–306.

Ramírez, A.R., W.A. Szelistowski and M. López. 1989. Spawning pattern and larva recruitment in Gulf of Nicoya anchovies (Pisces: Engraulidae). *Revista de Biologí Tropical* 37: 55–62.

Ramírez, A.R., M.I. López and W.A. Szelistowski. 1990. Composition and abundance o ichthyoplankton in a Gulf of Nicoya mangrove estuary. *Revista de Biología Tropical* 38 463–466.

Reise, K. 1977. Predator exclusion experiments in an intertidal mud flat. *Helgolände Meeresuntersuchungen* 30: 263–271.

Reise, K. 1985. *Tidal Flat Ecology*. Springer, Berlin.

Reise, K. 1991. Macrofauna in mud and sand of tropical and temperate tidal flats, pp. 211-216 *In*: M. Elliot and J.P. Ducrotoy (eds.), *Estuaries and Coasts. Spatial and Tempora Intercomparisons*. Olsen and Olsen, Fredensborg.

Rodelli, M.R., J.N. Gearing, P.J. Gearing, N. Marshall and A. Sasekumar. 1984. Stabl isotope ratio as a tracer of mangrove carbon in Malaysian ecoystems. *Oecologia* 61: 326-333.

Rojas, J., C.E. Villalobos, F. Chartier and C.R. Villalobos. 1988. Tamaño, densidad y reproducción de la barba de hacha *Tagelus peruvianus* (Bivalvia: Solecurtidae) en e estero de Puntarenas, Costa Rica. *Revista Biología Tropical* 36: 479–483.

Shaffer, G.P. and M.J. Sullivan. 1988. Water column productivity attributable to displace benthic diatoms in well-mixed shallow estuaries. *Journal of Phycology* 24: 132–140.

Szelistowski, W.A. 1990. Importance of mangrove plant litter in fish food webs and as temporary, floating habitat in the Gulf of Nicoya, Costa Rica. Ph.D. Dissertation. University of Southern California, Los Angeles.

Thistle, D. 1981. Natural physical disturbances and communities of marine soft-bottoms. *Marine Ecology Progress Series* 6: 223–228.

Vargas, J. A. 1979. Predation and community structure of soft-bottom benthos in Rehoboth Bay, Delaware. M.Sc. Thesis. University of Delaware, Lewes.

Vargas, J. A. 1987. The benthic community of an intertidal mud flat in the Gulf of Nicoya, Costa Rica. Description of the community. *Revista de Biología Tropical* 35: 229–316.

Vargas, J.A. 1988a. Community structure of macrobenthos and the results of macropredator exclusion on a tropical mud flat. *Revista de Biología Tropical* 36: 287–308.

Vargas, J.A. 1988b. A survey of the meiofauna of an Eastern Tropical Pacific intertidal mud flat. *Revista de Biología Tropical* 36: 541–544.

Vargas, J.A. 1989a. A three year survey of the macrofauna of an intertidal mud flat in the Gulf of Nicoya, Costa Rica, pp. 1905–1919. *In*: O. Magoon, M. Converse, D. Miner, L.T. Tobin, D. Clark (eds.), *Coastal Zone 89*. American Society of Civil Engineers, New York.

Vargas, J.A. 1989b. Seasonal abundance of *Coricuma nicoyensis* Watling and Breedy (Crustacea, Cumacea), on an intertidal mud flat in the Gulf of Nicoya, Costa Rica. *Revista de Biología Tropical* 37: 207–211.

Vargas, J.A. 1995. The Gulf of Nicoya estuary, Costa Rica. Past, present, and future cooperative research. *Helgoländer Meeresuntersuchungen* 49: 821–828.

Virnstein, R.W. 1978. Predator caging experiments in soft sediments: caution advised, pp. 261–273. *In*: M.L.Wiley (ed.), *Estuarine Interactions*. Academic Press, New York.

Voorhis, A., C.E. Epifanio, D. Maurer, A.I. Dittel and J.A. Vargas. 1983. The estuarine character of the Gulf of Nicoya, an embayment on the Pacific coast of Central America. *Hydrobiologia* 99: 225–237.

Warwick, R.M. and Ruswahyuni. 1987. Comparative study of the structure of some tropical and temperate marine soft-bottom macrobenthic communities. *Marine Biology* 95: 641–649.

Wehrtmann, I. and A.I. Dittel. 1990. Utilization of floating mangrove leaves as a transport mechanism of estuarine organisms, with emphasis on Decapod Crustacea. *Marine Ecology Progress Series* 60: 67–73.

Part 3

HUMAN ALTERATIONS AND MANAGEMENT

Human impacts are addressed in nearly every chapter in this book. The following chapters treat human alterations more explicitly and in a management context. Garniel and Mierwald (Chapter 16) look at long-term effects in an estuary that has been exposed to human alterations for millennia and will be subject to even greater modification in the future as funds become available. They place the channel fluctuations introduced by van den Berg and colleagues in a hazards context, underscoring the cycles of shoreline change related to reclamation and destruction–rebuilding cycles. Jackson (Chapter 17) looks at estuarine evolution from a smaller spatial and temporal scale but indicates that many of the aspects of evolution at larger scales, such as displacement of harbors down the estuary due to shoaling, construction of buildings and shore protection and alteration of shoreline environments to accommodate human activities in the upland. Doody (Chapter 18) places the chapters on shoreline evolution in Part 1 in a planning context, noting that the conversion of estuarine environments from natural to human-altered forms can be reversed by reinstating tidal conditions in formerly reclaimed land. Doody's chapter points to the link between science and policy, a topic that is examined further by Burroughs (Chapter 19) and Phillips (Chapter 20). Burroughs, concentrating on the discharge and control of pollutants, identifies the stages of the policy process where scientific input is likely to be most effective. Phillips concentrates on linkages at the upper boundary of the estuarine shore, identifying the spatial dimensions of scientific and management issues related to buffer zones and underscoring the difficulty of resolving conflicting demands for use and management.

Chapter Sixteen

Changes in the morphology and vegetation along the human-altered shoreline of the Lower Elbe

ANNICK GARNIEL and ULRICH MIERWALD
Kiel Institute of Landscape Ecology

16.1 INTRODUCTION

Originally the estuary of the Elbe was an ecotone of high diversity shaped by interconnected marine, terrestrial and fluvial dynamics. The only exchange medium, and the medium of energy flow, was water. Today, after 2000 years of human interference, the estuary has lost much of the character of its ecotone. Since early settlements, changes in population and steadily increasing technology resulted in heavy impacts on the ecosystem. The major aim of human alterations was to control the water by embanking, draining and channeling the stream and its tributaries. The estuarine environment was compartmentalized, and the alluvial marsh was cut off from river flow and tidal flooding as well as from surface runoff and subsurface seepage from the surrounding hills.

The following sections give a short synopsis of landscape history and a reconstruction of the historical loss of the character of the ecotone of the Elbe estuary. We emphasize the historical perspective of anthropogenic impacts on the estuarine marsh since early settlements and their consequences for the vegetation. Implications of a restoration project of a tidally influenced tributary of the Elbe near Hamburg are then discussed.

Estuarine Shores: Evolution, Environments and Human Alterations. Edited by K. F. Nordstrom and C. T. Roman.

16.2 THE ELBE ESTUARY

The Elbe has its source in the 'Giant Mountains' at the border of Poland and the Czech Republic. After 1165 km, passing through heavily industrialized areas, it opens into the southeastern part of the North Sea (German Bight) at Cuxhaven. The drainage basin extends to 144 000 km^2. The average discharge at Hamburg is 700 m^3 s^{-1} (survey by Buchwald (1990)). The estuarine sector of the river is the Lower Elbe, extending from Geesthacht near Hamburg to the North Sea at Cuxhaven (Figure 16.1).

16.2.1 Morphology

The estuary of the Elbe between Geesthacht and Cuxhaven is about 150 km long. At Hamburg the riverbed is 500 m wide while at Cuxhaven it opens to 15 km. Under natural conditions, the average depth of the Lower Elbe at Hamburg should be around 4.7 m. At the moment, the navigable channel is at least 13.5 m deep with plans for further dredging. In the vicinity of Hamburg, stream velocity is checked by opposing tidal currents. Here the stream divides into several channels and forms a delta 30 km long, 10 km wide, within the estuary. Downstream from Hamburg the stream shows a slightly meandering course. Estuarine sediments extend as far as Hamburg. Upstream marshes are the results of fluvial processes only. The marshes are built up mainly of fine-grained organic deposits that have been deposited behind protective natural levees. At Cuxhaven, the river flows into the Wadden Sea, a tidal flat area of 10 000 km^2 (Embleton 1984). The mouth of the estuary is characterized by subtidal sand shoals and channels (Reineck and Singh 1980).

16.2.2 Geology

The Lower Elbe is embedded in the Fennoscandian glacial deposits of the North European Lowlands (Figure 16.2). Its valley is surrounded by hills of an average height of 50 m and a maximum height of 150 m. According to Behre and colleagues (1979) a large depression emerged north of the high outwash plains of the Harburg Hills during the retreat of the ice after the second major advance of the Saalian Glaciation. A valley was formed inside this depression during the final phase of this glaciation (Warthe-Stage). A gorge-like valley was eroded as a consequence of the extremely low sea level during the glacial maximum. The deep interglacial sea level caused considerable erosion. At Hamburg the bottom of this valley is found at −30 m relative to mean sea level (MSL). During the Eemian Interglacial, this valley was filled up with fine clastic sediments to a level of −20 m to −15 m MSL.

The continental ice-sheet of the Weichselian Glaciation did not extend to the Hamburg area. The actual valley was formed as the major meltwater channel (the

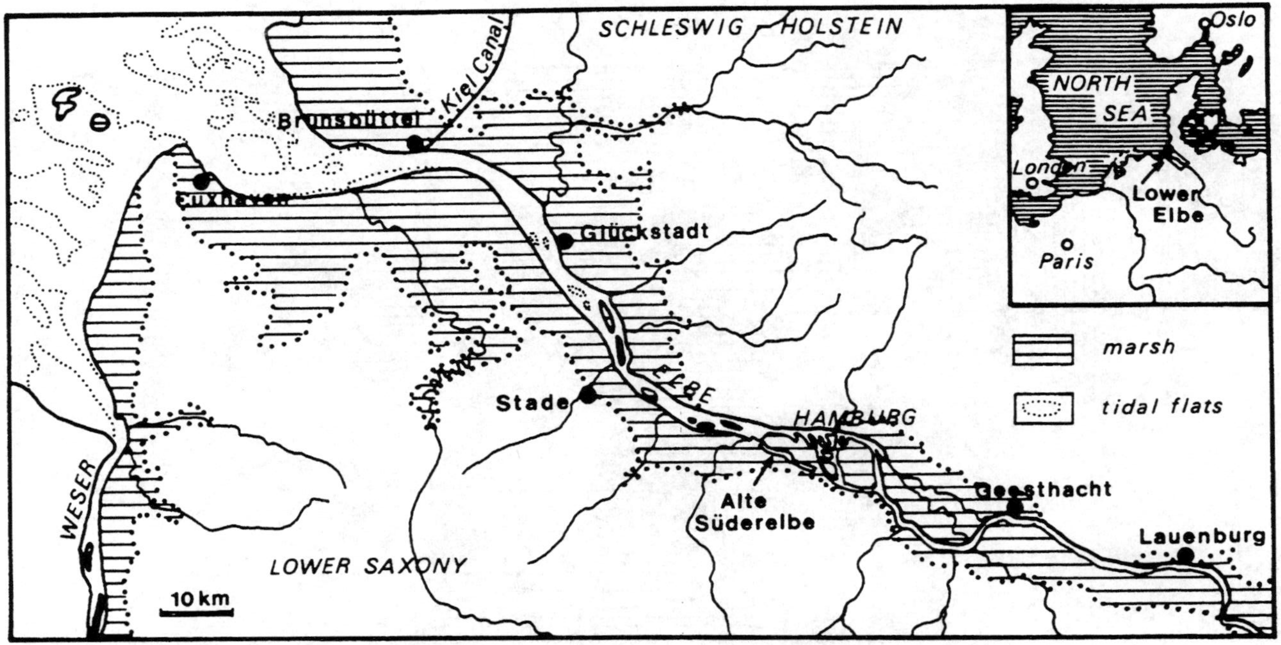

Figure 16.1 The Elbe from Lauenburg to Cuxhaven

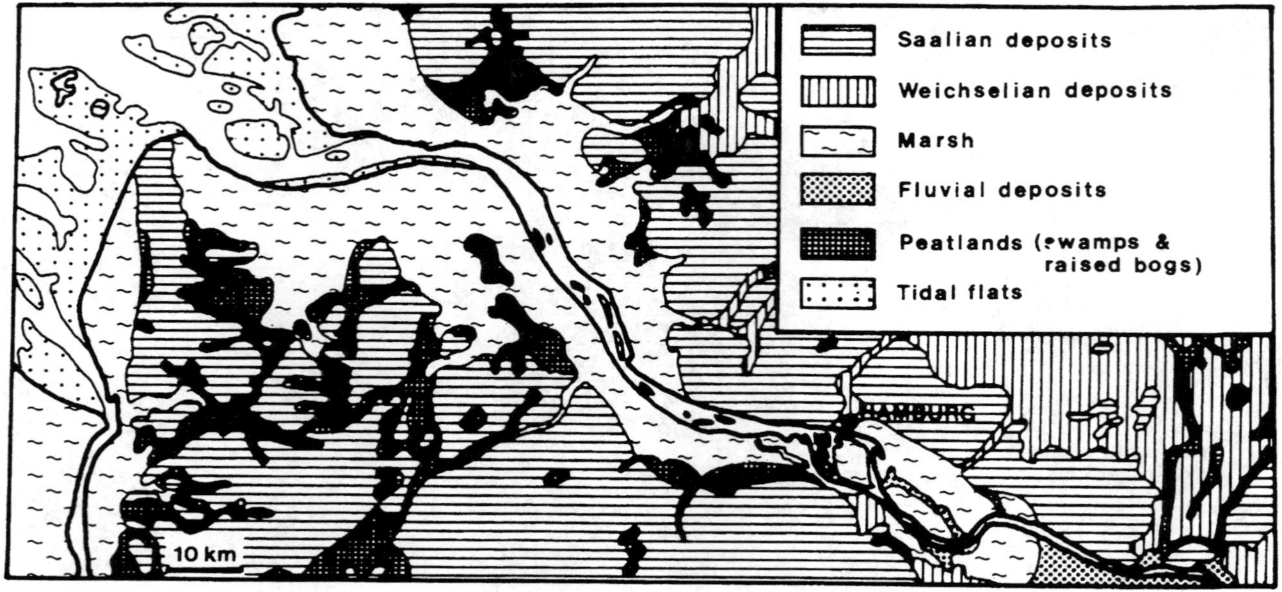

Figure 16.2 Depositional environments along the Lower Elbe

'urstromtal') escaping to the west along the ice margin. The valley of the Warthe-Stage became integrated into the urstromtal and was filled with coarse deposits. With the decrease of river flow and sediment load at the transition to the Holocene, the Elbe started to meander on the wide valley floor.

16.2.3 Climate and hydrology

The average yearly precipitation is 771 mm at Glückstadt. The mean temperature of the coldest month is +0.5°C (February), the warmest month +16.7°C (July). The average tidal range is 2.95 m at Cuxhaven and Glückstadt and 3.11 m at Hamburg-St Pauli (Dahl and Heckenroth 1983). The estuary can be divided into three transitional zones based on salinity, including: (1) a mesohalinic zone (1.65–0.3%) from Cuxhaven to Brunsbüttel; (2) an oligohalinic zone (0.3–0.05%) from Brunsbüttel to Glückstadt; and (3) a limnetic zone (0.05–0%) upstream of Glückstadt.

16.3 HISTORICAL DEVELOPMENT OF THE SHORELINE

Until early medieval times, the Holocene transgression of the sea, river flow and the influence of heavy storms determined the coastal development of the Lower Elbe. Settlers became the dominant morphogenic factor from the 11th century onward.

16.3.1 Postglacial movements of sea level

During the first phase of the postglacial transgression until the Atlanticum at about 6000 BP (Figure 16.3), the sea level rose by around $2\,\mathrm{cm\,a^{-1}}$. In the Preboreal (9000 BP) the sea extended to the area of Scharhörn off Cuxhaven. At the end of the Boreal (7500 BP) the shoreline had reached the mouth of the river Oste at -15 m MSL, cutting deeply into the tributaries.

This phase was connected with considerable erosion of glacial sediment, especially on the south bank. On the north side, large brackish marshes developed with scattered fen and alder swamp peat. Brackish estuarine deposits of that period can be traced as far as Pagensand. Considerable accumulation of fine clastic sediments took place upstream under freshwater conditions. As a consequence of the resulting increase in elevation, the stream branched delta-like into many tributary channels between Hamburg and Harburg.

At the end of the Atlanticum the rise of the sea level slowed down considerably, leading to much silt sedimentation. In the 2nd millennium BC the sea level rose to the watermark of today, with a temporary maximum at about 3000 years BP. The tidal influence reached the braided area at Hamburg.

The mainland border took its course at the foot of the Saalian hills. The marshes

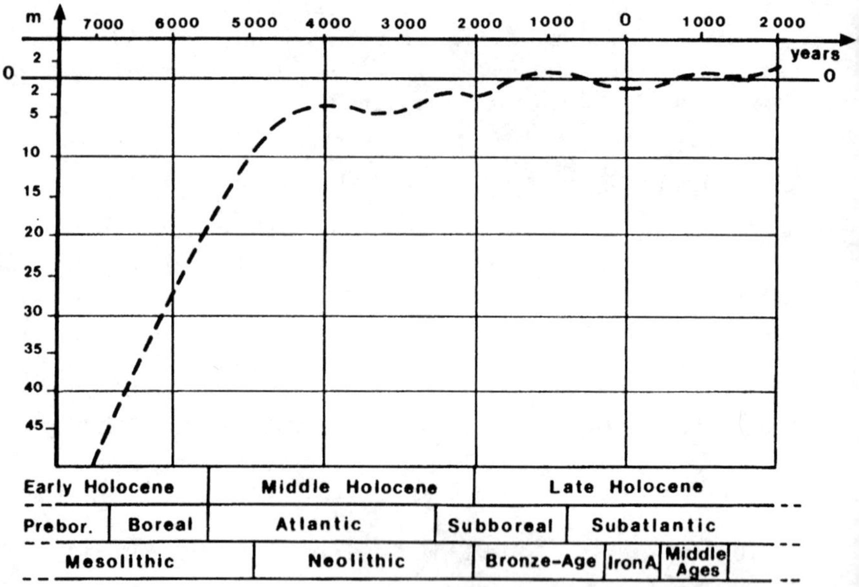

Figure 16.3 Rise of sea level at the southern North Sea coasts during the Holocene (from Menke 1976 and diverse authors)

became semi-terrestrial wetlands. Along the main tidal creeks and shorelines mainly minerogenic sediments accumulated. These deposits formed natural levees that protected the lower backswamps ('hinterland' or 'sietland') against tidal floods. Fine clastic sediment accumulated in this area only during high tides. Water supplies from the Saalian hills led to extended peat growth in the hinterland. The peat accumulation was interrupted only by rare floods, and resulted in the local development of raised bogs. Typical sequences of the sedimentary processes from bore holes are given by Schubert (1933).

Salt was washed out of the marsh soils during a period of regression after 2800 years BP. Sand supplies from the Upper Elbe and from the sea predominated over tidal flushing; the riverbed was filled with sand and branched into several channels. Permanent settlements were founded in the marsh that had transformed into mainland. The tidal influence did not reach Hamburg at that time.

About 1900 years BP the present phase of sea level rise began. During the first 800 years, the level rose 1.2 m, and settlements became acutely endangered. The tidal wave extended into the braided area at Hamburg again. A further sea level rise from about 1000 to 850 years BP resulted in the construction of the first embankments to fight land losses. Permanent sea level measurements started in the 18th century. Historical sources and archaeological finds reveal that, from the 16th century until today, the sea level in the southeastern North Sea has risen by an average of 0.25 cm a^{-1}. Now, near the end of the 20th century, the preliminary sea level maximum of 3000 years BP has been exceeded.

16.3.2 Fluvial morphodynamics

The meandering of the Elbe showed strong fluctuations in historical times. In the 15th century, the main active channel moved north, silting up the harbor of Stade, an important trading place in Lower Saxony. At the opposite shore, the erosion intensified and several villages were lost. At the end of the 16th century, a 2 km wide marsh zone accumulated on the south shore near Cuxhaven. At the same time, the northern shore underwent considerable erosion. During the 17th and 18th centuries, the stream shifted south again, and the new land off Cuxhaven was lost (inset, Figure 16.4).

It is assumed that the meander fluctuations of the Lower Elbe result from changes the subtidal sand shoals in the mouth of the river. These shoals are built up by sand shifting along the southern coast of the North Sea. According to Reineck and Singh (1980) the sand shoals are shifting at a rate of 27 m a^{-1}. As a result, the main channel of the Elbe slowly migrates from the shoreline of Lower Saxony (where it is at present) toward the shoreline of Schleswig-Holstein. An analysis of historical sources on land growth and loss in the Elbe area indicates a 100-year cycle of fluctuations (Simon 1965).

16.3.3 Morphodynamics through storm surge events

The coincidence of extremely strong landward winds from the North Atlantic Ocean, high or spring tide, and high discharge during periods of thawing or heavy rainfalls can lead to catastrophic storm floods in the estuary. The water rises more than 3 m above MHW during very high storm surges. The elevation takes place within only a few hours. Since the middle of the 12th century, 33 high and very high storm surges have been recorded. High storm surges occur once in 2–20 years on average; very high storm surges occur less than once in 20 years. No very high storm surge was recorded between 1825 and 1962, but since then there has been an increase of frequency.

During the great storm surge on 17 February 1962, 315 people died in the city of Hamburg. The water level was 6.5 m above MSL in the center of the city. Dikes broke at 120 places in the estuary, and an area of 4,300 km^2 was flooded, but this increase of retention space for flood waters prevented catastrophies from occurring in other locations.

In diked areas the consequences of storm surges can intensify because of the reduction of retention space and the initial resistance of dikes. When the breach in the dike finally occurs, the water pours rapidly out into the hinterland causing greater damages than during a gradual flood rise in the open marsh where the wave energy is dissipated.

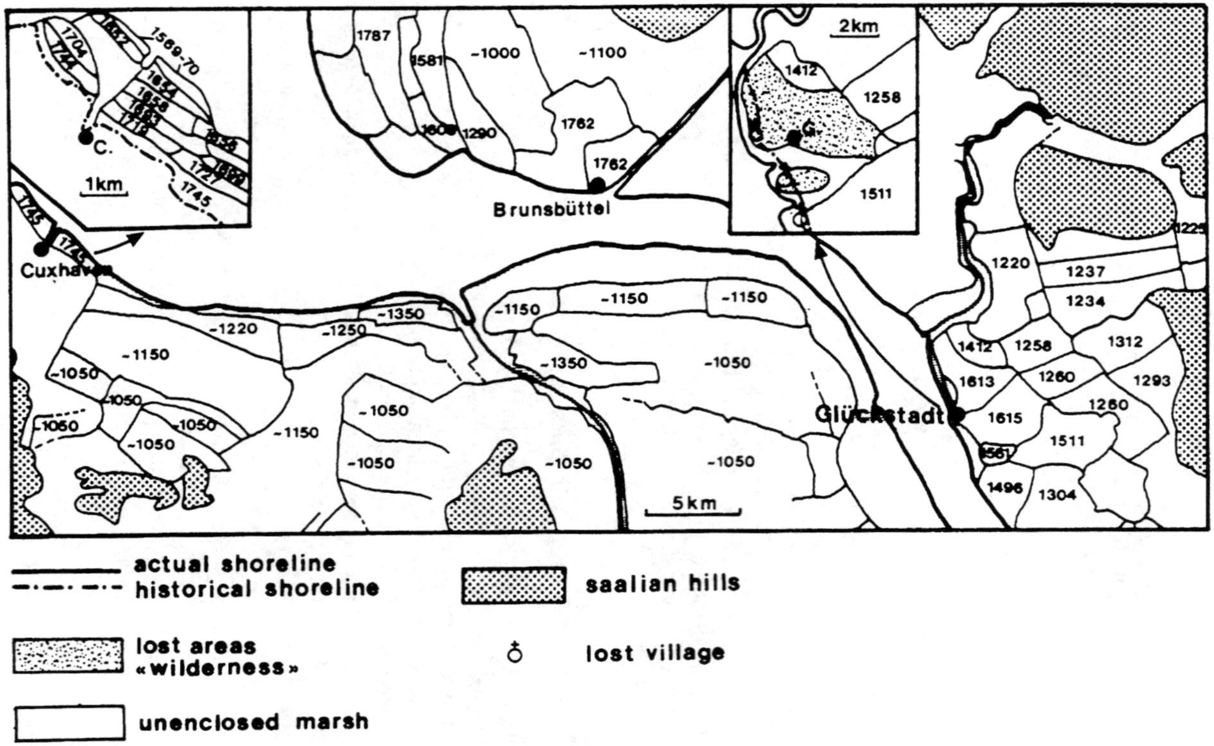

Figure 16.4 Enclosures at the Lower Elbe until the end of the 18th century

16.3.4 Human influences

Human impacts on the marshes originate from efforts to reclaim land for agriculture. Attempts to shelter the agricultural areas against the combined destructive effects of sea level rise, storm surge and the meandering of the river then follow. In this respect, settlers combined offensive and defensive strategies. Sometimes the consequences of offensive strategies required further defensive measures; i.e. soil compaction after drainage of marshes required digging deeper ditches for drainage and building stronger dikes.

The area south and east of the North Sea has been populated since the Neolithic Age. Estuarine marshes fit for settlement did not develop until 3000 years BP. The oldest remains of settlements can be traced back to the first century AD. (Behre and Menke 1969). The houses were built on slightly higher, sandy streamside levees. These places were arable because of their height; livestock grazed at lower areas. The human impacts on the estuarine ecosystems were of comparably low significance prior to the beginning of the early Middle Ages. From the second century AD onwards, tidal creeks were artificially deepened, and ditches were dug around fields to install a more effective drainage system. The backswamps and bogs remained untouched.

At this time, responses to storm surges and flooding were only defensive and included constructing low-level dikes of marshy clay around houses or placing the settlements on artificial dwelling mounts ('Wurten' or 'Warften'). These dwelling mounts consisted of alternating clay and habitation layers, revealing the necessity of repeated attempts at increasing elevation. This type of settlement is still found in the Elbe estuary.

The cultivation of the backswamps ('sietland') and bogs began during the 12th century. New settlers from Frisia introduced complementary techniques of dike construction and inland drainage.

16.3.4.1 Enclosures and land reclamation

The first dikes were constructed of marsh clay. At the sea side, the profile was too steep to withstand wave erosion, resulting in many catastrophies. A new or repaired dike was built according to the maximum level of the preceding flood. The long-term rise of sea level was not perceived until the beginning of the 20th century. From this date on, anticipatory construction became integrated in sea defence systems (Kramer 1992).

Extensive enclosures took place between the 12th and 15th century (Figure 16.4). Dikes were constructed along the seaward sections of the tributaries. During the next 500 years until 1962 almost no additional forelands were enclosed, which shows the efficiency of medieval diking.

Large areas were lost during storm catastrophies in the 15th century. These land losses are mainly due to shifting meanders of the river as well as rising water level.

As a consequence of land losses, the dikeline had to be transferred inland. Several villages fell victim to floods, as did arable land. The abandoned land was not resettled at that time, and these areas are marked as 'Wildniss' (= wilderness) on the old maps. Several of these wildernesses were re-enclosed in the 17th century (inset, Figure 16.4). Between 1618 and 1744 the dikes off Cuxhaven-Ritzebüll had to be taken back incrementally. It was not possible to win back these land losses until today (Figure 16.4).

From the 16th century onward, active land reclamation occurred along the Elbe. The undiked forelands are covered with a narrow system of ditches that enhances sedimentation. The sedimentation fields are surrounded by brushwood groins and are subdivided by earthdams in smaller units that are provided with an open drainage system (see Verhoeven et al. (1983) for detailed descriptions).

The maintenance of dikes was the duty of marsh farmers and meant intensive and long-term efforts. The dikes were often neglected during periods of war, epidemics or famine, which usually are connected with a loss of population. The resulting catastrophies led to further decrease of population. The shoreline of the Elbe changed little from the Middle Ages until the middle of the 20th century, but the maintenance of a status quo despite an average rise of the sea level of 0.25 m per 100 years occurred only because of constant improvements in engineering techniques.

The heavy damage during the great storm surge of February 1962 induced an extensive flood protection program, the principal aim of which was to shorten and straighten the dikeline. Between 1964 and 1978 all tributaries were provided with storm surge barriers that are closed only under storm flood conditions (Figure 16.5). In the name of coastal protection, a further 254 km^2 of estuarine marsh was enclosed, destroying nearly 80% of the semi-natural rivermarsh existing before 1962 (Dahl and Heckenroth 1983). Retention space has decreased considerably as a consequence of these enclosures. A further consequence was that the maximum water level during storm surges at Hamburg has increased by more than 0.4 m, a highly problematic development considering the continuous rise of sea level and the recent increase of intensive storm events.

16.3.4.2 Drainage of enclosed areas

The natural drainage system of the open marsh stayed untouched as long as settlements were restricted to dwelling mounds. After the construction of dikes, the enclosed land had to be provided with effective drainage for the now arable land and with an artificial discharge system for water from precipitation, rivers and backswamps. The surplus water was conducted into the foreland by a close-meshed system of ditches and channels passing the dikes by tidegates that were closed during high tides.

At the end of the Middle Ages the former tidal creeks had been deepened and channelized to improve the discharge. At times of insufficient drainage the land routes were often impassable, so that these channels became important commun-

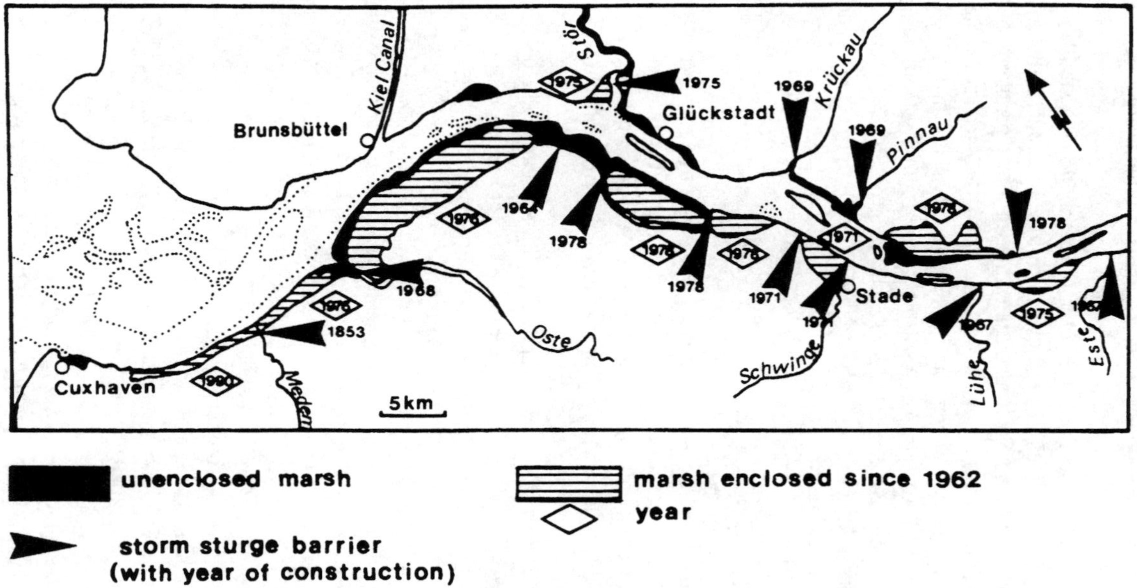

Figure 16.5 Storm surge barriers and enclosures since 1962 (from Grimm 1978)

ication routes. Severe difficulties were caused by water from bogs of the valle margin and the surrounding hills. The danger of floods during periods of heav rainfall led to the installation of a separate drainage system for the hinterland.

The end of the Middle Ages saw the beginning of the drainage and destruction o raised bogs – one of the most serious impacts on the ecosystem of the Lower Elbe Ditches were driven into the margin of the bogs and slowly extended to the cente By turf cutting, the surfaces of the bogs were lowered and finally cultivated. Th destruction of the raised bogs caused new hydrological problems because of th sudden lack of rainwater storage capacity. This led to accelerated runoff o rainwater into lower enclosures, and subsequently dikes had to be constructe around some bogs. The main period of turf cutting and bog cultivation was in th 17th century when large land losses and a growing population forced th acquisition of new land. Today only small degraded remainders of these bogs sti exist in the Elbe estuary.

16.3.4.3 Impacts on river flow

Since the beginning of the 20th century, deepening and widening of the navigatio channel led to considerable changes in flow velocity and tidal range. As one resu of increased flow velocity, the Elbe does not freeze, even in hard winters. Othe important effects are the augmentation of the average tidal range and the landwar shift of tidal influences.

The destruction of the river delta at Hamburg began with the first enclosure about 700 years ago. After 1850, when two arms (Norderelbe and Süderelbe) wer dredged to improve access to the harbor, the discharge concentrated in these tw channels. It was then possible to cut off other tributaries by sluices and to drain th enclosures of the delta. As a result, the tidal influence that had been diminished b the delta until then shifted to river sections upstream. At the same time, th navigation channel was deepened from 4.7 m to at least 13.5 m and widened fro 200 m to an average of 300 m. Consequently the tidal range at Hamburg increase by more than 1 m, to the present 3.11 m. The tidal wave shifted about 60 kn upstream and extended beyond Lauenburg (Figure 16.1). A barrage was constructe at Geesthacht in 1960 to counter the tide. This barrage is effective in cutting off th tide, but it also cuts off the ecosystems of the Upper Elbe from the estuary.

Today dikes are in close contact with the river edge almost everywhere and ther is no protecting foreland zone in between. The dikes have to be protected b revetments, causing further constriction of stream flow.

Even though the erosive power of the Elbe has increased considerably due to th activities identified above, extensive dredging is necessary to retain navigatio depths. The deepening of the navigation channel down to −13.5 m MLW produce about 50 billion m^3 of sand. The maintenance of the channel adds 10 billion m each year. One-third of the material is dumped in the North Sea while two-third are deposited as sandflats in the estuary. Deposition initially took place directly i front of the undiked marsh to provide a base for further embankments. Now, efforts are made to attach the deposits to existing sandflats or islands. The former islands of Schweinsand and Hanskalbsand grew from +0.8 m MHW into a large coherent sandflat that is now +5 m MHW (Dahl and Heckenroth 1983). Further plans included a complete barrage of the whole estuary off Brunsbüttel and the transformation of the Hamburg harbor to an artificial sandflat in the mouth of the Elbe. These have been postponed, mainly for economic reasons.

The present state of the river and surrounding marshes represents the results of human alteration of the ecosystem of the Elbe without regard to expected global change (see Schirmer and Schuchardt (1993) for more information on climatic changes and estuaries).

16.4 THE VEGETATION OF THE LOWER ELBE

16.4.1 Natural vegetation

An attempt to reconstruct the primary vegetation in the Lower Elbe area is made difficult by the long-term severe impacts on the ecosystem. Detailed paleobotanical investigations from the Elbe and other estuaries (Behre 1970, 1985) and the interpretation of historical sources, such as landscape descriptions and old maps, provide only fragmented information. The history of morphological, hydrological and anthropogenic parameters and investigations of the few remaining natural and semi-natural environments along the Elbe form the basis for a reconstruction of the natural vegetation cover without human influence (Figure 16.6a).

The vegetation of a locality depends mainly on its distance and elevation relative to the river bed, which determine the average number of flood events, the mechanism for sedimentation, and the salinity, which decreases from shoreline to the valley margin and from the mesohalinic mouth of the estuary to the upstream limnetic zone (Behre 1994). The following description gives a short survey of the most important natural estuarine environments along the Elbe and their characteristic vegetation types. Data on the remaining natural vegetation are taken mainly from Kötter (1961), Raabe (1982) and Wolf (1979, 1988).

16.4.1.1 Sublittoral zone

The submerged vegetation of the seaward section of the stream was probably characterized originally only by algae. The population density of algae is dependent on the turbidity of the Elbe, which is related to the amount of suspended load. Eelgrass (*Zostera* sp.), which is abundant in the Wadden Sea, does not intrude into the estuary. Only diatoms can be found in high densities. According to Eschenburg (1928) freshwater macrophytes were growing in the limnetic zone upstream of Glückstadt until the early years of this century.

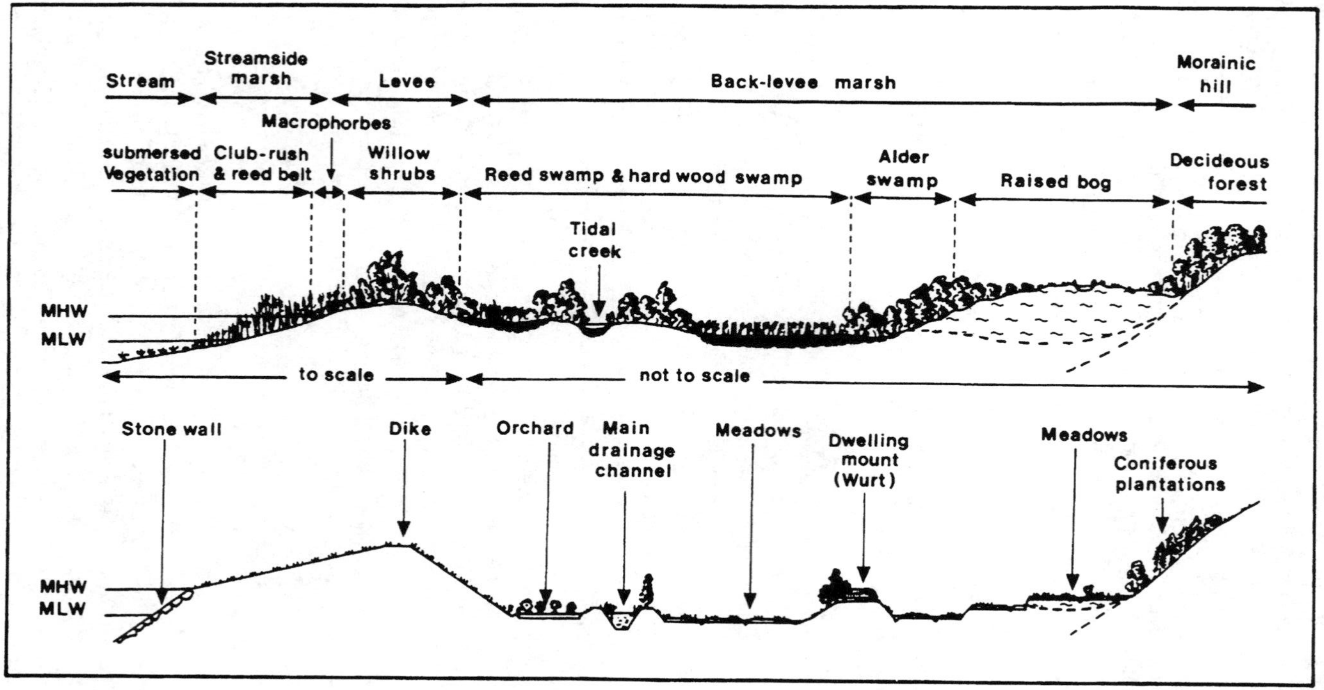

Figure 16.6 Vegetation along a transect from river edge to the valley margin

16.4.1.2 Littoral and upper littoral zones

The littoral zone is between −2 m below MHW and MHW. Under brackish conditions, the sand and mudflats of the riverbanks were colonized by tidal reed belts, dominated by *Bolboschoenus maritimus, Schoenoplectus tabernaemontani* and *S. triqueter*. This tidal reed was replaced by small, stress-tolerant species like *Eleocharis uniglumis* at portions of the shoreline exposed to waves and currents. The reed was associated with salt-tolerant species like the sea aster (*Aster tripolium*) on brackish tidal flats. Freshwater tidal flats are dominated by stands of common reed (*Phragmites australis*) and bulrush (*Typha angustifolia*), which tolerate high water-table variations. Above −0.7 m MHW this reed was only dominated by *Phragmites* stands with some productive perennials like *Lythrum salicaria, Sonchus paluster, Senecio paludosus* and *Angelica archangelica*, forming belts of luxuriant flowers.

The upper littoral zone around MHW is affected by only half of the tidal inundations. In winter, drift ice often destroys the reedbelt, which then is usually replaced by perennial macrophorbs. At the transition to the supralittoral zone, large amounts of drifting material are accumulated every winter. Mineralization of this organic material provides a surplus of nutrients for annual nitrophytes such as the different species of Chenopodiaceae, or *Bidens tripartita* and *Xanthium albinum*.

16.4.1.3 Supralittoral

Usually the supralittoral zone is covered by water only during storm floods. The vegetation was dominated by woody plants that could not colonize salty soils. Mang (cited in Preisinger 1991), divides the freshwater areas of the Elbe into three zones according to their vegetation and surface altitude above MHW, including:

1. MHW to +0.5 m – willow-shrubs dominated by *Salix viminalis* and *S. triandra* (*Salicetum triandro-viminalis*).
2. +0.5 m to +1.0 m – woodland dominated by *Salix fragilis* and *S. alba* (*Salicetum albo-fragilis*).
3. +1.0 m to +1.5 m – woodland dominated by *Fraxinus excelsior* and *Ulmus glabra* (*Fraxino-Ulmetum*).

16.4.1.4 Backswamps ('sietland') and raised bogs

The backswamps were only flooded during very high storm surges when water broke through the natural levees. Freshwater from rainfall and runoff from raised bogs of the valley margin create the upper layer of the ground-water table. Tributaries meandered through these large sietlands, extending the influence of tides to the backswamp area. Intact remainders of plant species can only be recovered from deep layers because the backswamps have all been drained and cultivated. It seems plausible that the backswamps were dominated by large *Phragmites* stands, *Salix* shrubs and *Alnus glutinosa* woodlands.

The large bogs at the valley margins were fed by springwater from the neighboring hills. Gradual elevation of the bog surface by accumulation of undecomposed biomass led to the development of ombrogenic raised bogs dominated by *Sphagnum*- species, *Ericaceae*, and different *Cyperaceae*. These raised bogs were bordered by alder and birch swamps.

16.4.1.5 Implications

The distribution of natural vegetation units in an estuary is characterized by several natural gradients. The vertical (cross-shore) dimension (from river margin to valley margin) is distinguished by the transition from fully aquatic to terrestrial environments along with a change from brackish to freshwater conditions. In the horizontal dimension (upstream) the dominant factors are the salinity gradient and the decrease of tidal influence. From investigations of estuaries under less anthropogenic interference it can be deduced that the Lower Elbe, with its distinct ecotone character, was once an ecosystem rich in species and structures.

16.4.2 Human-altered vegetation environments

Before the first human impacts, erosive and accumulative processes between river tributaries and tides formed unstable and changing borders between sea and land. As soon as the settlers had developed the necessary technical methods, they began to set up stable boundaries and to reclaim land for cultivation. The result of 2000 years of continuous settlement in the Lower Elbe area is a disastrous loss of natural environments. In turn, new environments have been established that are inhabitated by secondary communities (Figure 16.6b).

16.4.2.1 Submerged and shoreline vegetation

Macrophytes have completely vanished from the Lower Elbe since the early 20th century due to increasing turbidity caused by increased nutrient content, augmentation of current velocities, and intensification of impacts associated with navigation (chronic oil spills, etc.). Potential habitats on the shoreline have been considerably reduced by embankments and barrages constructed in the tributaries. The remaining wetlands have been cut off from tidal dynamics by sea defense dikes. As a result, these areas are occupied by completely different vegetation communities.

Incipient marshes were enclosed after the flood of 1962. Long sections of the new dikes do not show any foreland growth, so the former reed belt has vanished

Long segments of the shoreline had to be protected by stone and concrete walls because of the high-velocity currents and wave energy. An endemic *Poaceae, Deschampsia wibeliana,* is often found established in the joints of these walls. According to Wolf (1988) there are only 680 ha of tidal marshes left in the part of the estuary that lies in Schleswig-Holstein. The tidally influenced shrublands are restricted to only a few small places like the protected area of Heuckenlock in Hamburg.

Heavy grazing has influenced the vegetation cover of the remaining foreland for a long time (Dierssen et al. 1991). The tidal reeds that are sensitive to stress by grazing have been replaced by wet grasslands. Saltmarshes with characteristic species like *Puccinellia maritima* and *Juncus gerardii* are only found seaward of Brunsbüttel. Upstream, heavy grazing of wet sites usually leads to the development of monotone flooding meadows (few species of *Poaceae*, such as *Agrostis stolonifera* or *Alopecurus geniculatus*, growing on degenerated soils).

The ecotone along the Elbe estuary used to include many different environmental conditions (saltwater, brackish water, freshwater, different tidal ranges, different currents and still-water zones, silt and sand flats, large meadows and tidal creeks). These environments formed a multitude of habitats for water fowl and Wadden Sea birds. Today there are only a few areas left where parts of this differentiated ecotone are still realized. These locations are of worldwide importance for bird species like Bewick's swan (*Cygnus bewickii*) and the barnacle goose (*Branta leucopsis*). For this reason, the estuarine shoreline of Lower Saxony between Stade and Cuxhaven is under formal protection as a 'wetland of international importance' according to the RAMSAR convention. Despite this importance, it was not possible to stop the recent enclosure (completed in 1976) of another 6000 ha of marsh in the center of these wetlands (Klän 1976). After the enclosure, most of the former grazing grounds of swans and geese were changed into fields.

Large flats of sand from the navigation channel could have offered new habitats for flora and fauna, but the sand was artificially stabilized to protect against erosion by sowing different species of *Poaceae* for cover, while *Bolboschoenus maritimus* was planted to establish reeds at the shoreline. Afforestations with foreign *Salix* species occurred on sandflats with an altitude of more than +5 m MHW. The natural habitats for *Salix* shrubs are between +0.5 and +1.5 m MHW, so all plantations on the island of Schwarztonnensand died from lack of water and nutrients. This island is now covered by a sparse grassland community dominated by *Agrostis tenuis, A. stolonifera, Carex arenaria* and *Festuca ovina agg.*

16.4.2.2 The vegetation of the embanked areas

Embankments and drainage projects have been changing the landscape since medieval times. Raised bogs were destroyed by drainage and turf cutting so that now some former sites can only be traced by soil investigations. The former backswamp areas have been cultivated for such a long time that it is difficult to reconstruct their natural vegetation cover.

Until World War II the enclosed areas were mainly dominated by meadows and pastures rich in species, and intersected by nets of ditches with an abundant flora of macrophytes (Lötschert 1963). As a result of the recent conversion to high-intensity cultivation, most of the pastures have been replaced by arable fields, destroying much of the wetlands of special ornithological interest. Extensive orchards (mainly apples and cherries) have been established in the marsh near Hamburg.

Modern cultivation has led to high input of nutrients and pesticides (e.g. herbicides, fungicides, insecticides, acaricides) to water and soils and to an extreme loss of species diversity and habitat structures, especially in the secondary environments of the traditional agricultural landscape. Only a few well adapted species prosper and are now growing almost everywhere. The vegetation of cultivated land as well as of waste land is dominated by a few stress-tolerant and very productive nitrophytes. The dominance of wind-pollinating taxa like *Poaceae* or *Urtica dioica* in many agricultural plant communities has strong negative effects on the faunistic diversity of the cultural landscape.

The destruction of the natural ecosystems of the Elbe estuary can be divided into two phases. During the first phase (from the Middle Ages until 1960), the natural or slightly influenced environments were replaced by secondary formations like large wetlands, pastures and ditches which were still rich in species and of high ecological value. The second phase started with a further intensification of cultivation and extension of industrial areas along the Elbe, especially at Hamburg, Stade and Brunsbüttel. The disastrous loss of typical elements of flora and fauna of the traditional landscape is still going on.

The embankment of large parts of the remaining marshes is especially alarming because it took place without economic pressure and at a time when the ecological significance of wetlands and marshes was already recognized and accepted. Following several severe floods in the estuary during the past decade, the loss of retention space has become an object of intensive discussion, pointing to a new perception of the human value of the marshes that were lost due to embankment.

Lists of rare and endangered plant communities (Dierssen et al. 1988; Westhoff et al. 1993) identify the remaining tidal reeds as endangered or very endangered vegetation types. The *Salix* woodlands of the estuaries are on the eve of extinction and are protected according to § 20c of the German Federal Nature Protection Law.

The whole ecosystem of the Lower Elbe has been obliterated, so even a small-scale restoration of natural conditions is very expensive – if possible at all. The next section provides a perspective on the restoration process by giving a short account of a project designed to revitalize a tidally influenced tributary of the Elbe at Hamburg that was dammed after the catastrophic flood of 1962.

16.5 RESTORATION OF THE ALTE SÜDERELBE

In the area of Hamburg, at Altenwerder, further harbor extensions are planned on ecologically vulnerable sites. According to the Federal Nature Protection Law, the resulting destruction of the environment requires extensive compensation. The compensation will take place at the area of one of the main tributaries in the delta south of Hamburg. It is planned to revitalize the Alte Süderelbe (Figure 16.1) by reinstating tidal influence. The Alte Süderelbe was dammed and cut off from the main stream after the flood of 1962. This area has a high ecological potential for the development of different tidally influenced habitats.

The section of the Alte Süderelbe to be restored is about 7.5 km long. The width of the surface of the actual lake-like stillwater, which used to be a main tributary, is now between 200 m in the north west and 2 m in the south east. The restoration area covers about 200 ha. In the future, two sluices will connect the former tributary with the main stream to reinstate tidal influence. One sluice will connect the northwestern part of the Alte Süderelbe with the Mühlenberger Loch, a renowned ornithological sanctuary. The other sluice will be built 4.4 km to the east at the Köhlfleet, a channelized part of Hamburg harbor. The future maximum tidal range in the Alte Süderelbe may reach 2 m, but final decisions have not been made. They depend on the necessity of further compensation for road construction and future development of industrial sites.

Today this area is characterized by intensive agriculture, mainly orchards and meadows, disposal sites for contaminated harbor sediments (Schlickhügel Francop), large industrial sites and a nature reserve. The former tributary is mainly used as a large drainage channel and a water reservoir for spraying orchards to eliminate frost in early spring. Detailed descriptions of this area are given by Winterfeld and Schulze (1978) and Mierwald (1994).

According to the scenario for the project (Freie und Hansestadt Hamburg 1993) the Alte Süderelbe will develop into a tidally influenced freshwater ecosystem with mudflats, tidal reeds, *Salix* shrublands and *Salix* woodlands. It is supposed to form the center of an undisturbed complex of semi-natural habitats with a distinct ecotone character.

The tidal range will be controlled by the two sluices that will be closed during storm surges and high floods as a part of the sea defense. The artificial tidal regime inside the enclosure may lead to the development of completely new semi-natural environments that differ considerably from the habitats of the outside estuary.

The main differences in ecological parameters between the future tidally influenced Alte Süderelbe and the undiked parts of the estuary are:

1. Artificial tide of constant range (no neap or high tide).
2. Lack of morphogenic events like storm surges and strong ice-drift.
3. Reduction of inundation in winter or early spring, which should be related to strong accumulation of organic material.

4. Continuous input of nutrients and pesticides from the surrounding orchards.
5. Uncertain sedimentation processes, especially in the narrow eastern part of the Alte Süderelbe upstream of the eastern sluice, which will only be indirectly connected to the tide.

The actual vegetation, with its stands of highly competitive nitrophytes, may have a strong influence on the future development of new environments on the future shoreline. Under natural conditions, storm surges and ice-drifts would destroy parts of the existing vegetation cover and allow for the establishment of new species. Without natural disturbance, the development of extended *Salix* woodlands on the prospective shore will be unlikely because of the competitive strength of the actual plant communities.

The realization of this project is planned to start in 1999. It will cost about DM 50 billion. An extensive monitoring program will be conducted. It will be of great scientific interest to investigate the changes that will occur in the vegetation cover under the artificial tidal regime.

16.6 CONCLUSIONS

The landscape history of the Elbe estuary reveals an ever-increasing impact of humans, resulting from offensive strategies to alter natural environments to produce a resource base of human value and defensive strategies to provide stability for this resource base. Human actions reveal a progression from passive measures to active measures, and then to anticipatory response. These actions include enclosing marshlands, draining bogs, cutting turf, cultivating formerly natural environments, digging ditches for drainage, building dikes and revetments, shortening and straightening the dikeline, constructing storm surge barriers, dredging navigation channels and disposing of dredged spoil. Secondary effects, such as contributing additional nutrients and pesticides to the water, also have major impacts on estuarine environments.

Consequences of these actions include a decrease in retention space for flood waters, increase in the maximum water level during storm surges, decrease in rainwater storage capacity, changes in flow velocity and tidal range, shift in location of tidal influence, destruction or relocation of the river delta, creation of artificial forelands through spoil disposal, conversion of ecosystems of the Upper Elbe from estuarine to freshwater, elimination of natural habitats and loss of species diversity.

There are now only a few areas left where parts of the highly differentiated natural ecotone exist. Restoration efforts have resulted in new semi-natural environments, but these differ considerably from the natural habitats of the estuary and have caused changes in the locations and levels of natural disturbance and competitive advantages of plant communities. Changes in the lower Elbe reveal the large scale of human alterations and the difficulty of reinstating natural environments after these environments have been eliminated.

REFERENCES

Behre, K.H. 1970. Die Entwicklungsgeschichte der natürlichen Vegetation im Gebiet der unteren Ems und ihre Abhängigkeit von den Bewegungen des Meeresspiegels. *Probleme der Küstenforschung im südlichen Nordseegebiet* 9: 13–48.

Behre, K.-H. 1985. Die ursprüngliche Vegetation in den deutschen Marschgebieten und deren Veränderung durch prähistorische Besiedlung und Meeresspiegelschwankungen. *Verhandlungen der Gesellschaft für Ökologie* 13: 85–96.

Behre, K.-H. 1994. Küstenvegetation und Landschaftsentwicklung bis zum Deichbau, p. 182–189. *In* Lozan, J.L., E. Rachor, K. Reise, H.v. Westernhagen and W. Lenz (eds.), *Warnsignale aus dem Wattenmeer*. Blackwell Wissenschafts-Verlag, Berlin.

Behre, K.-H. and B. Menke. 1969. Pollenanalytische Untersuchungen an einem Bohrkern der südlichen Doggerbank. *Beiträge zur Meereskunde* 24/25: 123–192.

Behre, K.-H., B. Menke and H. Streif. 1979. The Quaternary geological development of the German part of the North Sea, p. 85–113. *In* Oele, E., R.T.E. Schüttenhelm and A.J. Wiggers (eds.), *The Quaternary History of the North Sea*. Acta Univ. Ups. Symp. Univ. Ups. Annum Quingentesimum, 2, Uppsala.

Buchwald, K. 1990. *Nordsee – Ein Lebensraum ohne Zukunft?* Verlag Die Werkstatt, Göttingen.

Dahl, H.-J. and H. Heckenroth. 1983. Ornitho-ökologische Untersuchungen zu Baggerarbeiten in der Unterelbe und zu geplanten Aufspülungen im Bereich Brammerbank. *Naturschutz und Landschaftspflege in Niedersachsen* 6: 1–114.

Dierssen, K., H.v. Glahn, W. Haerdtle, H. Hoeper, U. Mierwald, S. Schrautzer and A. Wolf. 1988. Rote Liste der Pflanzengesellschaften Schleswig-Holsteins. *Schriftenreihe des Landesamtes für Naturschutz und Landschaftspflege Schleswig-Holstein*, 6: 1–157.

Dierssen, K., I. Eischeid, W. Haerdtle, U. Hamann, K. Kiehl, P. Koerber, F. Luettke Twenhoeven, R. Neuhaus and J. Walter. 1991. Geobotanische Untersuchungen an den Küsten Schleswig-Holsteins. *Berichte der Reinhold Tuexen-Gesellschaft* 3: 129–155.

Embleton, C. (ed.). 1984. *Geomorphology of Europe*. Macmillan, London.

Eschenburg, H. 1928. Gemarkungsflora von Holm. *Schriften des Naturwissenschaftlichen Vereins für Schleswig-Holstein* 18: 62–161.

Freie und Hansestadt Hamburg, Wirtschaftsbehörde, Strom und Hafenbau, Planungsabteilung (ed.). 1993. *Vorbereitende Herrichtung von Flächen für Hafenzwecke in Altenwerder. Teil IV: Landschaftspflegerischer Begleitplan*. Freie und Hansestadt Hamburg, Hamburg.

Grimm, R. 1978. Die Entwicklung der Niederelbelandschaft in den letzten 10 Jahren aus ökologischer Sicht. *Verhandlungen der Gesellschaft für Ökologie Kiel*, 1977: 351–355.

Klän, D. 1976. Die ornitho-ökologische Bedeutung des Niederelberaumes – zugleich eine kritische Flächenbilanz. *Schriftenreihe des Deutschen Rates für Landespflege* 25: 301–305.

Kötter, F. 1961. Die Pflanzengesellschaften im Tidegebiet der Unterelbe. *Archiv für Hydrobiologie, Supplementband Elbe-Ästuar XXVI/1*: 1-2: 106–184.

Kramer, J. 1992. Entwicklung der Deichbautechnik an der Nordseeküste, p. 64–109. *In* DVWK (ed.), *Historischer Küstenschutz. Deichbau, Inselschutz und Binnenentwässerung an Nord- und Ostsee*. Konrad Wittwer Verlag, Stuttgart.

Lötschert, W. 1963. Die Pflanzengesellschaften der Elbmarschgräben. *Natur und Museum* 93/7: 291–298.

Mang, F.W.C. no date. Massnahmen des biologischen Wasserbaus im Tidegebiet der Elbe. Unpublished manuscript. Behörde für Wirtschaft, Verkehr und handwirtschaft – Amt für Strom- und Hafenbau.

Menke, B. 1976. Befunde und Überlegungen zum nacheiszeitlichen Meeresspiegelanstieg (Dithmarschen und Eiderstedt, Schleswig-Holstein). *Probleme der Küstenforschung im südlichen Nordseegebiet* 11: 145–161.

Mierwald, U. 1994. *Biotopkartierung und faunistische Untersuchungen zum Projekt 'Öffnung Alte Süderelbe'. Wirtschaftsbehörde der Freien und Hansestadt Hamburg, Strom- und Hafenbau.*

Preisinger, H. 1991. Strukturanalyse und Zeigerwert der Auen- und Ufervegetation im Hamburger Hafen- und Hafenrandgebiet. *Dissertationes Botanicae* 174: 1–296.

Raabe, E.-W. 1982. Die Zerstörung der Urlandschaft an der Haseldorfer Marsch. *Die Heimat* 89: 261–269.

Reineck, H.-E. and I.B. Singh. 1980. *Depositional Sedimentary Environments, with Reference to Terrigenous Clastics.* Springer, Berlin.

Schirmer, M. and B. Schuchardt. 1993. Klimaänderungen und ihre Folgen für den Küstenraum: Impaktfeld Ästuar, p. 244–259. *In* Schellnhuber, H.-J. and H. Sterr (eds.), *Klimaänderung und Küste – Einblick ins Treibhaus.* Springer, Berlin.

Schubert, E. 1933. Zur Geschichte der Moore, Marschen und Wälder Nordwestdeutschland II: Das Gebiet der Oste und Niederelbe. *Mitteilungen der Provinzialstelle für Naturdenkmalpflege Hannover* 4: 1–148.

Simon, W.G. 1965. Geschichte des Elbe-Ästuars von der Überflutung der Nordsee nach der letzten Vereisung bis zur Gegenwart nach dem Stand der Kenntnisse von 1964. *Mitteilungen des Geologischen Landesamtes Hamburg* 53: 163–209.

Verhoeven, B., M. Jespersen, D. Koenig and E. Rasmussen. 1983. Human influences on the landscape of the Wadden Sea area, p. 112–135. *In* Wolff, W.J. (ed.), *Ecology of the Wadden Sea*, Vol. 1. Balkema, Rotterdam.

Westhoff, V., C. Hobohm and J.H.J. Schaminee. 1993. Rote Liste der Pflanzengesellschaften des Naturraumes Wattenmeer unter Berücksichtigung der ungefährdeten Vegetationseinheiten. *Tuexenia* 13: 109–140.

Winterfeld, H.v. and H.-D. Schulze. 1978. Erläuterungsbericht zum landschaftsplanerischen Gutachten Süderelbe. *Schriftenreihe Freizeit – Freiflächen – Planung* 14: 1–88.

Wolf, A. 1979. Vegetationszonierung im Brackwasserbereich der Elbe. *Kieler Notizen zur Pflanzenkunde in Schleswig-Holstein und Hamburg* 11: 13–19.

Wolf, A. 1988. Röhrichte und Rieder des holsteinischen Elbufers unterhalb Hamburgs. *Mitteilungen des Naturwissenschaftlichen Vereins für Schleswig-Holstein* 58: 55–68.

Chapter Seventeen

Stabilization on the shoreline of Raritan Bay, New Jersey

NANCY L. JACKSON
New Jersey Institute of Technology

17.1 INTRODUCTION

Human development has altered the function of estuarine shorelines by filling wetlands (Thom et al. 1994), changed the sediment characteristics of intertidal habitat by introducing artificial nourishment, reduced or eliminated sandy beaches by constructing buildings and shore protection structures (Nordstrom et al. 1989), and altered the hydraulics of estuaries by reclaiming land and dredging (Findlay and Kirk 1988; Orford 1988; Nichols and Howard-Strobel 1991). Shoreline erosion and flooding of low elevations associated with storms are two major threats to development along estuarine shorelines. Land use regulations (incorporating setbacks and building practices) and shore stabilization are two loss reduction measures for mitigating these threats. In developed estuaries, shore stabilization (including both hard structures and artificial nourishment) is more common and has been a conspicuous component of the shoreline since the early 1800s (Zabawa et al. 1981; Rosen and Vine 1995).

Beach environments are susceptible to physical modification where shore stabilization is used to protect human development, or elimination where resource values of the shoreline are enhanced that do not require a beach (Nordstrom 1992). Concern over the effects of shoreline stabilization on beach dynamics and ecological habitat in estuarine environments is increasing, and managers are now beginning to consider these relationships in coastal management strategies (Shipman and Canning 1993; Thom et al. 1994). Most of our knowledge on the effects of shore protection alternatives is based on studies from ocean environments

Estuarine Shores: Evolution, Environments and Human Alterations. Edited by K. F. Nordstrom and C. T. Roman.

(e.g. Kraus and Pilkey 1988; Schwartz and Bird 1990). These studies reveal that shore-parallel structures can alter the shoreline by truncating the beach profile, shore-perpendicular structures can restrict longshore sediment transport, and introducing beach nourishment can change the elevation and sediment characteristics of the beach profile. Studies exist that assess the short-term physical effects of shore stabilization on beaches in estuaries over a period of years (Shultz and Ashby 1967; Ludwick 1987; Jackson and Nordstrom 1989; Hardaway and Gunn 1991; Jackson and Nordstrom 1994). In estuaries where wave energy is low and beach change is small, long-term studies may be more informative in assessing the effects of stabilization. Few studies assess long-term changes over a period of decades (Zabawa et al. 1981; Chu 1985) or examine the changes to the physical form and function of beach environments as a result of stabilization.

This paper reviews the types of alterations that have occurred to estuarine shorelines in the presence of development and how they influence the physical dimensions and sediment characteristics of sandy beaches. The focus is on sandy beaches and the physical changes that have occurred to these environments over periods of years to decades. Historical examination of development activity, shoreline changes and stabilization efforts is conducted on the southern shoreline of Raritan Bay (Figure 17.1). Types of alterations to beaches are related to physical characteristics of shoreline reaches such as topography, shoreline orientation, sheltering by headlands, changes in the function of the shoreline, and vulnerability of human development. Raritan Bay was chosen because the shoreline is extensively developed and there is a spectrum of types of physical environments and human development.

17.2 RARITAN BAY

Raritan Bay is a funnel-shaped estuary located within the New York–New Jersey metropolitan area (Figure 17.1). The areal extent of the estuary is approximately 70 km^2 (Fray 1969). Tides in the estuary are semi-diurnal with a mean range of 1.5 m and a spring range of 1.8 m (NOAA 1994). The dominant wave energy reworking beach sediments is locally generated, although some segments of the shoreline are exposed to ocean waves that enter to the north of Sandy Hook spit (Figure 17.1). Locally generated waves during non-storm conditions are generally less than 0.14 m high with periods less than 3.0 s (Jackson and Nordstrom 1992). Locally generated waves during storm conditions approach 1.0 m in height during breaking (Nordstrom 1992). Storms that affect the bay include extratropical storms, associated with low-pressure centers that frequently pass over the area between the months of October and March, and hurricane activity. A history of storm activity, from 1635 to 1960, reveals that severe storms occur at a rate of 14 times in 100 years (US Army 1960). Hurricanes that resulted in severe damage to property include the hurricane of 14 September 1944 and Hurricane Donna of 12 September 1960.

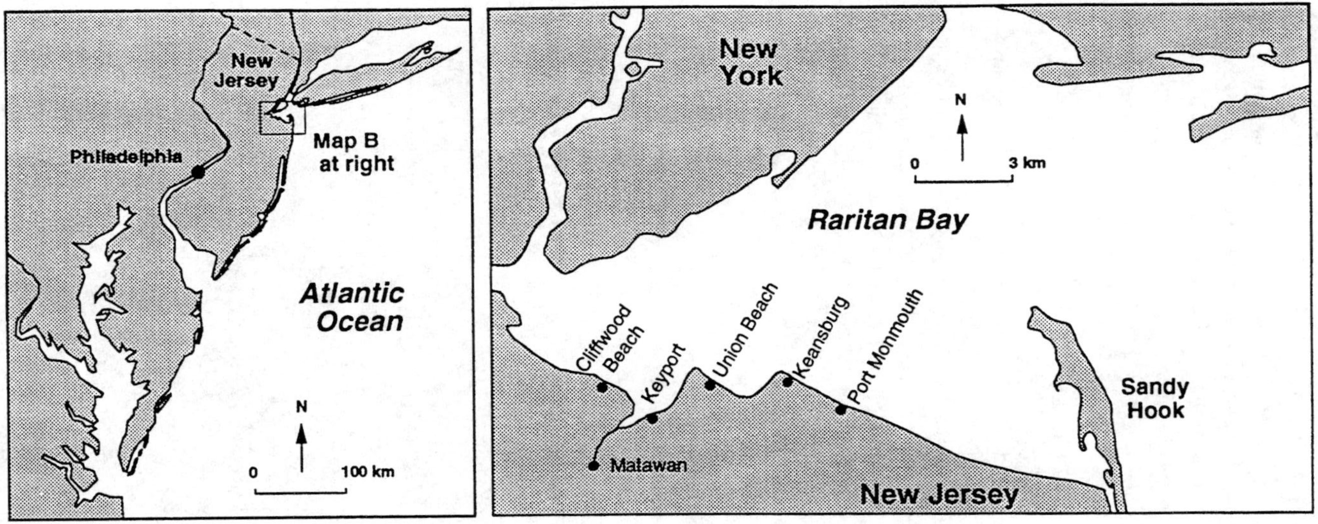

Figure 17.1 Locator maps of Raritan Bay

Storm surge can cause widespread damage to development situated on and near the shoreline of the Raritan estuary. In 1960, peak tides associated with Hurricane Donna reached 2.7 m above mean sea level. Tidal inundation occurred over more than 1600 ha and at depths as great as 1.5 m in the southern reaches of the estuary (US Army 1960). The highest peak tide recorded (3 m above mean sea level) occurred during the extratropical storm of 11–12 December 1992 and the shoreline was later included in a disaster area declaration (US Army 1993).

17.2.1 Study area

The area selected for detailed analysis extends from Whale Creek to Compton Creek for approximately 14 km (Figure 17.2). The area is representative of the spectrum of shoreline types and human alterations that can influence the physical characteristics of estuarine beaches. Shoreline topography in the study area consists of eroding bluffs fronted by beaches of up to 15 m wide; narrow beaches of low elevation front extensive marsh associated with the numerous creeks that compartmentalize the shoreline. Mean grain size in the study area in 1957, sampled at the location of mean high water, is 0.44 mm (US Army 1993). Much of the shoreline is intensively developed for both residential and commercial activities. Types of shore stabilization used in the study area include groins, bulkheads, and artificial nourishment.

Six shoreline reaches can be identified that represent differences in shoreline orientation and degree of sheltering (Figure 17.2). Reach 1 is 2.1 km in length extending from Whale Creek on the west to Matawan Creek on the east. The upland consists of an eroding bluff that was originally protected by a bulkhead and groins and is currently stabilized by a stone seawall and beach nourishment project. A

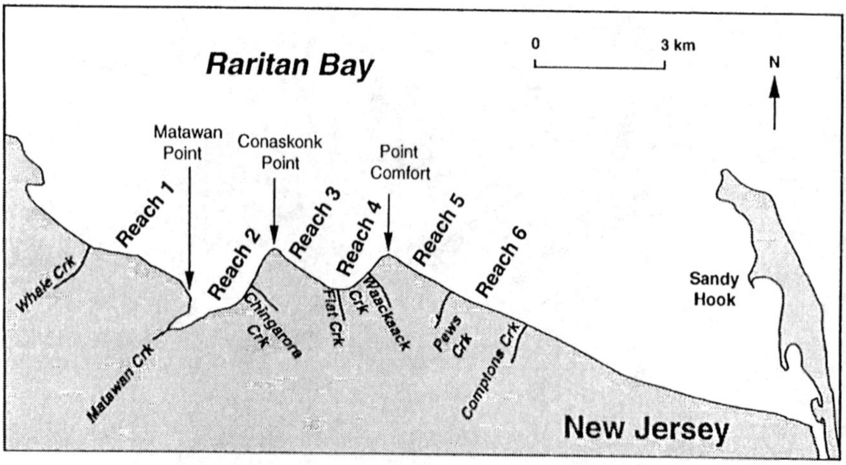

Figure 17.2 Study reaches located on the southern shoreline of Raritan Bay

linear artificial dune protects the low-lying area to the west of the bluff. A longshore drift divergence zone occurs near the midpoint of the seawall (Jackson and Nordstrom 1994). The segment to the west of Matawan Point is exposed to waves generated by the dominant winds from the northwest; ocean waves entering the mouth of the estuary reach the shoreline during strong winds from the east (Jackson and Nordstrom 1994). The shoreline to the east of Matawan Point is sheltered by Conaskonk Point, located in Reach 3; an extensive marsh is located adjacent to Matawan Creek. Residential development is located on the bluff and immediately to the east of the seawall.

Reach 2 (1.8 km in length) extends from Matawan Creek towards the west to Chingarora Creek towards the east. The reach consists of low-elevation marsh at the east end and higher elevation bluffs at the west end of the reach. The area is characterized as a low energy environment protected from the dominant northwest winds by Matawan Point in Reach 1 and from ocean waves by Conaskonk Point in Reach 3. Net longshore transport is to the southwest. The shoreline is stabilized by bulkheads and there are a few small pocket beaches located within and between bulkhead compartments (US Army 1993). The shoreline is used for recreational boating and fishing.

Reach 3 extends from Chingarora Creek to Flat Creek and is 2.7 km in length. Conaskonk Point is a marsh located to the west of a 1000 m long sandy beach in the community of Union Beach. The beach is protected from westerly winds by Conaskonk Point and from easterly winds by Point Comfort (located in Reach 5), but is located in a position to be affected by ocean swell waves that enter Raritan Bay north of Sandy Hook. The beach has a broad, low-tide terrace that helps dissipate waves at low water levels. There is no reported net longshore transport within the sandy shoreline compartment. The compartment is protected by bulkheads and artificial nourishment. Development, residential and commercial, is located in close proximity to the bulkhead.

Reach 4, 1.4 km in length, extends from Flat Creek to Waackaack Creek. The area is low in elevation and comprises an extensive marsh near Waackaack Creek (US Army 1993). The reach is sheltered to the east by Conaskonk Point but is exposed to winds from the northwest. There is no record of stabilization efforts in this reach.

Reach 5 extends from Waackaack Creek to Pews Creek (4.2 km in length) and includes the community of Keansburg located on the Point Comfort headland. The headland is exposed to waves generated in the estuary by northwest winds and ocean waves approaching from the east. Artificial nourishment has created a wide beach on the west and east sides of the headland. In addition, a continuous artificial dike was constructed landward of the backbeach to prevent flooding of inland development. An amusement area is located immediately landward of the dike; the artificial beach on the east side of the headland is used for public recreation.

Reach 6 is 1.8 km long; it extends from Pews Creek to Compton Creek, and includes the community of Port Monmouth. Jetties at both creeks compartmentalize the reach. The shoreline consists of a narrow beach and dune system fronting low elevation marsh. There is no dominant direction of longshore sediment transpo along the shoreline reach (US Army 1993). Development, primarily residentia structures, is located in close proximity to the extensive marsh associated with th drainage basins of the two creeks, and inland flooding occurs in connection wit major storms. This area has experienced more recent development than the othe reaches in the study area and there is an effort underway to revitalize this portion o the Raritan Bay shoreline.

17.3 METHODS

A history of shoreline development in the study area was derived from historica narratives (e.g. Ellis 1885; County of Monmouth 1922; Mustin 1929; Hunton an McCabe 1990). Information on topography and changes in location of marsh wa gathered from US Geological Survey maps (Cook 1887; Vermuele 1902; 1914 1938). An inventory of shoreline stabilization efforts was conducted using aeria photographs and US Army Corps of Engineers documents of the Raritan Ba shoreline (US Army 1960; 1993).

Rates of shoreline change were determined from US Army Corps of Engineer shoreline change maps from the Beach Erosion and Hurricane Survey for Rarita Bay and Sandy Hook, New Jersey, 1960 (scale 1:10 000). The data presented ar for the period 1836–1957. Shoreline change was determined by averaging shorelin position measured at regular intervals beginning at Whale Creek and ending a Compton Creek. Data within each reach were averaged; this average was the divided by the number of years of record to arrive at an annual average. Shorelin change adjacent to the creeks that bound each reach is reported separately. Recen shoreline change was determined by comparison of shoreline position on aeria photographs from 1963, 1972 and 1986.

Changes in the form of beaches in shoreline reaches are examined in the contex of shoreline development and types of stabilization employed. A shoreline segmen in the community of Union Beach (Reach 3) was selected for more detaile investigation to examine differences in the shoreline before and after constructio of a bulkhead. Profile data gathered before bulkhead construction (Nordstrom an Jackson 1989; Farrell et al. 1991) are compared with profile data gathered afte bulkhead construction (Farrell et al. 1994) to assess differences in elevation, slope and mobility of the profile. Additional data were gathered in 1995 at another sit along the bulkhead to assess differences in profile and sediment characteristics at a beach fronting the bulkhead and a pocket beach formed by two shore-perpendicula bulkheads (Figure 17.3). Beach elevation was measured at 2 m intervals from th bulkhead to the break in slope and from the landward limit of the backbeach to th break in slope at the pocket beach. Cores, 0.10 m in depth, were gathered at thes locations to determine differences in grain size populations on the surface and at a

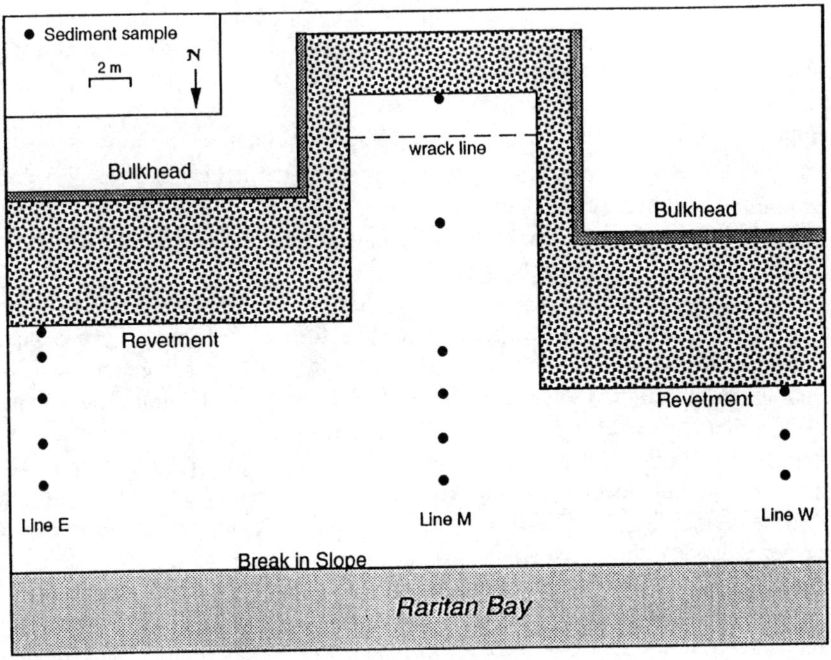

Figure 17.3 Union Beach showing a segment of the bulkhead and sediment sample sites

depth of 0.05 m. This depth was determined from an estimate of potential depth of wave reworking based on findings from previous investigations on estuarine beaches (Jackson and Nordstrom 1993; Sherman et al. 1994). Samples were washed, dried, sieved at 0.5ϕ intervals and analyzed using graphic measures (Folk 1974).

17.4 RESULTS

17.4.1 Development history

The Lenape Indians initially occupied the study area but left by the end of the 18th century (Hunton and McCabe 1990). Farming was an important economic activity in region and development of the communities along the shoreline was associated with transport of both seafood and agricultural products. During the 18th century, landings were established at Browns Point in Keyport (Reach 2), Point Comfort (Reach 5) and Shoal Harbor near Compton Creek (Reach 6) (Hunton and McCabe 1990). Development at the shoreline was limited due to the presence of marsh that was harvested for salt hay.

Increased industrial development, improved transportation and the beginnings of a tourist trade occurred during the 19th century (Hunton and McCabe 1990). Development during this time consisted of piers, docks and support infrastructure to transport products and passengers to New York City. Keyport, located near the mouth of Matawan Creek, was developed as a port in the late 1820s and served as an important transport route for agricultural produce to markets in New York City. The community increased in importance as a center for commerce and transport with the decline of the original port at Matawan (Figure 17.1) due to siltation of Matawan Creek from agricultural runoff.

The railroad was an important influence on the location and viability of the bayshore communities; once it became cheaper to transport goods by rail, many landings were abandoned. The first railroad connected Port Monmouth with the communities of Long Branch, Red Bank, Shrewsbury and Farmingdale (Hunton and McCabe 1990). Port Monmouth (located in Reach 6) was developed as a connection point between these communities and New York providing steamboat service and hotel accommodation. The community remained viable until the construction of the New York and Long Branch railroad, which provided rail transport between New Jersey and New York (Ellis 1885).

Union Beach was originally built as a port community in 1852 to establish a travel and trade connection between New York and the town of Florence (to be established on the Delaware River) (Mustin 1929). Community development ceased when permission to build the railroad was not granted by the legislature (Leonard 1923). The site was abandoned in 1872 and later incorporated in 1925 as a resort community (Hunton and McCabe 1990).

Resort development began in the late 19th century (US Army 1993). The Point Comfort headland (Reach 5) was developed as a resort community in 1906 for the working class of New York City (Hunton and McCabe 1990). The New Point Comfort Beach Company purchased approximately 1000 m of shorefront that later included a boardwalk, sandy beach and amusement area and adjacent wooded land for residential construction (Leonard 1923). The area, originally low and swampy, was developed with sediment dredged from Raritan Bay (Keansburg, New Jersey 1967). Published photographs reveal that residential and commercial development was located close to the backbeach; a recreation platform between the commercial development and a timber bulkhead provided a dry bathing beach at high water (Keansburg, New Jersey 1930). A passenger service between Keansburg and New York City was established with piers located on the west side of Point Comfort. The population of the area increased from 500 people in the winter months to 10 000 people in the summer. Published photographs of the area reveal that the summer population was housed in tents pitched on the beaches on the east side of the headland (Keansburg, New Jersey 1967).

The study area shoreline is now predominantly residential with some commercial development associated with shore-related activities. Most available land is developed. Many of the homes originally built as summer residences are now occupied year-round. Local residents continue to take advantage of shore-oriented recreational opportunities, with highest intensity use occurring in Reach 2 (boating and fishing) and Reach 5 (beach recreation).

17.4.2 Shoreline characteristics

17.4.2.1 Shoreline type

The most dramatic conversions in shoreline type (marsh, beach) occurred during the development of the bayshore communities. US Geological Survey maps depicting 1887 topography reveal that most of the shoreline was low in elevation with a few bluffs ranging from 3.0 to 5.0 m above mean sea level. These bluffs are located on the shoreline of Matawan Point (Reach 1), Keyport south of Chingarora Creek (Reach 2), on the west side of Union Beach (Reach 3), and Port Monmouth (Reach 6). Road development along the shoreline was originally confined to these higher elevations prior to 1887 but expanded into the marsh areas by 1914. Topographic maps from 1914 reveal that road construction was occurring in the marsh areas of Conaskonk Point and on Point Comfort, where fill was used to develop the resort area.

Point Comfort was predominantly freshwater marsh just landward of the shoreline on the west with pine swamp characterizing the shoreline to the east. A small water body, located in the middle of the headland, was surrounded by pine swamp. By 1914 the entire point was filled; marsh and pine swamps were replaced by a network of roads extending from Waackaack Creek to Pews Creek.

The bluff at Matawan Point (Reach 1) was 19 m above mean sea level in 1884; it was fronted by tidal marsh and there was extensive marsh to the east of the point. Marsh continued to dominate the landscape on the east side of Matawan Point in 1938; a network of roads existed on the bluff at this time. A published photograph from 1958 reveals severe erosion of the bluff, threatening an adjacent road (US Army 1960).

Recent fill activities have occurred in Reaches 1, 2, 3, 5 and 6. In Reaches 1, 3, 5 and 6 fill was used to increase the elevation and width of beaches and dunes. In Reach 2, fill was used to construct a parking area that is protected by a bulkhead (US Army 1993).

17.4.2.2 Shoreline stabilization

Figure 17.4 reveals the types of shore stabilization constructed through time and within each shoreline reach. A total of 77 groins have been constructed in the study area since 1924, with most construction occurring prior to the 1950s (US Army 1960, 1993). Groin construction has decreased since the 1970s. Groins, in various stages of repair, still exist along the more exposed reaches of the study area (Reach 1, Reach 5 and Reach 6) and along the shoreline at Keyport (Reach 2) where they

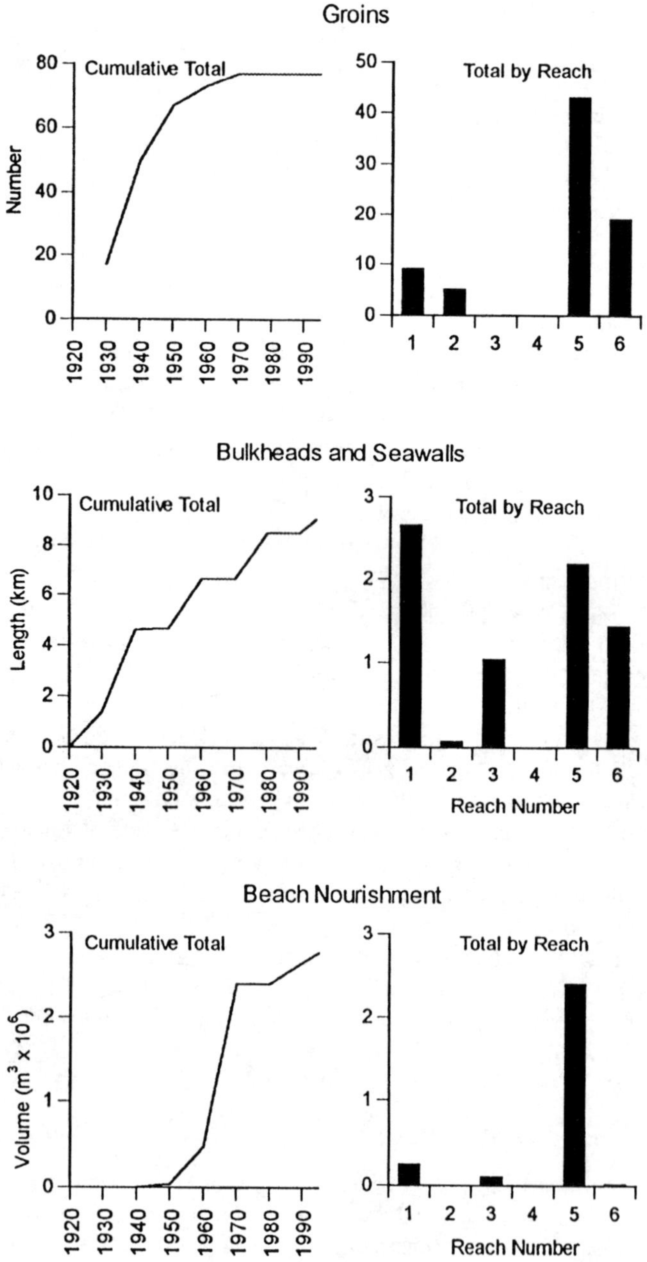

Figure 17.4 Cumulative totals and total by Reach of number of groins, bulkhead lengths and volume of beach nourishment

were constructed in front of bulkheads to retain sand in front of the structures. The groins in Keansburg (Reach 5) were constructed to maintain the beachfill slope during storms rather than as a trap for longshore sediment transport (US Army 1960).

Use of shore-parallel walls (bulkheads and seawalls) has increased steadily over time, and they are located in five of the six shoreline reaches (Figure 17.4). The total length of shore-parallel walls constructed is, in some cases, greater than the length of the shoreline reach, indicating that these structures have been rebuilt. The majority of the structures are privately owned. Most bulkheads are constructed of timber and most seawalls of concrete or stone. Average elevations above mean low water range from a low of 2.7 m (a bulkhead in Reach 5) to a high of 4.3 m (a seawall in Reach 1). The length of bulkhead segments ranges from 15 m to 730 m (US Army 1960).

An estimated total of 2.8 million m^3 of sand has been placed on the beaches in the study area. Of the estimated total, 87% was placed in Reach 5 between Waackaack Creek and Pews Creek (Figure 17.4). The earliest reported beach fill operation was in 1940 in Keansburg (Reach 5). This community was also the recipient of the largest fill operation in 1968 when approximately 1.9 million m^3 of fill was used for a beach and dike to prevent flooding. Smaller scale beach fill operations also have been conducted in Reach 1, Reach 3 and Reach 6. In Reaches 1 and 3, the fill was placed in front of a seawall and bulkhead respectively. In Reach 6 the fill was used for dune building operations, but most of this fill material was eroded during the storm of 1992 (US Army 1993).

Sources of beachfill in the estuary are from offshore and maintenance dredging of the adjacent creeks. For example, fill for nourishment operations in Reach 6 was both from a source 460 m offshore and from dredging of Pews Creek. There is a limited quantity of good quality beach sediment available from these creeks; most sediments dredged from the channels are predominantly silt and clay.

17.4.2.3 Shoreline displacement

Table 17.1 presents shoreline change rates between 1836 and 1957. The period between 1836 and 1855–57 represents shoreline conditions that most closely approximate conditions prior to development. The period between 1855 and 1932–34 represents conditions during development of bayshore communities and the period 1932–34 to 1957 represents conditions after their development. The greatest shoreline changes between 1836 and 1957 are associated with sand transfers at the mouths of creek channels (Table 17.1).

The shoreline receded in all reaches between 1836 and 1855–57. Rates of shoreline change are lowest in Reach 2 southwest of Chingarora Creek where Keyport was settled in the 1820s. This reach is more sheltered than the other five reaches from waves generated by northeasterly storms. Rates are highest in the most exposed reach (Reach 5) that was developed in the early 1900s as a shoreline resort.

Table 17.1 Average rates of shoreline change before and after development on Raritan Bay, New Jersey (m a^{-1})

	Before development (1836 to 1855–57)	During development (1855–57 to 1932–34)	After development (1932–34 to 1957)
Reach 1:			
Whale Creek	−5.0	+1.2	0
West of Matawan Point	−4.0	−0.2	+0.9
East of Matawan Point	−1.6	+0.5	+0.8
Matawan Creek	+9.9	+0.2	0
Reach 2:			
Matawan Creek	−6.2	+0.6	0
South of Chingarora Creek	−0.5	+0.01	+0.1
Reach 3:			
Chingarora Creek	−4.6	+1.2	+1.7
West of Conaskonk Point	−0.3	−0.5	−0.6
East of Conanskonk Point	−3.3	−0.5	−0.9
Flat Creek	−1.7	+0.8	−2.5
Reach 4:			
Flat Creek	−3.9	+3.0	−7.4
Between Creeks	−2.7	+1.6	−1.2
Waackaack Creek	−4.2	+2.8	−4.5
Reach 5:			
Waackaack Creek	−0.4	−0.4	−1.1
West of Point Comfort	−3.5	−0.4	+2.0
East of Point Comfort	−3.0	−0.7	+1.3
Pews Creek	−0.7	−0.2	−1.2
Reach 6:			
Pews Creek	−1.8	−0.2	−0.5
Between Creeks	−1.8	−0.1	−0.5
Compton Creek	−3.0	+1.1	+0.5

Negative values represent landward displacement of the shoreline and positive values represent bayward displacement of the shoreline.

By 1934 bulkheads, seawalls and groins were constructed in five of the six shoreline reaches. Shoreline change rates between 1855–57 and 1934 show a decrease in the magnitude of change (<1.0 m a^{-1}) in all reaches except for Reach 4 where there is an absence of shore stabilization.

Shoreline change between 1932–34 to 1957 reveals the effects of artificial nourishment in Reach 5 where there was a bayward displacement of the shoreline on both the west and east sides of the Point Comfort headland. This displacement of the shoreline does not include the nourishment operation in 1957 that moved the shoreline approximately 60 m bayward (Fray 1969).

Shoreline positions derived from aerial photographs in 1963, 1972 and 1986 are presented in Figure 17.5. The greatest changes in shoreline position are associated with beach nourishment operations in Reaches 1 and 5; the smallest changes are located in Reaches 2, 3 and 4.

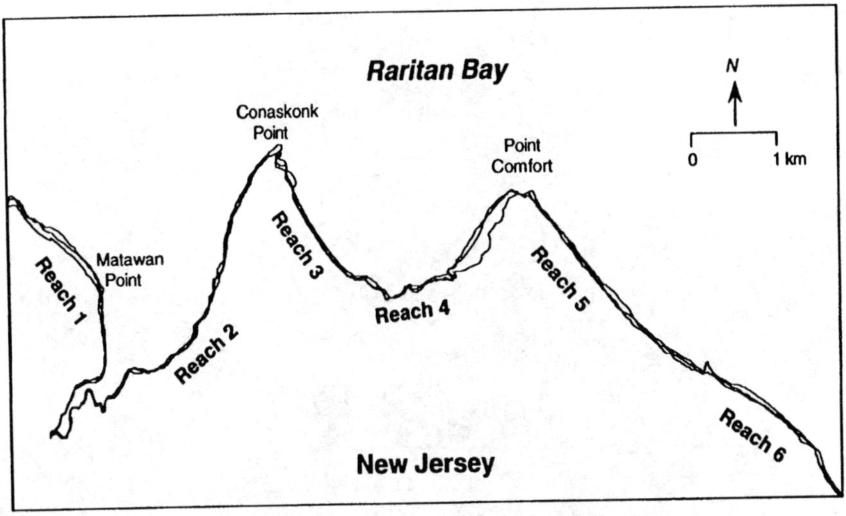

Figure 17.5 Position of the shoreline in 1963, 1972 and 1986

17.4.3 Beach characteristics

Natural beaches in Raritan Bay have average backbeach widths of 10 m and foreshore widths of approximately 20 m; the slope of the foreshore averages 6° and 0.5° on the low tide terrace (Nordstrom 1992). Dunes that are formed and maintained by local processes on unaltered segments of Raritan Bay are generally 1 m in height above the backbeach (Nordstrom and Jackson 1994). These dimensions can become considerably larger or smaller when the shoreline is stabilized for either shore protection or flood control and affect the recreational, ecological, and protective capacity of beaches. Types of adjustments made are a result of type of vulnerability, location of individual shoreline reaches within the estuary and development activity. Examination of beach characteristics reveals larger dimensions in both the elevation and width of beaches and dunes relative to natural conditions along more exposed reaches in the estuary. Beaches in the more sheltered environments have been reduced in size both longshore and cross-shore with increased stabilization.

17.4.3.1 Shore protection: artificial nourishment

Beach nourishment for shore protection has resulted in increased beach widths and changes in sediment characteristics. Introduction of fill material with grain sizes that are different from native beach sediments can result in dramatic changes to the beach environment. This situation is evident in Reach 1 where a narrow, coarse-grained beach backed by an eroding bluff and no dunes evolved into a wide, finer-grained beach backed by a seawall and large dunes.

Figure 17.6 Cliffwood Beach in Reach 1, looking northwest, before artificial nourishment in 1974 and after nourishment in 1989. The seawall is visible in the foreground

Approximately 249 000 m^3 of fill has been placed in front of the seawall and to the west of it in Reach 1 since 1982 (Figure 17.6). The distance between the seawall–beach contact and the break in slope between the foreshore and low tide terrace is 20 m on the west end of the seawall, 30 m in the middle of the structure and 5 m on the east end of the seawall. The fill sediments have a mean grain size of 0.35 mm

(Jackson and Nordstrom 1994); these sediments are finer than the 0.53 mm sediments sampled at the location of mean high water at Matawan Point in 1957 (prior to any nourishment activity) but are coarser than the grain sizes found on the low tide terrace at the location of mean low water (0.28 mm) (US Army 1960). The finer grain size of the artificial fill has led to increased mobility of the profile, and a considerable portion of the fill is now located on the low tide terrace near Matawan Creek (Jackson and Nordstrom 1994).

Aeolian activity has resulted in the accumulation of fill sediments on top of the seawall resulting in the development of dune forms that are separated from the beach profile. To the west of the seawall is an artificial dune that fronts a public road. No dunes existed in this reach under natural conditions. Dune elevation in September, just prior to the December 1992 storm, was 1.3 m above the backbeach (Jackson and Nordstrom 1994). These dunes were destroyed or severely scarped by that storm. The dunes were initially re-established by bulldozing and by 1993 dune elevations were 2.5 m above the backbeach (Nordstrom and Jackson 1994). Dune growth by aeolian processes is aided by the finer grain sizes and large source area created by the artificial fill. Dune elevations are sufficient at this time to prevent overwash and inland flooding during most storms.

Narrow beach widths fronting eroding bluffs limit the effectiveness of artificial dune construction as a stabilization alternative along some reaches. In Reach 6 sand fencing and vegetation was used for dune building at a bluff that is subject to wave attack during storms (US Army 1993). The dune building project in September 1988 displaced the beach–dune contact point approximately 4.5 m bayward (Nordstrom and Jackson 1989). Photographs taken after the December 1992 storm reveal that the dune was completely destroyed, exposing the eroded bluff (US Army 1993). The width of the beach fronting the bluff in 1995 is 8.0 m. Groins were constructed to maintain a beach but there is a lack of accumulation of sediment on either side of the structures (US Army 1993). A proposed shore erosion and flood control project for this reach will increase beach widths to 15.0 m and establish a dune 4.6 m in elevation above mean sea level (US Army 1993).

17.4.3.2 Shore protection: shore-parallel structures

Placement of shore-parallel structures occurs landward of and on the intertidal profile of estuarine beaches. Without the introduction of artificial nourishment, structures may become part of the active profile over time as erosional conditions persist. Bulkheads are common on the sandy shoreline segments in Reaches 2 and 3. Examination of profile characteristics in these reaches provides insight to long-term profile and sediment change with and without nourishment.

In Reach 2 bulkheads are located both landward of and in the active beach profile. Small beach compartments have been formed where shore-perpendicular and shore-parallel bulkheads connect. No artificial nourishment has been placed in this reach. Beach width is 3.2 m at the midpoint of a privately owned shoreline compartment bounded by bulkheads that were originally constructed before 1921 in Reach 2 (US Army 1960). The structures are currently fronted by a coarse-grained foreshore with gravel lag deposits; sediments sampled in the location of mean high water in 1957 reveal mean grain sizes of 0.4 mm.

Bulkheads in Reach 3 are confined to the 1000 m sandy segment of Union Beach. Examination of aerial photographs of Union Beach indicates that by 1977, bulkheads fronted more than 50% of the development located on the backbeach. By 1982, more than 75% of the shoreline reach was bulkheaded; structures that were evident on the east end of the shoreline reach in 1977 had no beach fronting them by 1982. A bulkhead and nourishment project undertaken in 1990 stabilized the remaining length of natural shoreline and replaced individual bulkheads with one structure (Figure 17.7). Previous investigations at this site reveal that the beach profile experienced low mobility before and after bulkhead construction in 1990 (Jackson and Nordstrom 1992; Farrell et al. 1994). The structure is subject to both wave attack on the bayward side and removal of upland on the landward side during storms, when waves and surge overtop the structure, thus increasing the potential for failure (US Army 1993). Areas of the bulkhead that were damaged during the storm of 1992 have been repaired. There is currently no dry sand beach fronting the structure at the east end of the compartment at high water. A small beach fronts the west end of the compartment that is 6.5 m wide; the foreshore slope is 5.7° at this location.

Figure 17.3 is a segment of the Union Beach shoreline showing the bulkhead at the west end of the shoreline compartment and the sediment sampling points. The width of intertidal beach between the stone revetment that protects the toe of the bulkhead and the break in slope varies from 8.3 m on the west to 11.5 m on the east. Figure 17.8 presents beach profiles fronting the bulkhead and in the middle of the pocket beach. The profile slopes fronting the bulkhead are 5.4° on the east side and 5.8° on the west side; the profile slope over the same distance at the site not fronting the bulkhead is 4.7°. Elevations are lower on the east and middle profiles compared with the west profile between 14 m and 24 m from the baseline.

Sediment characteristics for the segment portrayed in Figure 17.3 are presented in Table 17.2. There is little variation in mean grain size on the surface and at 0.05 m depth. Sorting values reveal higher values at depth than at the surface. This could either be a function of a lack of sediment reworking by waves or because the sediments composing the original upland and native beach sediments were sampled. Mean grain size for surface sediments gathered in the shoreline compartment in 1957 near the study site was 0.32 mm at the location of mean high water with a sorting value of 1.27 (US Army 1960).

17.4.3.3 Flood Control

Beach nourishment can provide protection from both erosion and flooding depending on berm elevations. Estimated berm elevation for natural beaches in the

Figure 17.7 Union Beach, in Reach 3, in 1988 before bulkhead construction (top) and in 1995 after bulkhead construction (bottom)

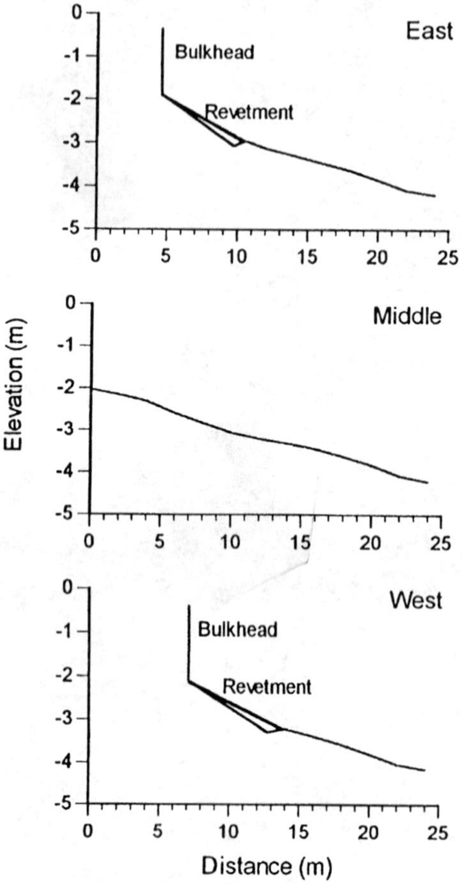

Figure 17.8 Profiles from Union Beach

estuary is 1.7 m above mean sea level (US Army 1960). In Reach 5, the berm elevation of the first large-scale project, funded by the state in 1957, was 3.4 m above mean sea level; berm width was 24 m with a foreshore slope of 4.5° (US Army 1960). A subsequent project by the federal government resulted in a berm elevation of 4.6 m above mean sea level and a berm width of 15 m (US Army 1993). Levees were constructed along both Waackaack and Pews Creek to reduce the interior flooding along with a sand dike located landward of an artificial backbeach. The elevation of the dike is 5.0 m above the backbeach (Nordstrom and Jackson 1994).

Keansburg has successfully maintained and enhanced the recreational aspects of the sandy beach and provided flood control and shore protection benefits. The dimensions of the beach are much larger than would be anticipated under natural conditions. The subaerial beach width is approximately 14 m on the west side of the

Table 17.2 Sediment characteristics of samples from the surface and at 0.05 m depth gathered at Union Beach, New Jersey

Transect	Distance	Mean (mm)		Sorting (phi)	
		Surface	Depth	Surface	Depth
East	10.5	0.361	0.359	0.477	0.527
	12.0	0.330	0.274	0.278	0.630
	14.0	0.293	0.607	0.441	0.805
	16.0	0.361	0.390	0.233	0.846
	18.0	0.319	0.463	0.465	0.624
Middle	0	0.262	0.264	0.303	0.448
	6.0	0.274	0.330	0.382	0.526
	12.0	0.366	0.332	0.373	0.478
	14.0	0.285	0.308	0.236	0.503
	16.0	0.310	0.373	0.469	0.812
	18.0	0.295	0.566	0.428	1.184
West	14.0	0.361	0.344	0.370	0.538
	16.0	0.328	0.369	0.327	0.877
	18.0	0.270	0.259	0.394	0.814

Point Comfort headland and approximately 38 m on the east side where the beach is used for public recreation.

17.5 IMPLICATIONS

Rates of shoreline change between 2.0 and 3.0 m a^{-1} occurred in the more exposed reaches of Raritan Bay before significant shoreline development. These rates decreased to <2.0 m a^{-1} when bulkheads and groins were used to stabilize the shoreline. Introduction of artificial sediment to shoreline reaches has translated the shoreline bayward and increased mobility in the more exposed reaches.

Beach nourishment is the preferred option for stabilizing eroding ocean shorelines in New Jersey (Nordstrom 1994). There is also an indication of this trend in Raritan Bay for both shore protection and flood control. The total quantity of fill placed on the shoreline of Raritan Bay is greater than the total volume reported for nourishment operations in Chesapeake Bay, Delaware Bay or Puget Sound (Nordstrom 1992). The total quantity of fill placed in Reach 5 in 1968 exceeds the volumes reported for most nourishment projects along the exposed ocean shoreline of New Jersey (Leonard et al. 1990).

Carter and Mitchell (1985) demonstrate the short-term stability of artificial nourishment along estuarine shorelines; the data from Raritan Bay suggest long-term longevity of beaches once nourishment operations are undertaken in estuaries, even along exposed reaches. Nourishment activities in Reach 5 have displaced the shoreline bayward; the beach originally designed for the Point Comfort headland was 15.0 m wide; beach width measured in 1995 on the west side of the headland has decreased by 1.0 m in 27 years and the recreational beach on the east side of the headland has increased by 13.0 m.

Shore-parallel structures, primarily bulkheads, remain the most favored form of shore stabilization along estuarine shorelines and, unlike the movement away from hard stabilization on the exposed ocean shoreline, there is no indication that the use of bulkheads will be abandoned. Low cost, ability to withstand wave energies in estuaries, and multiple uses for erosion and flood control, boat docking and land reclamation have contributed to the widespread use of these structures (US Army 1981). In Washington state, two counties on Puget Sound reported a total of 344 applications for bulkhead construction and repair over an 18-month period (Shipman and Canning 1993). In Raritan Bay, approximately 20% of the shoreline between Whale Creek and Compton Creek is bulkheaded.

The physical impacts of a shore-parallel structure on the fronting beach will occur when waves reflect or break on the structure. Increased energy in the swash zone associated with wave reflection off a shore-parallel structure may lead to increased scour (Zabawa et al. 1981), oversteepening of the foreshore slope (Slaughter 1967) and coarsening of beach sediments (Macdonald et al. 1994), reducing the suitability of these areas as habitat (Shipman and Canning 1993). Scour can also lead to emplacement of stone rip-rap placed on the bayward side of the structure to protect it from further erosion (Slaughter 1967). Scour was not evident at bulkheads in the study area during visual inspection in 1995. Rip-rap fronts the recently rebuilt bulkhead in Reach 3, but comparison of profile data 4 years before and 4 years after construction reveals no lowering of the profile or change in foreshore slope in the compartment (Nordstrom and Jackson 1989; Farrell et al. 1991; Jackson and Nordstrom 1992; Farrell et al. 1994).

Shore-parallel structures can be constructed on the low-tide terrace of estuarine beaches, without increased scour in front of the structure (Jackson and Nordstrom 1994). Nourishment in front of shore-parallel structures can result in beach widths that eliminate beach–seawall interactions (Macdonald et al. 1994). Beach widths fronting the west end of the seawall in Reach 1 are greater than would be found under natural conditions and the extra width reduces the frequency of wave uprush reaching the structure. In Reach 3, placement of the structure low on the profile combined with artificial nourishment reduces the physical effects of stabilization on the beach.

The net effect of the use of shore-parallel walls in estuaries can be significant reduction or elimination of sandy beach environments. Truncation of the beach profile by bulkheading has decreased the number of suitable spawning areas on the east side of Delaware Bay (Botton et al. 1988). Union Beach has recently lost at least 20 m of sand beach due to bulkhead construction low on the foreshore. Larger beaches have been created in other reaches through nourishment operations. Horseshoe crabs have been observed during spawning season on the artificial beach in Reach 1 but the suitability of these environments and the factors that may enhance these environments as habitat are unknown.

Increased investment and/or reinvestment along estuarine shorelines will result in proposals for construction of improved erosion and flood control measures as the value and extent of vulnerable properties rise. Alterations to the beach environment can be great, even when development is not located at the shoreline. Development history of the study area reveals that land use decisions have resulted in significant changes to beach characteristics. The road on the bluff in Reach 1 was vulnerable to undermining by wave action in 20 years after construction. Alterations have eliminated marsh environments and introduced artificial sediment that has increased the dimensions of the beach and dune system. Residential construction in the low-elevation marsh along the creeks in Reach 5 increased vulnerability to flooding and resulted in physical alteration of the shoreline to mitigate the threat to private property. A similar residential development pattern in Reach 6 has contributed to the need to reduce vulnerability to erosion and flooding. Plans are now underway to reduce the threat to inland property in Reach 6 with a flood control plan similar to the project implemented in Reach 5. These flood control measures have increased both the horizontal and vertical dimensions of the shoreline. The effect of these changes on shoreline or inland marsh habitat is unclear. Evaluation of 14 estuaries that are part of the National Estuary Program in the United States reveals that only four have identified land use change and intensification as a priority management area (McCreary et al. 1992). Results of this study show that significant changes to beach environments (in the cross-shore direction) have occurred due to land use decisions than from interactions between structures and the beach.

17.6 CONCLUSIONS

Humans have been a persistent process in the evolution of the shoreline in Raritan Bay. Stabilization measures to protect private property have resulted in a reduction in erosion rates and, in some cases, have reversed the net direction of shoreline change. Bulkheads are successful along sheltered shorelines provided the structure is of adequate height to prevent waves and surge from overtopping the structure. Beach nourishment has resulted in the most dramatic changes in the physical characteristics of shoreline reaches, including construction of dunes where none existed previously and an increase in the elevation and width of beaches. Nourishment along more exposed shoreline segments including the western segment of Reach 1 and the west facing shoreline of Reach 5 are more mobile than along less exposed segments in those reaches.

The scale of response to shoreline erosion and flooding is shifting from actions of individual property owners to broader scale responses that incorporate shoreline reaches and are planned and funded by higher government agencies. With increasing land development in close proximity to the creeks, vulnerability of private property to flooding will increase. Human activities will alter the physical characteristics of sandy shorelines to reduce flooding hazards. Berm elevations wi be increased through beach nourishment, and shore-parallel barriers will b constructed at elevations sufficient to prevent surge penetration.

ACKNOWLEDGEMENTS

The author gratefully acknowledges the assistance with aerial photo interpretatio given by L. Pace and K. Jagger.

REFERENCES

Botton, M.L., R.E. Loveland and T.R. Jacobsen. 1988. Beach erosion and geochemica factors: influence on spawning success of horseshoe crabs (*Limnulus polyphemus*) i Delaware Bay, USA. *Marine Biology* 99: 325–332.

Carter, L. and J.S. Mitchell. 1985. Stability of an artificially nourished beach, Balaena Ba Wellington Harbour, New Zealand. *New Zealand Journal of Marine and Freshwate Research* 19: 535–552.

Chu, Y.H. 1985. Beach erosion and protection, a case study of Lincoln Park, Washingtor *Shore and Beach* 53: 26–32.

Cook, G.H. 1887. *Map of the Sea Coast of Monmouth and Ocean counties, New Jersey*. G.W and C.B. Colton, New York.

County of Monmouth. 1922. *History of Monmouth County, New Jersey 1664–1920. Volum II*. NY: Lewis Historical Publishing Co., New York.

Ellis, F. 1885. *History of Monmouth County, New Jersey*. R.T. Peck and Co, Pennsylvani

Farrell, S., B.A. Sullivan, F. Oliver, S. Hafner, A.P. Meggison and T.S. Lyons. 1991. Fin report for 1990 on New Jersey beach profiles network: a series of FEMA monitorin survey stations. Contract C-29050. New Jersey Department of Environmental Protectior Division of Coastal Resources, New Jersey.

Farrell, S.C., B. Sullivan, S. Hafner and T. Lepp. 1994. New Jersey beach profile networ analysis of the shoreline changes in New Jersey coastal reaches one through fiftee Raritan Bay to Delaware Bay. New Jersey Department of Environmental Protection Office of Regulatory Policy, New Jersey.

Findlay, R.H. and R.M. Kirk. 1988. Post-1847 changes in the Avon-Heathcote estuary Christchurch: a study of the effect of urban development around a tidal estuary. *Ne Zealand Journal of Marine and Freshwater Research* 22: 101–127.

Folk, R.L. 1974. *The Petrology of Sedimentary Rocks*. Hemphill, Texas.

Fray, C.T. 1969. Final report: Raritan estuary sedimentation study. Contract No. 14-12-40 Department of Interior, Federal Water Pollution Control Administration, Washingtor DC.

Hardaway, C.S. and J.R. Gunn. 1991. Headland breakwaters in the Chesapeake Bay, 1267–1281. *In*: O.T. Magoon, H. Converse, V. Tippie, L.T. Tobin, and D. Clark (eds. *Coastal Zone 91: Proceedings of the Seventh Symposium on Coastal and Ocea Management*. American Society of Civil Engineers, New York.

Hunton, G. and J. McCabe. 1990. *Monmouth County Historic Sites Inventory: Summar Report*. The Office of New Jersey Heritage, the Monmouth County Park System, and th Monmouth County Historical Association, New Jersey.

Jackson, N.L. and K.F. Nordstrom. 1989. Estuarine beach management: shore processes and protection methods, p. 3360–3376. *In*: O.T. Magoon, H. Converse, D. Miner, L.T. Tobin, and D. Clark (eds.), *Coastal Zone 89: Proceedings of the Sixth Symposium on Coastal and Ocean Management*. American Society of Civil Engineers, New York.

Jackson, N.L. and K.F. Nordstrom. 1992. Site specific controls on wind and wave processes and beach mobility on estuarine beaches in New Jersey, U.S.A. *Journal of Coastal Research* 8: 88–98.

Jackson, N.L. and K.F. Nordstrom. 1993. Depth of activation of sediment on a steep sand beach. *Marine Geology* 115: 143–151.

Jackson, N.L. and K.F. Nordstrom. 1994. The mobility of beach fill in front of a seawall on an estuarine shoreline, Cliffwood Beach, New Jersey, USA. *Ocean and Coastal Management* 23: 149–166.

Keansburg, New Jersey. 1930. *The Story of Keansburg: Monmouth County's Fastest Growing Community*. Bureau of Publicity, Keansburg, New Jersey.

Keansburg, New Jersey. 1967. *Fiftieth anniversary of Keansburg, New Jersey 1917–1967*. The Advisor, Inc., Keansburg, New Jersey.

Kraus, N.C. and O.H. Pilkey. 1988. The effect of seawalls on the beach. *Journal of Coastal Research* SI4.

Leonard, L.R., K.L. Dixon and O.H. Pilkey. 1990. A comparison of beach replenishment on the US Atlantic, Pacific, and Gulf Coasts. *Journal of Coastal Research* SI6: 127–140.

Leonard, T.H. 1923. *From Indian trail to electric rail*. The Atlantic Highlands Journal, New Jersey.

Ludwick, J.C. 1987. Mechanisms of sand loss from an estuarine groin system following an artificial sand fill. Technical Report 87-2. Old Dominion University Department of Oceanography, Norfolk, VA.

Macdonald, K., D. Simpson, B. Paulson, J. Cox and J. Gendron. 1994. Shoreline armoring effects on physical coastal processes in Puget Sound, Washington. Volume 5. Report 94-78, Washington Department of Ecology, Shorelands and Coastal Management Division, Washington.

McCreary, S., R. Twiss, B. Warren, C. White, S. Huse, K. Gardels and D. Roques. 1992. Land use change and impacts on the San Francisco estuary: A regional assessment with national policy implications. *Coastal Management*. 20: 219–253.

Mustin, M. (ed.). 1929. *A Sketch of Monmouth County, New Jersey 1683–1929*. M. Mustin Co., New Jersey.

Nichols, M.M. and M.M. Howard-Strobel. 1991. Evolution of an urban estuarine harbor: Norfolk, Virginia. *Journal of Coastal Research* 7: 745–757.

National Oceanic and Atmospheric Administration. 1994. *Tide Tables: East Coast of North and South America*. Rockville, MD.

Nordstrom, K.F. 1992. *Estuarine Beaches* Elsevier Applied Science, New York.

Nordstrom, K.F. 1994. Developed coasts. p. 477–509. *In*: R.W.G. Carter and C.D. Woodroffe (eds.), *Coastal Evolution: Late Quaternary Shoreline Morphodynamics* Cambridge University Press, Cambridge.

Nordstrom, K.F. and N.L. Jackson. 1989. Processes and landform changes affecting management decisions on the Raritan Bay and Delaware Bay shorelines. Technical Report 1043. Rutgers University, Center for Coastal and Environmental Studies, New Jersey.

Nordstrom, K.F. and N.L. Jackson. 1992. Two dimensional change on sandy beaches in estuaries. *Zeitschrift fur Geomorphologie* 36: 465–478.

Nordstrom, K.F. and N.L. Jackson. 1994. Aeolian processes and dunefields in estuaries. *Physical Geography* 15: 358–371.

Nordstrom, K.F., N.L. Jackson, K.A. Jagger and J.R. Allen. 1989. Backbarrier beaches on a developed shoreline, p. 23–32. *In* D.K. Stauble (ed.), *Barrier Islands: Processes and Management*. American Society of Civil Engineers, New York.

Orford, J. 1988. Alternative interpretation of man-induced shoreline changes in Rossalare Bay, southeast Ireland. *Transactions of the Institute of British Geographers, New Series* 13: 65–78.

Rosen, P.S. and D.B. Vine. 1995. Evolution of seawall construction methods in Boston Harbor, Massachusetts. *Proceedings of the Institution of Civil Engineers – Structures and Buildings* 110: 239–249.

Sherman, D.J., K.F. Nordstrom, N.L. Jackson and J.R. Allen. 1994. Sediment mixing-depths on a low-energy reflective beach. *Journal of Coastal Research* 10: 297–305.

Shipman, H. and D.J. Canning. 1993. Cumulative environmental impacts of shore stabilization on Puget Sound, p. 2233–2242. *In* O.T. Magoon, W.S. Wilson, H. Converse and L.T. Tobin (eds.), *Coastal Zone '93. Proceeding of the Eighth Symposium on Coastal and Ocean Management*. American Society of Civil Engineers, New York.

Schultz, L.P. and W. Ashby. 1967. An analysis of an attempt to control beach erosion in Chesapeake Bay, at Scientists Cliffs, Calvert County, Maryland. *Chesapeake Science* 8: 237–252.

Schwartz, M. L. and E.C.F. Bird. 1990. Artificial beaches. *Journal of Coastal Research*, SI7.

Slaughter, T.H. 1967. Vertical protective structures in the Maryland Chesapeake Bay. *Shore and Beach* 35(2): 7–17.

Thom, R.M., D.K. Shreffler, and K. Macdonald. 1994. Shoreline armoring effects on coastal ecology and biological resources in Puget Sound, Washington. Coastal erosion management studies, Volume 7. Report 94-80. Shorelands and Coastal Zone Management Program, Washington.

United States Army, Corps of Engineers. 1960. *Raritan Bay and Sandy Hook Bay New Jersey. Cooperative Beach Erosion Control and Interim Hurricane Study*. US Army Engineer District, New York Corps of Engineers, New York.

United States Army, Corps of Engineers. 1981. *Low-Cost Shore Protection, Final Report on the Shoreline Erosion Control Demonstration Program (Section 54)*. Office of the Chief of Engineers, Washington, DC.

United States Army, Corps of Engineers. 1993. *Raritan Bay and Sandy Hook Bay, New Jersey Combined Flood Control and Shore Protection Reconnaissance Study*. US Army Corps of Engineers New York District, New York.

Vermeule, C.C. 1902. *Topographical Map of the Monmouth Shore with the Interior from Metuchen to Lakewood*. US Geological Survey.

Vermeule, C.C. 1914. *Topographical Map of the Monmouth Shore with the Interior from Metuchen to Lakewood*. US Geological Survey.

Vermeule, C.C. 1938. *Topographical Map of the Monmouth Shore with the Interior from Metuchen to Lakewood*. US Geological Survey.

Zabawa, C.F., R.T. Kerhin and S. Bayley. 1981. Effects of erosion control structures along a portion of the northern Chesapeake Bay shoreline. *Environmental Geology* 3: 201–211.

Chapter Eighteen

Management and use of dynamic estuarine shorelines

J. P. DOODY
Joint Nature Conservation Committee

18.1 INTRODUCTION

Estuaries are dynamic places where a series of interrelated habitats provide food and shelter for an abundance of plants and animals. In the absence of human interference an individual estuary develops in response to a complex set of interactions including the geology of the hinterland, relative sea level change, tidal range and pattern of inundation, sediment inputs/outputs and movement, and freshwater flows. Habitats occur in a sequence from the sea to the upper shoreline and include successional and transitional characteristics which reflect the dynamic interaction of the component habitats.

Relative change in sea level is a key factor in determining the limits of an estuary,and estuarine shorelines are particularly responsive. Shennan (1987) showed that in the case of the Wash (eastern England) a change from a net landward movement of the sea to one where relative sea level falls, can take place over periods measured in hundreds of years. A similar picture has been described by Steers (1981) for the coastal lowlands of Essex (south-east England) and six periods of transgressions and five regressions by the sea have been identified since the end of the last glaciation (Greensmith and Tucker 1973).

Historically, estuaries have been exploited for their often abundant food resources (fish, shellfish and wildfowl), for the disposal of waste and use of the upper margins of the zones for agriculture and in more recent years infrastructure (including industrial) development. As the upper margins of the shore are destroyed by enclosure the size of the embayment is progressively reduced. The upper limits

Estuarine Shores: Evolution, Environments and Human Alterations. Edited by K. F. Nordstrom and C. T. Roman.

of tidal influence, including saltmarshes, tend to suffer most and the natural, soft and flexible margin of the estuary is often replaced by anunnatural, hard and inflexible barrier. As the estuarine shore is forced seaward the extent of the intertidal land is reduced and with it the natural ability to accommodate rising sea levels, storms or human perturbations. This chapter seeks to assess the consequences of this 'estuarine squeeze' for wildlife and human activity and reviews new approaches to habitat restoration.

18.2 IMPLICATIONS FOR MANAGEMENT (THE COASTAL SQUEEZE)

18.2.1 Human influences – habitat loss

Past human activities have taken advantage of accretion of the coastline, notably by the growth of saltmarsh, to create more favourable conditions for exploitation. Evidence from archival material in the UK indicates that marshlands in North Norfolk were used for grazing well before the Norman conquest nearly 1000 years ago. As early as the 12th and 13th centuries attempts were made to exclude the tide from the upper levels of the saltmarsh (Allison, personal communication) and early enclosure and subsequent use for arable land was commonplace in Kent in the 13th and 14th centuries (Smith 1940).

Enclosure and development for intensive agriculture have had a major impact on low-lying land around many estuarine systems. The process involves the exclusion of the tide by the erection of an earth bank or other seawall, enclosing mature saltmarsh. In historical times many enclosed saltmarshes became 'grazing marsh'. These areas were usually enclosed, unimproved saltmarsh which was left as rough grassland under traditional grazing management (Gray 1977). In more recent years these areas have themselves been improved through drainage and ploughing for intensive agriculture (Doody 1995). Modern techniques can bypass this intermediate stage and drainage and ploughing can convert the land in just a few years to intensive agricultural use.

The enclosure for agricultural development is often piecemeal, the timing of individual works being dictated mostly by prevailing economic circumstances. Examples of the cumulative nature of the process are shown in Figure 18.1 for the Ribble estuary in north-west England where recent enclosure for agriculture has taken place. As can be seen from Table 18.1, significant enclosure of tidal land can occur.

Losses have also resulted from the development of ports, industry, housing and recreational use. Documented examples in Great Britain include the Tees Estuary (north-east England) (see Davidson et al. (1991)) and Cardiff Bay (south Wales) (see Doody (1995)). Many other estuaries around the world have been affected in the same way.

Table 18.1 Loss of saltmarsh to enclosure for agriculture, major estuaries in south-east and north-west Britain

Estuary	Area (ha)	Dates	Reference
The Humber (Humberside)	4663	17th to mid-19th century	Stickney 1908
The Wash (Lincs/Norfolk)	29 000	From the 17th century (3000 in 20th century)	Dalby 1957
Suffolk estuaries	10 817	From Roman times	Beardall et al. 1988
Essex and North Kent estuaries	4340	Mainly pre-18th century	Macey 1974
The Ribble (Lancashire)	1960	Past century	Royal Commission on Coastal Erosion 1911
Morecambe Bay (Cumbria/ Lancashire	1300	13th century to 19th century	Gray 1972

18.2.2 Loss of habitat – erosion from the sea

The direct loss of habitat through enclosure can readily be seen by the inspection of a sequence of historical maps. It is more difficult to assess losses that have taken place to seaward. Surveying land periodically covered by the tide, the absence of fixed and prominent features and the seasonal variability of the pioneer vegetation, made accurate representation of the shoreline on older maps very difficult. However, with the use of aerial photographs, suitably calibrated to make comparisons valid between different dates linked with field survey, measurement of change is possible.

The saltmarshes in south-east England (Essex and north Kent) have been enclosed by embankment for centuries with the creation of considerable areas of agricultural land. At the same time there has been a relative rise in sea level resulting from the combined effects of eustatic and isostatic adjustments that are taking place in south-east England (Shennan 1987). Because the upper limit of much of the saltmarsh is constrained by a seawall the intertidal habitat is ‘squeezed’ between it and the rising sea level (Figure 18.2). It is thought that these two effects may be partly responsible for the loss of saltmarshes observed between 1973 and 1988 (Table 18.2). In some localities the saltmarsh has been eroded completely in front of the seawall, which has led to the undermining of the toe of the sea defence structure. If the rate of sea level rise increases as a result of global warming, then the saltmarsh zone could ultimately disappear over a much wider area.

18.3 MANAGEMENT OPTIONS

In many low-lying coastal areas the estuarine shoreline is defined by a straight inflexible barrier. In a zone which is recognised as being subject to the forces of

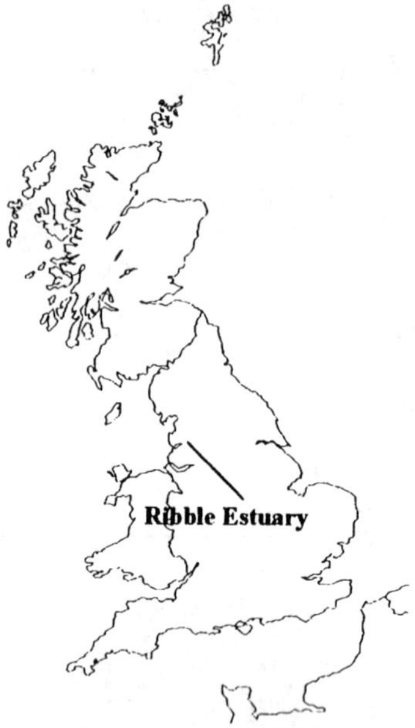

Figure 18.1 Cumulative loss of saltmarsh to enclosure for agriculture, Ribble Estuary, north-west England

nature, it could be argued that too much effort has been devoted to the 'battle with the sea'. As the coastline is squeezed into an ever narrower zone its ability to respond to the unpredictable forces of storms, tides and sediment movement, as well as human influence such as pollution, is compromised. This may have grave consequences not only for the survival of the wildlife which inhabits these areas but also for their future sustainable human use.

The results of past exploitation of the estuarine environment and the implications for future management leave a number of policy options for the future. These options can be considered under several headings including:

1. Strengthening of existing line of defence.
2. Accretion enhance in front of existing lines of defence.
3. Control inundation by tidal barrier.
4. Exclude tide through the erection of a tidal exclusion barrage.
5. Accommodate change by redefining management options within the zone of hazard.

Each of these is summarised below.

18.3.1 Strengthening the existing line of defence

Maintenance of the current line of defence by building a higher, wider or stronger

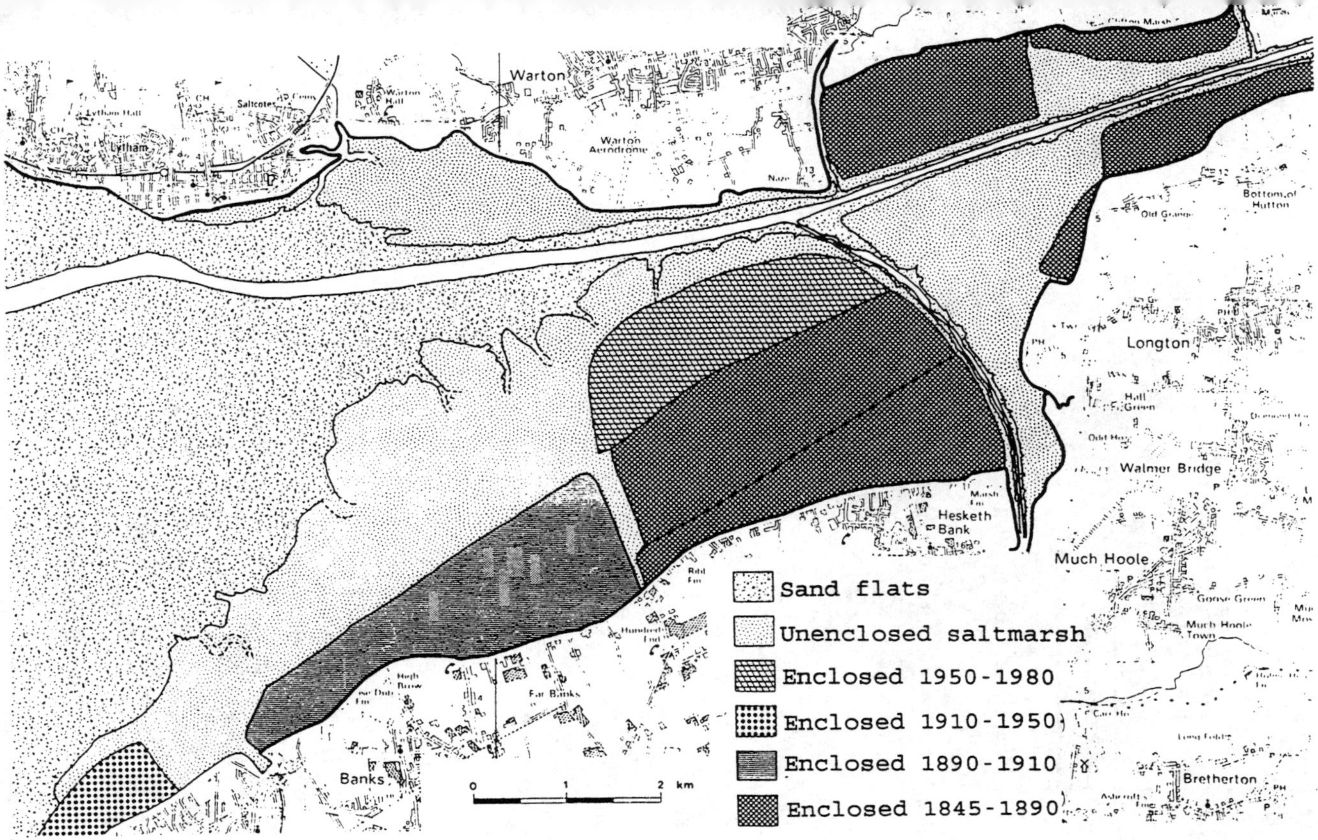

Figure 18.1 *(continued)*

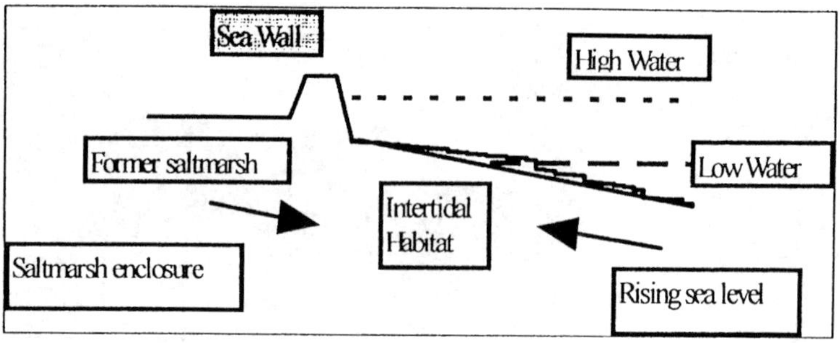

Figure 18.2 Loss of estuarine shoreline habitats – 'The coastal squeeze'

Table 18.2 Loss of saltmarsh in the estuaries of Essex and Kent, 1973–1988 (Burd 1991

	Original area (ha)	Total area lost (ha)	Loss (ha) to reclamation	Loss (ha) to erosion	% of original area eroded
Orwell	99.5	39.9	7.5	32.4	32.6
Stour	264.2	129.5	13.3	116.2	44.0
Hamford Water	876.1	170.6	1.2	169.4	19.3
Colne	791.5	97.7	5.2	92.5	11.7
Blackwater	880.2	200.2	–	200.2	22.7
Dengie	473.8	46.7	–	46.7	0.9
Crouch	467.1	146.1	22.1	124.0	26.5
Thames (Essex)	365.9	105.6	22.3	83.3	22.8
Thames (Kent)	77.8	17.5	3.2	14.3	18.4
Medway	843.8	198.3	18.2	180.1	21.3
Swale	377.0	61.6	3.4	58.2	15.4

sea defence or flood protection barrier, provides the tried and tested engineerin solution. The protection given to the coast can be measured in the time interval ove which storm or other extreme conditions might be expected to overwhelm th defences. In the Netherlands the design life for the series of dams and barrages o the Delta Project is set to withstand a 1 in 5000 year flood event. In the Wash th height of the sea walls is expected to be overtopped once ever 40 years or so an the Thames barrier is designed to protect London from tidal floods 1.2 m greate than any so far recorded.

18.3.2 Accretion enhance

The second approach involves encouraging the growth of saltmarsh through th enhancement of accretion. This has been a major activity, extensively used i Denmark, Germany and Holland (Kamps 1962) and is being introduced in the UK as saltmarshes are increasingly being recognised for their contribution to se defence (Brampton 1992). The technique for saltmarsh recreation involves the partial enclosure of tidal flats with brushwood fencing, groynes or other material to encourage siltation. Occasionally on estuarine shores 'beach feeding' through the artificial nourishment of the shoreline may also be used to help protect an existing seawall.

18.3.3 Storm surge/energy generation barriers

Storm surge barriers may have very different effects on the development of an estuary shoreline, depending on the frequency with which they are used to prevent tidal inundation. Since they are normally designed to operate only during extreme events their overall impact on the estuarine shoreline is likely to be small. By comparison, energy generation schemes using tidal barriers will, potentially at least, have a greater effect. They may cause major change to the tidal regime both inside and, perhaps more importantly, outside the barrier. In its turn this will modify the natural hydrological, sedimentological and ecological systems (see Carter 1988 pp. 406–416, for a summary of the likely impacts).

18.3.4 Tidal exclusion barrages

Tidal exclusion barrages such as those built as part of the Delta project in the Netherlands are massive engineering structures and very costly. These are likely to be used in only a few limited areas where the significant threat to human life and property justifies the investment. This well documented example shows that where they are built they can have a major impact on the estuarine shoreline (Hooghart and Posthumus 1989). These and the much smaller amenity barrages, (often justified to 'cover unsightly mud' in order to 'create a superb waterfront environment' as in the case of Cardiff Bay in south Wales) represent the final stage in the loss of the estuarine shore, which is usually replaced by a freshwater lake.

18.3.5 Set back and managed retreat

In those areas where enclosure of saltmarsh has been followed by agricultural use it may be possible to reverse the process. The value of saltmarsh for coastal defence and the inevitability of shoreline retreat as sea level rises has led to the adoption of a policy of 'managed retreat' in the UK as one of a suite of approaches to shoreline management. This seeks to reverse the enclosure of tidal land to recreate saltmarsh and by so doing improve the sea defence capability of the coast at the same time as improving its wildlife habitat potential. This policy attempts to define new coastal limits for the estuarine shore and is discussed in more detail below.

18.4 DISCUSSION

The majority of the human interventions on the estuary shoreline either destroy c greatly modify its natural dynamic state. As the upper margins of the shore ar enclosed the sequence of embankment results in cumulative losses within the tid basin. In their turn these activities influence the nature of the estuary and the tid waters that enter it, the full implications of which are not understood. The sequenc of events is shown in Figure 18.3a and 18.3b.

There is no doubt that in the short term a variety of techniques, designed t maintain the position of the artificial shoreline can be effective, see 18.3.1–18.3. above. However their long-term viability is not clear, particularly in areas whe sea level is rising relative to the land. The loss of saltmarsh recorded in south-ea England suggests that as the shoreline steepens under the influence of a rising se level, activities which seek to maintain the current line of defence may have seriou consequences for the 'natural' sequence of habitats. Elsewhere, in the Dutc Wadden Sea for example, it was found that enhanced sediment accretion b 'poldering' could keep pace with rising sea level. However, in contrast, in th eastern and western Shelt it could not (Dijkema et al. 1990). Thus in many areas is possible that the intertidal area has been so diminished as to bring into questio its long-term viability, and the loss of saltmarsh in south east England may be onl a foretaste of the future for other areas.

This has important consequences not only for the 'natural' shoreline habitats bu also for the viability of the sea defence structures. Saltmarshes provide a significar contribution to buffering the effects of storms and changing sea level and hence t the cost of the sea defence structure. The National Rivers Authority (NRA) in Eas Anglia have estimated the cost of seawall maintenance to meet agreed standards fc protection against flooding in Essex. They show a clear economic advantage i having an intact and wide saltmarsh in front of a seawall, see Table 18.3.

If an increase in the rate of sea level rise consequent upon global warmin becomes a reality then the problems become even greater and more widesprea Given the serious environmental consequences associated with the 'coasta squeeze' and the fact that the maintenance of the sea defence structures may nc

Table 18.3 Saltmarshes and sea defence costs (NRA 1992)

Saltmarsh width (m)	Seawall height (m)	Cost of seawall (UK £ m^{-1})
80	3	400
60	4	500
30	5	800
6	6	1500
0	12	5000

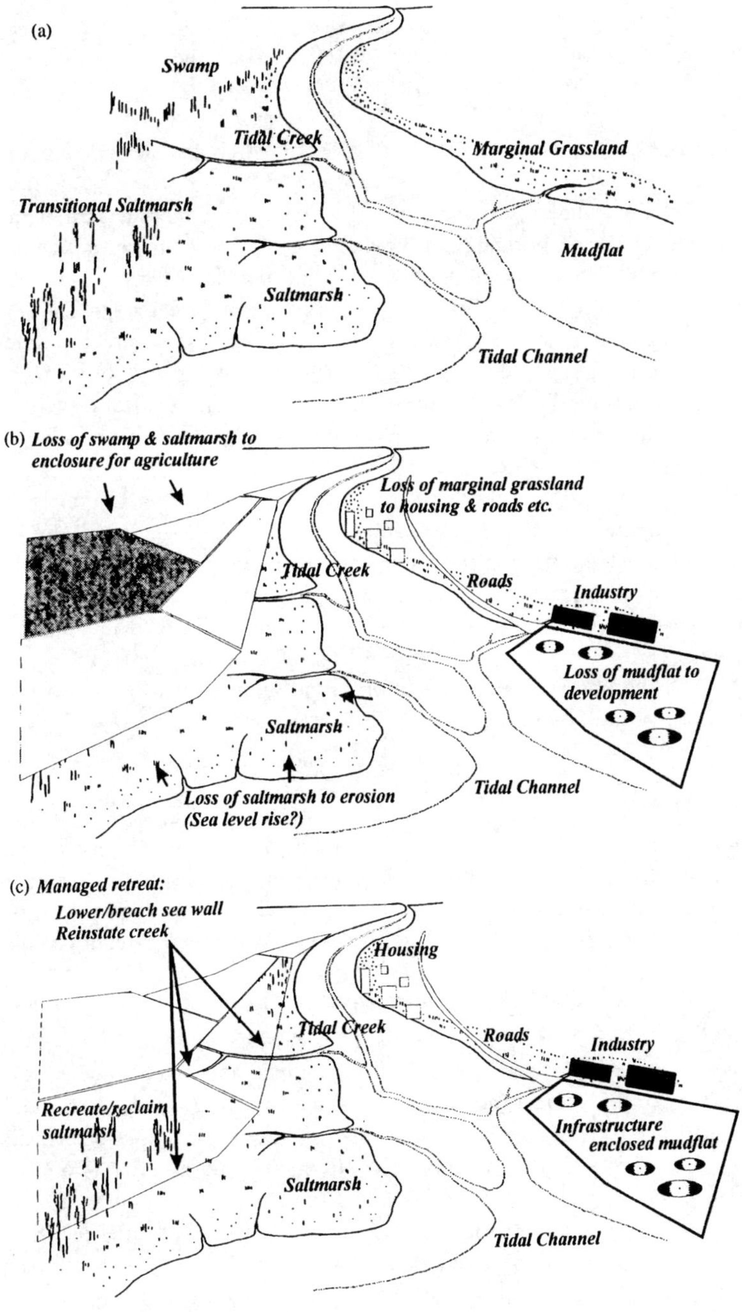

Figure 18.3 Sequence of enclosure of estuarine habitats and their subsequent 'reclamation'. (a) Original estuary; (b) habitat loss due to enclosure; (c) habitat recreation (managed retreat)

be sustainable in the long term without considerable investment, in some areas the case for retreating from the current line of defence becomes compelling.

18.5 CONCLUSION – THE CASE FOR MANAGED RETREAT

Managed retreat, which is depicted in Figure 18.3c, may have a number of advantages over the traditional hard 'engineering' solutions to coastal erosion and flooding. It will create opportunities for the development of new and the reinstatement of 'old' wildlife habitats. By reducing the design standards needed for 'new' inland defences it may also prove to be less costly to protect the hinterland. It is tempting to conclude that widening the intertidal zone may help re-establish its natural ability to accommodate changes in the sea level, the increased incidence of storms and storm surges which may be amongst the consequences of global warming.

Some examples of the reinstatement of tidal conditions on former saltmarsh ('natural reclamation') do exist, such as those in south-east England. Here sections of older seawalls have been left to deteriorate, allowing the tide and tidal water to return to the land. These show that new habitats (notably saltmarsh) can develop and that they are, superficially at least, not dissimilar to those which are thought to have existed prior to enclosure (Figure 18.4). Whether this provides a new way forward is uncertain. There is scant information upon which to build a policy of managed retreat and as yet the outcome is untried in all but a few small locations.

18.5.1 Research needs

Research must play a significant role in promoting any new policy, particularly one which seeks to replace the traditional human preoccupation with land claim and coastal protection, with one which accepts a reversal of this process. Of particular importance is the need to show the benefits for both nature conservation and human use.

Most examples which do exist are the result of 'accidental' rather than 'managed' retreat; however, short-term studies provide an indication of the likely outcome. A review of sites in Essex, where unplanned abandonment has taken place similar to that shown in Figure 18.4 above, showed that areas of land at or near high water were successfully colonised by saltmarsh. Older enclosures well below current high water progress more frequently to mudflat (Boorman and Hazelden 1995). A small 0.8 ha 'set back' site on Northey Island in the Blackwater estuary, one of the first experimental sites in the UK, showed good recolonisation by saltmarsh species within three years (Dagley 1995). However, these examples are small and may not provide a sound basis upon which to predict the outcome for larger areas.

These and other studies must be extended to other locations, notably in the

Figure 18.4 'Natural' saltmarsh reclamation on the River Deben, Suffolk. Note dead trees at edge of saltmarsh, suggesting inundation of former agricultural land

eastern USA and the Wadden Sea where more extensive examples are availabl However, although they will help establish that the approach has merits and can b effective, they will not provide a prescription for managing retreat at a singl location. Much more rigorous studies will be required.

Initially it will be important to establish the original extent of tidally influence land and the nature and scale of human interference, to help set the context for th managed retreat. This information should be augmented by data on the tidal regim sediment load and salinity gradients of tidal waters and height of the land i relation to the potential inundation. Predicting the likely response of plan according to their ecological tolerances will help establish the expected sequence c colonisation and provide a means of monitoring the outcome of the process fc nature conservation. Some or all of these studies will be needed if effectiv monitoring of both existing and planned examples of managed retreat are to b adequately assessed for their effectiveness.

The effect of the 'coastal squeeze' reminds us of the implications of interferin with coastal systems without understanding the consequences of our actions ove the long term and in relation to the estuary as a whole. Simply recreating forme tidal land may not result in a straightforward re-establishment of dynamic stabilit Evidence from the Netherlands suggests that an increase in the tidal prism withi the estuary of the Oosterschelde, which resulted from the loss of land initiall inundated by a storm surge in 1421, caused the rate of saltmarsh erosion to increas elsewhere in the estuary (Mulder 1989). Other evidence suggests there may also b a shift from a tide-dominated estuary to a river-dominated one, with all th implications this has for the development of the estuary and its wildlife as salinit gradients change (Carter 1988). Thus there is considerable scope for research int the long-term implications of re-establishing tidal land on areas of forme enclosure. Amongst the most obvious is modelling the long-term implications o increasing tidal volume on erosion rates and estuary tide levels.

To some extent it is an act of faith that in the longer term a wider, naturall functioning estuarine shore will be more resilient to change than a constrained an narrow one. The problem of pursuing a more enlightened approach to th management of a dynamic shore may therefore lie in reconciling the time-scales c human memory with those over which an estuary adjusts to the prevailin environmental conditions. It is unlikely that the concept of 'managed retreat' wi be as acceptable as the tried and tested engineering approach for some time t come. For this to be achieved some of the rhetoric associated with 'working wit nature' must be put into practice. In the context of coastal defence this implies a acceptance that natural structures (beaches, saltmarshes, sand dunes and shingl bars) may provide a more sustainable and cost-effective protection against erosio and flooding in more than the few rare examples where it is currently applied.

English Nature, the statutory body responsible for nature conservation i England is actively promoting such a policy in its 'Campaign for a Living Coast' At the same time the Ministry of Agriculture, Fisheries and Food, which i responsible for grant aiding coastal defence schemes, includes this as an option in their *Guide to Good Practice* (Ministry of Agriculture, Fisheries and Food 1993). In July 1995 an experimental 'set back' site of 21 ha was initiated by the Ministry when the existing seawall was lowered and a breach made to allow flooding of the formerly enclosed saltmarsh.

Given that the statutory agencies, and perhaps more importantly, the owners of the land are beginning to accept that managed retreat has some merits, more attention should be paid to providing evidence for its long-term benefits. Thus the results of scientific studies should be more obviously directed towards this aim. They should also be more widely disseminated and tailored for policy-makers. Equally important are the economic and cultural implications which will need to be included if landowners, taxpayers and policy-makers are to be persuaded that managed retreat is acceptable. This is essential if a reversal of the present trend of land claim is to be achieved which will benefit wildlife and the sustainable use of the estuarine resource.

REFERENCES

Beardall, C.H., R.C. Dryden and T.J. Holzer 1988. *The Suffolk Estuaries*. Suffolk Wildlife Trust, Saxmundham.

Boorman, L. and J. Hazelden 1995. Saltmarsh creation and management for coastal defence, p. 175–184. *In* M.G. Healy and J.P. Doody (eds.), *Directions in European Coastal Management*. Samara Publishing Limited, Cardigan, UK.

Brampton, A.H. 1992. Engineering significance of British saltmarshes, p. 115–122. *In* J.R.L. Allen and K. Pye (eds.), *Saltmarshes: Morphodynamics, Conservation and Engineering Significance*. Cambridge University Press, Cambridge.

Burd, F. 1991. *Erosion and Vegetation Change on the Saltmarshes of Essex and north Kent between 1973 and 1988*. Research Survey in Nature Conservation No 42. Nature Conservancy Council, Peterborough.

Carter, R.W.B. 1988. *Coastal Environments*. Academic Press, London.

Dagley, J.R. 1995. *Northey Island: Managed Retreat Scheme. Results of Botanical Monitoring 1991 to 1994*. English Nature Research Reports No. 128., English Nature, Peterborough.

Dalby, R. 1957. Problems of land reclamation. 5. Saltmarsh in the Wash. *Agricultural Review*, London 2: 31–37.

Davidson, N.C., D d'A. Laffoley, J.P. Doody, L.S. Way, J. Gordon, R. Key, C.M. Drake, M.W. Pienkowski, R. Mitchell and K.L. Duff. 1991. *Nature Conservation and Estuaries in Great Britain*. Nature Conservancy Council, Peterborough.

Dijkema, K.S., J.H. Bossinade, P. Bouwsema and R.J. De Glopper. 1990. Saltmarshes in the Netherlands Wadden Sea: rising high-tide levels and accretion enhancement, p. 173–188. *In* J.J. Beukema, W.J. Wolff and J.J.W.M. Brouns (eds.), *Expected Effects of Climate Change on Marine Coastal Ecosystems*. Kluper, The Netherlands.

Doody, J.P. 1995. Infrastructure development and other human influences on the coastline of Europe, p. 133–152. *In* A.H.P.M. Salman, H. Berends and M. Bonazountas (eds.), *Coastal Management and Habitat Conservation, Proceedings of the 4th EUCC conference, Marathon, Greece*. EUCC, Leiden, The Netherlands.

Gray, A.J. 1972. The ecology of Morecambe Bay. V. The Saltmarshes of Morecambe Bay. *Journal of Applied Ecology* 9: 207–220.

Gray, A.J. 1977. Reclaimed land, p. 253–270. *In* R.S.K. Barnes (ed.), *The Coastline*. John Wiley, Chichester.

Greensmith, J.T. and E.V. Tucker. 1973. Holocene transgressions and regressions on the Essex coast and outer Thames Estuary. *Geology en Mijnbou* 52: 193.

Hooghart, J.C. and C.W.S. Posthumus. 1989. *Hydro-ecological Relations in the Delta Waters of the South-west Netherlands. Proceedings and Information No. 41*. TNO ommittee on Hydrological Research, The Hague.

Kamps, L.F. 1962. Mud distribution and land reclamation in the eastern Wadden shallows. Rijkswaterstaat Communications, 4: 1–73.

Macey, M.A. 1974. Report 1d. Survey of semi-natural reclaimed marshes. *In* Aspects of the ecology of the coastal area in the Outer Thames Estuary and the impact of the proposed Maplin Airport. Report to DoE by NERC.

Ministry of Agriculture, Fisheries and Food. 1993. *Coastal Defence and the Environment. A Guide to Good Practice*. MAFF, Environment Matters, London.

Mulder, J.P.M. 1989. The changing tidal landscape in the Delta area of the south-west Netherlands, p. 71–88. *In* J.C. Hooghart and C.W.S. Posthumus (eds.), *Hydro-ecological Relations in the Delta Waters of the South-west Netherlands. Proceedings and Information No. 41*. TNO Committee on Hydrological Research, The Hague.

National Rivers Authority (NRA). 1992. *East Anglia Saltmarshes*. NRA, Peterborough.

Royal Commission on Coastal Erosion and Afforestation. 1911. Third (and final) Report Vol III, Pt V. The reclamation of Tidal lands. HMSO, London.

Shennan, I. 1987. Impacts on the Wash of sea level rise, p. 77–90. *In* P. Doody and B. Barnett (eds.), *The Wash and its Environment*. Research and survey in nature conservation No. 7. Nature Conservancy Council, Peterborough.

Smith, R.A.L. 1940. Marsh embankment and sea defence in medieval Kent. *Economic History Review* 10: p. 29–37.

Steers, J.A. 1981. *Coastal Features of England and Wales*. The Oleander Press, Cambridge.

Stickney, W.M. 1908. Plan showing reclamation from the River Humber Holderness. Appendix XX to evidence given at the Royal Commission on Coastal Erosion and Afforestation Second Report.

Chapter Nineteen

A policy orientation toward estuarine shores

RICHARD H. BURROUGHS
University of Rhode Island

19.1 INTRODUCTION

Shortly after the Second World War a new approach to understanding the affairs of government was introduced. It is called a 'policy orientation' and includes analysis of policy formation and execution (Lasswell 1951). The purpose here is to use this orientation to examine estuarine watershed development and water quality impacts. Policy content is familiar to many estuarine scientists. They observe the results of government actions and the presence or absence of scientific information in them. The processes responsible for the formation and execution of policy are not commonly found in the estuarine literature. However, these processes have a dominant role in influencing the uses of estuarine science. This chapter explains activities in the policy process and through specific examples shows how estuarine science affects policy.

The policy process is a chronological series of activities through which an estuarine issue proceeds as it is considered by society and ultimately acted upon by government. For background, the policy process will be briefly described. Through subsequent case studies, two parts of the policy process will be emphasized. The topic for both cases is water quality and sewage treatment. The first case focuses upon problem definition in the Thames estuary, UK. The second case examines program evaluation in the upper Narragansett estuary in the US. Throughout, the focus is on activities in the policy process and two questions related to them. First, in which policy activities is estuarine science most important? Second, what is a useful scientific contribution as seen from a policy orientation? An objective in

Estuarine Shores: Evolution, Environments and Human Alterations. Edited by K. F. Nordstrom and C. T. Roman.

answering both questions is to make scientific research more useful in the protection of estuaries.

19.2 THE POLICY PROCESS

The policy process is a way of conceiving the chronological activities that an issue passes through as society and government at any level deal with the issue. Key activities are identified in Table 19.1. From the perspective of idealized policy analysis, a single issue will move from top to bottom in the table. However, in reality functional activities may be omitted or be considered in an order different from the one shown. Often there are overlaps between activities. A first look should make it clear that in fulfilling the activities related to estuarine environmental issues, there are numerous occasions where environmental science has a

Table 19.1 Policy process. Ideally, estuarine water pollution control will pass through each functional activity if a government program to address it is established

Functional activities	Illustrative estuarine water quality questions
*Definition of problems in society	Is there a problem of degraded estuarine water quality? What aspects of it should be addressed?
Aggregation of concerned individuals	How many people are affected by low-quality estuarine waters?
Organization and representation	Are these people well organized? Do they know how to gain access to decision-makers?
*Agenda setting	How does estuarine water quality get on the agenda for government action?
*Formulation of proposals	What are the possible ways of solving this problem?
Legitimation of program by government	Which proposal is selected? Who supports it?
Budgeting for government program	Is adequate estuarine water quality program funding available?
*Implementation of government program	How and by whom is the estuarine water quality program administered?
*Evaluation of program	Has estuarine water quality improved as a result of the program?
Adjustment/termination	How should the program be changed to improve it? Should the program be terminated? Has a new problem been defined?

*Significant scientific input for this activity.
(Source: Modified from Jones 1984.)

predominant role (see asterisks in Table 19.1). Furthermore, analysts believe that science can contribute to policy through technological inputs and through an attitude or way of thinking about problems (Dror 1983).

Different analysts specify alternative structures for the framework, but all adopt similar schemes (Hogwood and Gunn 1984; Jones 1984). The first activity of the policy process is problem definition. It may be initiated by an event such as flooding or eutrophication, which is perceived as subject to amelioration through governmental intervention. Problems are not given and immutable, but rather they are subject to formulation by the policy analysts (Lindblom 1980). A scientific understanding of the consequences of human interactions with estuaries is fundamental to perceiving and defining a problem effectively.

If the number of persons in the general public affected is large, and they organize well, then their issues come before government. The activities of aggregation, organization, and representation are instrumental in making this happen. While scientific information is used in these activities, the scientists themselves are not directly involved in most instances. If that representation is successful, then the issue is recognized by many as one that requires public action. It is placed on the agenda for government action. For these activities (definition through agenda setting) to bring a problem to government, a scientific consensus about the nature of the problem and broad public support that something should be done are necessary. Scientific information at problem definition and agenda setting can catalyze the public. Scientists' testimony before government bodies is an important confirmation. These activities are responsible for delivering the issue to government.

Initial action in government includes formulation of proposals, legitimation of programs, and budgets (Table 19.1). Proposals record the results of the analysts' formulations. Effective formulations will specify the problem or problems to be treated, will identify which aspects of the problem are covered, and will consider effects if the proposal is adopted (Jones 1984). Thus, a proposal for a dissolved oxygen problem in an estuary may address point sources with the predicted effect that fisheries will improve. The same dissolved oxygen problem may also trigger an alternative proposal, such as the control of non-point pollutant sources. Proposal formulations appear in a variety of forms which can include executive branch pronouncements and interest group position papers.

Formulation of a proposal for a solution requires relatively certain predictions about what will happen in the estuary if specified governmental actions are taken. As such, formulation of proposals is a particularly important focus of science and scientists. Scientific diagnoses of an estuarine environmental problem and the likely consequences of technical or other solutions in addressing the problem are mandatory for the design of effective proposals. For example, proposals to address impacts of polychlorinated biphenyls (PCBs) on biota (Limburg et al. 1986) or of power plant cooling water on striped bass (Barnthouse et al. 1988) are scientifically detailed and can be influential.

If a proposal for a program gathers sufficient support, it is selected as the preferred program. It is legitimated in the sense that large numbers of people agree to a new way of arranging their affairs. Examples of legitimation include federal legislation for sewage treatment or local ordinances for setbacks around estuarine shores. Other forms of legitimation may occur through direct citizen participation. Beach clean-ups and citizen water quality monitoring efforts are examples. A final test of intent for programs housed in government is budgetary support. If forthcoming, it signifies that a proposal has even broader public support.

Next is the implementation of a solution (Table 19.1). At this point goal achievement becomes paramount. Government resources (personnel, budget, institutions) have been applied. Actions and results are expected. Jones (1984) identifies three components to implementation. First, an organization with adequate resources and methods needs to be in place. Second, the law or other goal needs to be translated into specific plans and directives. Third, the program instruments such as services, payments, or permits must be applied.

The actions associated with implementation of a program dealing with the environment can be identified as resource management. In more specifically defining this for the ocean, Juda and Burroughs (1990) observed that management is directed toward human utilization of the resource consonant with politically determined values and with the desire to increase benefits and avoid deleterious impacts. As this policy orientation implies, initial resource management activities may be modified as additional information becomes available. The concept of change can be manifested through conscious experimentation with proposal design and implementation. In those situations, adaptive management of natural resources is being practised (Holling 1978).

Ultimately, this application of governmental effort must be evaluated and the program adjusted or terminated. The results of the program are considered through evaluation (Table 19.1). Did the program accomplish what it set out to do? In estuaries, determining the outcome almost inevitably involves scientific measures of environmental quality (Burroughs and Lee 1988). In evaluation, it is particularly interesting to consider whether failures resulted from the inability of the cognizant government agency to deliver what was demanded of it, or whether the failure was due to an ineffective design embodied in the legislation.

All of the functional activities in Table 19.1 are candidates for scientific input. A few (see asterisks) are loci for significant scientific input, as has been indicated in the discussion above. Two, problem definition and program evaluation, have been selected from those highlighted to illustrate the connection between estuarine science and an activity in the policy process.

19.3 PROBLEM DEFINITION

For a policy analyst, problem definition 'is a medium through which we discover what we realistically want and how we may go about obtaining it' (Dery 1984). If the goal is beyond current means of achieving it, then it becomes an issue of limited interest to the policy analyst. Without relative certainty of obtaining objectives through known means, the policy analyst is unable to proceed. For the analyst problem definition starts when feasible scientific solutions are joined to acceptable governmental process. Thus, for the analyst, the policy problem begins when a scientific consensus is evident. A policy problem is an issue that people are concerned about and for which there is a solution. When decisions are to be made, policy analysts prefer certainty as do their clients, political decision makers.

A problem definition is non-useful when it is not possible to have a solution or where other problems arise. Good policy problems meet three requirements (Dery 1984). The values must be attainable; the values embodied in the solution are worth achieving; and the actions necessary to reach the solution are worthwhile in the context of the values advanced by the solution. To illustrate this third point, if protection of land results in sharply declining water quality because wastes have been transferred from one medium to another, then actions are not promoting an overall goal of environmental quality.

To illustrate scientific contributions to problem definition, the Thames estuary in England is an ideal case. The policy problem has been persistent, and the types of information available for problem definition have been documented.

19.4 PROBLEM DEFINITION IN THE THAMES ESTUARY

The human population of Greater London increased from around 1.5 million people in 1801 living in about 200 000 houses to 8 million people in 1931 (Wood 1982). For a while the tidal Thames served as a source of drinking water because, with normal winter flows, the river is fresh to central London (Morris 1988), and it was also used for human waste disposal. Two different problem definitions with contrasting solutions are apparent in this case.

19.4.1 Displacement as a problem definition

This problem definition is based on moving human waste away from the people who created it. As onsite capacity was exceeded at individual dwellings, human waste moved through open sewers or was directly discharged to tributary streams of the Thames which were gradually covered over by the growing city (Wood 1982). By 1830 water closets were widely used in the city, and through this primitive stream–sewer system large volumes of human waste were displaced from dwellings to the Thames in central London. In 1858, the Year of the Great Stink, sheets with disinfectant were hung in windows at Houses of Parliament to combat the odors emanating from the river (Wood 1982).

After the Great Stink, construction of interceptors ensued. In less than a decade, completed interceptor sewers north and south of the river displaced waste over

15 km downstream and seaward before discharging it into the river (Gross 1976; Wood 1982).

By initially defining the problems as the displacement of human waste from households to the river and later via the interceptors seaward, new problems had been created. These were cholera and associated declines in estuarine water quality. The cholera outbreaks of 1831–1832, 1848–1849, 1853–1854, 1865–1866, and 1871 resulted in tens of thousands of deaths. To address this, in 1852 water withdrawal for drinking from the fresh tidal reaches of the river was prohibited (Wood 1982). Estuarine water quality and odor remained an issue.

19.4.2 Treatment as a problem definition

Displacement downstream had moved but not removed the problem. Treatment emerged as an alternative problem definition. By 1878, organic-rich mudbanks were alleged to have formed in the vicinity of the interceptor sewer outfalls and, in the eyes of some, threatened navigation. Chemical precipitation, removing 17% of the putrescible matter, was reported from 1880–1915 at the treatment works where the interceptor sewers discharged (Wood 1982).

Even though treatment had been attempted, by 1950 hydrogen sulfide odors from the river were prevalent during summer adjacent to the main interceptor sewer outfalls. This widely acknowledged public nuisance became the subject of a 15-year study that began in 1949. At this point, the skills of estuarine scientists were fundamental for devising solutions. Their study identified organic pollution as the cause of the low oxygen levels and for the first time predicted what future conditions would be if changes in loading were made (Gameson 1964). In this context, the scientists had diagnosed the problem and predicted the future state of the estuary if changes were made to meet a goal of more than 10% saturation for dissolved oxygen (Gross 1976). The former, a problem definition activity, was directed toward understanding the factors affecting distribution of dissolved oxygen. The latter, formulation of a proposal for treatment, was a modeling effort. Ultimately, implementation targeted organic loading and reduced it through a series of treatment improvements. Pretreatment was also instituted. In 1957 the river was devoid of fish over a 60 km segment, but by 1978 over 50 species of fish had been caught on intake screens of a power plant in the same area (Norton 1983).

19.4.3 Problem evolution

As noted above, the test of a good problem is that it has feasible solutions. The evolution of scientific knowledge about estuaries coupled with innovations in sewage treatment technology over a century provided new solutions, and hence new problem definitions. This case history has been divided into displacement and treatment to illustrate how knowledge affects the evolution of problem definition. Realistic objectives and the means to reach them change appreciably as more information becomes available. Between the Year of the Great Stink, 1858, and the release of the report documenting the causes of declining water quality, 1964, scientific knowledge had changed substantially. By 1964 the source of the hydrogen sulfide was tied to the river, and its relationship to dissolved oxygen had been modeled. In the earlier period, displacement of waste seaward was a feasible means to reduce odors in central London. Treatment was not feasible because it was not sufficiently well known until early versions of chemical precipitation were instituted in 1885 (Wood 1982).

Problem definitions evolve because the results of the solutions they provide are evaluated. The second case study describes how estuarine program evaluation is done within a policy orientation.

19.5 PROGRAM EVALUATION

Program evaluation consists of a systematic assessment of program results and an explanation of the extent to which the program caused those results (Newcomer et al. 1994). An ultimate objective is to modify current operations to improve the program. Evaluation of the effectiveness of social programs began with consideration of Great Society programs (Wholey et al. 1970), and the methodologies for these analyses continue to evolve. They rely on the identification of program goals, the instruments to achieve them, measurement of results, and consideration of statistical significance. The purpose of this section is to explain how estuarine science may contribute to program evaluation.

The focus on evaluating program outcome is an important one to consider. Outcome emphasizes actual changes achieved to reach the goal desired as opposed to intermediary measures or outputs. In the terms of estuarine management a program output may be, for example, a regulation, the number of permits processed, or an allocation of funds. An outcome in an estuarine program is an improvement in water or sediment quality brought about by governmental action. The consequences of this outcome may also be considered in terms of living resource and environmental use changes (EPA 1994).

For the purposes of evaluation, the intended outcome in the estuary is a dependent variable (Burroughs 1987). The government program or output is an independent variable that is assumed to be causally prior to it. Experimental manipulation, which in this case is program implementation, occurs when the independent variable is intentionally altered. Assuming rival explanations can be eliminated through a control or other means, then, if repeated manipulations produce predictable changes in the dependent variable, a causal relationship may be established.

Estuarine programs use natural science information in a social program setting. They do not include all elements of an ideal experiment. Social programs are at best quasi-experiments (Caporaso 1973). Pollution control evaluations fit this quasi-experimental category (Roos and Bohner 1973). One variable is manipulated through the apparatus of government with the hope that the desired change will ensue. The task for the analyst is to design an approach that compensates for the lack of an ideal experiment but still leads to valid causal inferences. Primary among the obstacles in a quasi-experimental evaluation of estuarine water quality programs is the fact that the random assignment of units and the assignment of treatments are beyond the control of the investigator. Furthermore, information concerning events that have occurred must be extracted from records that were not necessarily kept for that purpose, since few programs are implemented with a requirement for evaluation data to be collected.

More specifically, for the relationship between sewage treatment and estuarine water quality, the independent variable is the initiation of plant functioning. The dependent variables are amounts of pollutants in the estuary. Testing the hypothesis requires pertinent data on water quality and on influences beyond the cause hypothesized. The latter can be important because the strategy of a quasi-experiment is to identify correlation and to distinguish correlation from causation (Hoole 1978). The approach requires the systematic testing of rival explanations before judgments about cause may be rendered.

A time series design is one of the many that are applicable to social program evaluation (Cook and Campbell 1976). In the estuary, data concerning pollution are collected before and after the initiation or enhancement of the treatment plant. With this series of before-and-after measurements, the question becomes clear. Did the sewage treatment plant affect environmental outcome? One wants to know not only whether there was a change, but whether it can be attributed to the proposed cause, and whether effectiveness can be generalized to other circumstances. In reality, well functioning treatment plants limit the discharge of specific contaminants. However, and most importantly, sewage treatment alone may not solve estuarine water quality problems because there are other sources of contaminants. Water quality program evaluation assesses the effectiveness of the solutions in place and identifies other sources of contaminants requiring attention.

Pollution has been referred to in a general manner. It will be helpful now to look at it as specific water quality parameters such as dissolved oxygen, nutrients, metals, and bacteria to specify what variables should be measured before and after a program is initiated. Secondary treatment removes 85–90% of biological oxygen demand (BOD), 10–30% of nitrogen and phosphorus, and 15–65% of the metals from the effluent (Mueller and Anderson 1983). Changes in these or associated parameters as observed in estuarine waters, organisms, or sediments are candidate outputs to evaluate.

Note that the latter evaluation differs from considering the functioning of the plant alone. To make calculations about the effectiveness of the plant, influent versus effluent concentrations are compared. To evaluate the effectiveness of sewage treatment in improving estuarine water quality, the objective is to note and attribute changes in contaminants at the estuarine level, not just at the plant level.

Understanding the system-wide impact of a specific governmental initiative is a complex undertaking. Such an analysis has been conducted for upper Narragansett Bay.

19.6 PROGRAM EVALUATION IN NARRAGANSETT BAY

Water quality in the upper Narragansett estuary has been degraded by population growth and industrialization along the Blackstone–Seekonk–Providence River complex (Burroughs and Lee 1988). Organic materials, bacteria, nutrients, hydrocarbons, chlorinated organics, and metals are presumed responsible, but only dissolved oxygen has attracted potentially useful long-term monitoring. By 1913 deteriorating dissolved oxygen was apparent, and by 1947 the bottom water was anoxic (Petruny-Parker and Robadue 1985).

A recent response to the low oxygen conditions was the upgrade of a sewage treatment plant to secondary treatment in 1972. The objective of this section is to consider what scientific information is useful; how it is used in a program evaluation; and whether the upgrade had an impact on the estuary given that there are a variety of other sources of oxygen demand. The scientific activities particularly germane to evaluation are monitoring and inferring causal explanations for change.

19.6.1 Environmental monitoring in program evaluation

For environmental monitoring to be useful in program evaluation, contaminants affected by the program under consideration must be adequately measured in time and space. In upper Narragansett Bay dissolved oxygen levels in the receiving waters were selected as the parameter in this evaluation for three reasons. First, the BOD concentration of typical wastewater may be reduced by 85–90% through secondary treatment (Mueller and Anderson 1983). Hence, the treatment process may have an impact on dissolved oxygen in the estuary. That, after all, is one objective of treatment in the first place. Second, data on oxygen concentrations in influent and in effluent at treatment plants were available. Third, sporadic data on dissolved oxygen in the waters of upper Narragansett Bay were available (Burroughs and Lee 1988).

Monitored discharges to the estuary from the plant show an 86% decrease of BOD loading from the plant where treatment was upgraded (Burroughs and Lee 1988). When including the BOD loading from the other treatment plants on the upper estuary, a BOD decrease of 73% to the upper Narragansett estuary was noted. Effluent BOD has declined precipitously. Clearly, the upgrade to secondary treatment at the plant has had an important impact on effluent BOD load to the upper estuary. But, did that affect water quality?

To complete that portion of the program evaluation, a careful monitoring of the receiving waters of the estuary is necessary. The design and execution of environmental monitoring are an applied science endeavor that is seldomly done for estuarine program evaluation. Therefore, the upper Narragansett evaluation had to operate within the constraints of available data. With monitoring data collected for other purposes, Burroughs and Lee (1988) demonstrated that an increase in bottom water dissolved oxygen correlates with operation of the upgraded sewage treatment plant. This increase in oxygen was observed over a 3.5 mile-long stretch of the estuary (Burroughs and Lee 1988). Dissolved oxygen in much of this area doubled but still remained below 4 ppm, a desirable goal for all marine organisms.

Correlation of this increase in dissolved oxygen with upgraded treatment is not causation. Ultimately, the relationship can only be interpreted as causal after rival hypotheses have been considered and rejected. In this case several alternatives were evaluated.

19.6.2 Cause in program evaluation

Inferring cause for monitored change in environmental variables is a scientific enterprise that is particularly important to program evaluation. It encompasses identifying alternatives that might better explain the monitored changes than the government program being evaluated. Four alternatives are considered here.

First is the possibility that influent BOD loads to the plant have declined. If that were so, then improvement in bay waters would be due to population and manufacturing decline in the watershed. Direct evidence from sewage treatment plant influent data rules out this possibility. Inputs remain steady over the time period discussed above.

Second, industrial BOD loadings discharged directly to the estuary could influence dissolved oxygen. However, throughout the period the total amount of manufacturing discharges are significantly smaller than sewage treatment plant discharges and a surrogate, manufacturing employment, remains steady (Burroughs and Lee 1988). Thus, a decline in manufacturing discharges directly to the estuary does not explain the overall increase in dissolved oxygen because variation in it has little impact on total load to the estuary.

Third, changes in the river loadings could be a source of the alteration of the dissolved oxygen levels in bottom waters of the estuary. Rivers include: the Blackstone ($21.3\,m^{-3}s^{-1}$), the Ten Mile ($4\,m^{-3}s^{-1}$), the Moshassuck and the Woonasquatucket ($3.2\,m^{-3}s^{-1}$) (Pilson 1985; White 1988). Effluent flow from the plant in question is about 3–4% of the amount contributed by the Blackstone where the plant discharges.

The impact of the main river affecting the upper estuary, the Blackstone, may be evaluated through covariance analysis (Burroughs and Lee 1988). A control is selected outside the influence of the sewage treatment plant under consideration. Dissolved oxygen data in the bottom waters before and after the initiation of upgraded sewage treatment are plotted for the control and for segments of the estuary. The ideal control station for this analysis would be located at the head of the estuary. Such a station would sample river water upon its entering the estuary. Unfortunately, such data are not available. None the less, a covariance analysis with available data showed improvements over 3 miles of the estuary at or below the discharge point (Burroughs and Lee 1988). If there were improvements upstream of the plant in the major river, they appear less significant to estuary waters than the impacts of the plant.

Fourth, some segments of the estuary show little or no improvement. Spatial coincidences indicate that this may be related to flows from two sources. One is the confluence of two small rivers (combined totals $3.2\,m^3\,s^{-1}$), or combined sewer overflows (CSO) (Burroughs and Lee 1988). While these assessments are based on limited data, they indicate that a substantial BOD load may enter the estuary from CSOs and possibly from small river flows near the location where dissolved oxygen levels show little or no improvement.

Thus, alternative causes for improvement in dissolved oxygen when examined with the available data did not provide better explanations for the observed changes. Increasing the level of treatment was successful in improving dissolved oxygen levels. For areas of little or no improvement in dissolved oxygen, CSOs and small rivers were identified as potential problems.

Determination of cause for monitored changes in estuarine water quality is fundamental to both scientific understanding of estuarine systems and to program evaluation. Scientifically robust causal explanations lead to the design of better programs. However, the agency responsible for the program may be hesitant to support an evaluation because of the consequences negative findings may have on program continuity. Thus, while estuarine science is important, it may not be sought for program evaluation.

19.7 CONCLUSIONS

This application of a policy orientation to water quality shows that five activities in the policy process are likely to be particularly dependent upon estuarine science (Table 19.1). For each of the science-dependent activities, the nature of the policy questions asked leads to scientific data and scientists as a means of their resolution.

Each of the following is a nexus for scientific contributions. Problem definition, the first functional activity in the policy process (Table 19.1), requires monitoring data about the state of the environment. Agenda setting often relies on scientists to validate a problem definition through testimony before government or other actions. Proposal formulation for the solution of environmental problems requires modeling of how a proposed intervention would alleviate the problem. Implementation of a governmental program can benefit from scientists' involvement in operational details, such as monitoring programs. Finally, evaluation of the program is dependent upon scientists' understanding of how to measure change in the environment and how to interpret cause.

The case studies in this analysis consider two of the five activities where estuarine science is particularly valuable. In the first case study, two problem definitions for the Thames estuary were considered. In the initial problem definition, displacement, wastes were moved from dwellings to the river and ultimately via interceptor sewers down river. The primitive scientific understanding of over a century ago meant that science was not influential at the time. In the redefinition the water quality problem was more directly addressed through increased treatment prior to release. An environmental objective, a minimum of 10% of the dissolved oxygen saturation, and modeling of dissolved oxygen responses to different levels of treatment were the central scientific contributions to the problem definition and proposed solution. Policymakers and the policy process are highly receptive to information about a problem for which there is also a feasible solution.

The second case examined a water quality program evaluation in upper Narragansett Bay. To evaluate a water quality program, monitoring is mandatory. Yet in this case contaminant distribution over space and time was inadequate for all but dissolved oxygen data. Those data were used to show a change in bottom water dissolved oxygen coincident with the plant upgrade. Considering the cause for dissolved oxygen had two primary results. Alternatives to the plant upgrade as a cause were inadequate. Therefore, the government program to upgrade the plant was successful in improving dissolved oxygen over a significant area of the estuary. Second, the spatial distribution of dissolved oxygen showed continuing problems in one geographic area, which implied that additional work should be considered for the CSO and small river flows that enter the estuary at that location. Evaluation of estuarine program environmental outputs is clearly a scientific endeavor. However, policy actors may choose to avoid scientific information by not facilitating evaluations for fear that the consequences might be negative.

In summary, considering water quality from a policy orientation alerts scientists to where and how their skills are likely to be most influential in governance of estuaries. The greatest mutuality of interest is in problem definition and proposal formulation. Scientific contributions in agenda setting and implementation are also potentially significant. Evaluation is problematic. On one hand, scientific information is needed. On the other, its inclusion, if negative, may jeopardize program continuity.

REFERENCES

Barnthouse, L.W., R.J. Klanda, D.S. Vaughan and R.L. Kendall (eds.). 1988. *Science, Law, and Hudson River Power Plants*. American Fisheries Society Monograph 4, Bethesda, MD.

Burroughs, R.H. 1987. Estuarine water quality program evaluation: a quasi-experimental approach, p. 639–644. *In* M.P. Lynch (ed.), *Estuarine and Coastal Management – Tools of the Trade*. The Coastal Society, Bethesda, MD.

Burroughs, R.H. and V. Lee. 1988. Narragansett Bay pollution control: an evaluation of program outcome. *Coastal Management* 16: 363–377.

Caporaso, J.A. 1973. Quasi-experimental approaches to social sciences: perspectives and problems, p. 3–38. *In* J.A. Caporaso and L.L. Roos (eds), *Quasi-Experimental Approaches to Social Sciences*. Northwestern University Press, Evanston, IL.

Cook, T.D. and D.T. Campbell. 1976. The design and conduct of quasi-experiments and true experiments in field settings, p. 223–336. *In* M. D. Dunnett (ed.), *Handbook of Industrial and Organizational Psychology*. Rand McNally, Chicago, IL.

Dery, D. 1984. *Problem Definition in Policy Analysis*. University Press of Kansas, Lawrence, KS.

Dror, Y. 1983. Basic concepts in policy studies, p. 3–10. *In* S. S. Nagel (ed.), *Encyclopedia of Policy Studies*. Marcel Dekker, New York.

Environmental Protection Agency (EPA). 1994. *Measuring Progress of Estuary Programs*. Office of Water, Washington.

Gameson, A.L.H. 1964. Pollution of London's river. *New Scientist (London)* 22: 295–298.

Gross, M.G. 1976. Estuarine cleanup – can it work? p. 3–14. *In* M. Wiley (ed.), *Estuarine Processes Volume I: Uses, Stresses, and Adaptation to the Estuary*. Academic Press, New York.

Holling, C.S. (ed.). 1978. *Adaptive Environmental Assessment and Management*. John Wiley, New York.

Hoole, F.W. 1978. *Evaluation Research and Development Activities*. Sage Publications, Beverly Hills, CA.

Hogwood, B.W. and L.A. Gunn. 1984. *Policy Analysis for the Real World*. Oxford University Press, Oxford.

Jones, C.O. 1984. *An Introduction to the Study of Public Policy*. Brooks/Cole Publishing, Monterey, CA.

Juda, L. and R.H. Burroughs. 1990. The prospects for comprehensive ocean management. *Marine Policy 14: 23–35.*

Lasswell, H.D. 1951. The policy orientation, p. 3–15. *In* D. Lerner and H.D. Lasswell (eds.), *The Policy Sciences*. Stanford University Press, Stanford, CA.

Lindblom, C.E. 1980. *The Policy-Making Process*. Prentice-Hall, Englewood Cliffs, NJ.

Limburg, K.E., M.A. Moran and W.H. McDowell (eds.). 1986. *The Hudson River Ecosystem*. Springer-Verlag, New York.

Morris, A.W. 1988. The estuaries of the Humber and Thames, p. 213–224. *In* W. Salomons, B.L. Bayne, E.K. Duursma, and U. Forstner (eds.), *Pollution of the North Sea: An Assessment*. Springer-Verlag, Berlin.

Mueller, J. and A.R. Anderson. 1983. Municipal sewage systems. p. 37–92. *In* E.P. Meyers and E.T. Harding (eds.), *Ocean Disposal of Municipal Wastewater: Impacts on the Coastal Environment*. MIT Sea Grant College Program, Cambridge, MA.

Newcomer, K.E., H.P. Hatry and J.S. Wholey. 1994. Meeting the need for practical evaluation approaches, p. 1–10. *In* J.W. Wholey, H.P. Hatry and K.E. Newcomer (eds.), *Handbook of Practical Program Evaluation*. Jossey-Bass, San Francisco, CA.

Norton, M.G. 1983. Experiences in the United Kingdom on the control of discharges of sewage and sewage sludge to estuarine and coastal waters, a case study, p. 947–1023. *In* E.P. Myers and E.T. Harding (eds.), *Ocean Disposal of Municipal Wastewater: Impacts on the Coastal Environment*. MIT Sea Grant College Program, Cambridge, MA.

Petruny-Parker, M. and D.D. Robadue, Jr. 1985. The Seekonk River: a case history c estuarine management, p. 884–888. *In Oceans 85: Ocean Engineering and th Environment*, vol. 2. Marine Technology Society, Washington.
Pilson, M.E. 1985. On the residence time of water in Narragansett Bay. *Estuaries* 8: 2–14
Roos, L.L. and H.J. Bohner. 1973. Compliance, pollution, and evaluation – a research desigr p. 271–280. *In* J.A. Caporaso and L.L. Roos (eds.), *Quasi-Experimental Approaches Testing Theory and Evaluating Policy*. Northwestern University Press, Evanston, IL.
White, S. 1988. Personal communication, US Geological Survey, Providence, RI.
Wholey, J. et al. 1970. *Federal Evaluation Policy: Analyzing the Effects of Public Programs* The Urban Institute, Washington, DC.
Wood, L.B. 1982. *The Restoration of the Tidal Thames*. Adam Hilger, Bristol.

Chapter Twenty

Natural and legal shoreline buffers

JONATHAN D. PHILLIPS
East Carolina University

20.1 INTRODUCTION

The shorelines of estuaries are a transition zone between what are typically important and fragile public trust waters and valuable, coveted real estate. Land use in this shoreline zone may be critical to the quality and ecological integrity of, and public access to, the water resource. It is thus often in the public interest to establish estuarine shoreline buffers to provide for some means of influencing land use and environmental impacts in this critical zone.

A key question and chronic point of debate is the width of the shoreline zone to be included. Buffer widths should be variable to accommodate the spatial variation in geology, topography, soils, hydrology, vegetation, and land use along the shoreline, but the political and practical realities of government are such that a single number is required for application across a range of conditions. At best, a single-number maximum or minimum buffer width that can be adjusted upward or downward according to site conditions is the most flexibility a pragmatist can hope for.

The purpose of this paper is to compare buffer widths based on 'natural' criteria (i.e. based on hydrological and ecological considerations) with those used or implied in legal, regulatory, or incentive programs applicable to estuarine shorelines. The focus will be primarily, though not exclusively, on hydrological criteria. Legal buffers are generally single-number minimum widths. Owing to the variability in shoreline environments on the one hand, and the political and economic pressure for minimum buffer widths on the other, it is likely that the

Estuarine Shores: Evolution, Environments and Human Alterations. Edited by K. F. Nordstrom and C. T. Roman.

widths of legal buffers will be narrower than the widths implied by natural criteria. This contention can be tested by comparing buffers delineated according to hydrological or ecological criteria to those mandated by law, regulation, or accepted practice. Because there is so much geographical variation in both the natural and legal frameworks, a meaningful comparison of natural and legal estuarine shoreline buffers must be location-specific. In this study the comparison is based on Carteret County, North Carolina, where there are considerable existing buffer delineation data. Carteret has been included in earlier work investigating the role of specific environmental variables in determining shoreline buffer effectiveness (Phillips 1989a), comparing buffer widths necessary to protect water quality within the estuarine shoreline regulatory jurisdiction in the state's coastal management program (Phillips 1989c), and examining the implications of riparian buffer establishment on land use (Phillips 1989d). I have also studied the buffer effectiveness of riparian forests along coastal plain rivers (Phillips 1989d), and hydrological and biogeochemical processes in wetland buffers in eastern North Carolina (Phillips 1995).

This study differs from the earlier work in four main ways. First, the focus is on comparing 'natural' buffer widths established by already developed models, and in the literature, with a broad array of legal buffers and setbacks, rather than on developing and testing buffer delineation models, or on hydrologic processes. Second, I present typologies of legal and regulatory buffers. Third, the earlier studies included hydrologic buffer delineation criteria based on transport capacity of overland flow, and on detention time of runoff. This paper adds additional criteria based on groundwater travel time. Finally, this paper integrates the overland flow hydraulics, storm runoff detention time, and groundwater flow considerations to determine minimum buffer widths.

20.2 TYPES AND PURPOSES OF BUFFERS

Legal and regulatory implementation of shoreline buffers has three general forms. One is mandated setbacks, that restrict certain activities or facilities within a fixed distance of the estuarine waters. Examples include separation distances between septic fields or sludge application areas and surface waters, and property line setbacks (Phillips and Phillips 1988). These setbacks are rarely unique to estuarine shorelines; they apply to any water body (or in the case of property line setbacks to any property boundary). A second form of implementation is a regulatory buffer, within which some activities may be prohibited and others require a permit. These may be designed specifically for estuaries and other critical waters; examples include the Maryland Critical Areas program (Chesapeake Bay Critical Areas Program, adopted 1987). Others, such as wetlands dredge-and-fill permit programs, are widely applicable in estuarine environments but not designed for or restricted to these areas (Yocom et al. 1990). The third general form is assistance and incentive programs to promote conservation or environmentally friendly land use. An example is the US Department of Agriculture's Conservation Reserve program, which specifically allows for enrollment of naturally vegetated riparian buffer zones. The latter is applicable to any riparian area, but some programs, such as Virginia's agricultural best management practice cost-share program, are specifically targeted to critical coastal shorelines (Dodd et al. 1993).

Establishing and maintaining vegetated riparian buffer zones provides a number of benefits and serves several purposes. While specific legal programs are generally designed with respect to a specific function or value, other benefits accrue as by-products.

One general purpose is that of a water quality buffer or filter. Vegetated buffers are known to be effective and cost-effective filters of both runoff and land-applied wastewater and are a widely accepted tool in coastal water quality management (Clark 1977; Vanderholm et al. 1979; Lowrance et al. 1985; Phillips 1989a,c,d; Dodd et al. 1993). Water quality buffers work in two general ways: by spatially displacing potential pollution sources away from the shoreline and by filtering polluted runoff and effluent (Phillips 1996).

Maintaining ecological integrity is a second general purpose. Estuarine shorelines are an important transition zone between aquatic and terrestrial environments, and a membrane for matter and energy fluxes between them. They provide critical habitats for species adapted to this zone (PTRF 1991; Bren 1993; Keller et al. 1993). An obvious example is shoreline wetlands and their biota, but higher, well-drained shorelines also provide specialized habitats for organisms such as beach grasses and bank-dwelling shorebirds. For example, dunes and other well-drained sands provide specialized habitats for a number of plants adapted to nutrient-poor, xeric conditions, and kingfishers often make their nest-holes in shoreline bluff exposures. Undeveloped riparian corridors may also be important for wildlife movement and shallow-water biota.

A third general purpose is to protect natural or cultural resources unique to or concentrated in estuarine shoreline areas. Important outcrops of the Flanner Beach formation in the Carolinas, a classic and important stratigraphic record of a marine transgression, for example, all occur in estuarine shoreline settings (Colquhoun et al. 1991). Important archeological sites are also often concentrated in coastal riparian areas; such is the case in North Carolina (Claasen 1982; Phelps 1983), where both native Americans and subsequent European settlers chose high, well-drained shoreline bluffs which provided both water access and the best agricultural soils in the region.

The fourth general purpose for estuarine shoreline buffers is to provide an erosion setback. While these are more common on ocean coasts, rapid estuarine shoreline erosion does pose management problems in some areas (Nordstrom 1992). North Carolina's estuarine shoreline regulatory zone was originally established as an erosion setback, for example (Phillips 1989c).

Shoreline buffers may also serve the purpose of allowing public access to recreational resources and public trust waters, and/or preventing encroachment of private land uses on public trust resources. Property-line setbacks prevent private encroachment on public trust property when applied to shores, just as they prevent encroachment of neighbors on each other when applied to land boundaries. In North Carolina and most other states the shore below mean high water is in the public trust. Maintenance of a buffer behind this area provides for more natural, pleasant, and safer public access to this public resource, with less potential for conflict. Access may also be a critical issue in efforts to acquire public shoreline land. Undeveloped shoreline areas, even though privately held, have traditionally been available in many coastal areas for fishing, hunting, hiking, beachcombing, swimming, and other recreational activities. Shoreline development often precludes this traditional access, prompting efforts to acquire public land to preserve public access to public trust resources (Phillips and Phillips 1988; PTRF 1991).

Finally, estuarine shoreline buffers serve aesthetic purposes. Independently of other environmental considerations, dense and uncontrolled development to the water's edge can be ugly, unpleasant, and unsafe, just as dense, uncontrolled highway strip development can be. Buffers may serve to prevent or mitigate such developments.

Whatever the purpose, the designated buffer would ideally coincide in some way with the natural riparian zone of the estuary, defined (following Bren 1993) as the area in close proximity to the estuary and distinctly influenced by that proximity. That definition is not easily implemented, and it varies dramatically according to the type and degree of influence. This ideal also ignores the fact that buffers are generally intended to mitigate the influence of land-based activities on the estuary, not vice-versa.

20.3 NATURAL BUFFERS

The focus is on hydrology and water quality, but a section on ecological considerations is also included. Two basic criteria for evaluating water quality buffer effectiveness have been identified: the ability to reduce the stream power of overland flow; and the contact time of both surface and subsurface flow within the buffer (Phillips 1989a). To these may be added a third criterion: the contact or travel time of subsurface flow within the buffer. The first two have been extensively examined in earlier work (Phillips 1989a,b,c,d) and will be only briefly reviewed here.

20.3.1 Hydraulic model

The Riparian Buffer Delineation Equation (RBDE; Phillips 1989a) was developed to produce a buffer effectiveness ratio: an index of the relative effectiveness of a buffer for reducing the transport capacity of overland flow or for increasing the contact time of flow through the buffer. Given a particular desired, reference buffer condition, any existing or proposed buffer can then be evaluated with respect to the reference. Guidelines for choosing a reference buffer are discussed by Xiang (1993a) and Phillips (1989e). In general, the reference buffer will be based on engineering design criteria, or on a local reference soil (with associated topographic characteristics) known to be suitable for onsite wastewater treatment, grassed waterways, and other soil and water conservation features (see SCS 1985). If the buffer effectiveness ratio (BER) exceeds unity, the buffer being evaluated is more effective than the reference buffer; BER <1 indicates an inferior buffer. The equation can be manipulated under various assumptions (i.e. rearranged to solve for any variable) to produce recommendations on specific buffer characteristics, most often width (Phillips 1989c,d; Xiang 1993a,b).

The hydraulic version of the RBDE is based on unit stream power theory and follows the work of Moore and Burch (1986). By combining standard equations for unit stream power as the product of velocity and slope with the Manning equation, unit stream power is seen to be a function of buffer width (slope length; L), slope gradient (S), and hydraulic roughness (Manning's n). The ratio of the infiltration rates in the proposed and reference buffers can be assumed to be the same as the ratio of their saturated hydraulic conductivities (K; Phillips 1989a, 1995). The RBDE-hydraulic version is then obtained as

$$\text{BER} = B_b/B_r = (K_b/K_r)(L_b/L_r)^{0.4}(S_b/S_r)^{-1.3}(n_b/n_r)^{0.6} \qquad (1)$$

where the subscripts b and r refer to the buffer being evaluated and the reference buffer.

The hydraulic model is most applicable where transport of the pollutants of concern is controlled or limited by the transport capacity of overland flow, such as sediment and other larger particulates. The RBDE-H is most sensitive to slope gradient. Example applications are presented elsewhere (Phillips 1989a,e). A similar equation could be derived from the D'Arcy-Weisbach flow equation, which has a superior physical basis. However, the D'Arcy-Weisbach friction factor cannot be directly estimated in the field, as can the Manning roughness coefficient.

20.3.2 Detention model

The detention version of the RBDE is based on travel time of imposed flows through the buffer. Surface flow is based on the Manning equation; subsurface flow is assumed to be Darcian. The hydraulic conductivity ratio is assumed to govern the partition of surface and subsurface runoff, and the ratio of water table slopes between the proposed and reference buffers is assumed to be the same as the ratio of their ground surface slopes. Under these conditions one obtains (Phillips 1989a)

$$B_b/B_r = (K_b/K_r)^{0.4}(L_b/L_r)^2(S_b/S_r)^{-0.7}(n_b/n_r)^{0.6}(C_b/C_r) \qquad (2)$$

where C is soil moisture storage capacity above the water table or a confining layer.

The RBDE-detention is applicable to dissolved pollutants and is most sensitive to buffer width. Applications of the RBDE (both versions) include the development of guidelines for coastal water quality protection Phillips (1989a;c); delineation of streamside buffer zones in geographic information systems (Xiang 1993a; Stratton 1993); identification of riparian conservation areas in land use planning (Xiang 1995); and environmental policy evaluation (PTRF 1991; Xiang 1993b). A 'cookbook' guide to its application to estuarine shorelines is given by Phillips (1989e).

20.3.3 Subsurface travel time model

The RBDE is concerned primarily with stormwater runoff (or effluent applications). The hydraulic version is concerned solely with overland flow and the detention version with surface and subsurface storm flow. However, delayed subsurface flows can have significant and insidious effects on surface waters, especially because this flow may comprise much of the local flow (Bonell 1993).

Mean water level in the estuary is the local base level, and is usually equivalent to mean sea level. Assuming Darcian flow, the travel time T from any given point on the land to the estuary is given by

$$T = D/[K(H - d_{wt})/D] \tag{3}$$

where D is distance to the waterline, H is elevation of the point above base/sea level, and d_{wt} is depth to the water table. If the water table is perched or controlled by a confining layer which outcrops at the shoreline, H should be measured relative to the elevation of the top of the confining layer. The equation assumes uniform permeability and a linear water table slope, and D refers to horizontal distance. In practice the permeability or saturated hydraulic conductivity of the most permeable horizon or layer at the site should be used.

For any desired travel time or flow delay T^*, equation (3) can be rearranged to solve for the distance necessary, given K, H, and d_{wt} at the site, to provide for T^*:

$$D = [T^* K(h - d_{wt})]^{0.5} \tag{4}$$

Once T^* is selected, equation (4) can be used to determine recommended or necessary buffer widths.

T^* was selected in this case based on criteria for evaluating soil suitability for onsite wastewater treatment (septic tanks). US Soil Conservation Service (SCS) criteria for evaluating soil suitability for septic tanks specify that only permeabilities in the range of 2–6 in h^{-1} (50–150 mm h^{-1}) and slope gradients of 2–8% pose 'slight' limitations. This implies a maximum saturated hydraulic conductivity of 6 in h^{-1} (150 mm h^{-1}) and a maximum hydraulic potential gradient of 0.08 (because the water table slope is generally less than or equal to ground slope). Most regulations (including those of North Carolina) require a minimum separation distance between septic fields and surface waters of 50 ft (15 m). Using equation (3) and $K = 150\,\mathrm{cm\,h^{-1}}$, $(H - d_{wt})/D = 0.08$, and $D = 15$ m, $T^* = 1250$ h is obtained. One can then determine, for example, that for a riparian site with a water table 1.0 m deep, an elevation of 5 m and moderately rapid permeability, a buffer width of 16 to 28 m is required.

20.3.4 Ecological buffers

There are no quantitative models or guidelines for establishing buffer widths based on ecological criteria. In a study of stream corridor management in the Pacific Northwest, Budd and colleagues (1987) suggest that when a variety of ecological criteria are considered, a minimum buffer width of about 15 m is desirable. Keller and colleagues (1993) examined stream corridors in Delaware and Maryland with respect to avifauna. They recommend a minimum corridor width (not counting stream width) of 100 ft, implying a minimum estuarine shoreline buffer width of 15–30 m. Bren (1993) reviewed buffer width requirements throughout Australia and found a median figure of about 20 m for well-defined streams. Owing to the value of buffers as wildlife habitat, Bren concluded that ‘attempts to reduce buffer widths below current levels would be strongly resisted on non-hydrologic grounds even if they did meet hydrologic criteria’ (1993: 299). The US Forest Service (USFS) guidelines for riparian buffers recommend a tripartite system. This includes a developed but low-intensity use zone for water spreading and deconcentration immediately adjacent to logging, farming, or development activities, a fully vegetated filter and uptake zone, and a natural, undisturbed streamside zone to provide ecological integrity. Minimum width of the latter two zones is recommended to be 75 ft (23 m; USFS 1992). The state of Kansas developed criteria for stream corridors from an ecological perspective; an ‘excellent’ corridor should have a minimum of 20 m of forested buffer along any bank or shoreline (Dodd et al. 1993; appendix 2).

20.4 NATURAL BUFFERS IN COASTAL NORTH CAROLINA

20.4.1 Study area

Carteret County, North Carolina (Figure 20.1) is generally representative of the coasts of the whole state, in that it includes all general types of estuarine shoreline environments: extensive marsh, fringe marsh, swamp fringe, low and high sediment banks, and sandy backbarriers. The climate is humid subtropical and the topography generally flat to gently rolling. Carteret County is in the embayed section of the North Carolina Coastal Plain, and lies on the late Pleistocene Pamlico and the mid-to-late Pleistocene Talbot marine terraces. The major estuaries in or

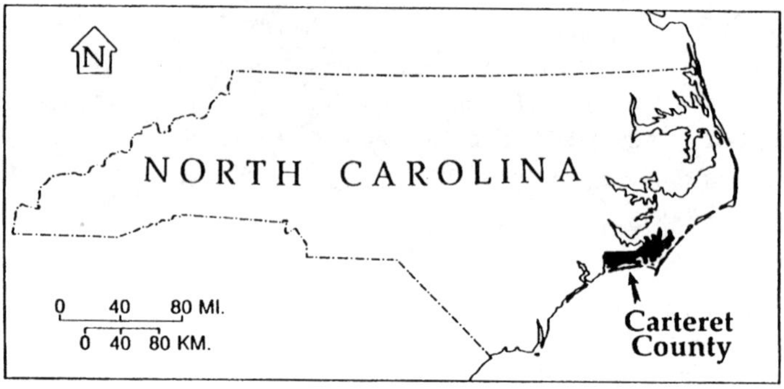

Figure 20.1 Carteret County, North Carolina study area

bordering the county include an embayed system dominated by wind tides (Pamlico Sound and Neuse River), tidal lagoons (Bogue and Core Sounds), and embayed tidal rivers (White Oak and Newport Rivers).

In all cases below it is assumed that regularly flooded riparian wetlands are part of the zone to be buffered rather than part of the buffer itself (i.e. the buffer zone begins at the wetland–upland contact), and that the buffer has a 100% ground cover of vegetation or litter which is not submerged by overland flow.

20.4.2 Hydraulic buffers

The hydraulic model was applied to Carteret County by Phillips (1989a) in a study of factors determining the effectiveness of water quality buffers in limiting the stream power of overland flow through the buffer. Given the range of shoreline conditions found in the county, slope gradient accounted for about 66% of the maximum expected variation in buffer effectiveness, and permeability for about 27%. Buffer width and surface roughness (Manning's n) accounted for 2 and 5%, respectively. Buffer width is so unimportant in the hydraulic version of the RBDE as compared with other factors, particularly slope, that it is almost meaningless to recommend buffer widths beyond a minimum recommended value of about 8 m. With sediment buffers, it is far more important to avoid steep slopes (greater than about 8%) and to maintain a complete vegetation cover that is not submerged by imposed flows or smothered by accumulated sediment.

The 8 m minimum applies to slopes of 0.02 or less. Where slopes are greater, buffer widths should be increased by the square of the ratio $S_b/0.02$, where S_b is the slope gradient of the area being evaluated (Figure 20.2). Where the shoreline is a cliff or bluff, the buffer zone is delineated from the top or shoulder of the slope. When these steeper slopes are treated thus, the maximum buffer slope in the study area is about 0.1. By the formula above, that implies a five-fold increase in the minimum buffer width, to 40 m.

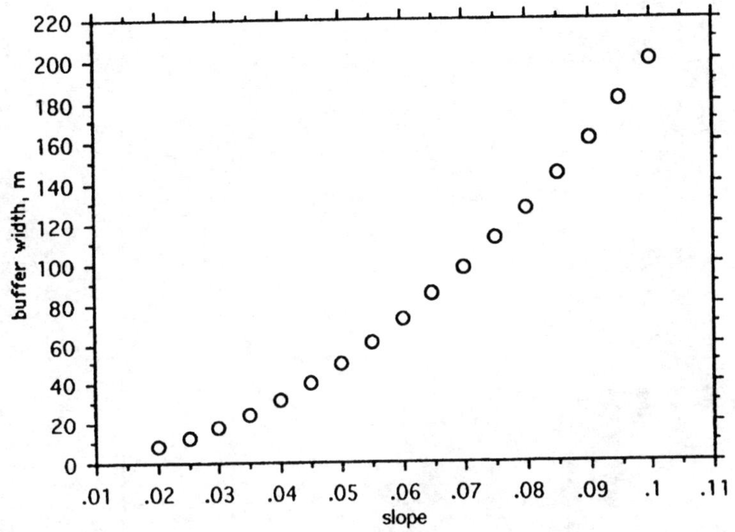

Figure 20.2 Graph showing the relationship between shoreline slope gradients and minimum sediment buffer widths. The minimum is 8 m for all slopes ≤ 0.02; for slopes >0.1 the buffer should be delineated from the top of the slope

20.4.3 Detention time buffers

There are 15 soil types found in estuarine shoreline areas in Carteret County, according to the county soil survey (Goodwin 1978). Using a reference buffer designed to provide an effective filter for urban stormwater runoff, necessary buffer widths (to achieve BER = 1) for all 15 soils were calculated, based on the optimal, least optimal, and median combination of hydrologic and slope factors for each soil, using the RBDE-detention (Phillips 1989c). The reference buffer was based on a 30 m, 2% slope of the Norfolk series (fine-loamy, mixed thermic Typic Paleudult), a common upland soil in the area. A typical grass cover ($n = 0.24$) with non-submerging flow was assumed. Buffer widths ranged from 31 to 259 m for least optimal, 39 to 142 m for median, and 19 to 106 m for optimal conditions. The overall range between worst-case, worst-soil and best-case, best-soil conditions is thus 19 to 259 m. Based on the fact that the least effective soils requiring the widest buffers are typically wetlands and thus subject to a separate set of regulations, buffer widths of 61–91 m were recommended (Phillips 1989c).

20.4.4 Subsurface travel time

Graphic solutions of equation (4) with $T^* = 1250$ hours are shown in Figure 20.3 for water table elevations (surface elevation – depth to water table) of 0–10 m, and US Soil Conservation permeability classes ranging from moderate to very rapid. These represent permeabilities or saturated hydraulic conductivities of 15–1020 mm h^{-1}.

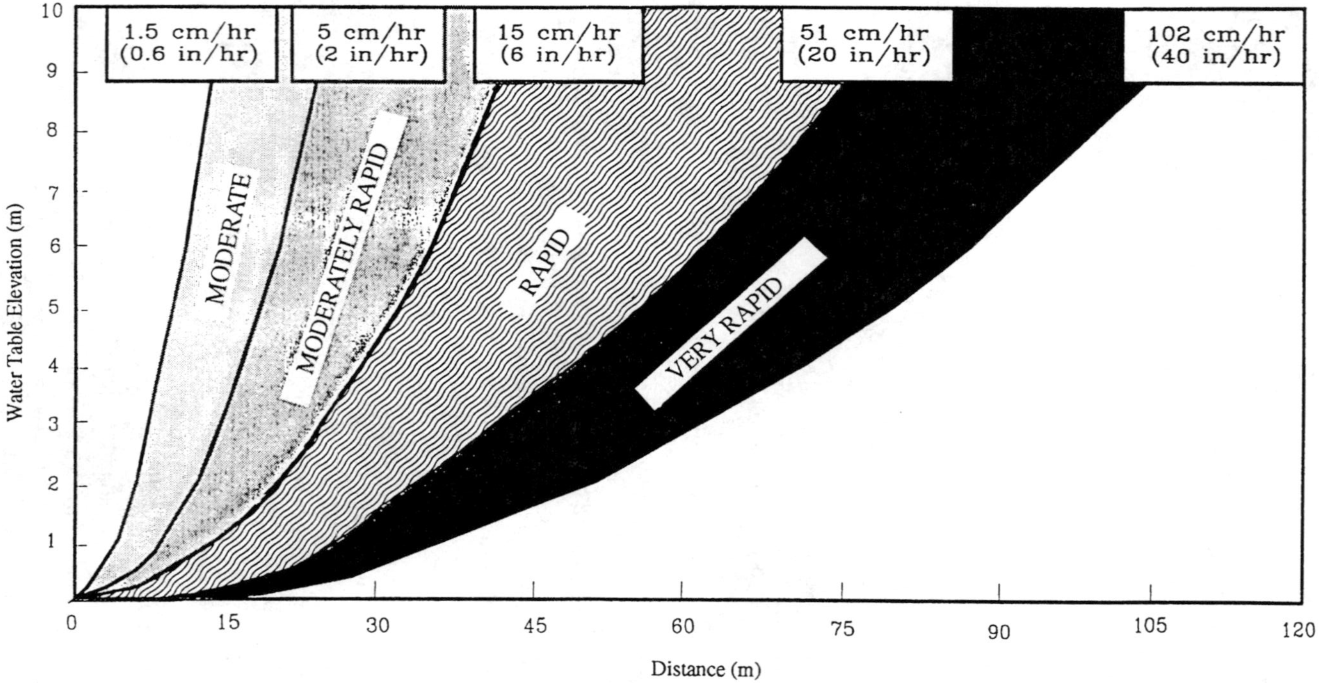

Figure 20.3 Graph showing the relationship between shoreline slope gradients and minimum sediment buffer widths. The minimum is 8 m for all slopes 0.02; for slopes 0.1 the buffer should be delineated from the top of the slope

More slowly permeable classes are not included because they are not found in the estuarine shoreline soils of the study area, and because in practice the maximum K of any horizon or layer at the site is used. For water table elevations of 0.5 m or more (less than 0.5 m is likely to be a wetland), necessary buffer widths range from about 3–6 m for a minimal hydraulic gradient in moderately permeable soils to 80–113 m for a steeply sloping water table in very rapidly permeable soils.

In the study area, water table elevations of more than 5 m above base level are rare, as are perched water tables. Adopting 8 m (25 feet) as a minimum setback, the range of recommended buffer widths based on this criterion would be 8–80 m. The former would occur in moderately permeable soils with water table elevations of less than 1 m, the latter in very rapidly permeable soils with water table elevations of 5 m. The former situations are likely to be wet, low-lying areas which would require much wider buffers under the detention time criterion. The latter are likely to be backbarrier sand dunes.

Note that these figures are minima, based on a reference time T^* derived from assuming the minimum setback distance, maximum water surface slope, and maximum permeability associated with a well-sited septic system. The range above is thus more properly expressed as 8 m to >80 m.

20.4.5 Water quality buffers: synthesis

Sites that perform well under one criterion above might fare poorly under others. A highly permeable, sandy soil might be a good hydraulic buffer due to the rapid infiltration of surface water, but a poor subsurface buffer due to the rapid throughflow or groundwater flux. To synthesize the three criteria, the minimum buffer width situations for the hydraulic, detention, and subsurface models were evaluated according to each of the other models.

The best hydraulic buffers have both minimal slope (0.0025) and rapid permeability (0.50 m h^{-1}). Assuming identical vegetation cover and surface roughness, and the maximum soil moisture storage capacity associated with soils meeting these criteria, a minimum buffer width of 19 m would be required using the detention time model, and 21 m under the subsurface travel model.

The best detention time buffer is the beaches/Newhan soil complex, a sandy soil found on barrier islands, under optimum conditions (maximum K and C, minimum S). A buffer width of 29 m would be required under the hydraulic model to achieve a BER of unity using the reference buffer given by Phillips (1989c). Depending on the elevation of the water table, a buffer width of 26–80 m would be required under subsurface travel time criteria.

The best subsurface buffer is a soil with permeability at the lower bound of the moderate range (i.e. 150 mm h^{-1}) and a minimal water table slope (0.0025). A soil having these properties would require a 132 m buffer under the detention model and a 33 m buffer with the hydraulic model.

When the three hydrologic criteria are combined, a minimum buffer width of

21 m is established, and an absolute range of 21–132 m. The higher figures are associated with the best subsurface buffer soils, which are characterized by low slopes and permeability. The latter properties cause these same soils to require much wider buffers according to the hydraulic or detention time criteria. Shoreline areas in Carteret County with minimal permeability and slopes are poorly drained hydric soils which pose other limitations for development (Goodwin 1978; Phillips 1989d). In light of this, an upper bound of 91 m (after Phillips 1989c) is more realistic.

20.4.6 Ecological buffers

Unless one is concerned with specific organisms or communities for which the habitat requirements can be identified (such as Keller et al. 1993 for avian communities), ecological buffer delineation is subjective and geared mainly toward maintaining an adequate amount and connectivity of riparian habitat. The major considerations are to maintain continuous corridors, avoid habitat fragmentation and allow sufficient space for riparian species (Clark 1977; Budd et al. 1987 Brinson 1991 and Bren 1993).

The Pamlico-Tar River Foundation (PTRF) considered buffer widths in eastern North Carolina. They recommended a minimum 15 m width for aesthetic and ecological purposes, regardless of hydrologic characteristics (PTRF 1991). Dodd and colleagues (1993) examined buffer issues throughout the Albemarle-Pamlico Estuarine Study region, which includes Carteret County. While their primary concern is water quality, they also consider ecological values and conclude that a buffer width of 8 m (25 ft) does not provide adequate riparian habitat for the region. Based on the literature above and in Section 20.3.4, a minimum buffer width of 15-20 m can be recommended for Carteret County on ecological grounds.

20.5 LEGAL BUFFERS

Carteret County's shorelines are potentially subject to local, state, and federal programs to mandate, control, or encourage waterfront setbacks or buffer zones. These vary considerably in the width specified. Dodd and colleagues (1993) review programs with the potential to protect or restore riparian buffers in eastern North Carolina; those programs directly applicable to this case study are outlined below.

North Carolina's Coastal Area Management Act (CAMA) requires permits for any land-disturbing activities in certain designated 'areas of environmental concern' (AEC) in Carteret and 19 other coastal counties. The estuarine shoreline AEC is defined as the strip between the mean high water mark and the point 75 feet (23 m) inland of that point. The estuarine shoreline AEC was originally proposed in 1974 as an erosion setback. Efforts to expand the AEC's width to address water quality concerns in the mid-1980s were unsuccessful. Phillips' (1989c,d) analyses showing that a 23 m buffer is inadequate to address water quality concerns and that wider buffers would have minimal land use impacts have also not been persuasive enough to inspire any CAMA revisions. Note that the estuarine shoreline AEC does not mandate vegetated buffers or prohibit development in the 23 m shoreline zone. CAMA simply requires permits within the zone, allowing some control (subject to considerable political pressure) over shoreline land use. Permit requirements also provide a strong incentive to avoid major land-disturbing activity in the AEC, if feasible, to avoid regulation.

North Carolina also has stormwater rules which apply to new development in the coastal counties which exceed a threshold in terms of impervious surface. These provide for setbacks of 25–30 feet (8–9 m) from surface waters and allow the use of 30–50 foot (9–15 m) buffers to treat stormwater for new development. Rules do not specify forested buffers but do imply vegetated buffers. Waters designated as High Quality Waters (HQW) in North Carolina also require a 30 ft (9 m) buffer in natural or established vegetation between built-on areas and surface waters. Only a fraction of the waters in Carteret County, and the coastal zone as a whole, carry the HQW designation.

The federally funded Albemarle-Pamlico Estuarine Study (APES) included hotly debated provisions for vegetated buffers along all flowing water bodies in the region, which includes Carteret County. The requirements do not specify the type of vegetation, and all surfaces except built-upon and severely compacted soils in the region quickly develop vegetation cover after any disturbance. Despite arguments of APES-funded researchers (Dodd et al. 1993) and others that it is inadequate to protect water quality or riparian habitats, the final APES management plan includes an 8 m (25 ft) vegetated buffer.

The North Carolina Divisions of Environmental Health and Environmental Management enforce state regulations on the siting and design of onsite wastewater systems, land application of treated wastewater, and land application of wastes such as sewage sludge and livestock manure. When such facilities occur near estuaries, mandated setbacks from surface waters apply to estuarine shorelines. These setback requirements are 50 ft (15 m) for septic fields and land application of treated wastewater, and 100 ft (30 m) for application of sewage sludge and animal waste. Requirements are based on poorly understood rules-of-thumb supposed to prevent water supply contamination by microbial pathogens. Applicants can petition for smaller buffers in specific situations; the state can also require larger buffers as permit conditions. Onsite wastewater disposal permits are handled by the Carteret County health department, and counties have statutory authority to set standards stronger than those of the state. Thus far Carteret County has shown no interest in exceeding the 'lowest common denominator' standards set by state law.

The county subdivision ordinance includes property line setbacks for structures. The intent was to prevent encroachment and to maintain access for fire and emergency vehicles. However, the required setback from shoreline property lines is 75 ft (23 m), which provides a de facto buffer in many cases (Phillips and Phillips

1988). The ordinances apply strictly only to permanent structures and do not mandate a vegetated setback.

The US Department of Agriculture's Conservation Reserve Program contains provisions specifically designed to encourage naturally vegetated riparian buffers. Agricultural land adjacent to water bodies can qualify for a 50% cost share plus rental payments to establish a permanent vegetative buffer. This program can be used in the coastal environment as an incentive to provide estuarine buffers in agricultural areas. The program allows enrollment of buffers 66–99 ft (20–30 m) wide.

20.6 A COMPARISON OF NATURAL AND LEGAL BUFFERS

Table 20.1 summarizes the natural and legal buffer widths discussed above. The incongruence is clear, especially when the median and high ends of the natural buffer ranges are considered. Legal buffers are wide enough to accommodate natural buffers only with respect to the most hydrologically favorable sites. Sites with steep slopes, rapid groundwater movement, low moisture storage capacities, shallow water tables, and slow infiltration require wider buffers, well beyond the range provided in legal buffer and setback programs. If wetlands were included in the natural buffer recommendations, the discrepancies would be even greater. The high water tables, low soil moisture storage capacity, and generally low permeability of wetlands make them poor buffers of surface runoff or subsurface storm flow (Phillips 1996), and most Carteret County wetlands would require buffer widths of 100 m or more. However, as some wetlands are protected under state and local dredge-and-fill permit programs, and virtually all are unsuited without hydrologic modification for any type of development (many of which are regulated), a maximum buffer width of 91 m was adopted here.

There are several reasons for the discrepancy. First, in North Carolina and elsewhere there is strong grassroots resistance to any form of land use control or regulation. This makes it extremely difficult to establish or expand any form of buffer program. Landowners, developers, and other vested interests exert strong political and economic pressure to resist, eliminate, or minimize the width of regulatory buffers. Second, for purposes of regulatory simplicity a single-number buffer width is required. This is undesirable from the scientific perspective due to the wide variation – even over a relatively small area – of hydrological and ecological conditions found in the shoreline zone. Political and economic pressures are such that establishment of single-number widths almost invariably err on the side of narrower buffers. Third, few legal buffers are established specifically for estuarine shorelines. Of the legal buffers listed in Table 20.1, only two apply exclusively to estuarine shorelines or to the coastal zone, and one was not designed to cope with hydrological or ecological issues. Finally, it is difficult to establish or agree upon 'natural' buffer widths. While most hydrologists and ecologists working

Table 20.1 Natural and legal estuarine shoreline buffers, Carteret County, NC. The natural buffers are assumed to be exclusive of the width of any riparian wetlands. See text for details

Type of buffer	Width (m)
Natural	
Hydraulic – to control sedimentation and other insoluble particulates	8–40
Detention time – to control delivery of dissolved pollutants and pathogens: absolute range	19–159
Recommended width of regulatory buffer	61–91
Subsurface travel time – to control subsurface pollutant delivery	>8->80
Hydrologic buffer considering all three criteria above	21–91
Ecological – to maintain riparian habitat integrity	>(15–20)
Legal	
CAMA estuarine shoreline AEC	23
APES vegetated buffer	8
NC coastal stormwater regulations	
surface water setbacks for new development	8–9
vegetated buffers for stormwater treatment	9–15
NC High Quality Waters, mandatory buffers for built-on areas	9
NC/Carteret Co. surface water separation for septic systems	15
NC surface water separation for treated waste-water land application	15
NC surface water separation for animal waste or sludge land application	30
Waterfront property line setback, county sub-division ordinance	23
Riparian buffer area, USDA Conversation Reserve Program	20–30

in riparian areas seem to agree that a minimum width in the range of 15–20 m is necessary, there is no apparent consensus of mean, typical, or maximum widths. While there are a number of models, such as those used here, for buffer evaluation, translating those into a recommended width depends on what goal, reference, or standard is used.

20.7 CONCLUSIONS

In Carteret County, North Carolina, hydrologic considerations suggest that vegetated buffer zones along estuarine shorelines, and inland of any shoreline fringe wetlands, should generally be 21–91 m wide. From an ecological perspective, a minimum width of 15–20 m is desirable. Regulatory and incentive programs which mandate, regulate, or encourage buffers or other waterfron setbacks call for or imply buffer widths of 8–30 m. The 'legal' buffers are thus adequate only where the most advantageous hydrologic conditions exist, and in some cases are inadequate altogether.

Both the environmental and legal frameworks will vary elsewhere. However, the general phenomenon – legal buffers of a width barely equal to or less than desirable buffer widths based on hydrological and ecological criteria – is undoubtedly widespread.

ACKNOWLEDGEMENTS

This chapter would not have been possible without Lynn Phillips' contribution o knowledge of coastal land use planning and management, and without Nate Phillips' willingness to walk, slog, and paddle many kilometers of shoreline with the old man.

REFERENCES

Bonell, M. 1993. Progress in the understanding of runoff generation dynamics in forests. *Journal of Hydrology* 150: 217–275.

Bren, L.J. 1993. Riparian zone, stream, and floodplain issues: a review. *Journal o Hydrology* 150: 277–299.

Brinson, M.M. 1991. Landscape properties of pocosins and associated wetlands. *Wetland* 11: 441–465.

Budd, W.W., P.L. Cohen, P.R. Saunders and F.R. Steiner. 1987. Stream corrido management in the Pacific Northwest. *Environmental Management* 11: 587–605.

Claasen, C.P. 1982. Shellfishing patterns: an analytical study of prehistoric shell from North Carolina coastal middens. Ph.D. Dissertation, Harvard University, Cambridge Massachusetts.

Clark, J. 1977. *Coastal Ecosystem Management*. John Wiley, New York.

Colquhoun, D.J., G.H. Johnson, P.C. Peebles, P.E. Huddleston and T. Scott. 1991 Quaternary geology of the Atlantic Coastal Plain, p. 629–650. *In* R.B. Morrison (ed.) *Quaternary Nonglacial Geology: Conterminous United States*. Geology of North America Vol. K-2, Geological Society of America, Boulder, Colorado.

Dodd, R.C., M. McCarthy, W.S. Cooter, W.H. Wheaton and S. Stichter. 1993. Riparian Buffers for Water Quality Enhancement in the Albemarle Pamlico Area. Albemarle Pamlico Estuarine Study Report, Raleigh, North Carolina.

Goodwin, R.F. 1978. Soil survey of Carteret County, North Carolina. US Department o Agriculture, Washington.

Keller, C.M.E., C.S. Robbins and J.S. Hatfield. 1993. Avian communities in riparian forest of different widths in Maryland and Delaware. *Wetlands* 13: 137–144.

Lowrance, R., R. Leonard and J. Sheridan. 1985. Managing riparian ecosystems to contro nonpoint pollution. *Journal of Soil and Water Conservation* 40: 87–91.

Moore, I.D. and G.J. Burch. 1986. Physical basis of the length–slope factor in the Universa

Soil Loss Equation. *Soil Science Society of America Journal* 50: 1294–1298.

Nordstrom, K.F. 1992. *Estuarine Beaches*. Elsevier, London.

Phelps, D.S. 1983. The archeology of the North Carolina coast and coastal plain: Problems and hypotheses, p. 1–51. *In* M.A. Mathis, J.J. Crow (eds.), *The Prehistory of North Carolina. An Archaeological Symposium*. North Carolina Division of Archives and History, Raleigh.

Phillips, J.D. 1989a. An evaluation of the factors determining the effectiveness of water quality buffer zones. *Journal of Hydrology* 107: 133–145.

Phillips, J.D. 1989b. Nonpoint source pollution control effectiveness of riparian forests along a coastal plain river. *Journal of Hydrology* 110: 221–237.

Phillips, J.D. 1989c. Evaluation of North Carolina's estuarine shoreline area of environmental concern from a water quality perspective. *Coastal Management* 17: 103–117.

Phillips, J.D. 1989d. Effect of buffer zones on estuarine and riparian land use in eastern North Carolina. *Southeastern Geographer* 29: 136–149.

Phillips, J.D. 1989e. Evaluating estuarine shoreline buffer zones for nonpoint source pollution control, p. 399–411. *In* Magoon, O.T. (ed.), *Coastal Zone '89*. American Society of Civil Engineers, New York.

Phillips, J.D. 1996. Wetland buffers and runoff hydrology, p. 207–220. *In* Warner, B.G. and G. Mulamoottil (eds.), *Wetland Environmental Gradients, Boundaries and Buffers* Lewis Publishers, Chelsea, Michigan (in press).

Phillips, L.R. and Phillips, J.D. 1988. Land use planning techniques for estuarine shoreline buffer zone establishment, p. 635–640. *In* Lyke, W. and T. Hoban (eds.), *Coastal Water Resources*. American Water Resources Association, Bethesda, Maryland.

PTRF (Pamlico-Tar River Foundation). 1991. A river of opportunity: A pollution abatement and natural resource management plan for the Tar-Pamlico Basin. Pamlico-Tar River Foundation, Inc., Washington, North Carolina.

SCS. 1985. National Soils Handbook. US Soil Conservation Service, Washington.

Stratton, W. 1993. Generating variable environmental buffer zones in the grid environment. Environmental Systems Research Institute, Inc., Charlotte, North Carolina.

USFS (US Forest Service). 1992. Riparian forest buffers: Function and design for enhancement of water resources. US Forest Service, Radnor, Pennsylvania.

Vanderholm, D.H., E.C. Dickey, J.A. Jackobs, R.W. Elmore and S.L. Spahr. 1979. Livestock feedlot runoff control by vegetative filters. US Environmental Protection Agency EPA 600/2-79-143, Washington.

Xiang, W.-N. 1993a. A GIS method for riparian water quality buffer generation. *International Journal of Geographic Information Systems* 7: 57–70.

Xiang, W.-N. 1993b. Application of a GIS-based stream buffer generation model to environmental policy evaluation. *Environmental Management* 17: 817–827.

Xiang, W.-N. 1995. GIS-based riparian buffer analysis: Injecting geographic information into landscape planning. *Landscape and Urban Planning* (in press).

Yocom, T.G., R.A. Leidy and C.A. Morris. 1990. Wetlands protection through impact avoidance: A discussion of the 404(b)(1) alternatives analysis. *Wetlands* 9: 283–298.

CONCLUSIONS

Chapter Twenty-one

Emerging research needs related to interactions on estuarine shores

KARL F. NORDSTROM
Rutgers University

and

CHARLES T. ROMAN
National Biological Service

21.1 INTRODUCTION

The preceding chapters highlight many of the linkages between environments resulting from interactions and exchanges of sediments, nutrients and biota along and across estuarine shores. Space limitations forced us to be selective in the coverage of topics and sites presented in this book. Many aspects of estuarine shores are mentioned but not assessed, including hard shorelines (e.g. rocky fjords) and biogenic landforms (including reefs and shell beaches). In other cases, topics are discussed but not extensively, such as nutrients and pollutants (Howes et al. Chapter 12; Burroughs Chapter 19), mangroves (Woodroffe Chapter 5; Vargas Chapter 15) and birds (Pye Chapter 2; Vargas Chapter 15; Garniel and Mierwald Chapter 16). Most chapters are reviews, so little attention is devoted to sampling techniques, despite their importance in quantifying characteristics of estuarine environments and specifying interactions. The exclusion of these topics from this book is not an indication of their perceived lack of importance.

Several authors call for additional site-specific explanations of principles that already have been applied or already are documented in other locations, while other authors call for new generic studies of shore processes that are poorly understood. Data on many estuaries are totally lacking (Healy et al. Chapter 6), and there is

Estuarine Shores: Evolution, Environments and Human Alterations. Edited by K. F. Nordstrom and C. T. Roman.

uneven coverage of specific topics in some locations (McIvor and Rozas Chapter 13). Critical data, such as reliable sea level or sedimentation data, are missing for some areas (Anthony Chapter 3). The need for site-specific data is evident at the scale of climatic zones (McIvor and Rozas Chapter 13) or distances measured in meters (Phillips Chapter 20).

There are far more studies that describe characteristics within estuarine shoreline environments than there are studies of the linkages between environments (e.g. beach and saltmarsh). Even fewer studies provide data-based process explanations of the linkages (Healy et al. Chapter 6; McIvor and Rozas Chapter 13). In this chapter, we concentrate on generic research needs for explaining the complex environmental linkages, while acknowledging the continuing need to document specific characteristics within environments and conduct site-specific studies in little-known areas. We have tried to refrain from repeating research needs identified in previous books dealing with specific environments on estuarine shores, such as saltmarshes (Allen and Pye 1992) and beaches (Nordstrom 1992) or specific topics such as estuarine habitat restoration (Thayer 1992), except where the results would contribute substantively to an understanding of the linkages between environments.

21.2 HYDRODYNAMIC, MORPHODYNAMIC AND BIOTIC LINKAGES

Greater specification of linkages between shoreline environments and deep portions of the estuary is required. Detailed investigation of the relationship between large basin-scale estuary form and processes (e.g. basin characteristics and effect on wave generation and surge heights) and smaller-scale or local morphodynamics (e.g. beach change) is still lacking (Pethick Chapter 8; Nordstrom et al. Chapter 9). The role of inputs from the open ocean to the estuaries was not the focus of this book, but several chapters point to the significance of this linkage in determining water levels at the shoreline (e.g. van den Berg et al. Chapter 7; Nordstrom et al. Chapter 9), morphologic and sedimentologic characteristics of the shoreline (Anthony Chapter 3) and fish movements (Able et al. Chapter 14). Distance from the ocean entrance plays a role in determining the magnitude of ocean wave effects (Pye Chapter 2), but site-specific controls, such as the effect of shoreline orientation, also can play a great role in determining the effect of ocean influences (Jackson and Nordstrom 1992; Jackson 1995).

The linkage between tidal channels and adjacent shoreline environments in terms of water flow, sediment transport, chemical exchange and use by fauna is acknowledged in this book and in previous literature (Stumpf 1983; Roman 1984; French and Stoddart 1992; Peterson and Turner 1994; Anthony Chapter 3; van den Berg et al. Chapter 7; Howes et al. Chapter 12; McIvor and Rozas Chapter 13). There is such a difference in the scale of these channels (size, order in tributary hierarchy), the effects of these scale differences and the methods used to gather data at these different scales that universal models are difficult to develop. Better specification of linkages at each channel order will greatly expedite these modeling efforts. Hydrodynamic studies that examine the linkage between channel flows and morphology at the estuarine-basin scale often concentrate on changes within the channels and banks without examining the effects beyond channel margins. Assessments could place associated shoreline changes in perspective by linking changing hydraulics to tidal current flow away from the banks. Greater specification of the relationship between channel density, dimensions of channels and bank characteristics to flow into adjacent areas will help identify the reason for spatial differences in the probability and magnitude of flooding and sedimentation rates, and wave generation and shoreline erosion.

Several authors note the relative lack of hydrodynamic studies that relate measurement of processes to morphologic, sedimentologic and vegetation change on estuarine shorelines (Anthony Chapter 3; Healy et al. Chapter 6; Pethick Chapter 8; Nordstrom et al. Chapter 9; Fonseca Chapter 11). Morphodynamic studies of some types of shorelines, such as mesotidal beaches (Nordstrom 1992) are well represented in the literature. Other types, such as microtidal beaches and mudflats, are still poorly understood (Pethick Chapter 8; Nordstrom et al. Chapter 9).

The lack of significant changes resulting from waves during light intensity storm conditions (wind speeds of $13\,m\,s^{-1}$) on low-energy beaches (Nordstrom et al. Chapter 9), the resistance of mudflats to erosion, and the significance of storm-related sedimentation to higher regions of saltmarsh (French and Spencer 1993) implies that meaningful process-oriented studies for some environments will require monitoring under energetic storm conditions. Greater investment in equipment may be required to make it more resistant to storm damage, and longer field deployment times may be required to ensure occurrence of rarer high-magnitude events.

Fundamental linkages and interactions of biota among tidal channels and adjacent shore environments have been identified in some detail. We know that fishes and crustaceans (i.e. nekton) utilize subtidal marsh creeks, the intertidal marsh surface, mangrove systems, adjacent seagrass beds and mudflats during periods of their daily feeding activities for avoiding predators and for spawning (McIvor and Rozas Chapter 13; Able et al. Chapter 14; Vargas Chapter 15; Kneib and Wagner 1994), but our appreciation of how temporal and spatial scales affect habitat linkages for individual species and guilds is just beginning. Research is needed to evaluate faunal habitat linkages over tidal, diurnal and seasonal periods and investigate the functional ecology of these habitats during these time-scales. Utilization of subtidal marsh creeks by saltmarsh fishes during low tide and feeding on intertidal marsh surfaces during high tides appears to be an essential linkage for maintaining viable populations (e.g. saltmarsh mummichog, *Fundulus heteroclitus* – Able et al. Chapter 14). Other critical linkages need to be identified and examined.

Regarding spatial scales, seagrass beds are identified as an important habitat when compared with adjacent unvegetated shore environments (e.g. Heck et al.

1989), but we do not know the critical threshold size for seagrass beds or other habitats (i.e. marshes, mangroves, mudflats) to serve their function effectively as predator refugia, feeding areas, spawning sites or for other life history requirements. Also, we have little appreciation of how proximity of habitats to each other affects faunal interactions – a relationship that is tightly coupled to the mobility of the organism.

Literature linking the upland with the shore is largely lacking, although this linkage has important implications for controlling nutrients and pollutants (Howes et al. Chapter 12; Valiela et al. 1990) and establishing management zones (Phillips Chapter 20). Activities in the adjacent upland at some distance inland can also result in alterations to the shoreline (Jackson Chapter 17). The spatial dimensions of linkages differ with different variables; wave effects may be highly localized (Nordstrom et al. Chapter 9), whereas socio-economic linkages extend well landward of the coastal zone.

There is greater representation in this book of cross-shore linkages than of longshore linkages, although longshore variations in shoreline environments are evident at all scales of investigation, including basin scale (tens of kilometers, Pye Chapter 2; Anthony Chapter 3), reach scale (Jackson Chapter 17) and site scale (tens of meters, Nordstrom et al. Chapter 9). Specification of sources and pathways of sediments and biota, calculations of the actual amounts transported and the relationship of longshore exchanges to cross-shore exchanges are required to identify the dependencies of shoreline habitats on longshore exchanges.

The interrelatedness of processes (e.g. wave climate and sea level change) and form (e.g. channel and beach characteristics) through simultaneous co-adjustment makes prediction of long-term morphologic changes difficult (Pethick Chapter 8). Prediction is especially difficult when the processes that can be monitored in the field are not the processes that are responsible for the major changes in sediments, morphology and biota (Nordstrom et al. Chapter 9). Some spatial linkages have a strong temporal component. Geomorphic inheritance, for example, plays a role in influencing present day processes by determining water depth, wave energy and exposure of individual shoreline reaches (Lampe Chapter 4; Jackson Chapter 17). There is an ongoing need to reconstruct how estuarine shorelines have evolved in the past, especially in response to sea level rise, in order to interpret existing features and anticipate future changes.

Future sea level rise has the potential to cause increased storminess and higher waves and storm surges, leading to increased frequency of inundation and changes in rates and locations of erosion and deposition (Tooley 1992; Barth and Titus 1984). These changes can cause drastic reduction in intertidal area and loss of associated habitat in developed estuaries (Doody Chapter 18) and may introduce cascade effects in physical and ecological processes that may continue independent of sea level changes (Stevenson and Kearney Chapter 10). Predicting future evolution is difficult, even in the better studied estuaries (Stevenson and Kearney Chapter 10). Care must be taken in the application of analyses of past changes to prediction of future changes, because there have been dramatic changes in tidal prisms, current flows and impacts of human activities (Stevenson and Kearney Chapter 10; van den Berg et al. Chapter 7; Garniel and Mierwald Chapter 16; Jackson Chapter 17).

The effect of climate change, other than its effect on sea level, has only been alluded to here (Anthony Chapter 3). The climate of an estuary affects the kinds of habitats and areal distribution of habitats (Fonseca Chapter 11; McIvor and Rozas Chapter 13; Vargas Chapter 15). Prediction of climate change requires greater specification of the climatic controls, greater knowledge of thresholds determining these climatic differences, and greater knowledge of the likelihood that these thresholds will be exceeded.

21.3 HUMAN IMPACTS AND HUMAN-ALTERED ENVIRONMENTS

Human impacts are conspicuous on nearly every estuarine shore in the world, whether they are generated within the estuary or outside it. There are numerous types of impacts, and many have had profound effects for many centuries. The impacts mentioned in this book include: loss of estuarine environments due to reclamation; increased deposition resulting from clearing of vegetation from an estuarine watershed; erosion of mine tailings; collection of peat and plant material; grazing and trampling by domesticated animals; salt production; construction of shore protection structures; construction of closure dams and storm surge barriers; conversion of non-tidal landscapes to tidal landscape as a result of flooding for military purposes or restoration; dredging and fill to improve navigation; alteration to enhance aesthetics or recreational use; introduction of flora and fauna; groundwater withdrawal; and addition of wastes, nutrients and pesticides. These activities cause direct changes to the portions of shoreline where they occur, and they cause indirect changes resulting in alterations to linkages between environments, such as shifts in channel locations and configurations that affect tidal amplitudes, maximum landward extent of tidal influences, current velocities and locations of accretion and erosion (Pye Chapter 2; van den Berg et al. Chapter 7; Garniel and Mierwald Chapter 16).

Human alterations are so ubiquitous in some estuaries that we have already lost the template for defining what is natural on the shoreline (Lampe Chapter 4; Garniel and Mierwald Chapter 16). Sea level rise, normally a natural mechanism for creation of new estuarine shoreline habitat, may now be a negative process where threatened upland environments are protected, because protection structures prevent re-creation of estuarine environments through flooding of low-lying land behind them, resulting in the 'coastal squeeze' (Pethick Chapter 8; Doody Chapter 18).

A large number of technical reports on specific human alterations to estuarine shore habitats is added to the literature each year. Usually, these studies are in the form of environmental assessments of planned projects, some of which can be large in scope (Garniel and Mierwald Chapter 16). The ongoing need to assess the effects of shorefront development, shore protection strategies, pollution and its control measures and dredging and dredge spoil disposal activities results in large numbers of reports, many of which are site-specific and unpublished. Synthesis of these disparate studies will help provide the basis for comprehensive generic models of habitat loss, conversion or re-creation.

Most studies of planned human projects predict their effects over a short time horizon. Models of change for human alterations should take into consideration the long-term effects of the designed projects and the effects of subsequent actions that will inevitably be taken, including: (1) defensive measures as the new use is threatened by an unplanned environmental hazard (Garniel and Mierwald Chapter 16); and (2) initiatives for new economic investment that result in conversion of land to a form perceived as more desirable for new types of human activities.

Natural estuarine environments undergo cyclic changes at a variety of temporal and spatial scales, including basin scale changes over millennia (Lampe Chapter 4) and centuries (Garniel and Mierwald Chapter 16) and local or regional scale changes related to storm and seasonal cycles (Pethick Chapter 8) and diurnal and semi-diurnal (tidal) cycles (Nordstrom et al. Chapter 9). There is evidence to suggest that cyclic changes are eliminated by human activities, such as dredging and other engineering improvements (van den Berg et al. Chapter 7; Garniel and Mierwald Chapter 16), but it is not always clear whether human alterations eliminate cycles, truncate the spatial dimensions of cycles or create their own cycles that are perpetuated by human action. Models of estuarine evolution under developed conditions are needed to complement the large number of short-term descriptive studies of effects of structures or pollutants that consider human alterations as an aberration rather than an integral, predictable component of environmental change.

Research on perception, cognition and behavior is required to identify the fundamental reasons for the inexorable conversion of estuarine shores to human artifacts. Observations by scientists that shorefront development is incompatible with natural processes have done little to dissuade people from destroying natural environments, occupying hazardous areas or building larger structures in reconstruction and protection projects. Recent acknowledgement of the long-term rise in sea level has resulted in anticipatory construction of sea defenses (Garniel and Mierwald Chapter 16), and additional hard shore protection structures are likely to be built by local shorefront managers, resulting in elimination of additional dynamic shore environments (Shipman and Canning 1993). Most pre-construction planning for shore protection is devoted to assessing the effect of the natural system on humans. Efforts could be devoted to evaluating the effect of protection structures on the natural system and to using technology to devise ways of achieving protection while allowing for exchange of water sediments, nutrients and biota across developed areas to maintain natural linkages.

21.4 ENVIRONMENTAL REGULATIONS, PRESERVATION AND RESTORATION

Environmental regulations can help reduce the rate of natural losses, but ideal regulatory programs may not be implemented because of difficulties of resolving socio-economic needs with scientific criteria (Garniel and Mierwald Chapter 16; Burroughs Chapter 19; Phillips Chapter 20). Compromise solutions may have to be found to ensure that natural values are maintained while human use is accommodated. Economics plays a major role in the feasibility of any human action. Economic input is crucial to the acceptability of regulations to preserve or enhance environmental values, and applied environmental research must include economics and other aspects of social science as integral components of research projects.

In some cases, only small degraded remainders of former natural environments exist, such as the raised bogs identified by Garniel and Mierwald (Chapter 16). Research is required to determine the extent to which small natural enclaves are viable as functional habitats and thus viable preserves. Specification of bottom-line requirements and allowable tradeoffs will be required to establish minimum effective dimensions and minimum numbers and kinds of species for management zones and preservation zones.

Restoration activities, including reconstructing substrate as well as revegetating substrate, are addressed in portions of the chapters (e.g. Garniel and Mierwald Chapter 16; Stevenson and Kearney Chapter 10), as is reinstatement of natural tidal conditions and effects of managed retreat (Doody Chapter 18). These factors may have increased importance if sea level rise is so rapid and widespread that it makes existing use of low-lying land untenable. Although it has been argued that restored habitat should mimic natural morphology and hydrology to maximize functional interactions (e.g. Peterson and Turner 1994), many restoration attempts will result in habitats that can, at best, be described as semi-natural (Garniel and Mierwald Chapter 16). Knowledge of how a system functions under natural conditions may be less important than knowledge of how natural functions can be maintained in a human-altered system. Restored environments are artifacts, and there are many questions beyond considerations of hydrologic processes, morphologic changes and biologic functions that are related to intrinsic value, aesthetics and other human perceptions and values, pointing to the need for humanistic studies.

21.5 SPECIFICATION OF RESEARCH QUESTIONS

The questions presented here evolve largely from the need to examine linkages between habitats that are highlighted as the principal theme of the book. The list is meant not to be exhaustive but an example of issues that can be examined in research programs with multidisciplinary structure or application.

21.5.1 Hydrodynamic and morphodynamic studies

What is the role of tidal channels in supporting critical environmental linkages between the open water, marshes, mangroves and other habitats?
How do linkages between channels and adjacent environments differ with differences in basin size and drainage density?
What is the role of bank profile or other geomorphic characteristics, such as scarp height or levee height, in transfers of water and biota between channels and adjacent marshes?
What are the differences in sediment characteristics and resulting habitat in locations where inundation is caused by different mechanisms of inundation (e.g. due to tidal flooding versus flooding by storms)?
What are the relative roles of waves and water levels due to ocean forcing compared to local estuarine processes?
To what extent are sediment exchanges between the upper and lower portions of the shore profile (upper foreshore–low tide terrace, mud flat–marsh) cyclical and to what extent are they unidirectional?
How do relationships between wave and current processes and shoreline morphology differ with differences in wave energy and tidal range? Are different models required for different energy regimes?
How do linkages between environments differ under weather events of different frequency and magnitude?

21.5.2 Biotic linkages

To what extent is habitat fidelity related to species mobility?
What is the primary function of estuarine shoreline habitats (e.g. feeding area, refugia from predation, spawning) and can other habitats perform those same functions and thus maintain linkages?
How do faunal linkages vary over temporal scales (tidal, diurnal, seasonal)?
How important is physical proximity of one environment to another in optimizing biotic linkages?
How does spatial scale affect habitat function and linkages?
Are specific habitat linkages essential to the maintenance of a population?
Which habitat linkages are universally applicable across latitudinal zones?

21.5.3 Identifying the changing nature of linkages

How will sediment supplies and budgets be altered by climate change and sea level rise?
How will the resulting storm frequency/magnitude relationships and the wind-wave climate be affected and in turn affect the dimensions and distributions of environments and their linkages?

What will the effect of these alterations be on habitat dimensions and juxtaposition with human structures?
How do we predict the location and type of shoreline in the future, considering that storm surge and wave and current energy may be increased in addition to still-water flooding?

21.5.4 Human impacts and human-altered environments

What are the many potential scenarios caused by the 'coastal squeeze'?
What is the effect of beach nourishment and associated changes in sediment size, and sedimentation rates on biota both within and adjacent to the nourished area?
How can technology be used to reduce human impacts on adjacent natural shoreline environments?
How can social science research contribute to identifying the process leading to human alterations and find ways of guiding human actions to environmentally compatible uses?
To what extent should hydrodynamic processes and rates of morphologic and biologic change (e.g. historic erosion rates) be used to assess future change in human-altered estuaries?

21.5.5 Environmental regulations, preservation and restoration

What are the policy and regulatory strategies required to support restoration and managed retreat?
What compromises can be made to accommodate environmental and developmental needs?
What will be the effect of habitat conversions related to eutrophication on sediment, nutrient and biotic linkages?
What setback distances are needed between upland development activities and adjacent estuarine shore environments to protect the functions and natural processes of these environments?
How can a restored shoreline environment be made to approximate more closely a natural environment in appearance, function and interaction with adjacent environments?
What are the fundamental hydrologic, sedimentologic and geomorphic criteria for habitat site selection of habitat restoration sites?
What are the optimum dimensions and minimum practicable size of a restored habitat (or entire preserve) to allow it to perform its designed function?
To what extent can restored habitats survive as isolated enclaves?
Which adjacent habitats must be preserved or restored to allow a target habitat (e.g. marsh) to perform its design function?

21.6 CONCLUDING STATEMENT

The contributions in this book are paralleled by other studies seeking to identify the relationships between estuarine shore environments in terms of the processes that link them into an integrated, complex and dynamic system. The focus on linkages requires an extension of the spatial and temporal scales of evaluation beyond short-term intraenvironmental studies. Although we underscore the desirability of comprehensive, multidisciplinary approaches to identify the complex inter-relationships, we realize that much research will be limited in scope, both temporally and spatially. Regardless of scale, we hope this volume will encourage future investigations of the transfers of sediment, nutrients and biota across the environments that define estuarine shores and give them their special value.

REFERENCES

Allen, J.R.L. and K. Pye (eds.). 1992. *Saltmarshes: Morphodynamics, Conservation and Engineering Significance*. Cambridge University Press, Cambridge.

Barth, M.C. and J.G. Titus (eds.). 1984. *Greenhouse Effect and Sea Level Rise*. Van Nostrand Rheinhold, New York.

French, J.R. and T. Spencer. 1993. Dynamics of sedimentation in a tide-dominated backbarrier salt marsh, Norfolk, UK. *Marine Geology* 110: 315–331.

French, J.R. and D.R. Stoddart. 1992. Hydrodynamics of salt marsh creek systems: implications for marsh morphological development and material exchange. *Earth Surface Processes and Landforms* 17: 235–252.

Heck, K.L. Jr., K.W. Able, C.T. Roman and M.P. Fahay. 1989. Fishes and decapod crustaceans of Cape Cod eelgrass meadows: species composition, seasonal abundance patterns and comparison with unvegetated substrates. *Estuaries* 12: 59–65.

Jackson, N.L. 1995. Wind and waves: influence of local and non-local waves on mesoscale beach behavior in estuarine environments. *Annals of the Association of American Geographers* 85: 21–37.

Jackson, N.L. and K.F. Nordstrom. 1992. Site-specific controls on wind and wave processes and beach mobility on estuarine beaches. *Journal of Coastal Research* 8: 88–98.

Kneib, R.T. and S.L. Wagner. 1994. Nekton use of vegetated marsh habitats at different stages of tidal inundation. *Marine Ecology Progress Series* 106: 227–238.

Nordstrom, K.F. 1992. *Estuarine Beaches*. Elsevier Science Publishers, London.

Peterson, G.W. and R.E. Turner. 1994. The value of salt marsh edge vs interior as a habitat for fish and decapod crustaceans in a Louisiana tidal marsh. *Estuaries* 17: 235–262.

Roman, C.T. 1984. Estimating water volume discharge through salt marsh tidal channels: an aspect of material exchange. *Estuaries* 7: 259–264.

Shipman, H. and D.J. Canning. 1993. Cumulative environmental impacts of shoreline stabilization on Puget Sound, p. 2233–2242. *In Coastal Zone 93*. American Society of Engineers, New York.

Stumpf, R.P. 1983. The process of sedimentation on the surface of a salt marsh. *Estuarine, Coastal and Shelf Science* 17: 495–508.

Thayer, G.W. (ed.). 1992. *Restoring the Nation's Marine Environment*. A Maryland Sea Grant Book, College Park, Maryland.

Tooley, M.J. 1992. Recent sea-level changes, p. 19–40. *In* J.R.L. Allen and K. Pye (eds.), *Saltmarshes: Morphodynamics, Conservation and Engineering Significance*. Cambridge University Press, Cambridge.
Valiela, I, J. Costa, K. Foreman, J. Teal, B. Howes and D. Aubrey. 1990. Groundwater-borne inputs from watersheds to coastal waters. *Biogeochemistry* 10: 177–198.

Index